FREUD AND THE AMERICANS
The Beginnings of Psychoanalysis in
the United States, 1876-1917

FREUD IN AMERICA:

Volume I
FREUD AND THE AMERICANS
The Beginnings of Psychoanalysis in the United States, 1876–1917

FREUD AND THE AMERICANS
The Beginnings of Psychoanalysis in the United States, 1876-1917

NATHAN G. HALE, JR.

New York
OXFORD UNIVERSITY PRESS
1971

Quotations from *James Jackson Putnam and Psychoanalysis: Letters Between Putnam and Sigmund Freud, Ernest Jones, William James, Sandor Ferenczi, and Morton Prince, 1877–1917,* edited by Nathan Hale, Jr. (Cambridge, Mass.: Harvard University Press, 1971), copyright © 1971 by the President and Fellows of Harvard College, are reprinted by permission. Translation of German texts by Judith Bernays Heller.

For my mother, Harriett Gerber Hale,
and in memory of Bernice Engle and Robert Lynwood McCoy

ACKNOWLEDGMENTS

The accumulating debts of this study to friends and colleagues are by now staggering. Judith Bernays Heller provided continuing insight into the world of her uncle, Professor Freud. Professor Henry F. May encouraged this study from its inception, especially during its apparently interminable shift of focus back toward nineteenth-century origins. Norman Reider read the manuscript time and again, in all its drafts. From his profound knowledge of psychoanalysis I have profited deeply. Professor Ilza Veith combed each chapter, saving me from many faux pas, medical and other. Dorothy Ross contributed searching criticism based on her extensive studies of Granville Stanley Hall and the origins of American psychology. To Mrs. Ross I also owe important material concerning Hall and Freud. To Samuel Haber and to Charles Rosenberg I owe invaluable criticism. Anne Sherrill read the manuscript with unceasing care for style and coherence. Ann Bradsher provided astute help with fact and interpretation. For all this patient, painstaking help, I am more grateful than I can say.

To the generosity of Marian C. Putnam, this study is deeply indebted. The papers of her father, James Jackson Putnam, provided insight, step by step into the development of the first influential American psychoanalyst. An edition of the Putnam correspondence, of which excerpts are included here, has been published by Harvard University Press. To the Harvard Press, and to Ann Orlov, editor, this study also owes a heavy debt. Edmund Brill provided important information and insight into his father's work as Freud's translator and leader of the American psychoanalytic movement. Eunice Winters has made available important portions of the Adolf Meyer papers and, with Anna Mae Bowers, criticized several chapters. Hanna Decker supplied information on Freud's reception in Germany. Others, who have criticized the text in one or more versions include Lewis Feuer, Hunter Dupree, Norman Bell, Lucille Birnbaum, Eric Carlson, and Barbara Sicherman. Versions of several sections were presented at meetings of the San Francisco Psychoanalytic Society and Institute, and I am grateful for the criticism and suggestions of its members. Sheldon Meyer has been a patient cicerone and a solid moral support.

My introduction to psychoanalysis in the pre-war period came first from Frederick Matthews's insightful master's thesis, "Freud Comes to America" (University of California, 1957), and then from John Burnham's pioneering doctoral dissertation, "Psychoanalysis in American Culture before 1918" (Stanford University, 1958). I have also profited from their later articles. I have come to disagree with some of Professor Burnham's conclusions, notably on the American interpretations of Freud. My opening chapter on the Clark Conference owes much to Professor Matthews's similar treatment, yet, I disagree with his attribution of the origins of American "adjustment" psychology in part to James Jackson Putnam.

During the revision of this study as a doctoral dissertation, Thomas Kuhn's *Structure of Scientific Revolutions* offered penetrating insights into material that already lay at hand. Because neurology and psychiatry in this period had so heavy an admixture of non-scientific elements, however defined, my use of Kuhn's conceptions hardly can be considered an application of his theories, least of all a rigorous one. Yet, many of them have proven stimulating and helpful. I am also indebted to Professor Peter Cominos's remarkable papers on English Victorian sexual morality and its collapse, and to the sensitive studies of the American character and conscience by Allen Wheelis.

Many people kindly contributed interviews or important information over the years: Karl Menninger, Gardner Murphy, and the staff of the Menninger Foundation, Topeka, Kansas; Phyllis Blanchard, the late Jacob Loewenberg; David Riesman; the late Anne Parsons; Gerald Grob; Helen Witmer, Joseph Freeman, Harry Solomon; Mrs. Morton Peabody Prince; Norman Bell, the late Henry Murray, George Wilbur, Ives Hendrick, Henry Viets, Walter Lippmann, Mrs. Smith Ely Jelliffe, Mrs. Carel Goldschmidt, Margaret Mead.

The staff of the library of the University of California, Berkeley, especially Jeannean Myles and Margaret Uridge; Mr. Richard Wolfe, curator of rare books, and the staff of the Francis A. Countway Library of Medicine, Boston; and the staff of the Stanford Medical Library provided invaluable help. Phyllis Fidiam patiently typed the manuscript.

A grant from the University of California, Berkeley, and a Haynes Foundation Fellowship facilitated the early stages of research and writing.

For permission to quote I am grateful to Edmund Brill, Basic Books, Inc., The Hogarth Press and the London Institute of Psychoanalysis, The Liveright Co., Random House, Inc., The Macmillan Co., Good Housekeeping, and the Johns Hopkins University Press.

Those to whom this book is dedicated, above all Bernice Engle, made possible its inception and completion.

INTRODUCTION

Any study of Freud's influence in the United States must begin with two myths, the first created by the founders of American psychoanalysis, the second, by their enemies. The first legend holds that Freud transformed neurology, psychiatry, the social sciences, education, and child raising; that before Freud there had been mostly darkness, inhumanity, a narrow constraint which ignored sexuality and childhood. This was the myth created by Freud's first militant followers, especially those who were in their mid-twenties or early thirties In 1909 when he visited America, and who became the first historians of psychoanalysis.

The enemies of psychoanalysis constructed their counter-legends. These began, perhaps, with the neurologist Francis X. Dercum, the humanist Paul Elmer More, and reached their apogee in Oscar Cargill, the literary historian, and the sociologist Richard La Pierre. "Freudian doctrine," they held, prescribed primitivism, sexual liberation, pessimistic determinism, permissiveness, and decadence.

Both legends agreed that Freud's influence had been decisive. The Clark University Conference in 1909, which, Cargill asserted, brought Freud to America where he delivered a significant series of lectures, had created an "earthquake" in public opinion. Both myths agreed that Freud's work had criticized and replaced a medical and a moral order.

This study was innocently undertaken to test these assertions and to explore Freud's influence in the popular and professional culture of the 1920's. After several years' work, certain problems seemed increasingly insoluble, certain paradoxes increasingly outrageous. For example, psychoanalysts wrote many tracts that touched on mental hygiene, child guidance, criminology, and education. Their opinions seemed at once remote from Freud's, yet close to aspects of his work. Where, for example, did their insistence on the importance of early childhood, environment, and parental influences come from? From John Watson's popular Behaviorism or from Freud, or from both and in what proportions? Did their emphasis on psychological factors in insanity come from the influential Swiss-American psychiatrist, Adolf Meyer, or from Freud and Jung and Eugen Bleuler, or all four, and again in what proportions?

On a popular level, the issues became more puzzling still. The 1920's were presumed by most historians to be a high point of popular interest in Freud. Yet a study of the nation's pulp magazines showed no trace of psychoanalysis. Even more peculiar, most articles about psychoanalysis were unfavorable. What did these paradoxes mean?

The problems of the 'twenties inevitably led to an attempt to untangle their earlier beginnings. The pre-World War period was merely to have supplied background for the 1920's, yet a closer study of the earlier period led to contradictions that, if possible, were more fundamental than those of the later one. Some historians had asserted that in the innocent Progressive Era psychoanalysts taught chiefly the sublimation of sexual impulses for the sake of social progress. Yet in 1920 in a symposium on sex education, A. A. Brill, Freud's chief American representative, noted that years before—in the fullness of the Progressive Era—he had given up prescribing intercourse to patients with unspecified problems. Horace Frink, another American psychoanalyst who had written extensively in the pre-war period, stated that absolute continence should never be insisted upon. Some psychoanalysts insisted on premarital chastity; others refused to take a stand. What then became of the psychoanalytic message of sublimation? Did these opinions bear any relation to Freud's? Was psychoanalysis so ambiguous?

Historians also had argued that before 1900 neurology and psychiatry were pervaded by gloom and fatalistic hereditarian notions. Yet, a cursory survey of American neurology and psychiatry from 1870 to 1900 indicates that many physicians gave a role to psychological factors and to environment. Indeed they asserted that proper training in childhood could overcome serious hereditary taint. Moreover, literature was filled with references to sexuality, and even to childhood.

Several studies of psychoanalysis in America, some of them excellent, had failed in one major respect. They had not clarified the state of American psychiatry, neurology, and sexual morality before Freud became important. Without analysis of the pre-Freudian order, the nature of Freud's influence could not be determined. The legends would remain unchallenged.

All these problems involved one central question. Where should a study of Freud in America begin? The psychoanalysts themselves hit upon the natural beginnings of their movement in 1924. Smith Ely Jelliffe, who edited the *Semi-Centennial Volume of the American Neurological Association,* included verbatim a debate in 1876 between James Jackson Putnam, professor of neurology at Harvard, and George Miller Beard, the neurologist who popularized the term "neurasthenia." Putnam had denied Beard's claims to have cured illness by "mental" means, because they were, perforce, unscientific.

Of all the first American psychoanalysts, Putnam had had the widest experience with other methods. He had begun as an "electrician," using as a tonic for the nervous system that mysterious force Beard had studied with Thomas Alva Edison. Yet, by 1900 Putnam himself was using "mental treatment" and by 1909 had become interested in psychoanalysis. What could have determined his dismissal of Beard and his later change of heart? Putnam presented a final paradox. He was descended from a long line of Puritan ancestors and came from the heart of prudish Boston's medical and social life. How could he have become interested in psychoanalysis?

Perhaps a study that began with the nineteenth-century moral order would illuminate the nature of Freud's influence. Like all legends, those concerning Freud in America have their share of truth as well as illusion. It is to be hoped that this volume will contribute to the process of sorting and clarification.

This study will begin with the Clark Conference where Freud presented his views to Putnam and to other influential Americans in September 1909. It will then explore the professional, social, and moral order that determined the American reception of psychoanalysis.

The young intellectuals of the pre-war period have been omitted from extended consideration here. They represent the beginning of the culture of the 1920's rather than the culmination of that of the nineteenth century. The symptoms of those among them who consulted psychoanalysts resemble less the classical neurotic complaints of "civilized" morality than those newer disorders of character and identity that became characteristic of a changing moral order. The young intellectuals of 1912, the Great War, and the influence of psychoanalysis in the 1920's will be taken up in a second volume.

CONTENTS

PART 1

The Clark Conference
and the
Nineteenth-Century Order

I

The Clark University Conference
September 1909

Freud's Psychoanalysis

In the autumn of 1909, the last year an American President rode to his inauguration in a carriage, Sigmund Freud paid his only visit to the United States. At fifty-three he was still a relatively obscure Viennese neurologist, known chiefly to physicians and psychologists on both sides of the Atlantic. His popular vogue in America had not yet begun.

After years of working in what Freud called his "splendid isolation," he received his first formal academic honor in the United States. He and two disciples, Carl Jung and Sandor Ferenczi, had been invited to Worcester, Massachusetts, to celebrate the twentieth anniversary of ambitious, unorthodox Clark University. Feeling ignored by Viennese academic circles, Freud gratefully accepted an honorary doctorate of laws for his contributions to psychology. Sixteen years later he recalled: "In Europe I felt as though I were despised, but over there I found myself received by the foremost men as an equal. As I stepped on to the platform at Worcester to deliver my *Five Lectures on Psychoanalysis* it seemed like the realization of some incredible daydream: psychoanalysis was no longer a product of delusion, it had become a valuable part of reality." [1]

Freud himself thus posed the problem of this study. Why did America welcome psychoanalysis more warmly than any other country? What was there in the nature of psychoanalysis and what in American conditions that created this affinity?

At the time, the trip aroused Freud's hope that there might be a future for psychoanalysis in the United States. He made lasting friendships with a few Americans. Yet he was puzzled and somewhat distrustful, amused but not pleased, by what he had seen—Worcester, the Adirondacks, Coney Island, his first movie, full of wild chasing. He admired Niagara Falls—it was grander and larger than he had expected. He was charmed by a porcupine and by the Greek antiquities at the Metropolitan Museum. Yet the American cooking irritated his stomach; the free and easy informality irked his sense of dignity.[2] He learned of a popular mania for religious mind cures, and he detected a distressing potential lay enthusiasm for his hard-won discoveries.

The Clark Conference was a decisive event in the history of psychoanalysis in America for several reasons—the moment at which it occurred, the personal relationships it established, the impression Freud's vision created in a few strategically placed Americans. The Conference brought together many of the leaders in American professional life who determined Freud's reception. Some members of his audience were disposed to listen sympathetically and others to reject his views. One of America's most famous psychologists, Edward Bradford Titchener, represented the introspectionist, experimental tradition at its most rigid. He detested theories of the subconscious and the application of psychology to the "cure of souls." [3] A few of his listeners had worked together informally over the preceding two decades in the criticism and reconstruction of neurology, psychiatry, and psychology. This had aroused their interest in psychoanalysis, which, they realized, still was largely known through the "gossip of prejudice and misconception." They had looked forward with deep interest to Freud's visit, so they could judge the man himself and assess what he might have to say. There were also partly admiring, partly hostile critics, such as William James, then fatally ill, who had come just to "see what Freud was like"; Franz Boas, who was demolishing doctrines of racism and transforming American anthropology; Adolf Meyer, the short, precise Swiss immigrant who would dominate American psychiatry between the first and second World Wars.

Some became enthusiasts—Hall himself, the president of Clark University, an expansive, somewhat eccentric psychologist who cultivated "zests" such as walking barefoot, and who had fostered American interest in child psychology; Edwin B. Holt, a pupil of William James, who would write one of the first and most successful popularizations of psychoanalysis; A. A. Brill, loyal and pugnacious, an immigrant from Austria-Hungary to the lower East Side, who had traveled with Freud's party from New York. Brill already had begun his controversial translations of Freud's work into English. Freud's future biographer, the young, brilliant, and cantankerous Ernest Jones, talked to William James and stored up impressions. Jones had become a friend of James Jackson Putnam, one of America's foremost neurologists, professor at the Harvard Medical School, with a reputation for uncompromising idealism.

The *Boston Evening Transcript* sent a special correspondent. There was one notorious uninvited guest, Emma Goldman, whom the *Transcript* referred to as "Satan," and who happened to be in Worcester on a speaking tour. Forbidden by the police to rent a hall, she had talked, perhaps about her favorite subjects, anarchy and free love, to some three hundred people on a sympathizer's front lawn.[4]

Freud wrote his friend the Swiss pastor, Oscar Pfister, that he had been tired and hadn't felt like preparing anything for America. So during a half hour's walk around Worcester he organized each morning's lecture for ex tempore delivery in German. Following a suggestion of Ernest Jones, who had spent some weeks in Boston, Freud decided to survey the whole field of psychoanalysis for the "practical Americans."

The Clark lectures, Freud's first extended synthesis, exaggerated certain elements present in his system up to about 1914, and these were singularly appropriate in the American environment. Freud emphasized the practicality, the optimism, the comparative simplicity of psychoanalysis, and he included one of his few discussions of sublimation. He also displayed his literary style, personal charm and appeal to laymen at their most beguiling. He condensed almost to the point of caricature the major theories he had worked out in his first great works, *The Interpretation of Dreams, Three Contributions to a Theory of Sex, The Psychopathology of Everyday Life,* and *Studies in Hysteria.* By concentrating on examples of trauma and catharsis, he seemed to emphasize the finality of psychoanalytic treatment and the ease of rational choices. He took a bold yet highly ambiguous stand for the reform of conventional sexual morality.

Five years later Freud's outlook changed drastically. Not only new theoretical and clinical considerations but the schisms within the psychoanalytic movement and the Great War deepened his disillusioned stoicism. But in 1909 the vicissitudes of sexuality, narcissism, the death instinct, the Superego and the Id remained to be formulated. If by some anachronistic miracle he had presented a summary of these final views, his American reception at Clark might have been very different. Scarcely a month after the lectures, Hall asked for them in a form in which they could be printed.

They "seemed to the best men who heard them: e.g., Putnam, Meyer etc., so admirably calculated to introduce American physicians and psychologists to your system; and your lectures were such masterpieces of simplification, directness, and comprehensiveness that we all think that for us to print them here, giving you reprints of course, would greatly extend your views at a psychological moment here and would do very much toward developing in future years a strong American school." [5]

Freud reluctantly complied, and the lectures were published in Hall's *American Journal of Psychology.*[6] He spoke each morning at eleven from September 7 through September 11. Before the first lecture, Hall had introduced him as one of the "foremost students of psychiatry," and he was greeted with "much applause." [7]

Freud opened with disarming humility, a humility he later regretted because it denigrated his own original achievements. He was honored to be associated with psychoanalysis, he said, but its first discoverer was a Viennese colleague, Joseph Breuer. It pleased him that the majority of his hearers were not physicians, because a medical education was not necessary to follow what he had to say. Anyone, he would reiterate, could test for himself some of the important principles of this new science.

He turned to Breuer's celebrated case, which he unraveled like a novel. Anna O., twenty-one, a girl of superior mind and character, presented the most serious symptoms: a paralyzed right arm and leg, impaired sight, nausea, mental confusion, delirium, and a changed personality. A layman might conclude she had had a brain injury that could prove fatal. Yet her vital organs were normal. This and other signs led the trained physician to one conclusion: she suffered from that "enigmatical state, known since the time of the Greek physicians as hysteria, which can simulate a whole series of symptoms of various diseases." [8]

Breuer hit on the idea that her symptoms were not arbitrary or senseless, but embodied the memory of one or more acutely painful psychic shocks when she had suppressed powerful feelings. As Anna O. recalled under hypnosis the first onset of each symptom and vented her emotions freely, the symptom permanently disappeared. She called this "chimney sweeping" or the "talking cure."

"No one," Freud said, "had ever cured an hysterical symptom by such means before, or had come so near understanding its cause." [9]

Freud confirmed Breuer's conclusions with his own patients. They, too, could not escape the past, and neglected present reality in its favor. Their symptoms were but the "remnants and memory symbols" of past traumatic experiences.

For a symptom to occur, strong emotions had to have been denied a normal outlet in appropriate words or actions. Then, the experience itself or its connection with the symptom would be forgotten. Only if the whole experience was recalled and the pent-up feelings fully discharged did the symptom disappear.

His second lecture offered a hopeful, practical approach to the origin and treatment of nervous disorders. He began by flatly contradicting Pierre Janet, the admired French authority who had lectured at Harvard three years before. Janet had explained a patient's failure to remember a traumatic event as the final result of hereditary degeneracy. Hysterics inherited a weakened ability to synthetize their experience, and tended to split off, or dissociate parts of it. But, Freud argued, hysterics actually showed no signs of impaired mental ability. Anna O. was even more fluent in English when she had forgotten her mother tongue.

His special technique and much of his subsequent theory came from trying the talking cure without hypnosis. He learned that patients fiercely resisted remembering. Why?, he wondered. It was as if a powerful force kept the memory from becoming conscious, the same force, he guessed, that originally had caused the patient to forget. What was this force? He concluded that it was made up of the "ethical, aesthetic and personal pretensions" of the patient. Each experience had aroused a wish that conflicted violently and irreconcilably with just these other forces. To protect himself from the wish and from the intolerable pain of conflict, the patient, after a struggle, forgot the experience.

Anna O. had not exhibited this struggle because hypnosis had suppressed the resisting forces. These were revealed with stark clarity in one of his own cases. A young girl, shortly after her father's death, had fallen in love with her sister's husband. Her sister died suddenly, and the girl thought to herself, "Now he is free and can marry me." She repressed this wish at once, but severe hysterical symptoms replaced it. Finally, without hypnosis and after agonizing resistance, she remembered the forbidden wish with "every sign of intense emotional excitement, and was cured." [10]

Freud offered his therapeutic alternatives to repression—sublimation and conscious choice. A wish merely thrust out of consciousness, like a rowdy tossed out of a lecture, could continue to raise havoc. But the therapist could offer a happier outcome than repression and symptoms. A patient could be led to accept his wish wholly or only in part; or he could "sublimate" it, direct it to a "higher goal." Or he might decide that he had been entirely justified in rejecting it. But now he could reinforce the faulty, unconscious mechanism of repression by his own better judgment. "One succeeds in mastering his wishes by conscious thought," Freud insisted, displaying his invincible faith in rational solutions.

Freud's third lecture described his technical innovations—free association and the analysis of dreams and the bungled acts of everyday life. Freud argued that these techniques could reveal repressed wishes because mental processes were strictly determined; nothing was trifling, nothing arbitrary or lawless. A patient must learn to say whatever entered his head, no matter how trivial, painful, or seemingly irrelevant. Without quite establishing the grounds for his belief, Freud argued that if the patient concentrated on seeking out the wish, any idea that occurred spontaneously to him would be a disguised substitute for that wish.

Laboratory experiments had confirmed his faith in the precise determination of associations. The Zurich psychiatrists Carl Jung and Eugen Bleuler had asked patients for their associations to a set of stimulus words. Prolonged or unusual reaction times and the associations themselves furnished clues to what Jung had christened the patient's "complexes," constellations of ideas highly charged with emotion, usually unconscious.

The "old and despised" art of dream interpretation, however, provided the most daring and unlikely *via regia* to the unconscious. Dreams demonstrated an astonishing similarity between normal and abnormal. Dreams of healthy adults showed outer similarity and inner relationship to the fantasies of the insane.

Most people—including all psychiatrists—regarded dreams as

absurd and senseless, full of "unrestrained shamelessness and immoral longings." Yet children's dreams were easy to interpret. They always fulfilled wishes aroused the day before but not satisfied. The dreams of adults, in fact similar, were distorted by the same psychic conflicts that caused symptoms. By associating freely to each element of a remembered dream, one could elicit its underlying meaning. Once more Freud expressed his rationalist assumptions by arguing that this "real sense of the dream" was "always comprehensible." It was associated with impressions of the day before and fulfilled an unsatisfied wish. Dreams also revealed the hitherto unsuspected importance of early childhood. For adults lived their childhoods over again in dreams, and invariably free associations led back to the emotions of those early years.

"With irresistible might," Freud remarked, "it will be impressed upon you by what processes of development, of repression, sublimation and reaction there arises out of the child, with its peculiar gifts and tendencies, the so-called normal man, the bearer and partly the victim of our painfully acquired civilization." [11] Freud added a hypothesis about dreams that supplied an attractive theoretical unity. Dreams were partly individual; yet they also were stereotyped and could represent the symbols that "we suppose to lie behind our myths and legends." The Oedipus complex, as dream and as drama, offered just such a universal symbol.

Then Freud set a dangerous precedent for psychoanalytic polemics. He insisted that one could test the results of psychoanalysis only by mastering its techniques. Some Europeans had scorned psychoanalysis without doing this. Yet these critics hardly would reject research with a microscope because it could not be confirmed with the naked eye. Still other critics resembled neurotic patients: their judgment was warped by their own uneasy repressions.

Freud broached his most explosive subject in his fourth lecture, the existence of sexual impulses in the child. Some members of his audience, including a future psychoanalyst, were shocked and incredulous. The *Boston Evening Transcript* omitted this disturbing theory except for a terse reference to "infantile inclinations" in dreams. The Worcester *Bulletin* was somewhat franker but also brief. It reported that "Dr. Freud showed in much detail the form which sexuality takes in young children . . . how it formed a love and fondness in a child for rocking, thumb sucking, and similar childish habits. He touched also on the natural curiosity of children, especially in the matter of the origin of babies and how this curiosity was suppressed and perverted by stock [sic] and other legends." [12] Even sympathetic colleagues had thought at first that he overstressed the role of sex. They had wondered why no other mental excitation could not also cause symptoms. He did not know why, but other factors simply did not do so.

Then he denounced current sexual taboos. One could judge the importance of sexuality only by the psychoanalytic method, he argued, because patients usually made every effort to conceal from their physicians any information about their sex lives. Moreover, ordinary doctors, like other "civilized" people, approached sex with an unpalatable combination of "prudery and lasciviousness."

"They do not show their sexuality freely, but . . . wear a thick overcoat—a fabric of lies—to conceal it, as though it were bad weather in the world of sex. And they are not wrong; sun and wind are not favorable in our civilized society to any demonstration of sex life." [13] Yet only if the memory of early sexual wishes were made conscious could treatment succeed, because these early experiences established sensitivity to later traumata.

"It is not very difficult to observe the expressions of this childish sexual activity," Freud noted. "It needs rather a certain art to overlook them or fail to interpret them." [14] His own colleagues had substantiated all his claims. A Clark University fellow, using what Freud called the "American method," presumably quantification, had amassed 2500 observations of the emotion of sex love before adolescence.

Freud defined sexuality as inclusively as possible—and as ambiguously—in terms of a developmental process. He began with the child's impulses, divorced from the reproductive act and made up of independent drives, whose aim was to produce pleasure from sensitive body areas, the mouth, anus, urethra, skin, genitalia. The child's drives, impelled by his helplessness, turned outside himself to other people, regardless of their gender, in active and passive pairs of feelings, pleasure in inflicting and in receiving pain, in exhibiting and in being seen. Then education, shame, disgust, morality, and repression blocked these early impulses. At adolescence the genitals assumed primary importance, and all the old component impulses became satisfied in adult sexual love.

This complex process could go awry and cause neuroses, or perversions: auto-erotism, homosexuality, or sexual infantilism. Too strong an expression of sexual impulses could cause a fixation at any early stage; later, if normal sexuality were hindered, the repressions broke down at exactly that point of fixation. Only this wide, genetic view, Freud asserted, could explain perversions, neuroses, and normal adult sexuality by relating them to the early mental and bodily sexual life of the child.

Flouting domestic pieties, he argued that every neurosis stemmed from the child's sexually toned attachment to his parents, a mixture of tenderness and hate. Parents stimulated this "nuclear complex" by their own love for their children, which was sexual in tone but inhibited in aim. Usually the father preferred his daughter, the mother, her son. The son reacted by wishing to take his father's place; the daughter, her mother's. This Oedipus complex was quickly repressed, but from the unconscious continued to exert a lasting influence. Every child had to be able to break away from this ambivalent attachment to his parents and transfer his feelings to other people.

Freud's final lecture combined a denunciation of civilized repressions with a defense of the power of conscious control over impulse. The neuroses, he resumed, represented a return to the gratifications of childhood, a retreat from reality which patients had found intolerable because of their own inner rigidities and because of external hindrances.

Curiously, however, neuroses also were directly related to the "most valuable products of human mentality." Civilization's high demands made reality unsatisfactory. The energetic and successful man transformed his wishes into actuality. Others constructed fantasies to fulfill their desires. The artist by a mysterious process turned his fantasies into artistic creations and thus, like the practical man, also returned to reality. Neurotic and normal people struggled with the same complexes. The difference between them lay in the relative strength of the clashing forces. For the weak or the disillusioned, the neuroses provided the special refuge, the cloister of our time.

Freud offered what he considered the most convincing piece of clinical testimony to those who had observed it, the "transference." Every patient experienced toward his therapist a tenderness often mixed with enmity that was shaped solely by his unconscious wishes. These could be dissolved or transformed by living them over again in relation to the psychoanalyst.

With his customary, artful anticipation of objections, Freud dismissed two as unworthy of consideration. The first was a prejudice against believing that mental life always was strictly determined. The second, more serious he thought, was a fear that repressed impulses, if brought to awareness, would overpower a person's moral and cultural forces. However, making a patient aware of his repressed wishes could do no harm for two reasons, Freud argued. First, an unconscious wish was incomparably stronger than a wish of which one was fully aware, for one could condemn the latter with all one's strength and maturity of judgment. Second, sexual impulses could be sublimated. The act of repression itself required effort. But when a wish became conscious, the energy previously needed for repression could be directed toward a socially more valuable goal.

Freud also suggested a highly ambiguous alternative: part of the "suppressed libidinous excitation" of the wishes had a right to direct satisfaction: "The claims of our civilization make life too hard for the greater part of humanity, and so further the aversion to reality and the origin of neuroses without producing an excess of cultural gain by this excess of sexual repression. We ought not to go so far as to fully neglect the original animal part of our nature, we ought not to forget that the happiness of individuals cannot be dispensed with as one of the aims of our culture."

He closed with a cautionary German folktale to illustrate the results of too much sexual repression. The inhabitants of the town of Schilda begrudged a work horse his expensive oats. They broke him of the bad habit of eating by gradually cutting down his rations; he finally learned to work on one stalk of oats a day. Then he died, to the astonishment of the townspeople. "We should be inclined to believe," Freud warned, "that the horse had starved and that without a certain ration of oats no work could be expected from an animal." [15]

Freud's ambiguities gave rise to conflicting interpretations that would plague the history of psychoanalysis. Some of these ambiguities may never be fully resolved, because Freud did not construct a final, closed system, but worked from cases and problems. Often he failed to discuss assumptions he considered self-evident. Sometimes he never quite reconciled earlier views with later ones. He early foreshadowed many of his later conclusions. Yet his emphases changed with his search for theories that would do justice to the stubborn harshness of the facts he observed in his patients and in the history of his time.

In a real though limited sense, Freud rebelled against conventional sexual ethics. Nowadays, when he is being interpreted as a dour ascetic rationalist, backward, so to speak, from the perspective of his final papers, it is important to recall the *élan* of his earlier protests. Precisely what was the moral code Freud so roughly attacked? What kind of revolutionist was he? A year earlier he had registered his complaints more tartly and specifically than he did at Clark.[16]

The "civilized" code as Freud described it, enjoined total sexual abstinence outside marriage. It presupposed repression of the infantile, partial drives—oral, anal, sadistic, masochistic, homosexual. As a physician, Freud had observed that the code encouraged neuroses and perversions, impotence and frigidity. Marriage was unsatisfactory, postponed too long during passionate youth when sexual desires were strongest. Marriage was vitiated by worries about conception and by inconvenient contraceptive techniques; marital intercourse usually was carried on for a few years only, for a limited number of acts. Women were forbidden to show sexual curiosity or even to know about their own sexual role. Freud doubted that premarital chastity for men after age twenty was the best preparation for marriage; and he denied that abstinence was either simple or healthy, as many physicians asserted. For most people abstinence was constitutionally impossible, and sublimation, difficult and intermittent.

People developed neurotic symptoms because they could not acknowledge sexual desires the moral code forbade; and women were the principal sufferers. "Civilized" sexual morality was the special ornament of the "cultured classes"; the "lower classes" were not bound by it. Only the "weak submitted to the code in all its rigor." One's approach to sexuality was paradigmatic of how one approached all other matters. Abstinence conduced to the making of weaklings, good followers, not "energetic self-reliant men of action . . . original thinkers, bold pioneers and reformers." He repeated this argument in 1917 and still later in *Civilization and Its Discontents.* Neurotics were ineffectual rebels, victims of their inability to repress, sublimate, or fulfill their powerful sexual drives.

He found the "most extreme" sexual morality, the American, "very contemptible," he later wrote Putnam. He himself stood for an "incomparably freer sexual life." But he added, "I myself have made little use of such freedom: only in so far as I myself judged it to be allowable." [17]

Freud's protest against "civilized" morality was grounded in his own experience as well as that of his patients. After his engagement at age twenty-five he had to postpone marriage for nearly five years to finish his medical training. Yet the moral code enjoined chastity. Freud developed what he later described as the "anxiety neurosis" of engaged couples. During the engagement and for several years of marriage, he suffered from that classical nineteenth-century syndrome, neurasthenia—moodiness, indigestion, sensitivity, a numbing feeling of deep tiredness. His late marriage was typical of the Central European intellectual who was rising from the middle or lower middle classes. In about one generation the marriage age of Austrian Jews rose from fourteen and fifteen to the late twenties, as they moved to cities and entered the professions. Freud's father, for example, a country wool merchant, had been married at seventeen.[18]

Freud's Jewishness left its own profound moral stamp. He emphasized effort, work, and achievement and insisted that the morality the individual absorbed as a child played an ineffaceable role in his psychic economy. Freud's theories of sexuality, as Henri Ellenberger and Stephen Kern have conclusively demonstrated, emerged from two decades of rich European work.[19] There were anticipations and parallels for nearly every element of Freud's views. Freud's uniqueness lay in the radical generalizations he made and in his organization of scattered insights into a coherent and brilliant synthesis.

Freud treated sexuality in a delicately aseptic spirit different from that of many other students of sexuality. Unlike Havelock Ellis, for example, he included no paeans to nakedness, or to the bliss of coitus under the open skies in bowers of pansies. Compared with some radicals, chiefly Europeans, who advocated the end of all social restrictions on sexual acts among consenting adults, Freud's view was conservative. If such proposals were adopted, he dryly observed, in one generation, "inferior individuals would automatically eliminate themselves in sterile indulgences in love." [20] Culture depended on the ability to suppress sexuality, and psychoanalysis made such conscious suppression possible. Although sublimation was a commonplace fact to most American and European students of sexuality, Freud gave it a uniquely important role in the model of therapy he presented at Clark. Yet difficulties attended this element in Freud's views: his elastic definition of sexuality and the mysterious nature of sublimation.

Precisely how sublimation was achieved Freud never fully explained. How much was conscious and rational, how much unconscious and involuntary? How much dealt with adult genital impulses? How much with infantile partial impulses? Freud did not make sublimation as simple as some of his contemporaries who believed that sexual energy, the most powerful animal instinct, was a force like any other in the canon of physics and therefore capable of being transmuted into other forms of energy. Yet in the Clark lectures Freud suggested that sublimation was partly a conscious process. But he also insisted there and elsewhere that the decisive formation of the sexual drives occurred in early childhood.

Freud's devotion to civilization was as clear as his complaints against the excesses of "civilized" morality. At Clark he voiced his full acceptance of the "higher" aims of culture. He wrote Putnam that "social morality" as distinguished from sexual, was "self-evident": the obligation to be just, honest, to consider others, not to make them suffer or take advantage of them. His own chaste life and family devotion were models of "civilized" morality. Yet the publicity with which moral demands were made, he added, made an "unpleasant impression" on him.[21]

Obviously the happier issue to neurotic repression Freud advocated at Clark rested on a belief in rational, conscious choice by the "higher . . . human mental facilities." Freud assumed a very real commitment to this goal, and to a standard of judgment as well. His protest against the sexual constraints of the later nineteenth century was undertaken from a deep loyalty to civilized values and to culture itself.

Yet the Clark lectures complicated ethical problems. For example, the most socially desirable actions might develop from the least valuable component sexual instincts. The greatest "friends of humanity" might have been in early childhood "little sadists and animal torturers," Freud wrote some years later. People were "good" or "bad" in relation to specific situations; they were never altogether either.[22]

At Clark, Putnam asked Freud if psychoanalysis would "rule out all moral estimates of whatsoever sort." If it were true that people are as they are because of "biologic evolution, personal experience and social education" rather than "spontaneous choice with reference to an ideal aim consciously conceived," could we then judge one person as better than another, his character "noble," another's "ignoble"?

Freud replied "with impressive earnestness," Putnam recalled, that it was not "moral estimates that were needed for solving the problems of human life and motives, but more knowledge." [23]

Freud made conflict about sexuality a central fact of human life. Sexual desire was inescapable. But morality was as coercive as desire. He believed each individual represented a unique balance of both these forces —sexual drive and internalized regulation. Naturally, he railed against the physicians who prescribed sexual intercourse as a cure for nervous disorders. His criticism of this prescription, the essay "On Wild Analysis," was promptly summarized in the *American Journal of Psychology* in 1910.[24]

He was also provokingly and deliberately imprecise about specific reforms. Many sexologists believed that current laws were damaging marriage itself. The Swiss psychiatrist August Forel wanted easier divorce and a close but chaste mingling of the sexes before marriage. Freud insisted that reform was urgent. He went so far as to say that "enlightenment of the many" by psychoanalysis, both as public knowledge and as therapy, and a consequently greater social tolerance ultimately might extinguish the neuroses. But he also cautioned that every social institution was intimately interlocked with every other social institution and that reform was not the physician's business.

Freud never relinquished his faith in science and in reason, the still small voice, whose mordant verities, once the frenzy of opposition had spent itself, finally would be heard. "I am no pessimist," he once said. "I permit no philosophical reflections to spoil my enjoyment of the simple things of life." [25]

But he did indulge in philosophical reflections, and he did descend from the comparatively optimistic peaks of 1909. His views about man and society became more complex; his doubts grew about the reformation of social customs. He began to explore the inherent difficulties of achieving normal sexuality, an intricate union of tenderness and sensuality. The Great War provoked an increasingly chary view of human nature. Yet in the black year of 1917 he wrote Maria Montessori that he was in "deep sympathy" with her "humanitarian and understanding" educational endeavors.[26] In his final period he usually appended to his gloomiest strictures a small footnote of hope. He continued to assert that changes in sexual customs and in attitudes toward property might allow for greater individual satisfactions. He even suggested that if religious education were eliminated, we would learn more about the potential intelligence of human nature. At the end of *Civilization and Its Discontents* he voiced the expectation that Eros would exert itself to counter hate and aggression. Then in 1931, he added a final sentence, "But who can foresee with what success and with what result?" [27]

Thus Freud could appeal both to optimists and to pessimists, to moral radicals and conservative reformers. This ambiguity was a major factor in his American reception.

American Issues

Freud came to America at a "psychological moment," as his host informed him.[28] Symptoms of crisis and change were coming to the surface in those aspects of American cultural and professional life that psychoanalysis touched most deeply—sexual morality and the treatment of nervous and mental disorder. This study will explore the complex relationship of psychoanalysis to the background and course of these mounting, almost simultaneous crises. A "crisis" in this context means a period of intensified change when social roles and fundamental principles are being redefined.

Reactions to Freud were partly determined by positions Americans already had taken concerning these growing crises, as the responses of participants at the Clark Conference indicate. Ultimately Freud came to symbolize a whole complex of changes that took on momentum in the years when psychoanalysis was becoming an important factor in American professional and cultural life. The Conference also offered a prophetic sample of the social interaction that accounts for what Lord Bryce considered to be the uniquely swift spread of new ideas in the United States. Present at Clark were three major agents of cultural diffusion—professionals, laymen, and the press. Traditionally, each group influenced the others. Professionals with innovations to introduce tried to win public support. Laymen took a keen interest in the latest scientific expertise. The press, with which professionals often eagerly cooperated, crystallized styles and disseminated information in hyperbolic and simplistic ways.

In the reception of psychoanalysis, neurology and psychiatry played the most important role. American neurologists and psychiatrists saw an apparent increase in the incidence of nervous and mental disease and a decline in recoveries. A few were increasingly disillusioned with the accepted somatic style of etiology and treatment. Two years earlier, the leader of the nation's institutional psychiatrists had declared, "Our therapeutics is simply a pile of rubbish." [29] Against this pessimism of the somatic style Adolf Meyer and a small advance guard had set themselves.

Meyer attacked the prevailing assumptions that insanity resulted from a brain lesion in those with an inherited predisposition. More vehemently than before, partly because of the influence of the psychoanalysts, he argued that insanity could result from learned dynamic habit patterns. He wrote to the Boston neurologist Morton Prince, shortly after the Clark Conference, that Freud appealed to his "inborn need for causal and dynamic chains." [30] In his own Worcester lecture, Meyer argued that the psychoanalysts' clever interpretations of complexes and symbolism had gone far toward explaining the apparently senseless symptoms of dementia praecox, the earlier term for schizophrenia, a variety of insanity then widely regarded as organic and largely incurable.[31]

Freud's therapeutic optimism appealed to James Jackson Putnam, one of the nation's foremost neurologists. His reaction to Freud linked the professional and moral crises. Before accepting views that tampered with aspects of the accepted moral code, Putnam had to be sure of Freud's integrity. His impression of Freud's sincerity as well as his brilliance led him to invite Freud, Jung, and Ferenczi to his camp in the Adirondacks. There for three days, around the campfire, they discussed psychoanalysis, and the personal ties they established led directly to the founding of the American Psychoanalytic Association less than two years later.[32]

For a number of important psychologists, Freud's system advanced the cause of a practical rather than a purely academic psychology. In 1900, Hugo Münsterberg, the Harvard psychologist, remarked that, in questions of behavior, psychology remained silent as a sphinx.[33] Neurologists and psychiatrists had looked in vain to scientific psychology for models of psychological functioning that might help to explain their clinical data. Yet, Titchener's school, still largely dominant, tended to disdain such practical considerations. The psychoanalysts' Clark lectures led Meyer to open this issue in an abortive correspondence with Titchener, who did not modify his views for some years.[34] Meyer also had drawn the psychologists' attention to Carl Jung's word association tests, which seemed to open up methods of uncovering repressed material and which had helped significantly to make Freud's views important.

The functional psychology of the psychoanalysts had aroused the interest of William James. At Clark he had discussed parapsychology with Jung, and, in parting, had put his arm around Ernest Jones's shoulder and remarked: "The future of psychology belongs to your work." [35] Yet James's reaction was mixed. Hall insisted that Freud and James had not been congenial and that James had told him Freud was a "dirty fellow." [36] Possibly James was irked because Jung and Freud had easily extracted from a girl who claimed mediumistic powers the fact that she had done so to win the man she loved. Hall himself was annoyed because the Europeans to whom he had referred the case had discovered so easily what he and two Clark colleagues had failed to find out after many interviews.

James was annoyed by Freud's scientific pretensions, and he soon wrote to his former student, the psychologist, Mary Calkins: "I strongly suspect, Freud with his dream theory, of being a regular halluciné. But I hope that he and his disciples will push it to its limits, as undoubtedly it covers some facts, and will add to our understanding of 'functional' psychology, which is the real psychology. The 'function' of Titchener's 'scientific' psychology (which 'structurally' considered, is a pure will-of-the-wisp) is to keep the laboratory instruments going, and to provide platforms for certain professors." [37]

James also wrote to his friend, the Swiss psychologist, Theodore Flournoy: "I hope that Freud and his pupils will push their ideas to their utmost limits, so that we may learn what they are. They can't fail to throw light on human nature, but I confess that he made on me personally the impression of a man obsessed with fixed ideas. I can make nothing in my own case with his dream theories, and obviously 'symbolism' is a most dangerous method." [38]

In 1925, two years after his first operation for cancer, Freud recalled James's courage at Clark: "I shall never forget one little scene that occurred. . . . He stopped suddenly, handed me a bag he was carrying and asked me to walk on, saying that he would catch me up as soon as he had got through an attack of angina pectoris which was just coming on. He died of that disease a year later; and I have always wished that I might be as fearless as he was in the face of approaching death." [39]

For Hall and his students, Freud and Jung had confirmed the importance of two of their deepest interests—sexuality and childhood. Lewis Terman, a former Clark graduate, wrote Hall some months later that the lectures had "stirred up" his thoughts more than anything he had recently read. If Jung and Freud were right, "their work is the biggest bomb that has struck the psychologists' camp in recent years. . . ." [40] It may have been Hall, a contributor to the *Nation* since his student days, who summed up his own impressions in an unsigned account on September 23.

> Far too little is known in America of either the man or his work. For twenty years he has been publishing memoirs and even volumes, based

on his vast clinical experience with cases of hysteria and allied phenomena. A pupil of Charcot and Westphal, he has developed what may now be called a new system of psychology, which seems to a growing number of workers in this field, of whom the writer is one, to be the best word yet spoken there. His views are now being talked of in Germany as the psychology of the future, as Wagner's music once was dubbed the music of the future. And yet, partly because so much that he taught was new and revolutionary, he has until lately had but scant recognition; and because he attempts to do justice to sex in his scheme, he was for years, socially ostracized. Happily, he is now coming into his own and a growing circle of very vigorous young men in all civilized countries are giving him due recognition and working out his ideas. It is difficult to give in brief and lucid phrases, any adequate conception of a system that has so many details and even technicalities. His system seems to go beyond Janet; and if it be confirmed, it plays havoc with many of the systems of both philosophical and of laboratory psychology. . . .[41]

In describing Freud as a lonely, ostracized pioneer finally coming into his own through a band of vigorous young workers, Hall projected something of his own self-image as well as Freud's, and at the same time evoked a popular American archetype. Hall's description was a perfect example of that sensitivity to popular culture and journalistic considerations that marked many American intellectuals.

In America popular and professional culture were closely bound together, and, beginning with the Clark Conference, the reception of psychoanalysis occurred on both levels. At Clark, Freud himself acted as a popularizer, and his lectures were intended for laymen as well as specialists. Perhaps encouraged by Hall, Freud also granted an interview to the correspondent for the Boston *Evening Transcript*, Adelbert Albrecht, a journalist and translator. Although Freud occupied a modest place in the *Transcript*'s accounts, Albrecht's partiality was clear. He dismissed Titchener as "critical, rather than creative," while to his mind Freud had gained the "eager interest" and in many cases "won the adherence" of the assembled scientists. Albrecht reinforced Hall's theme of the aggressive youth of Freud's followers. Not only was Freud "one of the most eminent neurologists of Europe," but he was recognized by "the most serious young scholars of the present generation as one of the greatest, if not the very greatest of psychotherapeutists." He described Freud affectionately in a way that Freud, sensitive about his age, might have disliked: "Students of Dr. Freud's books on psychic analysis have doubtless fancied him a cold and cheerless person, but that prepossession vanishes when one confronts the man, bent and gray, but wearing the kindly face that age could never stiffen, more than selfishness, and hears his own stories of his patients." [42] He dramatized and simplified what Freud already had shorn of considerable complexity, and he picked out salient features of Freud's lectures. He stressed Freud's opposition to heredity theories, and the effectiveness of psychoanalytic "cure" in recalcitrant cases.

In the atmosphere of the Conference, Freud functioned almost as if he, too, were an American educator. He denounced popular enthusiasm for religious mind cures. Yet, to the correspondent he made his points in appropriately "American" language, comparing suggestion therapists to Indian medicine men.

Freud's protest against "civilized" morality was omitted from the *Transcript* account, significant testimony to the operation of conventional moral norms. The revolt against "civilized" morality had taken on stylish momentum in Europe in the 1890's, of which the American *fin de siècle* was but the palest reflection. Although there were signs of a crisis in sexual morality among small groups in America then, it did not become acute until the years from 1912 to 1918. The main issues were discontent with the prevailing reticence and the exposure of contradictions between ideals and actualities. Already, by 1909, religious and cultural conservatives complained of a crumbling of moral codes, a new mass society bent on business and pleasure. Subjects that respectable families never would have mentioned a decade before now were being publicly talked about. Some were shocked at the academic nudes displayed in fashionable magazines. Darwinism, relativism, and pragmatism were "blasting the Rock of Ages" and destroying a reverence for moral truths once believed to be eternal. A few Americans asked whether their country were progressing or degenerating.[43]

Since the 1890's small groups of academic intellectuals had been questioning some of the tenets of American morality, and among them were founders of American psychology. G. Stanley Hall and James Mark Baldwin, for example, disliked what they called the "Puritan" suppression of joy and pleasure.[44] They had no wish to destroy the core of American morality. But they did believe that "dogmas" and "Mosaic" attitudes should be dispensed with. They repudiated, in particular, American prudery and reticence which prevented the open discussion of sexual problems and the exposure of moral evils.

Freud's ambiguous protest against "civilized" sexual morality provoked the most disparate reactions. In 1910 the Dean of the University of Toronto told Ernest Jones: "An ordinary reader would gather that Freud advocated free love, removal of all restraints, and a relapse into savagery." [45] A decade later, a protégé of Emma Goldman's, André Tridon, a lay analyst in New York, condemned Freud's doctrine of sublimation as high-minded romantic nonsense. Miss Goldman argued that Freud had proven that sexual repression had crippled the intelligence of women.

She was present at the Clark University gymnasium Friday evening, September 10, 1909, when Freud and the others received their honorary degrees. Visibly moved, Freud said in German: "This is the first official recognition of our endeavors." She recalled how Freud looked at that moment. "In ordinary attire, unassuming, almost shrinking . . . among the array of professors looking stiff and important in their university caps and gowns . . . he stood out like a giant among pigmies." [46] Emma Goldman's perception of Freud's stature probably was hindsight. The claim of G. Stanley Hall that the Conference "launched" psychoanalysis in America was hyperbolic, yet partly true. The attention paid to psychoanalysis in both the professional and lay press rapidly increased. Freud's Clark lectures became a widely read introduction that reinforced the best known version of psychoanalysis for the next decade.

Freud's optimism at Clark was a characteristic cultivated with special intensity by his American audience. Like the psychologist, J. McKeen Cattell, most were happy to be alive in an age of "transition and turmoil." They would have agreed with former President Theodore Roosevelt that, despite a lowering cloud of "evil forces," American prospects were clear. G. Stanley Hall, who had cultivated optimism to tide himself over the painful early days of Clark University, rejoiced in 1904: "Today more than ever, hard times and retrogression are relative and transient . . . good will and beneficence" are at the root of things; "man is the favored and protected being of a power that makes for his unfoldment"; he can trust that all "beyond his ken is well"; that "real evil can not befall the good man, living or dead; and that he can be glad and euphorious that he is alive." [47]

Nothing so reinforced optimism as evolution, Hall believed, and nothing so attested to evolution as the progress of science. Sanitation, the automobile, the wireless, the electric light, the airplane were some of its tangible manifestations. In 1908 a distinguished New York neurologist announced that medicine had wrought miracles and would continue to do so. Antitoxins had conquered diptheria, typhoid, cholera, and tetanus; anaesthesia had eliminated the pain of surgery. The possibilities were limitless. A venerable Supreme Court Justice wished to live long enough to witness the great inventions, the blessings of "peace, comfort and happiness of the next fifty years." [48]

Perhaps the distinguishing characteristic of these years was the matching of alarms and salvation, crises and remedies. The style of progressive calls for uplifting solutions were those which governed the tone of American sexual morality, a tone that echoed the endless sermons exhorting to righteousness on which most Americans had been raised. The reformer, Frederick Howe, later recalled that, when he listened to Woodrow Wilson, it was as if he "got hints of impressions received at home," impressions of "moral passion" when preachers lamented "lukewarmness to Christian ideals" and neglect of responsibility to the church.[49]

Solutions in neurology and psychiatry, as well as in sexual morals, were clothed in a similar rhetoric. William Alanson White, Theodore Roosevelt, the journalist Ray Stannard Baker, Harvard's president Charles W. Eliot described problems and their remedies in much the same terms. First, they exposed and denounced evils: the apathy of psychiatrists and the suffering of patients, the misery of slums, the corruptions of politics, the viciousness of conventional morals. Shame and guilt were intended to inspire redeeming change, to provoke action and good works. The immediate past was dark, but propects were bright.

The rhetoric, the optimism, the close ties between professional and popular culture were all reflected in Freud's American reception. To understand the American response requires a detailed examination of the late nineteenth-century order in sexual morality and the treatment of nervous and mental disorder.

II

American "Civilized" Morality
1870 - 1912

The self-appointed guardian of America's moral purity, Anthony Comstock, spoke at Clark University a few months before Freud. The appearance there of a representative of the nineteenth-century order at its most extreme and one of its severest critics symbolized the growing crisis in "civilized" morality. Where Freud argued that the repression of sexual thoughts could be pathogenic and drew a fundamental distinction between fantasy and behavior, Comstock warned of the baneful effects of "corrupt publications" on the young. Although many of his fellow countrymen ridiculed his zeal and his rhetoric, few would have disagreed with the fundamental premise on which then and earlier he based his case for censorship—the necessity for mental purity: "Once the re-imagining faculties of the mind are linked to the sensual nature by an unclean thought the forces for evil are set in motion which rend asunder every safeguard to virtue and truth." Impure thoughts were "pestilental blasts from the infernal region. . . ." [1]

Freud's host, and Clark's president, Granville Stanley Hall, agreed with Comstock's premise—until his acquaintance with psychoanalysis. His invitation to Comstock testified to his essential conservatism as well as his political astuteness, for his own books had been denounced as immorally frank and he himself was a member of Boston's Watch and Ward Society.

Despite enormous differences, Freud, Hall, and Comstock reflected nineteenth-century "civilized" morality and they agreed on certain fundamentals. They believed that civilization and progress depended directly on the control of sexuality and on the stable monogamous family. They believed in different ways that "mind" should govern the "sensual nature." Even Freud displayed some of the reticence Comstock would enforce by public censorship. It was only after great inner resistance that Freud brought himself to publish the sexual histories of his patients. These broad similarities between men so unlike testify to the strength of the common elements in the European and American version of "civilized" morality.

The American version of "civilized" morality, moulded the American conscience and thus prepared the social ground for the reception of psychoanalysis. This study will explore its formulation by community authorities, the social conditions that gave rise to it, the sanctions that preserved it, and later, the effect on some of those who absorbed it as a fully developed system—the first psychoanalysts and their patients.

Here and in Europe the ancient core of the code, religious in origin, was the same: on the level of adult sexual conduct it prohibited sexual intercourse outside marriage. But nineteenth-century "civilized" morality added prudery to virtue, cloaking sexuality in reticence and requiring that women remain ignorant of their sexual role until marriage. Americans added still more. They required purity of thought as well as of behavior from both men and women and insisted that continence was supremely important. They also described women as feeling little sexual passion.

"Civilized" morality operated as a coherent system of related economic, social, and religious norms. It defined not only correct behavior, but correct models of the manly man and the womanly woman, and prescribed a unique regime of sexual hygiene. These norms were instilled by the pillars of community authority: clergymen, physicians, parents, teachers. "Civilized" morality was, above all, an ideal of conduct, not a description of reality. In many respects this moral system was a heroic attempt to coerce a recalcitrant and hostile actuality.

Freud gave "civilized" morality a distinct social basis: he identified it with the rising, cultivated professional classes who supplied the bulk of his patients. Its purpose, he argued, was to discipline sexuality for the sake of work and cultural achievement. Freud's social analysis can be profitably extended.[2]

In Europe and America the equivalent social classes tried to live by it, and as will become apparent later, it structured similar neurotic symptoms. Probably "civilized" morality was highly prized by many of those Protestants, Catholics, and Jews who formed the middle classes in Europe and the middle and upper classes in America. The European bourgeoisie prided themselves on a strict discipline of work and sexuality because this set them off from the brutish masses and from lazy, wasteful aristocrats. By the early twentieth century those Americans who thought of themselves as self-consciously middle class believed that purity and industry distinguished them from the promiscuous and often alien poor, and from the idle rich, the American equivalent of the European aristocracy.[3]

"Civilized" morality remained the dominant American ideal until about 1912. Indeed, it grew progressively more stringent from the 1880's into the first years of this century. Yet the beginnings of the full-blown system are difficult to date. We know more about the breakdown of "civilized" morality than about its causes and its growth, and much of our information comes from hostile critics such as Freud, who wrote in the 1890's and early 1900's. In nearly every relevant field—the history of morals, sexual behavior, child raising, neurology and psychiatry— our knowledge still is sketchy. Much of this discussion therefore must be speculative, serving chiefly to raise problems for further exploration.

Surely the full system of "civilized" morality had not yet developed in the late eighteenth century. The Founding Fathers were not noticeably reticent or prudish; they did not describe women as passionless or belabor the values of mental and physical purity. One of their descendants, Charles Francis Adams, was acutely sensitive to this real difference in mores. In 1891 he accused New Englanders of almost systematically suppressing any evidence of grossness among their ancestors. Like most cultivated Americans, he deeply wished to believe that there had been "real moral improvement" since 1791. The improvement was twofold: in actual behavior as well as in "outward appearance and respect for convention." Adams believed that progress in conduct was paralleled by the progressive refinement of language from Shakespeare to Hawthorne.[4]

By the 1830's, when Tocqueville visited the United States, significant elements of the later system already were present—an emphasis on virtuous behavior, equally stringent demands on both sexes, reticence in literature. Tocqueville argued that American morality resulted from a coherent blend of religious and economic influences—evangelical Protestantism and middle-class egalitarianism. America was supremely bourgeois and lacked aristocratic or popular traditions of hedonism. Society took on the tone of the "commercial" and "manufacturing" classes, in which Tocqueville included American farmers because they so often produced for a market or held land for speculation. The pursuit of hard work and profit promoted a regular, sober life. Tocqueville's observations applied chiefly to settled communities and not to the frontier, to New England or parts of the Midwest dominated by the New England tradition. Moreover, they reflected the opinions of the middle- and upper-class Americans he knew best.[5]

Moral demands were stricter and more widely observed than in Europe; family life was purer. Democratic equality encouraged marriages for love. As a result, marital faithfulness was the rule, adultery and fornication exceptional. If prostitution existed, as it did, Tocqueville felt that it was less disruptive to family life than European gallantry and intrigue. He noted that lapses were equally condemned in men and women and that both had been punished by the Puritans for premarital intercourse. Paradoxically, sexual probity often went hand in hand with devastating financial laxity. Strict morals created a useful and orderly discipline that made risk-taking and success in business possible.

This American social discipline served a fierce passion for rising in the social order and for acquiring wealth. A contemporary of Tocqueville's, the American psychiatrist Amariah Brigham, also noted the American's restless ambition. Even if men were content to be "daily laborers," he wrote, they wanted their children to be "among the most distinguished of the learned, or among the rulers of the country." [6]

Equality and upward mobility created an appropriate pattern of relaxed parental authority and early autonomy for children. Tocqueville's typical European aristocratic father, like the father Freud described, linked the present with the past, and acted as the keeper of traditions, customs, and manners. He was "listened to with deference . . . addressed with respect," and the love felt for him was "always tempered with fear," Tocqueville wrote.[7] In America, however, paternal authority was weaker; sons displayed more affection than awe. The American went directly from boyhood to manhood without a protected adolescence; he learned early to judge and to fend for himself. American boys and girls associated freely from childhood.

American women, too, exhibited unique qualities. They were not "virginally soft," like the convent-bred French girl, but forthright, free, and "rational." They had to be resilient to withstand drastic shifts in place or fortune. Often women were scarce, and, partly for this reason, were treated with unusual chivalry. They were free to travel anywhere without fear. Yet they were far more bound to the home and to domestic pursuits and duties than European women. Men were careful not to offend women with rude, coarse speech. American literature, including novels, assumed women to be chaste.

Tocqueville's American women were not yet the angelic, innocent creatures prized by the later nineteenth century. American girls were frank in their speech, and far more conspicuous for "purity of manners than for chastity of mind." [8] Maternal, like fatherly discipline, was relaxed, and girls early learned to make their own decisions. They had a realistic and precocious knowledge of good and evil; their eyes were opened to vice in order to shun it. The ideal woman, then, was knowledgeable, chaste, hard-working, and dutiful, but also frank and independent. She occupied a "loftier position" than the women of other lands, and she set the moral standards.

Abundance reinforced morality, possibly encouraging Americans to marry earlier than Europeans. The pattern of age at first marriage in the United States still is uncertain for the entire period up to 1890. Demographers are just beginning to construct a more accurate overview and to demolish old myths.

One of the first legends to be destroyed was the romantic faith of the late nineteenth century that most colonial Americans married in their 'teens. This legend was in fact a protest against what observers thought was a much later age of marriage among their contemporaries of the professional and business classes. In *The Winning of the West,* Theodore Roosevelt wrote: "A young man inherited nothing from his father but his strong frame and eager heart; but before him lay a whole continent wherein to pitch his farm, and he felt ready to marry as soon as he became of age, even though he had nothing but his clothes, his horses, his axe and his rifle." [9]

It now seems likely that in the seventeenth century the New England pattern resembled the European, with men tending to marry in the late twenties and women between twenty-one and twenty-four. In the eighteenth century, however, the age of marriage in America may have declined so that ordinary men and women married in their early to mid-twenties. The educated upper classes, graduates of Harvard and Yale, for instance, married a few years later; only a tiny 2.5 per cent still were single at the age of forty-five. Marriage was an economic and social necessity in the conditions of American settlement, and children, an economic advantage. Infant mortality was high. Wives often died in childbirth; since it was almost a religious obligation to remarry, Americans often had several successive wives in a lifetime. Americans ridiculed Thomas Malthus, the English economist who fretted over the pressure that might be exerted on the food supply by the multiplying lower classes.[10]

Much that Tocqueville described as uniquely American continued a robust existence into the twentieth century: the passion for upward mobility; marriage for love; relaxed family discipline; the close agreement of religious and economic values. Yet American sexual morality became noticeably more severe and refined.

An exceptionally high value was placed on continence and purity. This more stringent system remained the dominant ideal of the middle and upper classes until about 1912 and of the men and women who became America's intellectual and political leaders in the Progressive Era. "Civilized" morality was honored in the rich aristocratic New York family of Theodore Roosevelt, the impecunious Presbyterian parsonage in the South where Woodrow Wilson was born, in the New England farming village where G. Stanley Hall grew up, in the Illinois home of William Jennings Bryan. These men were born from the 1840's to the early 1860's.

What explains the new genteel severity? Because analogous developments occurred in European "civilized" morality, common factors may have been responsible: longer education and professional training for more people; the desire for a higher standard of living and for smaller families; the growing influence of women; possibly a rise in the age of marriage.

By the 1890's Americans were marrying somewhat later than in the early years of the Republic. On the whole, old-stock Americans married later than immigrants, city people than country folk, professional men than farmers and workers. Marriages in the South and West generally were earlier than in the Northeast. The greater the amount of education, the later the age of marriage for both men and women—and perhaps most important, more Americans were getting longer education and professional training. On the whole, the well-educated upper classes tended to marry among themselves. Harvard and Yale graduates were marrying at around twenty-nine to thirty-one, somewhat earlier than professional men in England.[11]

The number of single people among the educated classes increased. Of the men who were graduated from Harvard between 1870 and 1879 nearly 28 per cent still were single between the ages of forty and fifty. More girls who attended women's colleges remained single or married later than girls who did not. At the top levels of American society there may have been growing numbers of spinsters, bachelors, and couples who married late. Conditions related to the economic system may help to account for this.[12]

Middle- and upper-class men felt obliged to support a family at an acceptable standard of living. Women were not expected to work outside the home because this might reflect on the husband as a provider. For those ambitious to rise, a suitable economic stake was essential. Young men were cautioned not to marry until their "prospects" were assured. "Penury," one writer admonished, was "prone to undermine wedded love"; a certain amount of capital or income was "almost indispensable." [13]

But this "indispensable amount" also increased as the standard of living rose. The expectation of a more comfortable life, often condemned as a "desire for selfish luxury," was considered the chief single cause for marrying late and for having fewer children. "To give $10,000 tastes and aspirations on $1000 incomes," G. Stanley Hall warned, "tends to delay and perhaps repress the desire for a family, and thus the best years for genesis are lost." [14]

The skills required to earn an adequate income grew more difficult to acquire; education lengthened. Training in law, medicine, and other professions became more rigorous, costly and protracted. More Americans began to attend high school and college. "For the majority of college men," G. Stanley Hall observed in 1903, "marriage is deferred much later than in former years because in the sharp competition of the present day, whether a man enters one of the professions or goes into business, a period of five to ten years and often more elapses before he is ready to take upon himself the support of a family." [15]

Cities contributed indirectly to late marriage because the urban cost of living was higher and large families were less desirable. In the city, children tended to be a disadvantage, unless parents were willing to have them work and forgo education. By the 1860's Harvard and Yale graduates were having about two children per family. Some members of the upper and middle classes began to regard the fecund poor as victims of unbridled lust. Indeed, sensuality, after drunkenness and misfortune, came to be regarded as a principal cause of pauperism. Around 1900 "cultivated and refined" young people were coming to believe that large families, presumably among their social equals, signified a "vulgar and sensual life."

As medicine and living standards improved, more children survived. If large families contributed to the downward movement, how were children to be prevented? Later marriage provided one answer, for people who married late had fewer children. Birth control was still another. Throughout the nineteenth century Malthusian practices were crude, and in the United States the transmission of contraceptive information through the mails had been forbidden by Anthony Comstock's obscenity law passed by Congress in 1873. Physicians were so bound by reticence and possibly by fear that they seldom discussed contraceptive techniques even at medical meetings. Nevertheless, birth control information enjoyed a wide underground circulation. Methods were risky—coitus interruptus, probably the most common, crude condoms, sponges for women, abstention during fertile periods, intercourse without ejaculation. Abortion probably was widespread despite its dangers.

The difficulty of obtaining contraceptive information, and the crudity of methods, the desire to limit families and the tradition that a wife must "submit" to her husband's demands presented women with agonizing problems. Sexual intercourse involved the risk of unwanted pregnancy, abortion, and, possibly, death.[16]

In the Progressive Era a woman physician observed that her sex had become "cruel and cunning" in its desperate desire to escape unwanted "propagation." [17] Many women came to regard marriage as little better than legalized prostitution. Sexual passion became associated almost exclusively with the male, with prostitutes, and women of the lower classes.

The influence of women in America, already strong in Tocqueville's day, grew stronger toward the turn of the century. More and more women worked outside the home, alongside men, and were exposed to new opportunities for sexual lapses. Middle- and upper-class girls began to pursue professional careers, sometimes in law and medicine, most often in teaching. They played a growing role in formulating both the norms of behavior and the manners of later nineteenth-century Americans. By 1900 women made up the majority of the nation's elementary-school teachers; some of them proudly hoped that the "vast affectional resources of the woman's heart [would be] brought out in every school room." [18]

By 1900 middle- and upper-class women were taking a more militant interest in enforcing a single standard of sexual morality. Possibly sexual morality became associated with the authority of the mother rather than the father. Here the city and the industrial system may have played important roles. Boys who grew up on farms, such as G. Stanley Hall, often worked with their fathers and absorbed information about sexuality and sexual morality from them. In urban families, men spent more and more time away from home and took a less active part in family discipline. One of the first sex education manuals written in the Progressive Era suggested that probably mothers would first impart sexual information to young sons and daughters. A popular etiquette book suggested that if mothers were their sons' confidants, well-behaved future husbands would be the result. Indeed, a mother's task was to keep her children pure and innocent during the decisive first ten years of life. Women should check "offensive speech" and conduct, set the tone of society, and control the manners and behavior of their fiancés. Around 1900 complaints increased that American culture was being feminized. Schoolmarms, G. Stanley Hall believed, were inculcating a "Sunday School" type of feminine sexual morality, overnice and prissy. Such "maternal codes," he charged, were making it more difficult to discipline boys. Another writer said that mothers were setting the "spiritual tone" of the household and had done the most to lift mankind from the "pit of animalism." [19] Lord Bryce, no doubt consciously echoing Tocqueville, considered the pervasive influence of women a purifying and ennobling force in American society.

Americans clearly realized that sexual passion was strongest during youth. Yet most young American men and women continued to have great freedom to be together, and it was assumed this would not be abused. Among the well-bred young, overnight excursions, conducted with light-hearted gayety and a total absence of impropriety, were fairly common.

As the age of marriage grew later, as more people remained single, as women came to work outside the home, sexual temptation may have seemed more dangerous than ever before. The alternatives to purity—prostitution, fornication, adultery—all broke the core of the code to which the middle and upper classes were deeply committed. Lapses from purity also risked illegitimate pregnancies and "excessive amativeness" in marriage, unwanted pregnancies. Sexual promiscuity fostered venereal diseases in a period when syphilis was still a loathsome and incurable scourge.

For all these reasons, it was entirely logical to place an unusual emphasis on continence and to condemn sexual pleasure outside of marriage as supremely evil. Ignorance, it was commonly believed, protected modesty, which in turn protected virtue. Children were systematically raised in total innocence. The model lady was angelic and relatively passionless. Sexual temptation had to be minimized by silence and censorship, by social ostracism and strict decorum. As Freud was plainly aware, "civilized" morality had a consistent internal logic.

Up to about 1912 standards of purity grew more rigorous and more refined. Throughout the development of "civilized" morality there had been close reciprocal influences between cultivated Americans and English middle-class leaders. In nearly every aspect of reform and culture, including sexual morality, English example set the tone. Doctors and clergymen, two important American community authorities, sometimes citing English precedent, developed appropriately stringent dogmas of sexual virtue and sexual hygiene. They stressed the economic and religious utility of sexual virtue. Not only salvation, but success could be expected from chaste living.

When and how the Protestant churches began to place a conspicuous value on sexual purity is unclear. No doubt complex religious developments, including the evangelical movement, were partly responsible. Henry May has suggested that as interest in theology declined, concern for personal virtue grew. By the late 1870's, according to the physician A. J. Ingersoll, an emphasis on absolute purity of thought and a condemnation of all sexual desire as low and sinful was widespread, preached with special fervor by theology students and by college and seminary teachers.[20] The agreement of religious and economic values that Tocqueville had noted can be observed in the sermons of two of America's most fashionable urban preachers. Their congregations, like those of most Protestant city churches, became more exclusively restricted to middle- and upper-class native Americans.

Phillips Brooks, rector of Boston's Trinity Church, assumed both a "divine hate" of impurity and an austere devotion to hard work in the truly "choice young man." Henry Ward Beecher associated sloth with sexual licence and animadverted on the evils of Solomon's "strange woman": "Thou hast forsaken the covenant of thy God. Go down! Fall never to rise! Hell opens to be thy home!" [21] Beecher warned against sexual thoughts. "The first purity of imagination," he thundered, "if soiled, can be cleansed by no fuller's soap." G. Stanley Hall recalled that his minister often dwelt on impurity, the "unpardonable sin." Many years later, Billy Sunday denounced sexual sin at his mass revivals; revivalist churches may have encouraged a sense of guilt in adolescent converts. Anton Boison, a founder of pastoral psychiatry, who grew up in the 1890's, complained that the churches and the Young Men's Christian Association aroused the sense of guilt and the fear of hell to control sexual conduct.

A Boston authority on the family commended the possibility of married celibacy. Only higher considerations and the mutual consent of husband and wife sanctified sexual intercourse. The body, readers were admonished, was the temple of the Holy Ghost, and "should under no condition be defiled." [22]

The double sanctions of economic probity and religious purity, morals and utility, were invoked in language suggestive of Max Weber in 1870 by an editorial in the *Nation.* Henry James, Sr., the father of William and Henry James, had complained that the *Nation* seemed to consider the sexual passion the most "untameable." The editor replied: "A man is morally ruined, in the eyes of the economist, when he runs about whoremongering, instead of working at his calling and supporting his wife and children; he is morally ruined, in the eyes of the theologian, when he disobeys God's law. But there is nothing necessarily incompatible in the two views." [23]

Some physicians taught a sexual hygiene that explicitly united economic and religious motives. The most devout developed the strictest standards of sexual discipline. Religion and medicine completely harmonized, one physician wrote, because religion was an "unrecognized branch of higher physiology." [24] Doctors, articulating the ideals of the industrious middle and upper classes, subordinated sexual gratification to family affection and social achievement. Physicians developed models for the manly man and the womanly woman that prescribed optimally healthy sexual behavior.

The central doctrine of this "civilized" sexual discipline held that energy must be carefully husbanded and that weakening excesses be avoided. Physicians often likened bodily energy to a stock of goods, or more commonly, to a bank account. Its fixed and limited quantity easily could be exhausted unless judiciously replenished. All otiose gratification therefore was undesirable, and sexual continence, or self-restraint, was a logical necessity.

Some physicians urged continence on social and economic grounds. Beginning in the 1830's and 1840's a new medical literature cautioned not only against masturbation, fornication, and adultery but against "excesses" within marriage.[25] Physicians assumed that the fewer the number of children the more vigorous they would be. William Acton, an influential English doctor, explained that only the boy with "abundant energy" could be successful in the "struggle with the world." He made the class and occupational implications of his argument quite clear. Continence was appropriate for the business and professional men of the new industrial and commercial urban world. He prescribed continence for the unmarried, and moderation for the married "hard-working intellectual" but not for the "strong, muscular countryman." An American physician observed that men engrossed in strenuous enterprises or arduous physical occupations could and did remain continent for long periods. With crass directness, Dr. James Foster Scott, who had been trained at Yale and Edinburgh, expressed the economic advantages of sexual purity in terms of Lamarckian social Darwinism. The pure would be the most vigorous and therefore their descendants would be the "most likely to survive in the struggle for existence." [26] The instincts of self-preservation and sexuality worked harmoniously together to create wealth and a stable family. Steady application to business or other pursuits resulted in the accumulation of property. Those who conformed to society's moral restraints came to occupy a "dignified position among the solid men of the community." G. Stanley Hall in 1911 summed up the advantages of husbanding sexual energy: "Thus the sex organs have two functions: the first is reproduction and the other is to give force and energy to all other parts and to character generally. All work involving great effort, either mental or physical, requires sexual temperance, and Delilah always robs Samson of his strength." [27]

Like Freud, influential physicians believed that sublimation of the sexual instinct was a primary source of the high achievements of civilization. Henry Maudsley, an English psychiatrist widely read in America, wrote that if man were "robbed of the instinct of procreation and all that arise from it mentally, nearly all poetry, and, perhaps, the entire moral sense, as well, would be torn from his life." [28]

Higher creations resulted from transmutations of physical sexual energy, which functioned like electricity. An American physician emphasized the similarity to electricity by quoting Meredith's *Ordeal of Richard Feverell:* "All at once an alarming delicious shudder went through her frame. From him to her it coursed, and back from her to him. Forward and back love's electric messenger rushed from heart to heart." [29] Sir Thomas S. Clouston, a Scottish psychiatrist also highly respected in America, insisted that sexual energy could be transmuted just as black coal was converted into luminous electricity.[30]

It was especially necessary to conserve energy during the years of growth. Physicians developed a theory of late maturity justifying the postponement of marriage that the middle and upper classes practiced. Men did not complete full physiological and moral growth before twenty-one and more usually at twenty-four or twenty-five. By then a man would have secured his place in the world and could afford to support a wife. Until marriage, all his energies must be concentrated on winning his way. Women reached maturity earlier, at about twenty. Sexual indulgence before development was completed could be disastrous physically, mentally, and morally; indeed indulgence might foster bodily weaknesses and ineradicable sexual obsessions.[31]

One influential group, led by Acton, and reinforced by the French-American neurologist, Charles Edouard Brown-Séquard, insisted on continence because of the ostensibily energizing properties of semen. Acton's *Functions and Disorders of the Reproductive Organs in Childhood, Youth, Adult Age and Advanced Life* went through eight American editions from 1857 to 1894. The fourth American edition in 1867 was described by *Medical Record* as "a classic" based on sound philosophy, science, and a "religious stand-point." Born in the Shillinton rectory in 1814, Acton opened his book with a catalogue of injunctions to continence from the Bible, the church fathers and the saints, who, he wrote, were far franker about sexuality than his own contemporaries. Ecclesiastical authorities universally enjoined mental and physical purity: sometimes they stressed the invigorating properties of semen, thus partly explaining the prowess of a celibate priesthood. The "life giving substance," Acton elaborated, should be carefully husbanded until full maturity. Even then, semen played a "most important part in the human economy" and could be "ill spared in the healthy, vigorous adult." [32] If retained in the body, semen was reabsorbed and invigorated the entire organism. Later a few physicians extended this doctrine to the female sexual secretions as well. The sexual substances created not only the primary and secondary sexual characteristics, but enhanced the finest mental, moral, and physical traits. One of the first American physicians to use Acton as an authority, John Cowan, added a phrenological justification for continence. The brain produced a "nervous fluid" that supplied the body with vital power. When a special organ, such as that of amativeness (located in the cerebellum), was "greatly employed," the fluid was diverted to that organ at the expense of other vital functions. Reabsorbed by the system, semen promoted the development of this vital brain fluid. Cowan's book, first published in 1869, still was widely advertised and apparently selling briskly in 1916. Dr. Alice Stockham, in *Tokology,* a popular compendium of vegetarianism, Eastern mind cure, and gynecology, relied on Acton and Cowan to justify the beneficent effects of continence in producing healthier children. Dr. J. H. Kellogg, the sage of corn flakes and Battle Creek, Michigan, based an elaborate doctrine of strict continence on Acton and on the research of Brown-Séquard, who at eighty, Kellogg testified, displayed remarkable "vigor and vitality" after an infusion from the sexual glands of a male rabbit. Kellogg's book of sexual counsel, *Plain Facts for Both Sexes,* sold 300,000 copies between 1879 and its fifth edition in 1910.[33]

Influenced by Acton or by his French mentor, Claude Lallemand, many physicians issued nightmarish warnings to both sexes about the evils of masturbation. Masturbation would destroy energy and unsex its victims. It would cause sickness and insanity, and by Lamarckian laws of heredity, result in offspring destined to become nervous and moral weaklings. From 1840 to 1880 a rash of semi-popular tracts spread this gospel. Young men were warned against involuntary losses of semen. In 1902 a professor of urology in the Chicago Polyclinic still could issue some of these warnings. Dr. Kellogg in 1910 repeated most of them. As late as 1912 New York had museums for "men only" depicting the horrible results of onanism. Quacks battened on these doctrines and advertised secret remedies.[34]

However, in the late 1870's influential British and American physicians disputed the extreme claims of Acton and Lallemand. Both were now described condescendingly as "older authorities," as indeed they might have been since Acton died in 1875, and Lallemand, in 1853. Probably the first major protest was delivered by Sir James Paget, Acton's London colleague, and Queen Victoria's physician. Paget and other physicians on both sides of the Atlantic were consulted by disturbed, anxious men who believed that their strength was draining away in nocturnal and diurnal pollutions. Sexual hypochondriasis and functional disorders were becoming an irksome medical problem, chiefly among the "sensitive and highly educated" classes. A distinguished surgeon and neurologist, Paget argued that masturbation, which he believed to be almost universal among both sexes, was no more harmful than sexual intercourse. Rather than causing nervous and mental diseases, it was a symptom of their existence. Nevertheless, nearly every physician continued to insist that excessive masturbation and sexual intercourse were debilitating and dangerous.[35] Masturbation must be cured, and physicians recommended hard work, rest, cold steel, padlocked canvas drawers, and, sometimes, blistering fluids.

As knowledge became more precise, it became clear that the internal secretions of the testes, not semen, created the secondary sexual characteristics. Gradually, the best-trained physicians developed a more moderate attitude.[36] The grounds for prohibiting masturbation shifted from physical to moral and psychological. Paget was one of the first to insist that guilt was more harmful than the physical act itself. The new attitude was best represented by Howard A. Kelly in 1908, probably America's most distinguished gynecologist and professor at the Johns Hopkins University Medical School. He argued that, although the physical effects of masturbation were "surprisingly small," the moral and psychological ones were baleful. "The undermining of self-respect, the tortures and the shame react on the general health surely and frequently and deeply." [37] He suggested that "confession" was a first step toward curative treatment by suggestion and will power. Yet, the tone of the older physical warnings lingered on in the first manuals of sex hygiene for the young written in the Progressive Era, and in pamphlets issued by the United States Public Health Service in the early 1920's. Perverse methods of intercourse, coitus interruptus, and homosexuality all violated the natural order and inevitably would bring mental and physical disaster. Physicians combined an avid interest in the details of these phenomena with frequent expressions of horror and disgust.

Sexual hygienists were especially troubled by two related problems, the desirable frequency of sexual intercourse and the sexual nature of woman. Extreme partisans of retaining sexual secretions hoped to minimize intercourse. Acton cautioned that three times a week was excessive.

Cowan taught that sexual intercourse should take place after the menses, and then, conception having occurred, not again until about eighteen or twenty-one months or even three years later. Kellogg argued that intercourse take place infrequently and then solely for the sake of procreation. Most physicians argued that the only true purpose of the sexual act was to beget children. Refusing to lay down laws of frequency on grounds that individual strengths and passions differed, they usually warned against marital "excesses," and one of America's most celebrated neurologists, William Hammond, an authority on "male and female impotence," flatly declared that daily intercourse was too much, even for the strongest.[38]

Physicians were clearly aware of the power of the sexual instinct, and most of them, no matter how genteel, fought a romantic sentimentality that would divorce body and mind, love and sexuality. Most medical authorities insisted that "temperate gratification" had a beneficial "sedative and tonic" effect. On this point neurologists, psychiatrists, gynecologists, and urologists agreed. A few even warned that "continence" could be bought at too high a price in nervousness and illness, and they tended to urge marriage as soon as possible after "full" maturity had been reached. All of them insisted that love and sexuality were inextricably linked. One function of sexual intercourse was to endear married couples to each other. "To ignore the bodily and secular aspect of it, . . ." Acton wrote, "would be as false and unwise, though not so degrading, as to forget the mental and spiritual." [39]

Some physicians, however, believed that the sexual nature of women made problematical this mutual physical basis of love. Physicians agreed that women, partly because of their training, had less sexual appetite than men, and there is some evidence that the physicians' stereotype of woman became progressively "purer" from the 1870's to about 1912. By 1911 at least one physician found the passionless woman the norm.[40] In 1879 a distinguished Philadelphia gynecologist, George H. Napheys, divided women into three categories: those with little or no sexual feeling, a group larger than generally supposed; a more numerous group tyrannized by "strong sexual passion"; and the "vast majority," with moderate sexual appetites. He denounced wives who plumed themselves on their "coldness and the calmness of their senses." [41] By 1906, however, some physicians regarded the asexual female as the norm: "It may be offered that the sexual appeitite in the majority of American females is evoked only by the purest love. In many the appetite never asserts itself and, indeed, the only impulse thereto is in the desire to gratify the object of affection." [42] A standard physiology text argued that for the "majority of women belonging to the more luxurious classes," after the age of twenty-two or twenty-three sexual intercourse from a physical point of view was a "mere nuisance." [43]

In some women this repugnance resulted from intense pain experienced in intercourse because of physical and psychological factors, including male ineptness. Physicians coined terms to describe this painful sensitivity—"dyspareunia," "vaginismus." Nearly every gynecologist, plagued by the questions of his patients, took a stand on one question. Did the absence of female pleasure in intercourse cause sterility? A few physicians believed that it contributed to barrenness, sometimes directly, sometimes by discouraging the husband's potency. Usually, however, gynecologists and urologists insisted that the woman's reaction made absolutely no difference whatever. They repeated time and again that women who were sexually anaesthetic bore children and often had large families.[44]

Some physicians, including Napheys, apparently had to muster a variety of arguments to persuade patients of the "dignity and propriety" of the sexual instinct in women. Nevertheless, many of them, including Tilt, stressed the importance of woman's satisfaction in intercourse. Tilt cited the evidence of spontaneous orgasms in "pure-minded" widows to show the power of the instinct.[45] A Philadelphia gynecologist and colleague of Weir Mitchell's, William Goodell, warned husbands that mutual pleasure was essential to the true function of marital intercourse. Such consideration had several important consequences. For one thing, a husband must be careful to "restrain" his ardor at times when his wife was indifferent. Nor must he subject her to indiscriminate pregnancies, or "excessive" intercourse: "Destroy the reciprocity of the union, and marriage is no longer an equal partnership, but a sensual usurpation on the one side, and a loathing submission on the other," he warned.[46]

Some physicians puzzled over why women tended to be sexually anaesthetic. Some ascribed coldness to training in "repression"; Alice Stockham observed that women believed that if they did not feel sexual pleasure they would not conceive. Coldness then was a method of birth control. When "womanliness and modesty are synonyms for repression," she wrote, when woman "lives in fear of maternity, and believes restraint on her part prevents vitality of life germs, when, too, erroneous habits pervert her every function, how can we tell what is natural for her." The lessons of "repression" were "graven" into woman's "very being by all the traditions, prejudices, and customs of society. . . ." [47]

Another physician protested: "A woman who would confess to feeling sexual desire would lose her reputation for virtue; and because under these circumstances, she will not often confess to it, we set it down in our text-books that an ideal mother has no sexual feeling, and only submits to her husband's embraces to gratify his passions. . . ." [48] A neurologist attributed this coldness to "hyper religious and hyper moral" ideals taught by preceding generations. [49]

Sexual repression, modesty, and innocence were associated with middle- and upper-class women. These traits were the sexual equivalents of social gentility and refinement. Women were expected to be "ministering angels," creatures of "more heavenly endowments." The words "suppression" and "repression" were often used interchangeably to describe the way females should control their passions. An etiquette book laid it down that ladies suppressed "undue emotion" and cultivated a modest reserve. [50]

Middle- and upper-class males were brought up to regard girls of their own class as future wives, to be won romantically, not to be used as objects of passing sexual desire. Often the struggle against temptation was fiercely consuming. The sexual passion, one eminent urologist wrote, was a "burning flame . . . a seething volcano without metes or bounds. . . ." [51] Mental chastity aided the struggle for physical purity. Manly men and womanly women kept their minds pure. A physician who campaigned for sex education in the Progressive Era argued for clean thoughts on grounds that precious energizing substances must not be lost. "If a man wants to avoid injury to the body and brain through the juices thrown into the blood stream by mental pictures and lecherous imaginations, he must be mentally chaste." [52] Pure minds protected men against that dangerous temptation, recourse to prostitutes. Some physicians condemned this not only because of disease but because sensual images "conserved in the cerebral centers" would prevent "pure sexual relations" with one's wife. [53]

Absolute chastity was a noble and necessary preparation for marriage. Physicians usually insisted that the strictest "continence" was perfectly compatible with health. "It is those who stand this long test successfully," G. Stanley Hall wrote, "that deserve and win in fullest measure the deep and abiding love of their mate, and merit all the respect and gratitude that children should owe to their parents." [54]

Mere sensuality, which drained energy from useful effort, or sexual intercourse divorced from love or the desire for children were the most despised violations of the "civilized" code. These transgressions, which reduced men to the level of beasts, were denounced with scatological epithets as "low," "filthy," "unclean," "defiling." Dr. Howard defined "normal maleness and femaleness," as "the desire for home and children . . . the everpresent watchfulness over family. . . ." [55]

The ideal marriage united the well-suited man and woman, who were both virgins, were strongly attracted sexually, and yet who felt deep mutual reverence and love purged of "excessive sensuality." Devotion, faithfulness, affection for children were to be expected from the creative channeling of sexual energies. Because the inheritance of acquired characteristics was accepted doctrine into the first decade of this century, it was commonly believed that these virtues would be passed on to future generations.

"Civilized" sexual morality aimed to foster the affectionate, industrious family. The closeness and affection, and probably most of the other requirements, seem to have been met in the marriages of many Americans prominent in public life—Theodore Roosevelt, Woodrow Wilson, William James, and their parents. After forty years of marriage, G. Stanley Hall's father and mother remained passionately attached to one another. During a brief separation, Hall's "wholly undemonstrative father" wrote often and voluminously, and "longed to see and even to kiss her, which we children cannot remember ever seeing him do. . . . My mother felt that she could only half live till he came home. . . ." [56] Hall's parents were poor yeoman farmers. The domestic happiness of the successful professional family was described by Dr. Weir Mitchell: "There are households in which the best qualities of heart and mind are so fitly joined together in a partnership of high aims and dutiful industries as to give at last a sense of that oneness of life which realizes the true conception of marriage." [57]

Social customs, manners, and literature reinforced "civilized" morality. Polite society rigorously condemned sexual lapses, and America's first woman physician, Elizabeth Blackwell, urged parents to ostracize all those who were known to lead immoral lives. A character in one of Dr. Mitchell's novels, published in 1900, observed: "In England one slip as to money matters will ruin the career of a politician—with us it is strangely passed over and soon forgotten. But let a man with us go notoriously wrong about women, and he is ruined." [58] Lord Bryce argued that in the best American society an immoral millionaire would be rigorously excluded, whereas in England a patent of nobility might assure his social acceptance. American etiquette books insisted that morals were as important as manners and that women, especially, must guard social purity.[59]

Since Tocqueville, most American literature had remained as reticent and as committed to "civilized" morality as polite society. William Dean Howells celebrated the New England ideal of sexual purity at its most exalted. His heroes and heroines agonized over the right to declare their passionate and romantic love. Should a girl not give up a suitor because her sister seemed more entitled to his love? Should a man not suppress his affection for a girl because she might have misconstrued his attitude toward her father as one of dislike? In *A Modern Instance* Howells examined, with irony but also with profound admiration, the nobility of a young Boston Brahmin who had resisted the temptation to suggest divorce to a woman he loved who was married to a blackguard. His renunciation, an act of the disciplined Will, expressed the highest standards of class, breeding, and religion. In the opinion of a relentlessly prim male friend of the hero, the fabric of social morals could be preserved only by such personal sacrifice.

"The natural man is a wild beast. . . . No, it's the implanted goodness that saves,—the seed of righteousness treasured from generation to generation, and carefully watched and tended by disciplined fathers and mothers in the hearts where they have dropped it. The flower of this implanted goodness is what we call civilization. . . ." [60]

The extent and depth of literary purity are difficult to exaggerate. Apart from a pornographic sub-literature, sexual themes usually were dealt with obliquely and there was none of the frank literature of passion, commonplace in France, for instance. Indeed, such literature was disapproved of.

Women made up the great majority of the reading public, and their preferences swayed authors and editors. Henry Ward Beecher complained that women were "sneering" at Shakespeare's coarseness and forgetting his nobility.[61] From the 1870's on, there were growing male complaints about female prudery and exaggerated modesty. In 1895, after a visit to the offices of the *Century* magazine, Weir Mitchell, as genteel a novelist as Howells, protested: "The monthly (sic) are getting so ladylike that naturally they will soon menstruate." [62]

The censorship of sexual themes was part of Anglo-Saxon good taste, as well as a more serious mission to uplift the masses. Anthony Comstock's career was based on the prosecution of what he considered to be salacious literature and art. The abnormal was censored even more severely than the normal. Theodore Roosevelt, whose literary tastes were fairly catholic, remarked that his own "atavistic puritanism" made him dislike a reference to Oscar Wilde in the poetry of George S. Viereck, an enthusiastic Bull Moose rooter.[63] Mabel Dodge Luhan, one of the first of a procession of analyzed women to write about their therapies, recalled how the Hearst editor, Arthur Brisbane, cut the word "pervert" from a story. He said it was understood by the "crowd" in one vile way, and "represents a thought which I do not want to put into the minds of millions of people even for the sake of truth." [64] In 1913 the Boston police threatened to act against the impeccable neurologist, Morton Prince, son of a mayor of Boston, for publishing psychoanalytic articles that dealt with sexuality. As late as 1917 some of the basic psychoanalytic texts were kept in a guarded room of the New York public library, to be read in a cage only by those with special permission. Solemn treatises on sex were available officially only to clergymen, physicians, and jurists.[65]

The social controls of ostracism, reticence, and literary censorship were paralleled on the level of personality by the inner psychological controls of Will and Conscience. These functions were derived from pre-Darwinian scholastic and Common Sense models of the mind. The emphasis placed on Will in the late nineteenth century by physicians and psychologists may have reflected in part the more rigorous demands of "civilized" morality. Often discussions of Will in its inhibiting aspects were illustrated by control of the animal passions, sometimes hunger, anger, or sex. The mission of the Will was to restrain the sexual "more than any other function," according to one manual of sexual hygiene. The Will's "highest qualities" were taxed in the struggle to master the "lower nature." Victory in this sometimes desperate battle guaranteed true manliness or true femininity. A "castrated" chastity, achieved without struggle, was worthless. The insistence on mental purity obviously raised new and profound problems of internal control over wishes and desires and may have required a far more sensitive conscience. Phillips Brooks, for one, hoped that Christians would preserve a conscience as tender as a child's, yet cultivate manly virtue in the midst of temptation, to keep themselves "unspotted from the world." Dr. Stockham approvingly cited Professor Huxley's insistence that both Will and Conscience together acted to regulate sexual passion: "That man has had a liberal education who has been so trained in youth that his body is the ready servant of his will . . . and who, no stunted ascetic, is full of life and fire, but whose passions are trained to come to heel by a vigorous will, the servant of a tender conscience." [66]

Socially, "civilized" morality reinforced this conscience by appropriate public manners and tastes. David Riesman has defined social character loosely yet usefully as a pattern of traits significant numbers of a given social group share in common. No doubt many middle- and upper-class Americans conformed to the demands of "civilized" morality and fulfilled the models of the manly man and the womanly woman. America's public leaders sometimes were chosen partly because they best exhibited these traits and reinforced them by example and precept.

In the Progressive Era, a growing note of defensiveness sounded in the calls to uplift. Appeals to moral standards were intended to bolster norms already under attack. Moral leadership received even stronger emphasis than before. G. Stanley Hall suggested that Americans were following men like Roosevelt because they were "ethical teachers." [67] Just as the stereotype of the asexual woman became more extreme, so, too, the emphasis on morality grew more clamorous. Native Americans opened campaigns to end the double standard. New methods of control were sought to enforce sexual purity.

Theodore Roosevelt best exhibited the truculent concern for norms and the identification of purity with manliness. In 1913 he easily won the title "greatest man in the United States" in a poll conducted by the muckraking *American Magazine,* which also published enthusiastic stories about mental healing and psychoanalysis. His character displayed superbly the social discipline of American "civilized" morality. He was sure of who he was and ought to be; his moral judgments were unconditional; his private conduct was governed by accepted public standards. He had what the psychoanalyst Alan Wheelis might describe as a "coherent sense of self," or identity.[68]

Like most Americans of the upper and middle classes, Roosevelt believed in Free Will. A man's fate was in his hands. Without being explicit about his definitions, he seemed to think of Will just as William James did—a capacity for effort against the line of greatest resistance. Will included steeling oneself to duty, regardless of pleasure or pain, and subsumed resolution, courage, energy, self-control, fearlessness in taking the initiative. Roosevelt respected most about a man what he made of himself. His own transformation, partly directed by his adored father, from the spindly, asthmatic child to the Harvard pugilist was a superb example of this therapeutic self-improvement.

Correct training, including swift, certain punishment, produced the manly man and the womanly woman. Men were at their most masculine when they worked and fought and kept themselves pure. Women were truly women if they were willing and able to bear children. For both the "prime end of life" was the family.[69]

Roosevelt expressed in starkest terms the American pride in sexual purity. Men and women should be chaste: sexual intercourse must occur only within lifelong, monogamous marriage. Adultery, he once remarked, was as heinous a crime as treason. Probably Roosevelt also believed that sexual acts should be directed solely toward the procreation of children, judging from his condemnation of birth control. No nation could be great, he insisted, which did not accept the principles of honesty and "moral cleanliness."[70] Before his marriage at twenty-two to Alice Lee, he thanked heaven that he was "absolutely pure. I can tell Alice everything I have ever done," he noted in his diary.[71]

By 1900 nearly all the assumptions about society and human nature that Roosevelt made so confidently already had been questioned. The economic and cultural factors that fostered later nineteenth-century "civilized" morality were changing.

The models of the human mind upon which Roosevelt based his assumptions and which provided the psychological controls of "civilized" morality also had been challenged. The faculties of Will, Conscience, and the concept of the unified and responsible Self were no longer adequate descriptions of what was known about human personality. A character in Dr. Weir Mitchell's autobiographical novel joked about the dilemma posed by the new knowledge: "If I am two people, and one can pop up like a jack-in-the-box, I may be six people, and how can I be responsible for the love-affairs of five? Have I six consciences?"[72]

The unconscious, as it was discovered toward the close of the century, bore an uncanny resemblance to the precise opposite of the values of "civilized" morality and the related disciplines of work and achievement. The unconscious was observed clinically in nervous patients, who displayed baffling symptoms that defied social norms and whose motives remained mysterious. Men failed to compete and achieve. Women acted out to the point of caricature the conventional stereotype of the passive, angelic female. The unconscious seemed to harbor murderous violence and primitive sexuality. By focusing attention on these aspects of personality, psychology and medicine raised problems that required radical new solutions.

III

The Somatic Style
1870 - 1910

Freud challenged not only "civilized" morality, but the somatic style in neurology and psychiatry with which it was historically associated in America. At Clark University he argued that when his Viennese colleague, Joseph Breuer, first saw his famous hysterical patient, Anna O., in the 1880's, no one could understand or relieve her condition. Freud's polemic partly distorted the past, helping inadvertently to create the myth his followers furthered of a pre-psychoanalytic period of therapeutic impotence and despair.

Yet, in the 1880's and early 1900's, some hysterical patients were recovering under a different system of diagnosis and treatment. Its success probably never can be exactly reconstructed, just as for different reasons the success of psychoanalysis remains uncertain today. The problem of recoveries will be taken up later, but certainly they occurred. In 1906, for example, a seventeen-year-old schoolgirl was treated on the neurological service of James Jackson Putnam at the Massachusetts General Hospital, one of the most sophisticated in America. She complained of severe head pains following tonsilitis. She had dreamed that a rooster was snipping off her right leg, just as a tailor might do with a scissors, and when she awakened her legs gave way. Then she dreamed that a physician thrust a knife blade to the hilt into her arms and legs and pinched her until she was black and blue.[1]

Her case illustrates the elements of the somatic style. Her symptoms were meticulously diagnosed as hysterical because they had no organic basis. Her dreams were left without interpretation, and orthodox physicians would have attributed both dreams and symptoms to disordered or maldeveloped brain function. Her relations to her family were not explored nor were the circumstances of her life. But she was assiduously treated solely by massage and exercise and in six days was discharged, the pain gone.

Psychoanalysis grew out of the somatic style in which Freud and all the first psychoanalysts were trained. Some of Freud's important assumptions were culled directly from it. Its unsolved problems and those of existing alternatives led Freud to create his psychological system and aroused initial interest in it. Both psychoanalysis and the somatic style accepted a medical model of nervous and mental disorder. Both systems, at least in youth, were vigorously hopeful. What distinguished each was not presence or absence of therapeutic zeal but different methods of eliciting and organizing data.

The somatic style limited the material physicians systematically investigated; often what they observed could not be logically accounted for in its terms. The norms of "civilized" morality, to which the somatic style was closely linked, also limited the medical approach.

The somatic style and "civilized" morality exhibited roughly the same historical pattern. They became dominant in the 1870's, rigid in the 1880's and 1890's, and were in a period of crisis by 1909. They asserted similar values and used the same hierarchical model of mental functions. Both were identified with the middle class and with a portion of that class. Both emphasized the determining role of outward form: the cranium indicated the quality of brain and mind, just as the decorum of reticence bespoke mental purity. Both systems were inaugurated and destroyed by a similar generational combination—a majority of the young, together with a few older authorities.

Because of the large proportion of non-scientific elements in neurology and psychiatry, "style" is an appropriate term for a consistent approach usually based on a distinctive model and a related set of hypotheses. However, no single style ever has been able to account for all the phenomena of nervous and mental disorder.

The American somatic style was set in the 1870's and 1880's chiefly by bright young neurologists such as James Jackson Putnam or Silas Weir Mitchell, who still dominated the profession in 1909. They shared new techniques and attitudes and created a new variant of the medical role that led directly to psychoanalysis.

Neurologists first specialized in somatic disorders of the nervous system as well as those illnesses that occupied the borderland between normality and insanity—morbid fears and compulsions, insomnia, fixed ideas, impotence, epilepsy, hypochondria, hysteria, nervous exhaustion. These disorders, neurologists argued, had been ignored or misunderstood. Alienists mistook them for types of madness. General physicians regarded them as gross physical diseases, as purely imaginary or as malingering. Only the most scientific diagnosis could establish their existence and lay the basis for sound treatment. By explaining all of them as somatic in origin, neurologists drew them within the medical model of involuntary sickness.

They were therapeutic enthusiasts who held out hope of recovery from the intensive private care not only of the borderline disorders, but of insanity during a period of decline in American public asylum care. Neurologists believed themselves unusually qualified to treat insanity because of their new knowledge of the brain and nervous system and of related organic diseases, such as poliomeylitis, tumors, multiple sclerosis.

Neurology grew out of general medicine and psychiatry between the 1850's and the 1870's. In America the Civil War brought rich experience with wounds to the brain and nervous system, malingering, and exhaustion. Many American neurologists had begun as general physicians or professors of anatomy and physiology. One of the first medical specialties to organize, neurologists founded their national association in 1875 and adopted the Chicago *Journal of Nervous and Mental Disease* as their official organ.

Neurologists struggled into the 1890's to establish their specialty not only against the opposition of alienists and general physicians but against mind cure practitioners who began to flourish, especially in Boston, in the 1880's. This brisk lay competition made neurologists insist all the more on the scientific purity of the somatic style. They saw themselves as young pioneers; their leaders included "excellent pathologists, sound logicians, acute clinical observers, men rich in every scientific attainment and of world wide fame." [2] This exalted self-image partly reflected the neurologists' social origins and training. Like their private patients, they tended to come from the rapidly growing ranks of the urban "comfortable classes." Neurology as a profession probably would not have existed without the swift increase of professional, clerical, and business groups in the major cities, which chiefly furnished its private clientele. A comparison between the presidents of the American Neurological Association and those of the Association of the Medical Superintendents of American Institutions for the Insane between 1876 and 1895 illuminates these contrasting origins. American alienists tended to come from small towns and farming communities; many of them did not attend college before medical school; fewer studied in Europe. The neurologists, however, often were born and raised in the great cities, and several were from old, professional families. They attended the major colleges and often received long additional training in Europe. They worked and taught at the important medical schools in the centers of American university and professional life—Boston, New York, Philadelphia, Baltimore, Chicago, St. Louis. Neurologists prided themselves on their wide culture. A surprising number wrote novels, plays, poetry, and journalism; a few achieved popularity and, like Silas Weir Mitchell, minor critical distinction.

The somatic style reflected the prestige of physics and chemistry, important general developments in medicine, and advances in neurological anatomy and histology. The major discoveries were European, and budding neurologists went abroad to study. The detailed history of these advances often has been told and is less important here than the spirit and axioms they engendered.

It is difficult to exaggerate the neurologists' sense of expectancy, their faith that within the foreseeable future every puzzle would be solved at last. Their confidence rested chiefly on brilliant new knowledge that began to illuminate that darkest Africa, the brain and nervous system. This once unknown continent, as neurologists called it, was being explored and mapped. The inhabitants, to alter the image, appeared to be intricate reflex mechanisms. If neurologists could learn exactly how they worked, how parts might be repaired or replaced and energy replenished, a science of cause and cure would be won.

The most spectacular discovery that confirmed this mechanistic view was the localization of functions in the brain. In the 1870's German and British investigators demonstrated for the first time, on animals, that the cerebral cortex responded to stimulation. If an electric charge of low intensity touched specific parts of the cortex, specific movements resulted. If an artificially induced lesion destroyed those parts, the movements would be abolished, at least for a time. The cortex seemed a mosaic of overlapping areas, each governing a definite function. Experiments not only on animals but nature's "tragic laboratory" of human nervous and mental disease provided the basic data. Occasionally investigators observed defects in human behavior, and then were able to relate these to lesions found postmortem. In 1861, for example, Paul Broca, a French physiologist, discovered a lesion in the third left frontal convolution of a patient who had laughed and cried, but had been unable to speak sensibly, although his vocal cords were intact. Broca concluded that the speech function must be located in that area of the brain. Some physicians saw aphasia as providing the key to the relation between mental and bodily functions. By the 1880's some physicians concluded that there must be acoustic and motor centers and even one for the "intellect proper." One day every function would be localized. Learning about the brain became à la mode among laymen as well as physicians. William James urged his Harvard students to arm themselves with a saw, chisel, scalpel, and forceps and dissect a sheep's head.[3]

The "nerve cell" powered the mechanism, evolving out of chemical activity "nerve force" or nervous energy. Although never clearly defined, it was like other unseen natural forces whose properties were known, such as gravity, light, heat, motion, and above all, electricity. Nerves and muscles transmitted electrical impulses, whose velocity and intensity could be measured, and presumably electricity produced "molecular changes" in nerve tissue.[4]

Lesions caused the nervous mechanism to break down or to function inefficiently and *ubi est morbis,* "where is the seat of lesion," became a major object of the neurologists' search. Ultimately, they hoped to relate each symptom to a specified pathological condition. A "lesion" in one part of the system might cause disease in another part by reflex action. Thus diseased ovaries might provoke epileptic fits. Yet, the apparently precise word lesion embraced a multitude of conditions, "structural," "nutritional," or "functional." Structural lesions represented permanent, ascertainable changes in tissue. Functional lesions could not be detected by known methods, although they were assumed one day to be discoverable. Nutritional or "trophic" lesions governed the ability of the system to generate energy. Although functional disorders could be serious and persistent, usually they carried a more sanguine prognosis. Neurologists placed within this category a number of complaints previously believed to be entirely structural in origin.

One mental disease, general paralysis, a syndrome of complex moral, mental, and motor disabilities, seemed to confirm the theory of somatic lesions. Postmortems seemed to disclose diffuse brain pathology, and in 1905 this disease was proved to result from syphilis. Because of new methods of staining tissue, better microscopes, the discovery of bacteria, it seemed natural to extend the model of general paralysis to every form of nervous and mental disease. Although assiduous research disclosed no lesions in most varieties, neurologists were not deterred. They argued that the very fact that patients still functioned in most important ways proved that the lesions must be subtle and complex and would be found by new methods. So they felt free to create entirely speculative pathologies. Paranoia, for example, resulted quite possibly from irreversible general atrophy of the brain; mania and melancholia from cortical irritation; hallucinations from abnormal subcortical functions.[5] The psychotic symptoms of nervous exhaustion, such as an inability to concentrate, must be a product of cerebral malfunction. Or perhaps neurasthenia came from a depletion of phosphorous in the nervous system. Hysteria sometimes was attributed to maldevelopment or malnutrition of the higher cerebral centers because of uncontrolled emotionalism, which it was the assigned task of such centers to govern.[6]

The major "predisposing" cause of nervous and mental disorder was also somatic—an inherited defective nervous constitution. Tangible morphological signs betrayed its presence—deviations from a presumed norm based on statistical averages, although it was emphasized that only tendencies to nervous and mental disease could be inherited, usually not the disease itself.

Well into the first decade of this century, belief in the inheritance of acquired characteristics made possible a close interaction between environment and heredity. Geography, social customs, stress, acquired vices or virtues could help determine the forms tendencies to inherited and nervous mental disease finally took. Details of hereditary theory will be examined later because they played a major role in the crisis of the somatic style.

It is important to emphasize that neurologists insisted that disadvantageous heredity could be overcome or neutralized by proper somatic hygiene. Early training in healthy habits, control of the passions, the elimination of stress or debilitating brain work effectively could forestall the outbreak of symptoms in the child or adult of tainted stock.

Hygiene conserved the energy of the weak constitution, and the principle of "nutrition," or the fostering of nervous energy, remained the fundamental principle of neurological therapeutics for half a century. "Nutrition" was never clearly defined. Diet, rest, massage, exercise, drugs, electricity were the neurologist's major weapons—because they improved the ability of the nervous system to manufacture energy. The editors of *Popular Science Monthly* neatly vulgarized the new neurological wisdom: "Mind force cannot come from nothing . . . and here as elsewhere one effect is at the expense of another. Thinking and feeling exhaust the mechanism, and we are involved with practical questions of waste and repair, exercise and rest, food, blood, nutrition, and the hereditary qualities of the nerve centers." [7]

To argue that "thinking and feeling" exhausted the nervous mechanism was at once to introduce psychological concepts into the somatic scheme. The raw data of nervous and mental disorder have always been human speech and behavior. The simplest somatic diagnosis required psychological assumptions. Consider, for example, a case of analgesia caused by typhoid fever presented by James Jackson Putnam to the American Neurological Association's first meeting in 1875.

Why, Putnam asked, could his patient no longer feel heat or pin pricks over a circumscribed area of the skin on the left side of his body, including the hand? Something interrupted the transmission of "impressions upon the skin" to the "seat of conscious sensation" in the brain. Did this interruption occur in the peripheral nerves of the spinal cord, or in the brain itself? By elimination of possibilities, partly theoretical, partly empirical, Putnam concluded that the interruption must occur in the peripheral nerves. To arrive at this conclusion he had had to ask what a "sensation" was. Did the sense of heat differ from the sense of pain? Were there separate conductors for such sensations, presuming them to be different? What was the relation of "sensation" to "consciousness"? [8]

At such points as these the interests of neurologists and of young psychologists such as William James and Granville Stanley Hall converged. Both exchanged ideas with Putnam, Beard, and Weir Mitchell, and at first hoped to ground traditional psychological conceptions in the operations of the nervous system. Yet, they were also keenly aware of the problems of the traditional psychology that most neurologists willy nilly had adopted.[9]

The first American neurologists and psychologists borrowed their models from a common tradition of philosophical psychology, chiefly

Scottish Common Sense, filtered through phrenology, and, more important, British Associationism. They discarded the innateness of the old "faculties of mind," yet retained them as categories of mental functions. Few realized that these categories might be less useful than a new psychological system empirically constructed to describe nervous and mental disease, such as Freud's psychoanalysis (with important exceptions) would attempt. In 1879 Hall observed: "Must we not have a radically new classification of mental faculties more fundamental than the almost mythic (as to its origins) one of feeling, intellect, will, and may it not be due to the falseness, or superficiality of our classification that so far it has not been possible to localize faculties in the brain?" [10]

Not only categories, such as feeling or will, but operations, such as judgment, were borrowed from the same tradition. Such operations were thought to be composed of simple elements, ultimately, sensations, associated in a particular way. Working directly from this psychological scheme, neurologists viewed the brain as made up of elements connected by fibrous tracts along which associations traveled. The fundamental importance that association played in Freud's psychology derived partly from its status in neurological theory.

A theoretical problem that had vital practical consequences was the relation of the psychological to the neurological systems. Did the "will" or the "emotions" alter the functions of the brain and nervous system? The prestige of physics with its law of thermodynamics partly dictated the commonest solution: the physical and psychological systems were closed; energy could not be lost or acquired from one to the other. John Hughlings Jackson, who profoundly influenced Freud and the Americans, argued that "volitions, ideas and emotions," i.e. states of consciousness, did not produce "movements or any other physical states." [11] The two occurred together in a parallelism, but were utterly distinct. For every mental state there was a correlative nervous state, but neither interfered with the other. This theory was the basis of all later nineteenth-century scientific psychology.

William James summarized the major implication of the mechanical view: "Feeling is a mere collateral product of our nervous processes, unable to react upon them any more than a shadow reacts on the steps of the traveller whom it accompanies. Inert, uninfluential, a simple passenger in the voyage of life, it is allowed to remain on board, but not to touch the helm or handle the rigging. . . . In a word, feeling constitutes the 'unscientific' half of existence. . . ." [12]

Parallelism, codified in the 1870's and 1880's, reinforced the already overwhelming emphasis on the physical symptom, tangible or speculative, and a concerted attack on parallelism was one of the first tactics adopted by some of the later psychotherapists.[13] In practice many, perhaps most physicians, ignored theory, comfortably switching from somatic to psychological and vice versa without discomfort. But, as will become clear later, lack of formal status granted to the emotions often meant, in retrospect, that they were discussed in an undisciplined and naïve way. Parallelism, like other elements of the somatic style, confined systematic attention to limited aspects of nervous and mental disorder.

Not only were the models physicians used covertly psychological, but descriptions of nervous and mental disorder were blatantly normative and still are. Social norms, usually implicit, then and now were invoked to describe an ostensibly medical reality. Patients were "alienated," "withdrawn in affection." Their "thinking," "feeling," and "acting" were "confused" or "disordered." They failed to perform their "social duties" or fulfill their "vocations." The degree of departure from the social norm usually measured the seriousness of the disorder. At the most benign level were listed what seemed to be intensifications of everyday problems—worry, anxiety over health, sleeplessness, depression, elation, fatigue, overactivity. At a more serious level occurred abrupt and marked "changes of character," "perception," or "judgment," ineradicable doubts, ritual obsessions, excessive, non-existent, or perverse sexual desires. At a still more serious level there were crippling periods of elation or depression or apathy, systematized delusions, and apparently motiveless rage. Well before Freud it was assumed that there was a continuity from normality to madness.[14]

In the later nineteenth century, the norms of the "civilized" moral order were enshrined in the traditional psychological models and given a somatic base. The rank order of the old divisions was preserved: intellect and will were "superior" to emotion and appetite. Judgment, reason, and control were located in the frontal lobes of the cortex, which inhibited the action of the "lower" centers where the primitive and childlike drives and instincts, including the sexual passions, were "located." Freud's later assignment of the enlightened control of sexuality to the ego reflects this classical outlook.[15]

Evolutionary theory reinforced this moral hierarchy by suggesting that conscious decisions and moral judgments represented "higher," more complex, and later stages of development. Inspired directly by Herbert

Spencer, Hughlings Jackson arranged the functions of the nervous system in an ascending order from the simplest, most organized and therefore most mechanical to the complex, least organized and most voluntary. Although Jackson explicitly refused to equate "higher" and "lower" with good and evil, he nevertheless assumed intellect to be more "voluntary" and "complex."

In disease, the evolutionary stages were reversed, and "dissolution" took place: the least organized, most voluntary, most recently learned function gave out first, leaving the next lower level of functioning intact. For example, in aphasia the most voluntary aspect of speech was the first to disappear while expletives might still be spoken. In insanity the "highest nervous centers" became diseased, conscious voluntary adjustment dissolved, leaving intact automatic and "emotional" conduct.[16] Freud may have derived his theory of regression partly from these views of Jackson.

Not only did the neurologist's hierarchy of mental functions reflect the moral norm, but the hygiene of sexuality and of nervous and mental disease were identical. Sobriety, hard work in moderation, thrift, and self-mastery protected sanity, just as they protected the family and ensured a desirable social status. Sensuality and excess destroyed alike mental and sexual health.

The "generative organs" were assigned two important roles. If natural laws of healthy sexual living were broken by masturbation, perversions, or excesses, nervous and mental disease might ensue. When sexual transgressions were not thought of as causes, they were thought of as frequent symptoms in the behavior of patients.

Scarcity economics and the theory of the limited bank account applied alike to nervous and sexual energy. Each person had limited resources at his disposal. One could overdraw one's account by excessive expenditure, by too much strain and too little rest, and thus become a nervous bankrupt. This theory had serious consequences. Well into the Progressive Era, people with "constitutionally" weak nervous systems, either inherited or acquired, were warned to curtail their activities. They should avoid too much sedentary brain work, which was thought to be peculiarly exhausting. Sometimes they were advised to avoid a professional career: the demands of law, medicine, or the pulpit might create excessive strain and consequent breakdown. The city, with its maddening, frantic pace, was to be avoided.[17]

These values expressed the outlook of a professional class. Like other professionals, specialists in nervous disorders were undergoing longer periods of apprenticeship that often necessitated the postponement of marriage. Medical leaders admonished physicians to shun the "earthly Aphrodite" and to cultivate habits of work.[18] Not all physicians accepted every aspect of "civilized" morality, but the most respected were its most loyal adherents.

The neurologist's class status also was expressed in beliefs about the incidence of nervous and mental disease and their criticism of the new commercial and industrial world. The incidence of insanity, they believed, was highest among the poor, who, on the whole, lacked the "vital force" of the upper classes, partly, it was admitted, because of harsh social conditions. Sometimes neurologists were willing to accept a frank class differential in the treatment of the insane, and warned against the folly of "palace hospitals" for those unaccustomed to luxury.[19] At the same time they could argue for the application of tougher medical standards to asylums.

Although they saw hysteria and neurasthenia in every social class, these diseases were usually identified with their own social groups, the upper classes. This identification was reinforced by the fact that neurologists themselves or members of their families often suffered from nervous disorders. Just as some of the most famous specialists in tuberculosis were tubercular, so neurologists, such as Putnam, Silas Weir Mitchell, or George Beard, were often "nervous." No wonder they treated these disorders with sympathetic respect.

To interpret them as entirely somatic robbed them of the stigma of breaking the norm of control over the emotions. Morton Prince, who married a nervous girl in 1884, superbly illustrated this logic a year later. The physician was too prone to dismiss a woman who complained of pain, but who exhibited no "coarse anatomical lesions" as "only hysterical." "How different all this becomes, when we regard every symptom, even the thought which prompts a patient to exaggerate her symptoms, as synonymous with disordered physiological activity of the cells and fibres, and every change of disposition and character . . . means an imperfect working of the brain. . . ."[20]

Patients responded to the somatic style with a luxuriance of physical symptoms, objective or subjective. Like the seventeen-year-old patient in 1906, they complained of headaches or of weakness and fatigue. Sometimes symptoms could be intensely mimetic and expressive, as in the "passionate movements" Jean Martin Charcot observed in his hysterical female charity patients at the Salpétrière, or the compulsive tics and grimaces S. Weir Mitchell observed in children. Whether the predominance of physical symptoms represented what patients expressed or merely what physicians chose to record is impossible to determine. Possibly patients described their symptoms in a somatic language they sensed would be acceptable to their physicians.

Neurologists frequently described nervous and mental disorder in a mode that bears resemblance to that middle-class favorite, the mystery novel. Physicians invariably began with an account of the patient's baffling symptoms and ended with a rational, quite unexpected explanation of them. This mode of the "mystery solved" was common to specialists as different as Freud, S. Weir Mitchell, and Cesare Lombroso.[21] Often other non-scientific factors shaped neurological observations—religion, institutional needs, personality, and crucial relationships with patients. Just as Freud learned the mechanism of conversion from Elizabeth R., so Mitchell devised the rest cure for a sympathetic, but debilitated New England matron.

Perhaps because most neurologists treated patients in the setting of their daily lives rather than in institutions, they were sensitive to environmental factors. Usually their patients' behavioral symptoms were milder, less systematized and symbolic than those of the chronic insane. Thus, the borderline cases seemed to emerge naturally from the stresses of everyday life. It was on these social sources of strain that some of the first neurologists focused their attention.

Today some of the problems the neurologists examined are becoming once more a meaningful part of psychiatric research.[22] As new nations are being changed by industrialization, psychiatrists and anthropologists are re-examining the relation of rapid change and social disorder to nervous and mental illness. When S. Weir Mitchell suggested that rapid change was unsettling, he was voicing a medical critique of modern, urban industrial culture that paralleled its development. When Beard suggested that steam power, the periodical press, the telegraph, the sciences, and the mental activity of women were fostering nervous exhaustion, he was describing the major agencies that transformed the nineteenth century.

Yet, of itself change hardly can be considered a "cause" of nervous and mental disorders. Rapid change also acted as a healthy stimulus to millions of Americans. The only careful study of American asylum statistics indicates no rise whatsoever in insanity in this period.

The neurologists' observations can be interpreted, however, as dealing with responses created by dysfunctional, internalized norms of behavior. Then the social diagnoses of Beard and S. Weir Mitchell become acutely meaningful. Like most physicians until quite recently, they worked from case histories they themselves recorded, without statistical controls or objective checks. Their views represent the shrewd, intuitive judgments of perceptive men, who were sensitive to their patients' complaints and to the social tension they observed.

Their descriptions of the commonest nervous disorders of women and men emphasized conflicts within individuals who could not fulfill social norms, yet, because they had internalized them, could not consciously reject them. Hysteria sabotaged the "civilized" norm of refinement in two ways, by "fits," outbursts of emotion that directly violated it, or by incapacitating physical symptoms, which, because they made a woman helpless, caricatured the very delicacy and softness she and American men had been taught to reverence. Something of the ambivalence this disease provoked is apparent in the most famous American studies by S. Weir Mitchell.

His well-to-do patients arrived in Philadelphia from all over the United States, sometimes borne on litters, their eyes bandaged to keep out light. Most suffered from nervous exhaustion combined with the baffling somatic symptoms of hysteria: uncontrollable vomiting; fits of tears and laughter; paralyses or unco-ordinated movements that resembled those of paresis; convulsions similar to those of epilepsy; swellings that resembled tumors; blindness; diarrheas without ascertainable organic cause, or stubborn constipation. Only missing were the multiple contractures and stereotyped convulsions observed by Jean Martin Charcot among his charity patients at the Salpétrière in Paris.

A few were malingerers, for example, a thirteen-year-old girl who looked as if she were dying yet deliberately vomited her food. "I could never extract from this child, when well, anything beyond the statement that she 'just could not help it,' " he wrote, "and if I pressed her further, she said she was sorry and took refuge in tears." [23]

In others stubborn paralyses might begin in an attack of hysterics, or what had been called the "vapors." A girl "cried and laughed by turns and then had a slight fit, on coming out of which she could no longer lift her right arm, or rather could lift it but a few inches." [24] Still another developed paralysis of both legs after seeing her father seriously injured in a fall from a horse.

Such cases, Mitchell argued, required the most acute diagnostic skill. Sometimes inept doctors doomed patients to lifelong invalidism because incapable of diagnosing their physical symptoms as hysterical. Yet, Mitchell's own observations were sharply limited by the somatic style and even more severely by the norms of "civilized" morality.

His rationale of cause and cure were both somatic, with, however, important and often unexplored psychological elements. Thus the causes of hysteria were "physical states or defects," ultimately the result of moral, i.e. psychological, causes or these combined with "constitutional conditions." [25] First would come a shock or trauma—loss of money or status, injury, death, prolonged strain, disappointment in love. This weakened health, which in turn altered the character, chiefly through loss of control. The once resolute woman now was indecisive; trifles were magnified until the slightest discomfort became unbearable. Then a train of physical symptoms ensued.

The way specific symptoms developed was clearest in children. Often they mimicked exactly the symptoms of others with astonishing accuracy or developed precisely the signs of an illness their anxious parents had warned them about. Much, Mitchell observed, that was ascribed to heredity came in fact from early imitation. The final result could be a stubborn hysteria, which, contrary to popular belief, seldom cleared up easily or quickly.[26]

Mitchell's rest cure, which often was highly successful, assumed that as physical strength returned, so, too, would moral and emotional control. Any severe nervous illness, it often had been observed, caused a rapid loss of weight. As fat left, so did red corpuscles. Mitchell aimed to "fatten" and "redden" his patient until she resumed her ideal aesthetic shape, "padded and protected" by a moderate layer of fat. He added daily massage to traditional diet and rest and regarded his treatment as a welcome departure from the indiscriminate drugging of the age. Too many nervous patients had become addicts of opium, chloral, bromide, and other sedatives, because too often doctors prescribed them.

The doctor played the "tranquil tyrant," like a Victorian *pater familias,* and exacted absolute obedience. He first changed the moral atmosphere by secluding the patient with a sensible nurse, allowing no visitors, reading, or writing. Patients were fed and massaged on a rigorous schedule. In part the rest cure was an exercise in deprivation calculated to make the doctor's visits important and women glad to become active again. Sometimes the taste for rest became obsessive. In one celebrated instance, Mitchell threatened to climb into bed with a patient who refused to get up and thus had her out in a trice.

The cure's final aim was to induce self-control and to create insight into the "moral failures" of "selfish-invalidism." "At last," he wrote, "you get her to promise to fight with every desire to cry, to twitch, or grow excited." [27] Women must learn to endure. The outward repression of emotions fostered their inner control.

Mitchell admired the feminine ideal of his time, yet detected its inner contradictions. Women were "pre-eminent for affection, sympathy, devotion, self-denial, modesty, long-suffering or patience under pain, disappointment, and adversity, for reverence, veneration, religious feeling, and general morality." [28] These virtues enhanced emotion, which, if uncontrolled, led straight to nervousness.

Yet, control, other neurologists observed, could become excessive and exact peculiar penalties, among them, perhaps, some of the symptoms of hysteria itself. A few neurologists believed they detected a growing restraint in women's expression of emotion, and this more rigid control may have been reflected in neurotic symptoms. In the 1860's and in most of Beard's writings, hysteria was characterized above all by "paroxysms" or "fits." [29]

A telling example of the conflicts engendered by rigidly refined controls can be seen in William James's sister, Alice, a spinster who suffered a hysterical breakdown in 1867 and lived on until 1892, a nervous invalid. She described her own life as a struggle between her inner controls and her "violent" impulses that issued now in a desire for suicide, now in further turns of hysteria:

"As I used to sit immovable reading in the library with waves of violent inclination suddenly invading my muscles taking some one of their myriad forms such as throwing myself out of the window, or knocking off the head of the benignant pater as he sat with his silver locks, writing at his table, it used to seem to me that the only difference between me and the insane was that I had not only all the horrors and suffering of insanity but the duties of doctor, nurse, and strait-jacket imposed upon me, too."

She did not have even the consolation of swearing. "What an awful pity it is that you can't say damn," said her friend, Miss Loring. "I agreed with her from my heart. It is an immense loss to have all robust and sustaining expletives refined away from one! At such moments of trial refinement is a feeble reed to lean upon."

"Conceive of never being without the sense that if you let yourself go for a moment your mechanism will fall into pie and that at some given moment you must abandon it all, let the dykes break and the flood sweep in, acknowledging yourself abjectly impotent before the immutable laws. When all one's moral and natural stock in trade is a temperament forbidding the abandonment of an inch or the relaxation of a muscle, 'tis a never-ending fight."

The English physician Charles Lloyd Tuckey relieved her briefly by hypnosis, and she learned the secret "of suspending for the time from his duties, the individual watchdog, worn out with his ceaseless vigil to maintain the sanity of the modern complicated mechanism." Dying of cancer, she was tormented still. The physical pain "falls away like dry husks from the wind, whilst moral discords and nervous horrors sear the soul. . . ." The diary had been a release, and on the day of her death, dictating to Miss Loring, "she could not get her head quiet until she had had it written: then she was relieved. . . ." [30] It is significant that Alice James's murderous rage was directed at her father and equally significant that, except for considerations of heredity, the neurologists largely ignored those intricacies of family life that novels of the period so painstakingly described.

Weir Mitchell could deal subtly in his novels with the passions that he could not treat "scientifically" in a neurological lecture or essay. He created heroines, such as the delicate hysteric in *Dr. North and His Friends,* who became a somnambulist and a split personality from thwarted passion. He described the overpowering attraction, clearly sexual, of another heroine. Yet his novels fulfilled the most exacting requirements of gentility. Owen Wister, a more popular but equally genteel writer, praised Mitchell for avoiding the "perversions" of Zola or Flaubert: "No matter what dark alleys his muse passed, what mud she had to cross, she comes to us always with her skirts clean and her heart warm." [31] Mitchell once glanced at a book about Freud. "Where did this filthy thing come from," he fumed and threw the book into the fire.[32]

Mitchell's case histories describe women as if they were totally asexualized. He knew that "deeply human" motives were at work in his patients, and he collected their autobiographies, hoping to shed light on the "natural history" of nervous diseases. Patients themselves usually did not know what their motives were. Some felt as if their wills had been paralyzed, others, as if they lived in a dream.

Many longed to confess. Mitchell's response was curiosity, but also reserve and secretiveness. In a famous passage, he wrote: To the physician "sad confessions come . . . more, indeed, than he may care to hear. . . . The priest hears the crime or folly of the hour, but to the physician are often told the long sad tales of a whole life, its far-away mistakes, its failures, and its faults. None may be quite foreign to his purpose or his needs. The causes of breakdowns and nervous disaster and consequent emotional disturbances and their bitter fruit are often to be sought in the remote past. He may dislike the question, but he cannot avoid it. . . . The moral world of the sickbed explains in a measure some of the things that are strange in daily life, and the man who does not know sick women does not know women." [33] Mitchell's reticence extended to himself. "No man can reveal his whole life," he wrote. "No man dares. The wisest are the most silent." [34]

In that disorder from which Mitchell and innumerable male and female patients suffered, neurasthenia, or nervous exhaustion, a similar normative conflict emerges: inability to fulfill an increasingly demanding standard of work and breakdown through overconformity. The energy "civilized" morality sought to preserve was notably absent in this disease. Patients complained of exhaustion and inability to concentrate or to work at all. The term "neurasthenia" was launched, if not originated, by a bombastic New York neurologist, George Miller Beard, in 1869.

Like the rest cure, neurasthenia was among the major American contributions to nineteenth-century neurology. Freud struggled with Beard's elusive description of the disease, just as, for a time, he experimented with a modified Weir Mitchell rest cure.

Just as the neurologist's attitude toward hysteria reflected ambivalence about the growing demands of refinement, so the attitude toward neurasthenia reflected criticism of the new industrial and commercial world. Too much "brain work" was thought to be peculiarly debilitating, and the new civilization demanded it in ever growing amounts. Brain work, Mitchell argued, gave no warning symptoms of exhaustion. At the end of a hard day, the brain remained "morbidly awake," repeating the day's work over and over. Perhaps research would one day prove that more nervous energy was expended in brain than in muscular activity.

Not only the type of work but the attitude toward it were now insidious. Differences from the more rural conditions Tocqueville described forty years before are illuminating. The pressures to succeed are fiercer; ruin and success occur with unprecedented swiftness. The very fact that more people were "joining the ranks of the comfortable classes" may have made failure all the more bitter.

The American had become utterly absorbed in business, not so much to make money as for the sake of work itself. He had neither the "time nor desire" for books and was too uncultivated to travel. For him the asylum gaped wide open. "The fiend of work he raised no man can lay," Mitchell warned.[35]

Worry and anxiety made overwork pathological and caused not only nervous exhaustion but "dyspepsia, consumption and maladies of the heart." Mitchell observed a correlation between occupation and liability to nervous disorders: first came manufacturers, railway officials, merchants, brokers, young men given business responsibility prematurely, overtaxed scientists, the overschooled young, and finally clergymen, lawyers, and physicians.[36]

Neurologists both distrusted and admired the new tycoon. Mitchell romanticized his energy and ruthlessness, and in the end, accepted him almost as a social equal. The neurasthenic, like the hysteric, represented the obverse of the period's positive values. He was an anti-hero whose characteristics were not pluck and strength but powerlessness and depression. Yet, these symptoms usually occurred only after work had been defined too narrowly and too slavishly pursued. Neurasthenia represented an involuntary alienation, a "sickness" rather than a conscious rebellion toward a counter-value of, for example, aristocratic leisure, or hedonism.

Sexual disorders associated with "civilized" morality, especially among men, often were treated by neurologists. Reticence, in particular, sometimes had dysfunctional results. There were married couples ignorant of copulation, husbands too well bred to entertain the thought of relations with their wives. Some were "overwhelmed with shame" at discussing sexual difficulties. "Sexual neurasthenia" was widespread, while unusually strong sexual desires were a source of embarrassment, especially to clergymen. One complained that his wife held exalted ideas of purity. "My desire to fondle and look at and admire her form, she regarded as 'signs of manly weakness.' She thinks yielding to these things only hurts me and excites my passions." [37]

Masturbation caused pandemic fears of loss of manhood. In 1874 a young man of twenty-three consulted James Jackson Putnam. The youth asked to be "fitted for marriage," because he believed his "generative powers" had been "almost entirely destroyed" by masturbation. For the most part, such cases were treated by drugs and special regimes. For cerebro-spinal exhaustion, the result of "sexual excesses and overstudy," this was prescribed: "the fat of beef and mutton to be eaten freely; butter and cream likewise. Sexual intercourse to be absolutely abstained from. Mental labor to be made as regular and unemotional as possible and sufficient sleep to be obtained: nerve tonics; phosphorus or zinc phosphate, oxide of iron, arsenic, strychnia, for relief of spinal pains, electricity and the emplastrum, belladonna, etc." [38]

Although most neurologists accepted both the somatic style and the canons of "civilized" morality, an important minority disagreed with aspects of both. George Beard argued that nervous diseases were increasing, partly because the "conventionalities of society" required repression of the emotions and secrecy about sexuality, the "most powerful passion of human nature." The attitude of secrecy extended even to physicians who had written little or nothing about the genital apparatus and who thus forced the public to seek advice about sexuality from quacks. Parents were equally reticent with their children, who consequently got their information from the gutter. Beard urged the sexual enlightenment of every child at an early age. Neurologists also saw cases of nymphomania and male and female homosexuality. A New Jersey physician, Andrew J. Ingersoll, argued that the voluntary repression of sexual life was a major cause of hysteria. A Philadelphia neurologist, James Hendrie Lloyd, who later would denounce Freud's Oedipus complex, suggested that emotions of which people were ashamed caused hysteria and required careful psychological investigation.[39]

Of the early neurologists, Beard most actively pursued psychology, psychotherapy, and the new medical studies of hypnotism and the unconscious, which he called the "involuntary life." Yet, Beard's psychology was undiscriminating and relied on a mythological somaticism. He considered "morbid fear" to be the most "overlooked, slighted and misinterpreted" sign of neurasthenia; its victims were "almost kicked out" of the average physician's office. Some feared high places, others dreaded the open streets, others thought they had committed disgraceful crimes or that they were going insane. Still others avoided dirt and wiped off chairs before they sat in them. Patients knew their fears were absurd but, Beard explained, the "emotional nature," under the influence of the "exhausted nervous condition," overcame "reason and will." [40]

The crudeness of Beard's psychology is apparent from a paper he delivered to the second meeting of the American Neurological Association in 1876, much of which probably was derived from the English alienist Hack Tuke's text on psychotherapy. Beard argued that emotions could cause disease to "appear and disappear without the influence of any other agency." Unrestrained "fear, terror, anxiety and care, grief, anger, wonder and expectation" were more pathogenic than the "most potent poison." But "reason, hope, joy, resolution, ambition, self-confidence" could cure acute and chronic, functional and organic diseases. Patients could be restored "from death itself."

Beard's claims aroused a furor. William Hammond, a founder of American neurology, replied that, if Beard's doctrines were taken seriously, "we should go back to monkery—give up our instruments, give up our medicines and enter a convent." Neurology would descend to humbuggery. The debate continued:

"Dr. Putnam regarded the experiments as unscientific because the emotions could not be isolated.

Dr. Mason objected to the term mental therapeutics and denied its existence.

Dr. Emerson regarded the experiments as unscientific because the exact nature of the elements experimented with was not known." [41]

Having rebuked Beard, the neurologists turned to more congenial pursuits. A physician exhibited "transverse sections through the human brain," made with the recently invented microtone that sliced sections as thin as .01 mm. to .005 mm. Dr. Putnam demonstrated a new English gas-cautery and advocated the standardized testing of galvanometers, which measured the intensity and direction of currents used in electro-therapeutics.

Toward emotion and hypnotism most neurologists were hostile, except as hope and faith in the physician soothed the way to recovery. In 1913 Mitchell still dismissed the new psychotherapy with its "hundred and twenty-one questions" and its "consultation" of dreams and the subconscious. [42]

The young psychologists, William James and Granville Stanley Hall, were more interested in Beard's hypnotic experiments than were his own colleagues. James led Beard to hypnotize away the sense of dizziness. Hall conceived of hypnosis as a result of the purely psychological function of attention on the basis of Beard's hypotheses. [43]

Beard continued to insist that "mental therapeutics" was an unwritten chapter of medicine that should be in every future textbook. He convinced a very few, including the New York neurologist Alan McLane Hamilton. Another was Charles L. Dana, who assisted Beard for a time, experimented gingerly with hypnosis, and sympathetically reviewed books on telepathy and clairvoyance. Later as professor of neurology at Cornell University Medical College he tolerated the enthusiasm of some of his students for psychoanalysis. Beard led Edward Cowles, director of the McLean Hospital, to explain insanity as a pathological intensification of fatigue, and in the 1890's much laboratory research vainly attempted to prove this.[44]

No tradition of active psychotherapeutics or investigation derived from Beard. When Morton Prince initiated the enduring American interest in hypnotism and the subconscious, he referred not to Beard but to French and British experimenters. Yet Beard's formulation of the common doctrine that modern civilization caused an increase of nervous and mental disorder persisted. Freud's criticism of "civilized" morality partly grew from this tradition.

Belligerent somaticism and an acceptance of "civilized" morality characterized most neurologists, many of whom, by 1900, were listed in the social register, that new index of American upper-class status. The first psychoanalysts, often from different social origins, confronted precisely this combination of a medical style united with social position and a moral stance.

Absorbed as the first neurologists were in extending knowledge of the brain and nervous system, with a definition of the scientific that ruled out the unquantifiable emotions, they devoted less energy and attention to the systematic study of social and psychological factors. Yet some of their most important observations were precisely in these areas which the somatic style did not easily encompass. Moreover, this style and "civilized" norms together prevented the exploration by both doctor and patient of important material that lay just below the surface.

The neurologists' high confidence that the somatic style would disclose final knowledge of the brain and nervous system and thus the scientific control of disease set the terms of their disillusionment. When dogged differential diagnosis and the most painstaking research failed to achieve expected results, an awareness of the style's limitations grew among a nucleus of professional leaders. With this growing awareness the crisis of the style began. Ultimately this crisis facilitated a systematic exploration of patients' feelings on a level that gave them serious etiological significance and the development of alternative therapeutic methods. Finally, a crisis in "civilized" morality brought to the surface existing discontents, conflicts within individuals, and contradictions between norms and behavior that had formed important undercurrents throughout the period.

PART 2

*Crisis and
Reconstruction
1885-1910*

IV

The Crisis of the Somatic Style
1895 - 1910

By 1908 Freud's theories were being damned and praised in America during debates over fundamental principles in neurology and psychiatry. Discussion raged over the relation of mind and body, heredity and environment, the nature of treatment and of the physician's role. This crisis of the somatic style had two simultaneous aspects—criticism of existing orthodoxy and the construction of alternatives, and each aspect reinforced the other. At Clark University, Freud unwittingly took stands on several issues of this crisis that were, by contemporary standards, radical. This chapter will examine the criticism, and later chapters the new alternatives.

Today, there are indications that classical psychoanalysis is undergoing a similar crisis of criticism and reconstruction for analogous reasons —conflicting theories, a search for more adequate models, therapeutic failures, an inability to distribute services widely enough, and, finally, alternatives belligerently advanced by neo-somaticists, behaviorists, and revivers of hypnosis.

At the Clark Conference, the young Swiss-American psychiatrist, Adolf Meyer, professor of psychiatry at the Johns Hopkins University, argued that nothing could be gained by continuing to insist that mental disorder resulted from brain lesions that were in fact vague and mysterious, fatalistic notions of degeneracy or mythical auto-intoxications of the nervous system.[1] Meyer and a small group of critics and innovators pressed embarrassing questions that advocates of the somatic style had failed to answer.

They could not explain the lack of brain lesions or other pathology in dementia praecox, an apparently incurable organic insanity. Why, if it involved a "profound deterioration of the cerebral cortex," did apparently hopeless cases suddenly recover?

"And what is thought?" the psychopathologist Boris Sidis asked. "Metaphysics! Cells and stains, that is real science." Adolf Hoch, another young psychiatrist, poked fun at the respectable speculation that hallucinations were irritative focal symptoms in the brain. Much of what floated "in the air as solid tradition" in psychiatry, Meyer concluded in 1904, was "far from a firm foundation. . . ." [2]

Such criticism of traditional somatic approaches had important consequences for psychoanalysis. It made some somaticists more cautious and aware of what they did not know. It justified a detachment of the psychological aspects of nervous and mental disorder from physiological processes. No longer was it necessary to speculate that hysteria involved some unspecified change in the "nutrition" of the "cerebral hemispheres," because hysteria was marked by "mental" symptoms.

Criticism struck at the heart of the crisis—the exhaustion, overextension, and failures of somatic theory and practice. The crisis began on a very small scale in the late 1880's, gathered momentum in the 1890's, but did not become truly significant until the years from about 1904 to 1909 when psychotherapy became a sensitive issue. A crisis is defined here as a period of accelerating change in fundamental assumptions and professional roles.

But the changes in neurology and psychiatry differed drastically from analogous shifts in the "hard" sciences.[3] For in neurology and psychiatry these changes were slow, piecemeal, and often inconsistent. The somatic style itself never had been exclusive, and its theorists always had accepted large, unsystematized currents of social, psychological, and normative data.

Moreover, the advance guard did not desire to discard somaticism out of hand, partly because the causes of much nervous and mental disorder remained unknown. Freud and Janet, for example, believed that ultimately a truly adequate somatic etiology and treatment would be discovered. Both developed energic theories that belonged in fact to the scientific bases of the somatic style, and Freud's psychoanalysis in many respects remained a theory rooted in the body and in the psychological consequences of physical functions and attributes.

What the advance guard *did* want was a shift in emphasis and role, the marshalling of energies for psychological observation and treatment. They argued that the somatic style, as then defined, had exhausted its actual possibilities. Its means did not yet match its ambitious ends. Physicians were concentrating their attention on a limited set of factors and ignoring others that seemed clearly important, but that occupied only a perfunctory place in accepted views.

During the crisis all the major postulates of the somatic style were challenged. Traditional views of diagnosis, classification, etiology, prognosis, and treatment bristled with unresolved difficulties. The theory of lesions and the correlative theory of brain localization came to seem irrelevant and naïve as explanations for much nervous and mental disorder. Classifications based on the structure of the nervous system came to seem highly artificial. Traditional groupings were replaced, then the validity of classification itself was questioned. Hereditary explanations of neurosis and insanity proliferated, became overextended, and were sharply attacked. Painstaking laboratory research failed to disclose pathology for many nervous disorders and some of the most serious and widespread varieties of insanity. Rates of recovery from insanity declined markedly, although more and better hospitals were built. Treatments based on restoring the nutrition of the nervous system—diet and rest, for instance—failed to produce the results claimed for them. In the hands of most physicians, the therapy of the nervous, and above all, of the insane, remained at a standstill.

Symptoms of this crisis can be observed in the provocative rhetoric of a small advance guard and in those retrospects and prospects by which the pulse of a profession can be taken. The brash and hearty confidence of the 1870's and 1880's was giving way to self-doubt and an awareness of complexities and unsolved problems. Perhaps most neurologists and psychiatrists admitted these difficulties. But the majority believed that a solution would come from more careful and fervent application of the somatic principles they had been using all along. Their outlook was expressed in 1897 by Bernard Sachs, an American pupil of Theodore Meynert, and a prominent New York neurologist: "The past of psychiatry has been full of discouragement; the present is involved in a maze of uncertainty, but the future is full of hope." [4]

The tiny advance guard shared social and temperamental characteristics. They represented, above all, a crossing of disciplines and professions. Drawn from neurology, psychiatry, and academic psychology they constituted an informal network of people, chiefly on the eastern seaboard, with similar interests, most of whom knew each other. They often worked in the same hospitals or clinics, attended the same medical meetings, and debated the same professional problems. As already noted, Granville Stanley Hall and William James, the nation's first psychologists, co-operated informally with neurologists such as George M. Beard and James Jackson Putnam. Men trained in neurology, such as Adolf Meyer, applied their skills in insane hospitals. Psychologists, such as Boris Sidis, trained by James, began investigating borderline cases of insanity and practicing psychotherapy.

New institutions fostered the interaction of all three professions. Some members of the advance guard emerged from that holy of holies, the laboratory, where somatic theories were pushed to their limits. Laboratories had been founded at public mental hospitals, partly because neurologists persistently had attacked their unscientific methods and lack of research.[5] Some laboratories were created by psychiatrists inspired by psychologists. Outpatient clinics, especially in larger cities, brought together neurologists and psychiatrists.

Most stimulating of all, especially for the younger men, were the new university-connected psychiatric institutes and hospitals for teaching and research on the European pattern. The prestige of these institutions facilitated the dissemination of innovation. They included the Boston Psychopathic Hospital, founded in 1912, the University of Michigan's Psychopathic Institute, in 1906, the Phipps Clinic of the Johns Hopkins University, in 1913, and most important, the Pathological Institute of the New York State Hospitals, organized in 1895. The Institute proposed comprehensive research not only in the somatic causes of insanity but, more revolutionary, in anthropology and psychology.[6] A. A. Brill was encouraged to study in Europe and first learned of psychoanalysis through an Institute training course under Adolf Meyer. New medical organizations, such as the Boston Society of Neurology and Psychiatry, were formed to bring together both professions. Finally, there were informal groups, such as the Boston circle of physicians and psychologists interested in psychotherapy and psychopathology.

If these new institutions provided positive inspiration, public hospitals, where the crisis was most apparent, furnished a more abrasive stimulus. Drugging, wild surgery, perfunctory case notes, and scant attention to patients angered some staff physicians and made them welcome alternatives to the accepted procedures of the somatic style.

The advance guard constituted a new professional generation. A very few were older psychiatrists, trained in the tradition of Moral Treatment, and a few older neurologists, such as Putnam. But the majority were born in the late 1860's and 1870's and were perhaps five to ten years younger than the leading somatic neurologists. Like the neurologists, the advance guard were urban and often college educated. They included a few members of the upper classes, such as Putnam or Morton Prince, but most were from the middle classes. Several were Europeans, including eastern European Jews.

They were open to new ideas and willing to champion unpopular changes and were especially close to European developments, still the major source of innovation. They had sufficient command of new views in science and psychology and of the somatic style itself to see its weaknesses. It was from this group that the first psychotherapists and the psychoanalysts were drawn.

The American advance guard set themselves against a declining optimism in neurology and psychiatry, as prognoses became more guarded and unfavorable. Neurologists trained during the high point of somatic orthodoxy, the 1870's and 1880's, took a less enthusiastic view of recovery for many nervous diseases than their elders, James Jackson Putnam, George Beard, S. Weir Mitchell, or the other founding fathers of the American Neurological Association. After 1900 neurasthenia, hysteria, and epilepsy seemed less curable than before.[7]

Alienists who worked in public hospitals for the insane became far more disillusioned than neurologists. The alienists' statistics showed a 50 per cent drop in recovery rates between the 1870's and 1910. In 1875, recoveries had ranged between 40 to 60 per cent of admissions; the over-all rate averaged about 46.8 per cent. By 1885 recoveries had dropped to between 20 and 40 per cent, averaging about 29 per cent. By 1910 recoveries had dropped still further to between 15 and 35 per cent. In New York they actually had risen a few points to about 25 per cent, with an additional 15 to 20 per cent discharged as able to live at home. Regardless of variations in the rates of different hospitals, however, the over-all decline was remarkable. Perhaps the majority view was expressed by a Midwestern asylum physician in 1913: "I believe that very few of our leading psychiatrists are now especially enthusiastic over the cure from insanity as a general proposition." [8]

Alienists provided several comforting explanations of the decline. They had become more scientific about statistics and set higher standards for recovery. The growing backlog of chronic patients reflected lives prolonged by better hospital care. Relatives often delayed sending patients to asylums, thus depriving them of early treatment when chances for recovery were best. The helpless aged were being unloaded on public mental hospitals because under modern urban conditions it was difficult to care for them at home.

Psychiatrists and neurologists assumed that nervous and mental diseases were growing more widespread and virulent through the combined action of heredity and environment. Accepting the inheritance of acquired characteristics into the first decade of this century, physicians argued that modern urban civilization was augmenting inheritable diseases by fostering alcoholism and syphilis and by weakening the physical constitution.

As recovery rates declined, the proportion of cases attributed to heredity rose from about 20 per cent in the late 1840's to 90 per cent by 1907. Heredity became the major explanation for the tragic paradox of nineteenth-century psychiatry: the apparent drop in recoveries despite more hospitals and better care.[9]

But as heredity became more important, it also became more and more widely and loosely defined, and from the 1890's on came under increasing attack. At first the major indicator of an inherited tendency was the presence of nervous and mental disease in relatives: the closer the relative and the more serious the disease, the worse the taint. Gradually the hereditary diathesis was expanded into an elaborate set of hypotheses that explained as variants of a single condition everything from hysteria to crime and insanity to genius. Heredity became almost identical with the theory of degeneration.

Full-blown degeneracy theory entered America from Europe, chiefly in the 1880's with the second generation of somatic neurologists, proliferated in the 1890's, and reached a high point under a Mendelian disguise from 1907 to 1917. The major theorists for Americans were Bénédict-Augustin Morel and Valentin Magnan in France; Henry Maudsley and Thomas Clouston in England; R. von Krafft-Ebing and Moritz Benedikt in Austria; and Cesare Lombroso in Italy.

It is difficult today to assess their views sympathetically. We have come to judge hereditary determinism as a pessimistic doctrine that rationalized inequalities in society and therapeutic nihilism in treatment. Yet most theorists of heredity were humanitarian radicals.[10] Because they believed that the environment could modify heredity, they often sought sweeping social reforms, such as the careful regulation of working conditions, as well as eugenic measures to prevent the birth of defectives.

They also hoped to discredit what they considered to be cruel and antiquated theological doctrines of free will and responsibility petrified in legal codes still being applied to human beings who were demonstrably sick and defective. By a careful evaluation of heredity in all its physical and psychological manifestations, it would be possible to discover those members of society whose behavior was determined by inherited forces beyond their control. This knowledge would in fact humanize the treatment of criminals and the insane. As William Alanson White observed in 1894, the theorists of heredity had made it possible to define the criminal in social terms, to see him partly as the "victim of circumstances." [11] Thus, well before the advent of psychoanalysis, traditional doctrines of free will and responsibility were being discarded by neurologists and psychiatrists on a purely somatic and hereditarian basis.

Much that today is attributed to environment and blamed on the behavior of parents was then assigned to heredity. For example, parents of adolescents suffering from the nineteenth-century precursor of schizophrenia often were observed to be bizarre. The parents' eccentricity and the child's mental disease were alike interpreted as manifestations of the family's inherited degeneracy. Similarly, it was observed that insanity broke out in the offspring of tainted families at crucial stages of physiological development, notably adolescence. Today some psychiatrists have interpreted schizophrenia in young adults as the result of defective maternal care or the result of stress at a critical stage of socialization.[12] Earlier views, like those now fashionable, seriously attempted to account for such apparent regularities.

Hereditary theory depended directly on a somatic view of cause. Morel argued that degeneration from an ideal human biological type caused phenomena ranging from the "slightest eccentricity" to idiocy and hopeless insanity. The underlying degenerative condition tended to become manifest in progressively more severe symptoms during a single lifetime or across generations. Hysteria might appear in a family, and within four generations it would die out in sterile idiocy. Thus nature mercifully preserved the integrity of the species.[13]

Environmental factors—disease, poisons such as alcohol or hashish, immorality, insalubrious climates and soils, unhealthy cities and factories, poverty, malnutrition—caused the original degenerative tendency. The degenerate could be recognized by a combination of physical, moral, and intellectual stigmata, some of which appeared in early childhood. These represented departures from a presumed norm. Degeneracy theorists energetically collected statistics and attempted to establish identifying traits for normal human beings, criminals, and the insane, then to draw meaningful comparisons among them. But their statistical techniques were usually crude, and they chiefly studied institutional populations in prisons and asylums, where chronic cases or recidivists tended to collect. Thus the distribution of traits among the general population remained unknown.

If standards were assumed for any trait, such as intelligence or shape of the head, too little or too much could be considered a sign of morbid deviance. Such opposite qualities as excessive tallness or shortness, brilliance or idiocy, callousness or squeamishness, overdevelopment or underdevelopment of sexual organs or appetites could be included among the stigmata. By the end of the century positive and negative deviance was pushed to amazing limits.

Physical stigmata were diagnostically important because of the close reciprocal relation that was believed to exist between somatic and psychological factors. Thus the outward morphological sign not only expressed, but caused, as it were, the inner somatic disgrace. Every advance in knowledge of the brain and skull, Moritz Benedikt argued, reinforced the ancient conviction that man "thinks, feels, desires and acts according to the anatomical construction and physiological development of his brain. . . . That a defective, atypically constructed brain, cannot function normally, is so evident as to leave no ground for discussion." Lombroso declared: "Is it possible that individuals afflicted with so great a number of alterations should have the same sentiments as men with a skull entirely normal?" [14]

The physical stigmata also came to include a variety of illnesses—rheumatism, alcoholism, tuberculosis, cancer, epilepsy and hysteria, or any disease that weakened energy for long periods. Krafft-Ebing argued that certain kinds of insanity, such as paranoia or what later was called manic depressive insanity, were symptomatic of degeneracy. Thus, some of the diseases degeneracy was supposed finally to cause, became, by circular logic, diagnostic of the state that caused them.[15]

Psychological stigmata assumed a growing importance. Cruelty, immorality, above all, lability and impulsiveness, which could appear in early childhood, marked the degenerate. Later theorists argued that precisely these psychological qualities were common to both the insane and the criminal, presumably because the inhibiting centers of the brain were insufficiently developed.[16] These same qualities also marked the degenerate but gifted artist, the eminent scholar or, finally, the man of genius. They were especially important because sometimes they were the only telltale signs in people with unblemished bodies.

The *reductio ad absurdum* of psychic stigmata was reached in an American text: "precocity or retarded evolution of intellect or instinctive propensities"; "exaggerated conscientiousness or absence of moral sense, with fanatic religious zeal, or great depravity" . . . ; "morbidly heightened imagination . . . ; eccentricity of ideas and of feelings, odd conceits and novel emotions" . . . ; "disproportionate development of mental faculties, one-sided talents, display of fantastic genius, and defect of higher rational processes . . . general lack of harmonious co ordination of the intellectual, emotional and volitional elements of the mind." [17]

Americans—until the eugenics movement around 1907—took a more optimistic view of heredity than most Europeans. The degenerative process was not inexorable and could be prevented by early measures and environmental amelioration.[18]

Several intellectual and social factors hardened American attitudes. The first probably was Francis Galton's insistence that nature was overwhelmingly more important than nurture. The second were attacks on the inheritance of acquired characteristics by the German biologist August Weismann and by those who revived Gregor Mendel's genetic laws around 1900.

Although some physicians remained Lamarckians and others paid little attention to theory, many psychiatrists and neurologists turned to Mendelianism as more becomingly "scientific." Yet the first psychiatric Mendelians were curiously uninventive. They retained the nineteenth-century diathesis and expanded it even further to include "loquacious rambling" and nervous fidgeting. The diathesis was a "unit character," transmitted as recessive to the normal. Manic depressive insanity and dementia praecox probably were passed on with some degree of directness. If both parents were neuropathic, the Mendelians argued, their children were likely to be neuropathic.[19] Logically, only eugenic programs could halt the apparently mounting incidence of insanity. From the 1880's on a few had proposed extirpating the sexual organs of criminals, epileptics, and the insane. By 1900 a lay enthusiast was advocating the painless destruction of "defectives" by carbonic acid gas.[20] Between 1905 and 1917 the impact of Galton's eugenics movement on American psychiatry reached a peak. The tone of eugenic proposals by responsible psychiatrists became increasingly harsh and punitive. In 1914 the leader of America's institutional psychiatrists proposed that the incurably insane be sterilized.[21]

Eugenic enthusiasm involved far more than theory. Class biases, institutionally fostered patterns of chronicity, and the new immigration played key roles. Public mental hospitals served chiefly the lower middle classes and the poor, and the percentage who became chronics was high —two-thirds of males admitted to the Pennsylvania Hospital for the Insane, for instance. Living in an atmosphere that destroyed the sense of identity, perhaps wearing makeshift clothes, institutional populations seemed to constitute a special class of human beings to those who observed them.[22]

These observers were overwhelmingly native Americans, often now with college as well as medical degrees. The social distance from their patients increased markedly with the new immigration; barriers of language and custom mounted higher than ever.

The new immigrants, especially Russian Jews, seemed at first to have a high incidence of "mental defect" and neurasthenia and to be unusually liable to dementia praecox. By 1911 the foreign born made up 29 per cent of the population of New York State, yet provided 48.02 per cent of first admissions to insane hospitals.[23] Some American psychiatrists began to identify dementia praecox, alcoholism, and syphilitic insanity with defective stock or un-American violations of morality and sobriety.

As eugenic proposals grew fiercer, the entire complex of hereditary theory came to seem more grotesque and vulnerable. Some physicians who had been proponents of degeneracy in the 1890's now attacked it —William Alanson White, for instance, and Smith Ely Jelliffe, his friend and collaborator who had once sketched thousands of ears of insane patients in a search for stigmata.[24] Critics argued that hereditary theories prejudiced therapeutic effort and outcome, and were overextended, contradictory, vague, and not based on scientific method.

Henry Maudsley warned Americans in 1896 that Morel's concept had been robbed of its scientific meaning and had become a "metaphysical something," stretched to "cover all sorts and degrees of deviations from an ideal standard of feeling and thinking, deviations that range actually from wrong habits of thought and feeling to the worst idiocy, and some of . . . which are no more serious marks of morbid degeneracy than long legs or short legs, long noses or short noses." [25] Others charged that diseases differing completely in form, prognosis, and treatment had come to be included as equivalent "stigmata." If "the same course were pursued with other diseases, 'confusion worse confounded' would soon reign in our general ideas of etiology and pathology," wrote the American alienist, Theodore H. Kellogg.[26]

The extension of the theory of degeneration to include genius aroused the deepest indignation. A reviewer in the *American Journal of Insanity* argued that it was preposterous to suppose that the "man of genius is brother to the epileptic and the lunatic, and first cousin to the criminal," [27] and that the highest products of the human imagination were symptoms of decay. In 1904 James J. Putnam complained to the New York neurologist Charles L. Dana that to consider neurasthenia as a sign of insanity and constitutional weakness, as Dana proposed, was senseless and could prejudice the outcome of treatment:

> Take for example, such cases as that of Parkman, the historian, who was never able to sleep . . . and could only work a few minutes at a time, yet who accomplished such a vast amount of first-class historical research, and was of such splendid will and courage. Surely he represents a kind of person whom it would be a mockery to speak of as mentally weak in any usually accepted sense. Is not a large share of the best literary, artistic and intellectual work of the world done by people who are more or less neurasthenic?
>
> . . . The fault, I think, with most conscientious physicians who have to do with persons whose mental makeup is a little peculiar is that they prejudice their treatment in advance by assuming that the case is a bad one and that treatment will be of no avail. For one such suspicious student who turned out badly there would be a dozen who were more or less suspicious of their neighbors . . . but in whom the symptom was really due to lack of proper bringing up or some unfavorable experience or loneliness, and the like, and it would be a great misfortune to such a person to let him find out that he was classified under a title that would imply to him that the asylum yawned for him. . . .[28]

During a symposium on heredity before the American Neurological Association in 1907, S. Weir Mitchell rejected pessimism as false and prejudicial to treatment. Like Putnam, he believed in the "therapeutics of hope"; his own experience contradicted pessimism, especially in the histories of families "in the upper social life of his own city." [29]

Objections accumulated to the methodology of the hereditarians. William James denounced Lombroso for an apparently "incurable incapacity" to reason accurately, and Franz Boas attacked Lombroso's statistical methods. His samples were too small; he had failed utterly to prove the existence of a "criminal type." He had not studied the prevalence of stigmata among the general population but had restricted himself to describing its prevalence among selected groups, such as criminals, epileptics, or the insane. Without a knowledge of how frequently stigmata appeared in the normal population, their significance could not be determined. Only in this way could a true "average type" be constructed. Adolf Meyer noted that many sober anthropologists assumed that the normal type was a purely arbitrary assumption and often embraced a "great number of physiological variations." [30]

The failure to find out how widely the stigmata were distributed in the general population created embarrassing anomalies. Research by a journalist among members of Congress, for instance, disclosed heads displaying "inequalities that would astound Lombroso." [31] Too often the respected businessman, the saintly clergyman, the honest worker, exhibited signs of degeneracy. What then, became of their diagnostic significance?

Perhaps the most devastating attack on theories of heredity reflecting advances in statistics and genetics was launched by the psychiatrist Abraham Myerson in 1917.[32] Hereditary theorists had collected only "positive instances" and ignored negative ones. They were unsophisticated about statistics and had never tested their hypotheses against control groups. Lombroso, for one, had worked from the loosest possible analogies.

In a *reductio ad absurdum* of the diathesis, Myerson asked, does "headache or miserliness in a great aunt have anything to do with dementia praecox in a patient? If it has, then it condemns all of us to a psychopathic career." [33] The arguments for heredity in nervous and mental diseases were often circular. Insanity, for example, was part of an "inherited predisposition" which caused insanity. It was preposterous to suppose that "there is a unity equivalence in the diseases of a list which starts from A, proceeds alphabetically to include apoplexy, alcohol, blindness, etc., down to tumor. . . ." Moreover, discriminating clinicians were carefully sorting out different kinds of nervous disorders and no longer lumped them together as variants of the same universal disease.

The "neuro-pathic constitution" could hardly be considered a "unit" character like the color of a pea. "We may be sure," the biologist Edwin G. Conklin indignantly noted, "that when the whole 'alphabet of degeneracy from alcoholism to wanderlust' is attributed to the lack of a single hereditary factor, there has been a pitiful failure to recognize the complexity of the phenomena in question." [34]

Other critics stressed the importance of environment, arguing that it was not racial stock but the shock of immigration itself, the alienation and isolation, the harsh competition in American slums that caused the apparently higher incidence of insanity among the foreign born. New medical psychologies, including psychoanalysis, that emphasized subjective factors helped to undermine faith in the diagnostic significance of outward form and behavior. For the long run the most important result of the controversy was the destruction of biological models for the social sciences and the emergence of a new cultural determinism. Men were being forced to seek in cultural and social traditions for the causes of those traits once attributed to the inheritance of acquired characteristics.[35]

"Classification Mania," 1898–1910

Without sound classifications of nervous and mental disease, studies of heredity were useless. Yet such classifications, despite attempts to fit them into somatic formulae, were often as backward as eighteenth-century medical nosologies, a chaos of multiplying systems and random observations.

A paradoxical development in classification brought to prominence an apparently incurable organic insanity, dementia praecox, later called schizophrenia. Psychoanalysis made one of its most brilliant contributions in the early attempts to understand it. That contribution, which brought Carl Jung to the Clark Conference, will be described in Chapter VIII. The provocative dilemma itself requires elucidation here.

Adolf Meyer and other members of the advance guard had been criticizing traditional American psychiatric classifications since the late 1890's—the division into acute and chronic types of mania, melancholia, circular insanity, and dementia. The chronic category, Meyer argued, was founded on ignorance and error because chronic patients sometimes emerged from years of insanity to become "self-supporting and rational citizens." [36]

Meyer pointed out statistical contradictions in conventional classifications. The eminent Sir Thomas Clouston listed 50 per cent of the patients in his Edinburgh clinic as suffering from "mania," or "morbid mental exaltation and excitement." In Paris, however, only 6 per cent of all cases were so classified and in Zurich only 8 per cent. Mania was considered a recoverable psychosis in American and British tradition. Yet, at New York State hospitals only 35.6 per cent of those suffering from acute mania recovered, while abroad 80 per cent got well. Were psychiatrists dealing with the same clinical entities? Did their treatments differ so radically? Did recoveries vary so hugely between Europe and the United States? These were uncomfortable questions.[37]

Satisfying answers seemed to be provided by the elaborate new system of Emil Kraepelin, the Munich psychiatrist. His method was introduced by Hoch and Meyer in the late 1890's and became accepted orthodoxy for the next thirty years, despite the growing objections of Meyer and others to its rigidity.

Kraepelin held out a tantalizing hope of "prognostic precision" and based his classifications on a thorough study of symptoms, course, and outcome, what he conceived to be the "underlying disease process." The intricacies of his system cannot be described here, but dementia praecox requires attention. An accomplished clinical lecturer, Kraepelin illustrated this new category with dramatic flair: "Gentlemen, you have before you today a strongly built and well-nourished man, aged twenty-one. . . . He broods, staring in front of him with expressionless features, over which a vacant smile occasionally plays . . . apparently speechless, and not troubling about anything." His memory was intact, but his judgment was "weak," he was "flighty," he displayed "mental and emotional infirmity," he was given to a "silly, vacant laugh," and displayed emotional dullness and "absence of independent impulses of the will." Above all, like other dementia praecox patients, he exhibited a marked apathy, a total lack of interest in his environment. When Kraepelin extended his hand, this patient would fail to grasp it. The prognosis was very unfavorable.

Despite occasional surprising recoveries in even the worst cases, as a rule dementia praecox victims never recovered completely. For, Kraepelin surmised, this disease entailed severe damage to the cerebral cortex.

Kraepelin expressed much of the pessimism erroneously associated with all somaticists. He concluded that the physician's power against the "mighty adversary" of insanity was "very narrow." [38]

The adoption of Kraepelin's system, and notably of the dementia praecox category, had curious effects. Some patients who might have been given the more hopeful prognosis of acute mania or melancholia now were diagnosed as suffering from dementia praecox. Kraepelin enlarged and contracted periodically the percentage of patients he estimated should belong in this category. At a time when he set the figure at about 14 per cent, New York alienists were diagnosing from 19 to 34 per cent of all new admissions as victims of dementia praecox. A very high rate was diagnosed at the Connecticut Hospital for the Insane at Middletown, where Kraepelin's English translator, A. Ross Diefendorf, was assistant physician.[39]

Once more, as with hereditary theory, medical fashion played an important role in the spread of dementia praecox. One critic observed: "It is a conspicuous fact that at certain intervals of time new nosographic entities appear on the psychiatric horizon, which have the misfortune to become popular. . . . In their passage they increase in bulk at the expense of less well defined pathologic conditions, especially by the absorption of numerous cases from that important category in all classifications—the undiagnosed group, which are often found to fit the new pattern with singular convenience." [40]

Some Americans who adopted dementia praecox enthusiastically widened its definition. The head of an Iowa asylum described this typical beginning for the insidious disease: "A son or daughter at home loses interest in study, drops out of school or neglects work, formerly well done, and lounges about the house. . . ." [41] A more serious initial symptom might be "silliness" or a "contrary disposition." Once the category was accepted, psychiatrists proceeded to fill it with patients. A candid, busy superintendent found dementia praecox a useful comfort: "The most difficult thing for me was to obtain some place where I could put my doubtful cases. I personally have not the time to work out any new schemes, either as to classification or scientific research. In going through the wards, there are many cases which puzzle me, and we do not know how to classify them. I have found dementia praecox a very comfortable, I may say vulgarly, dumping ground." [42]

What kinds of patients did this category so conveniently fit? The very term "dementia" provides an important clue. Kraepelin's dementia

praecox implied that the symptoms often observed in senile patients—
a loss of ability to think clearly, confusion, withdrawal—occurred
prematurely in the young; "deterioration" set in early in the illness. G.
Alder Blumer, a hospital superintendent, argued in 1906 that Kraepelin
had never bothered to define dementia closely and that he used it to
cover quite different conditions—senile dementia, in which perception
and memory were impaired, and dementia praecox, in which both were
intact. Possibly, Blumer suggested, dementia praecox was not an organic
illness at all, but a functional one.[43]

There is a close correspondence between the withdrawal and emotional flatness that was the *sine qua non* of dementia praecox and the
apathy a few psychiatrists believed to be the result of institutional routine and neglect. Thousands of inmates were "permitted to starve and
die a slow death for lack of proper psychic stimulation," a German immigrant alienist charged in 1901.[44] A sympathetic colleague responded,
"We do not, I fear, make sufficient effort to reach these minds in some
way. . . ." In some asylums, patients worked "in long lines, being driven
along just like oxen or mules." That spectacle, an alienist confessed, gave
him a feeling of "very great discomfort." [45]

Because cases considered hopeless of necessity received less attention,
the physician's attitude toward recovery was crucial. On that basis alone,
a few American neurologists and psychiatrists bitterly fought Kraepelin's
system and the introduction of "dementia praecox." Bernard Sachs
argued that many patients who might be classified as suffering from this
disease could function usefully for years. An alienist at Manhattan State
Hospital, who had been trained in the older, more hopeful American
tradition, attacked this "new menace" in psychiatry. The "gloomy prognosis" would do nothing but harm. Scores of patients with symptoms
that resembled those of dementia praecox actually recovered—but
only if they received zealous, constant, early treatment.[46]

Unremitting care was what most public asylums failed to provide.
At Kankakee, Michigan, there were six physicians for 2200 patients. A.
A. Brill, who worked for a time at New York state hospitals, recalled
the usual case note: "The patient is stupid, dull and demented." The
note would end, "patient died suddenly." [47]

During the years that Kraepelin's system became standard in many
American hospitals some psychiatrists began to reject rigid classifications. In 1906 William Alanson White argued that clinical types were
not clean-cut and that categories overlapped. He associated nosological

entities with the "old psychology," which looked on the mind "as composed of a number of cubby holes in each of which was pigeon-holed a faculty and each faculty—feeling, intellection, volition—was just as distinct from the other as the conception implied." [48]

Difficulties in classification also became acute for neurologists, who were moving in two opposite directions. Some of those experimenting with psychotherapy tended to enlarge the types of nervous disorders considered to be functional, that is, without known organic etiology.

A second group of neurologists was moving toward a more exclusively organic conception. Moses Allen Starr, for example, argued that all epilepsies resulted from organic cortical lesions. In general, the strict somaticists tended to espouse less hopeful prognoses and to regard hysteria, epilepsy, and neurasthenia as manifestations of degenerate heredity.[49]

The underlying assumption of somatic pathology was attacked in 1906, by the president of the American Neurological Association, Henry B. Stedman, a Boston neurologist. He argued, as did the French medical psychologist, Pierre Janet, that most psychoneuroses were not diseases of the nervous system at all. Rather, they were practically "mental" diseases of functional origin. In 1915 Morton Prince insisted that 75 per cent of all the ailments treated in nerve clinics were functional.[50]

The problem of the "traumatic neuroses" illustrates how disorders grouped together because of an ostensibly physical origin finally became differentiated into some that were genuinely organic and others that were functional. In 1871 the British physician John Eric Erichsen had argued that a wide range of injuries could be classified together as the results of "spinal concussion." These often occurred during railway accidents, and were principally characterized by an absence of obvious injury. The "lesions" ranged from meningitis and hemorrhage into the spinal cord to hysterical weakness or "anaemia" of the spine. The latter, Erichsen speculated, might be caused by a vibratory jar or concussion which disturbed the "molecular condition" of the sympathetic nervous system and the spine or the supply of blood to the cord. Some physicians cited Erichsen in behalf of the claims of railroad accident victims. In 1883 another British physician, Herbert Page, in the employ of the railroads, argued that few of the ailments Erichsen had grouped together were in fact organic; most were functional and of little importance, and he called them "traumatic neurasthenia." [51]

In 1884 James Jackson Putnam suggested, as had Page, that many

of the cases Erichsen had discussed were hysterical. Charcot, Putnam
noted, had established diagnostic "stigmata" of hysteria such as circumscribed anaesthesias, paralyses, and contractures that much resembled the
symptoms displayed by this class of railway accident victims.

Putnam cited the example of a man who was severely jolted against
the iron seat of a railway carriage. He was completely paralyzed at first;
then paralyzed on one side only. Half his body was totally anaesthetic to
pin pricks. Gradually, the paralysis disappeared. But the hemi-anaes-
thesia remained; this, Putnam argued, the patient could not possibly
have simulated. Symptoms might be hysterical, yet at the same time involuntary, serious, and lasting.

These symptoms were not caused by lesions, no matter how subtle,
such as minute hemorrhages or "vascular disorders and nutritive
changes." The causes, Putnam concluded, were "psychical," and presumably operated in those predisposed by heredity, although most cases
showed little evidence of inherited predisposition.

The differential severity of symptoms, Putnam argued, could be accounted for by differences in social class and training. Because of real insecurity and a dread of poverty, wage earners suffered the highest proportion of serious hysterical symptoms.[52] These careful differential
diagnoses by neurologists resulted finally in a marked increase in the
number of functional disorders.

At about the same time, the nosological entities established by Charcot and Jackson for hysteria and epilepsy, as well as Beard's vaguer
"neurasthenia," also were undergoing attack. Beard's former assistant,
Charles L. Dana, argued in 1904 that when mental symptoms such as
morbid fears and impulses predominated in cases of neurasthenia, these
probably were early stages or abortive types of insanity. Dana had reversed Beard's tactic of separating neurasthenia from insanity. Despite
Dana's protestations of hope and the fact that he included more symptoms in a strictly psychological category, Putnam, for one, saw the practical consequences as pessimistic.[53]

While neurasthenia could be viewed in large part as a kind of pro-
dromal insanity, sharp arguments were occurring over the nature and
prognosis of epilepsy. These conflicts illuminate the position of those
who wished to widen the psychological and functional aspects of neurological disorder and those who wished to make diagnosis more exclusively somatic. Between 1870 and 1890 Hughlings Jackson had established that the major kinds of epilepsy, "idiopathic" and what later was

called "Jacksonian," were essentially the same. They resulted from lesions, either structural or functional, in the cerebral cortex.

In 1904 Moses Allen Starr suggested that both types of epilepsy resembled each other so closely they must have the same cause: a circumscribed lesion anywhere in the brain. Where a few decades before the American neurologist William Hammond's prognosis for epilepsy had been guarded, but relatively favorable, Starr argued that epilepsy was a sign of degenerative inheritance and that recovery was unlikely.

This view provoked serious opposition. A few colleagues argued that Starr's pessimistic conclusions should be deleted before his paper was published lest epileptics and their relatives become unduly discouraged. Adolf Meyer noted that the treatment of epileptics indicated the existence of an irreducible "psychic factor," and that the therapeutic zeal and optimism of attending physicians had a marked effect on recovery. How, he asked, could Starr explain patients in whom a lesion had been removed by surgery, yet who failed to recover? [54]

A view of epilepsy the exact opposite of Starr's was taken by Putnam in a paper published that same year. He and his young assistant, George Waterman, suggested that physicians undertake careful investigations of the psychological state of the epileptic, such as already had been undertaken for the hysteric. Although epilepsy was assumed to be a result of inherited degeneracy, the stigmata were by no means always present in epileptic patients. Perhaps a painstaking investigation would show a connection between psychological processes and the symptomatic acts that occurred during an epileptic attack.[55]

Charcot's classical description of hysteria as a nosological entity—with such stigmata as circumscribed anaesthesias or narrowing of the field of vision—also was rejected. French physicians had been doubtful of Charcot's conclusions at least since the 1880's. Then in 1906 one of his pupils, Joseph Babinski, insisted that hysteria was not a nosological entity at all. Its symptoms were caused by "suggestion," and could be cured by "persuasion." Thus hysteria was reduced to a purely psychological disorder.[56]

Localization Dethroned, 1890–1910

Classical theories of the localization of functions in specific areas of the brain were questioned by a few neurologists in the 1890's, and demolished in the early 1900's. Although not every critic turned to psychotherapy and psychology, it was hardly accident that the first psychotherapists were among those who offered sophisticated objections to the classical theories.

Psychotherapy was developed in part because a few men were sufficiently in command of somatic theory to have a clear view of its weaknesses and extrapolations. Aware of how purely speculative and indeed "metaphysical" much localization theory was, these critics could view with equanimity the uncertainties of a psychological approach.

Some of the men who became psychotherapists established reputations as critics of localization. Among these were Sigmund Freud, Morton Prince, and Adolf Meyer. Localization had bolstered faith in the somatic pathology of nervous diseases and especially of insanity. Only the barest outlines of this highly technical and richly detailed controversy can be touched on here.

As William James had discerned, theories of localization, such as Theodore Meynert's, directly paralleled psychological theories of elements and association.[57] The critics of classical localization argued that localizers had assumed the mind to be a collection of separate units derived, perhaps unwittingly, from obsolete conceptions of mental faculties, such as those of phrenology. If the mind were composed of such separate elements, and if it worked like any other mechanism, damage to particular parts would damage particular functions. Motor functions had been localized with considerable certainty; sensory functions with far less. Extremists assumed that ultimately all major psychological functions could be located in discrete parts of the brain. A neurologist at the Johns Hopkins University, Lewellys F. Barker, presumed that when the "intellectual centers" were paralyzed, there would result an impaired inhibition of "lower centers concerned with the more primaeval instincts and emotions." [58] This flight of fancy, based in part on the theories of Hughlings Jackson, demonstrates the lengths to which neurologists allowed their imaginations to lead them. Aphasia, or the loss of the function of articulate speech, had been one of the keys to the construction of models of localization. A re-examination of aphasia led the way to the destruction of the more naïve aspects of the classical theories.

One important attack, read by Henri Bergson and noticed in the United States, came from Freud. Essentially Freud argued that the speech function could not be split up into distinct parts but was a "continuous cortical region within which the associations and transmissions underlying the speech function are taking place; they are of a complexity beyond comprehension." [59] He insisted, chiefly on the authority of Hughlings Jackson, that psychological and neurological events must be strictly separated. No meaningful attempt could be made to interpret the one in terms of the other.

The most serious and influential attack on classical localization theories came from the French clinical neurologist Pierre Marie in 1906. In a scathing review of slovenly neurological work he showed how few cases supported the elaborate theories erected upon them. According to the English neurologist Henry Head, Marie abolished the classical consensus and ushered in a period of "chaos" when each investigator saw aphasia from his own independent viewpoint. The result was a more open welcome to a psychological and "dynamic" interpretation of the speech function.[60]

Finally in 1910, Morton Prince, following Von Monakow, rejected the entire basis on which theories of localization had been based. He argued that even if a lesion of a specific area of the cortex impaired a specific function, this did not prove that the function itself was limited to that particular area. Clinicians, he charged, were "extremely prone to metaphysical psychology." [61] An earlier judgment of Adolf Meyer's still held good, that brain anatomy was a field of "wild semi-scientific speculation. . . ." The classical views were nothing less than a Cartesian anthropomorphism. "Nobody would, of course," Meyer wrote, "search the pineal gland for the spring of action, the 'soul'; but this same soul is, though split up a little, seated in the centers." The center was a "sort of homunculus, a mysterious Pygmy who acts his part as a little man would, accumulates images and energies and discharges them, sends them along the wires of fiber paths to his superiors and inferiors who discharge again on others. Such action is decidedly more anthropomorphic-electrical than physiological." [62] The criticism of localization as well as the discovery of the neurone in the early 1890's and the subsequent work of the English biologist Charles Sherrington made impossible the earlier, simplistic views of nervous function.

The New Unitary Principle, 1895–1910

Critics of localization had suggested an important new concept, toward which many different neurologists and psychiatrists were moving. Not only the brain, but the "whole nervous system and the whole man" should become the physician's concern. An approach that considered the patient in terms of his available energies and functional adjustments emerged from these debates over fundamental principles. In 1904 James Jackson Putnam launched an attack on the most sacred canon of neurology, Rudolph Virchow's insistence that all disease was the result of a localized anatomical lesion: "A man meets with an injury attended with great nervous shock, and the neurologist is ready enough to spend infinite pains on the study of the necrotic areas in his spinal cord, but is apt to overlook the fact that this localized process does not explain why he has at the same time lost flesh and strength and color, and has become the football of his delusions and fears."

"Imitation and fashion," he continued, "play a large part even in scientific investigation, and the almost universal tendency to bend all energies to the search for the physical evidence of localized lesions has led too often to a disregard of disturbances usually classified as functional."

As an alternative, Putnam suggested a "physiological principle" that emphasized the "function of the organism as a whole." Based on the theories of his teacher, John Hughlings Jackson, this principle assumed "interlocking organizations of energic functions, superposed one over another in ever increasing complexity." Putnam insisted that such a conception harmonized with the results of the investigations of dissociation in hysteria carried on by Pierre Janet, Sigmund Freud, Morton Prince, and Boris Sidis. Putnam also argued that more emphasis should be placed on the reciprocal relationship between "bodily processes and mental states." "Both of them," he wrote, "are manifestations of energy, and there must be some denominator common to them both." [63] Adolf Meyer's psychobiology later represented a similar attempt to find a principle that would unite mind and body into an indissoluble functioning unit. It was supported by the research of Walter B. Cannon, the Harvard physiologist, who demonstrated that "ideas" and "emotions" could alter bodily functions in striking and essential ways. The old doctrine of parallelism, with its separate pigeon holes for mind and body, seemed increasingly obsolete and unnecessary.

The Failures of Therapy and the Conflict of Roles, 1900–1908

Around the time of Freud's visit to the United States, the most acrid debates among neurologists, and to a lesser extent alienists, concerned therapeutics. Two presidents of the American Neurological Association, Silas Weir Mitchell and Henry Stedman, in 1906 and 1909, respectively, deplored their colleagues' slight concern with treatment.[64] They still were prescribing rest, diet, massage, discipline, and encouragement in most nervous disorders.

In 1907 and 1908 psychotherapy was bitterly discussed among New York, Boston, and Philadelphia neurologists, who tended to split into two camps, the orthodox somaticists and the proponents of systematic psychotherapy. The latter felt despised and rejected, classed among the charlatans and the faith healers. Yet they saw themselves as the protagonists of a new knowledge and technique which more hidebound authorities were too limited, and possibly too lazy, to learn.

Organicists argued that functional disorders were relatively unimportant; the true sufferers were those with organic diseases. Only the latter, Bernard Sachs insisted, were worth the neurologist's serious attention. Moreover, most psychotherapeutists, he argued, gave the preposterous impression that there was no organic basis for nervous diseases.

These debates over therapy precipitated a crisis over the neurologist's sense of identity and self-image and hardened the views of both sides. For a neurologist to call himself a psychotherapist was equivalent to saying he was not a neurologist.

"I'm a neurologist first," Sachs remarked. "I am not going to call myself a psychotherapeutist any more than I would call myself an electrician, although I do use electricity occasionally." A colleague argued that psychotherapy placed the "patient himself" in the role of the physician. "It leads to neglect of physical and real methods," insisted Francis X. Dercum, a Philadelphia neurologist.[65]

To neurologists from Putnam's generation forward, the very definition of their specialized role depended on a deep and thorough knowledge of the brain and nervous system. During the same period "mental methods" often were associated with the "mind cure" practitioner, the charlatan, or the vaudeville showman. As will be indicated in Chapter V, in the 1890's many physicians, including many neurologists, were hostile to hypnotism.

Yet, therapeutic failure involved neurologists in still another role conflict. Their function as physicians was to cure their patients. What were they to do? If they adopted non-neurological methods, they ceased to be neurologists in the sense their entire training dictated. If they did not, they ran the risk of failure as physicians. One way out of the dilemma was to attack the speculative and unscientific nature of their own extended somatic generalizations, and thus to make "mental methods" more intellectually respectable. It was also possible to incorporate systematic psychological methods into neurology.

Six years later, after psychotherapy had become more respectable, Charles L. Dana argued that it had saved the breadth of the neurologist's practice. No longer was there a danger that he might become only an esoteric specialist, consulted for the diagnosis of rare and incurable disorders, while general practitioners treated patients by psychotherapy. The neurologist had extended his practice to embrace "the art of investigating the character, temperament and social conditions of his patients, heredity, education and the fundamental traits of their character." A few years later, Dana would regard this new direction as a return to the principles of George Beard.[66]

The problem of therapy also was acute in psychiatry. Most public and especially most private mental hospitals officially continued "moral treatment." Yet, its content had changed profoundly since John Butler had described it in the 1880's. Then its essence had been the closest attention paid by doctors and staff members to the thoughts and feelings of patients, with whose daily lives they were deeply and deliberately involved.

The most striking example of this change in meaning came from Edward Cowles, the superintendent of the McLean Hospital. Judging from his meticulous case histories, he had a highly sophisticated understanding of his patients and spent long hours exploring their personal problems. Yet he failed to include this activity in his description of "moral treatment," which he defined as the "healthful occupation of body and mind," including recreation, exercise, manual labor, carpentry.[67] To some extreme somaticists, moral causes had come to seem the old wives' tales of demented patients and distracted relatives.

Nevertheless, in a few private asylums physicians were using methods of psychotherapy that probably went back to Hack Tuke's *Illustrations of the Influence of the Mind Upon the Body in Health and Disease.* First published in 1872 and reprinted in 1884, Tuke's manual had received the approval of even so ferocious a somaticist as J. G. Kiernan.

In 1903 a seventy-five-year-old alienist urged his colleagues not to ignore the therapeutic influence of the mind.[68] The prevailing emphasis on insanity as a brain disease was making physicians ignore careful, individualized mental treatment. Two years earlier, a New England alienist,

Charles W. Page, had argued for a revival of the therapeutic approach of his teacher, John Butler. Until his retirement in 1872 Butler had continued his acute and subtle psychological therapy of individual cases at the Hartford Retreat, and his "mornings on the lawn," when he would soothe the fears and plant the "seeds of hope in the hearts of many unhappy patients." [69]

Most alienists relied on somatic treatments such as hydrotherapy, drugs, and rest; and there had been decided therapeutic improvements in the late nineteenth century. Surgical procedures, such as asexualization, once the rage at Utica for epileptic insanity, for instance, had been given up. The heroic drugging of the 1870's and 1880's had been replaced by a greater emphasis on occupation, even if this meant working to keep the institution functioning. Yet the failure of somatic therapies was reflected in the low rates of recovery.

As yet no cerebral pathology and no consistent imbalances of the internal secretions had been found to explain such common kinds of insanity as dementia praecox or manic depressive insanity. This failure was not quite so devastating as sometimes has been suggested. To Putnam, for instance, it was not the failure of pathology that called existing views into question. It was rather that somatic theory failed to explain adequately the way nervous and mental disorders developed and failed even to describe them adequately.[70]

There was little neatness or uniformity in the history of these years of professional crisis. Some of the most devoted exponents of hereditary degeneracy were among the first to take up psychotherapy. Moreover, any necessary connection between somaticism and pessimism must be discarded. The organicists were not, as one early psychoanalyst implied, stupid men interested only in the custodial care of patients. The search for a physiological cure for insanity was highly optimistic, if not Utopian. A few men believed that an antitoxin would be found to cure insanity as easily as smallpox could be prevented by vaccination.

The advance guard's confrontation of somaticists with irksome questions and alternative hypotheses tended to harden for a time the attitude of their opponents. One Philadelphia neurologist, with Mary Baker Eddy and Christian Science in mind, insisted that any approach to insanity which failed to take the "organic brain" into account was "metaphysical." [71] In a final turn of the screw, as the orthodox physiological approach became more rigid, it also became more grotesque and vulnerable. In 1906, in a letter to the young psychiatrist Elmer Southard, Putnam observed that the "rather forced character" of the Philadelphia neurologist Charles K. Mills's attempts to classify mental disease on the "practically exclusive basis of anatomical data" had provoked Adolf Meyer's skepticism.[72] It is important to emphasize that the changes emerging during the period of crisis were gradual, contradictory, and eclectic.

The inability of neurologists and psychiatrists to solve theoretical and practical problems in the terms set by the somatic style proved significant in a negative way in this period of crisis. This failure, exacerbated by the confrontation of alternative, functional approaches, facilitated the reception of psychoanalysis.

The crisis began with a realization that rates of recovery had been steadily declining, with the result that pessimism among alienists deepened, especially from about 1900 to 1914. After 1900 less favorable prognoses also became the fashion among some neurologists.

The decline in recoveries was explained in large part by theories of heredity which became overextended between 1895 and 1917. The advance guard of critics picked out the contradictions of these theories and used them to question some of the assumptions of somatic etiology.

From about 1890 to 1910, it became increasingly apparent that neither gross lesions nor metabolic malfunctions could be found to explain the most important varieties of insanity. Morton Prince argued that it was useless to explain, for example, that "Smith hates Jones" because of a "congested optimic thalamus," or that the "melancholiac thinks he has committed the 'unpardonable sin' because of a hypothetical toxemia." [73] Theories of localization that had supported a somatic interpretation of nervous and mental diseases also were destroyed.

Problems of classification between 1898 and 1910 increased the difficulties. The adoption of Kraepelin's system, especially his category of dementia praecox, enlarged considerably the proportion of patients who could be viewed as suffering from an organic and relatively hopeless kind of insanity. Yet, the inconsistencies of Kraepelin's definition of the disease, the evidence of unexpected recoveries, and the already elaborate psychological explorations of hysteria, led a few physicians to suggest that dementia praecox might be functional.

The tendency of some neurologists to classify more and more disorders as organic and of others to emphasize their functional and psychological character exacerbated debate over underlying principles of treatment. What sense did it make to treat psychological cases by rest and diet? All this criticism made possible a methodological and provisional divorce between the psychological and the somatic levels of explanation. Freud had faced this problem, and solved it by abandoning his project of a psychology for neurologists and by developing the more purely psychological nature of his theoretical system for the *Traumdeutung*. Yet, all physicians looked finally to a solution from more sophisticated physiological knowledge. In the meantime, the advance guard knew more clearly what was sheer speculation and what was not. They were prepared to investigate systematically psychological and environmental factors without having to translate them into somatic terms.

All these factors together precipitated a crisis in defining the professional roles of neurologists and psychiatrists. Neurologists had been trained in knowledge of the brain and nervous system. Yet, the logical treatment for the functional disorders that made up the bulk of their practice was psychotherapy.

For alienists, the problem of role was more complex and more restricted by institutional requirements and routines. Psychotherapy required time. Yet, there were too few physicians in even the best public institutions to carry out this new method of approach. In defending psychotherapeutic experiments, William Alanson White had argued that alienists must attempt to cure their patients. By re-emphasizing this aspect of their professional role, they could justify the effort required by new, experimental departures. Every aspect of this crisis—the drop in recoveries, the problems of classification, localization and treatment—became acute between 1900 and 1909.

V

American Psychologists
and Abnormal Psychology
1885 - 1909

The psychoanalytic system Freud described at the Clark Conference was one of several dynamic medical psychologies created between 1885 and 1909 by a few innovators in Europe and America. This international reconstruction in neurology and psychiatry represented the positive aspect of the crisis of the somatic style. Many conceptions often regarded as Freud's original contributions were a common part of these international developments.

Systematic psychotherapy and the rediscovery of the unconscious were the major achievements of the new psychopathology. It resulted from three converging streams of professional activity—medical observations of nervous and mental disorder; new experiments in hypnosis; the research and theory of academic psychologists. This and the next two chapters will assess the relation of these parallel developments to psychoanalysis.

The major advances were European, and the quantity and quality of European research usually was higher. In many areas, but not in all, Europeans anticipated Freud's hypotheses more directly than the Americans. Nevertheless, the American contributions were significant and in several instances unique. American psychologists studied more assiduously than American neurologists or psychiatrists several elements that were fundamental to Freud's theories—childhood, dreams, and the development of the sexual instinct. Their interest in these topics motivated their interest in psychoanalysis, and Freud, an omniverous reader, used and criticized some of the American work.

Freud and the American psychologists had been profoundly influenced by evolutionary theory. The Americans had been led to study the child because they wished to discover what was original and truly instinctive in human nature. Discarding the innateness of such psychological functions as conscience, they observed their emergence from the child's experience. Evolutionary theory led them to assume that the sexual instinct could hardly appear at puberty without prior development. Evolutionary theory also led them to believe that civilized customs had developed from primitive beginnings and to suspect that the extremes of "civilized" morality might be unnatural and unhealthy. Several of the major psychologists—James and Hall, especially, suffered from conflicts over sexuality that marked their generation of American intellectuals, and they protested aspects of the accepted moral system, notably its prudery and reticence.

The contributions of psychologists to the new psychopathology were major and represent a debt seldom sufficiently acknowledged. By 1910 the interest of the "better psychologists" in Freud prompted some neurologists to give his theories more respectful attention.[1] A few psychologists, notably Hall, William James, and Josiah Royce, broke with the narrowly materialistic determinism that had formed the foundation of the neurology of the Gilded Age. They encouraged a number of physicians, notably James Jackson Putnam and Adolf Meyer, to explore nervous and mental disorder from a directly psychological outlook, without recourse to underlying somatic hypotheses such as the nutrition of the nervous system or the malfunctioning of the cerebral cortex.

Like Freud, some psychologists attempted to provide a new model of human personality that would obviate the old dichotomy between mind and body, emotion and thought. Working from neurological physiology and traditional categories of mental function, these psychologists attempted to create new systematic hypotheses about mental and emotional processes. They viewed the human being as a radically modifiable animal that learned by imitation, habit, and association. The term "mechanism," borrowed from somatic theories of the reflex, was applied to these psychological functions, and later by psychoanalysts to Freud's theories of condensation, projection, and displacement. The psychologists' new model of behavior was more subtle, comprehensive, and orderly than the crude changes on hope and joy, fear and terror, rung by the early neurologists.

American psychologists could make these contributions to the new psychopathology partly because neurology and psychiatry had profoundly influenced their own discipline. Both were as central as the experimentalism of Wilhelm Wundt to the first generation of psychologists—William James, G. Stanley Hall, and James Mark Baldwin. Some of them had been motivated by their own neuroses to take a lively, personal interest in "nature's experiments" in nervous and mental disorder.

The impact of the psychologists often was informal, operating through a network of friendships and acquaintances. Such personal ties were closest in New England, which became the American center of the new psychopathology, notably at Boston and Harvard, and, on the periphery of this genteel world, at Clark University in Worcester. William James, Josiah Royce, and Hugo Münsterberg, professors of psychology and philosophy at Harvard, worked with Boston neurologists and psychiatrists. James Jackson Putnam, Harvard's first professor of neurology, was one of five men whose "intellectual companionship" stimulated William James's *Principles of Psychology,* and the major achievements of the Boston school of psychopathologists came after the publication of James's *Principles.* James took his students to the psychiatric demonstrations by Adolf Meyer at the Worcester Insane Hospital, and the two men discussed psychiatry and scientific method. A brief survey of some of the relevant work of the American psychologists will illuminate the context in which Freud was invited to Worcester.[2]

G. Stanley Hall (1844–1924)

It is an ironic tribute to the success of Freud's theories that Granville Stanley Hall seriously underestimated the pioneering work of himself and his students when toward the end of his life he wrote his great *Life and Confessions of a Psychologist.* These efforts had become blurred in a tribute to the triumphs of psychoanalysis, yet, in rudimentary and fundamental ways, their insights resembled Freud's: the development of the sexual instinct beginning as early as the age of two and one-half; the pervasiveness and sublimation of sexuality in non-sexual areas; a theory that perversions resulted from arrests or fixations of normal development; the decisive role of instinctual, unconscious drives; the fragility of conscious, rational controls. It is hardly surprising that Hall's student, Lewis Terman, found Freud's Clark lectures electrifying, for he himself had been exploring some of the topics Freud discussed.

These achievements were part of Hall's grandiose plan to bring psychology close to the "throbbing life of the great world" and to inaugurate for "our practical age and land" the "long hoped for and long-delayed science of man." [3] Yet the science of man and Hall's attitude toward the traditions of his parents involved him in conflicts he spent his life trying to resolve. They were typical conflicts of those of his generation who confronted psychoanalysis. Hall disliked the "Calvinist" New England rejection of joy and art, yet he sought to preserve the New England virtues of hard work and a tender conscience. He wished to study sexuality scientifically, yet feared that such knowledge might subvert morality.

He came from a family of sturdy old New England farmers, and his parents' home life was marked by the sternest reticence. Almost every manifestation of affection was chilled by the "very atmosphere of Puritanism." Hall's mother was his chief confidant, and her outward placidity masked inner struggles she revealed in a tortured, religious diary she kept secret from her husband. At fourteen Hall rebelled against his father. He developed an almost mortal fear that his heart would stop, or, as the neighbors believed, that the world would come to an end. He dreaded bulls, rams, and lightning. Revivals filled him with "something of a fear fetish of hell." [4]

In his autobiography, written in his seventies, he professed to wish to be known as he really was, stripped naked of convention and pretense, despite possible censure and ridicule. Nearly all his life he experienced feelings of isolation—from his father, his schoolmates, his colleagues. He was graduated from Williams College, having soaked himself in romantic literature, then went to Union Theological Seminary, but could not accept its theology and turned to philosophy instead of the ministry.

Germany, where he went to study in 1868, became a spiritual and moral refuge. "It almost re-made me," he recalled. German philosophy opened an exhilarating world after the narrow, dry American denominational training. His experience of German mores inspired a rebellion against the constraints of American morality and culture. He fell in love for the first time, and came to realize that he was "a man in the full normal sense of that word." He confessed that this "initial fling" was followed by years of "chaste sublimation." He also took up beer drinking and dancing, and admired *gemütlich* German ways. When he returned to America in 1870, he found "staid old New England" stifling and provincial, narrow and inflexible. Its mores were settled and lifeless; he was revolted by the Puritan Sabbath and the "Calvinist's frown at earthly joys." The New England conscience, he later wrote, was a kind of "moral Fletcherism," an endless chewing of the cud of guilt.[5]

After teaching at Antioch College, he settled in Cambridge, Massachusetts, and in 1878 at the age of thirty-four received Harvard's first doctorate in psychology under William James. With Putnam he studied the blind girl, Laura Bridgman, and became interested in a school for defective children and the nearby insane asylum.

His interest in mental pathology deepened during a second trip to Germany in 1878, to study psychology. He was impressed by the psychiatric clinics of Karl Westphal, medical director of the Berlin Charité Hospital. On February 3, 1879 he wrote to the American classicist, Charles Eliot Norton, that Westphal's classifications were "so close that one comes to see mild traits of insanity in all one's friends." [6] He was fascinated by Helmholtz, and studied with Wundt in Leipzig, where hypnotic experiments were being carried on. He heard Meynert in Vienna, Charcot in Paris, and Liébault and Bernheim in Nancy. Back in Boston in 1881 he continued his studies of hypnosis. These investigations and long conversations with Charles Peirce convinced Hall that he could not solve epistemological problems in terms of physiological psychology.[7]

In 1883 Hall was appointed professor of psychology at the new Johns Hopkins University. He set up a psychology laboratory and organized and directed Baltimore's charity insane asylum. His students included John Dewey, Woodrow Wilson, J. McKeen Cattell, Joseph Jastrow, William H. Burnham. Earlier he had become acquainted with George Beard and Silas Weir Mitchell.

In 1888 Hall was asked to organize Clark University in Worcester, Massachusetts, and made it a remarkable community of students and teachers. Lewis Terman, the psychologist, described the "utterly unrestricted *Lernfreiheit*" he found there in 1904.[8] No attendance records were kept for the fifty full-time students; no marks were given. The only

examination was the four-hour doctor's oral. Students registered by noti-
fying Hall's secretary. Hall encouraged them to sample every course and
pursue only those that seemed interesting. Every Monday from 7:15
P.M., often till after midnight, Hall presided over his seminar, devoted
chiefly to minute criticism of students' reports. The seminar, like the
University, produced some original and distinguished work—and
sometimes a nervous crisis in students whose reports had not found
favor.

Hall had a flair, as had James, for predicting what would become
important and for working on new and unorthodox problems. Genetic
psychology and his interest in sexuality prepared him to welcome Freud
and to give psychoanalysis, after the Clark Conference, his helpful but
ambivalent support. From Clark came some of the first psychologists to
apply psychoanalysis to child guidance and educational psychology.

Hall revealed most clearly both his deep loyalty to the ideals of his
parents and yet his rebellion against them in his attitudes toward sexuality. An early sense of shame and guilt made him hate "Puritanism"; yet
he was all the more devoted to what he called "America's fine old standard of morality."

Hall suffered for years from an association of filth with sexuality and
from his father's cautionary misinformation. As a child he learned only
one name for a "certain part of the body: the dirty place." His father
warned him of a youth "who abused himself and sinned with lewd
women and as a result had a disease that ate his nose away . . . and
who also became an idiot. For a long time, if I had any physical excitation or nocturnal experience I was almost petrified lest I was losing my
brains and carefully examined the bridge of my nose to see if it was getting the least bit flat. I understood that anyone who swerved in the
slightest from the norm of purity was liable to be smitten with some
loathsome disease which I associated with leprosy and with the 'unpardonable sin' which the minister often dwelt upon." For years Hall struggled against masturbation and believed nocturnal emissions made him
"exceptionally corrupt and not quite worthy to associate with girls." One
school he attended before he was twelve was rife with "homosexuality,
exhibitionism, fellatio, onanism, relations with animals, and almost
every form of perversion described by Tarnowski, Krafft-Ebing or Havelock Ellis." [9] He was too young to be corrupted, except in thought. In
college guilt about masturbation and emissions led to a temporary religious conversion.

Freud's emphasis on the emotional elements of sexuality profoundly impressed Hall. He carefully read the other new European sexologists, but complained that they gave too little emphasis to the emotions and too much to the cruder aspects of sex. "Chemical theories of sex" were inadequate for human psychology, he argued. Albert Moll and Havelock Ellis were "too somatic, and therefore partial and inadequate, to explain the higher and normal manifestations of love." In 1904 Hall abandoned a new course in sex education at Clark because a few students developed a morbid interest and outsiders listened surreptitiously at the door. He also had found it unexpectedly difficult to speak plainly about perversions, although women were excluded from the room.[10]

At Clark, Hall constructed his historical and speculative genetic psychology out of evolutionary conceptions and new studies of childhood. Like James, Hall enjoyed expounding the closeness of man to the animal, the child, the savage, the criminal, and the insane. These precious subjects had been neglected because of a lust for life after death and because of a strong idealist tradition which, since Plato, had denigrated the naïve and the unconscious in favor of the rational. Yet precisely these subjects revealed clues to the mind's development. Clinical studies of the defective and insane were beyond price, for they illustrated the reverse of evolution—"devolution." The development of the individual recapitulated that of the race. For years Hall taught a course in "psychogenesis" that grandly surveyed progress from the cosmic mists through paleontology, geology, and the animal world to man. Perhaps, he even suggested, man's mind might be evolving toward a higher type.

A degree of irrationalism and worship of the primitive were among the lessons Hall derived from his evolutionary theology. Like other evolutionists he fought for the principle that mind and body were one. Evolution taught that "instincts, feelings, emotions and sentiments" were "older and more all-determining" than the consciousness of the idealists, or the intellect of the "homo academicus." What men learned from introspective study of their own rational consciousness was likely to be "narrow, provincial . . . and possibly even in some sense degenerative." Consciousness might even be merely a "wart raised by the sting of sin, or the product of alienation or a remedial process." To Hall reason was often as colossally mistaken as superstition. The controls of reason and of consciousness were fragile, easily destroyed by anger and passion, by the resurgence of "long for-gotten facts, and feelings and impulses of an immeasurable past. . . ." [11]

Child Study and Sexuality

Largely inspired by Hall, studies of the child, begun in the late 1870's, provided the groundwork for genetic psychology. Their participants regarded them as the result of a uniquely "American" impulse to take a "fresh, independent look at the primal facts of human nature and at growth itself." [12] Two methods were used in these studies, which, by the 1890's, had spread to the Midwest and the Pacific coast. The first was the direct observation of children, usually one's own. The second was the questionnaire, usually distributed and filled in by schoolteachers. Refining Hall's methods, Arnold Gesell, who studied at Clark in the early 1900's, became the nation's first professional expert on child development. Distrusting questionnaires because those who answered them might be incompetent observers, James Mark Baldwin based his important *Mental Development in the Child and the Race* largely on the study of his own daughter. He traced the growth of such functions as will and conscience that earlier philosophical psychologists had believed were innate.

Making extensive use of the French psychopathologists with whom he had studied, Baldwin argued that imitation, of which hypnotic suggestion was a variant, was the fundamental factor in learning. The child's sense of self first began to grow when he learned to distinguish things from people. Self-consciousness developed through imitation, and the sense of right and wrong emerged from the early experience of approval or disapproval from parents, nurses, and companions.

Although early experience left the deepest traces, character was not fixed in childhood, but developed over the years. Many traits, such as those requisite to the pursuit of a profession, were not fully formed until about the age of thirty.

Two conflicting attitudes toward the child are discernible in these early studies. The child was impulsive, self-centered, without a perception of the boundaries between reality and imagination, or between good and evil. At the same time, he was worshiped with Wordsworthian enthusiasm. The child's emotions, Baldwin remarked, were undistorted by convention and self-consciousness; children were as "spontaneous as a spring . . . pure and uninfluenced by calculation and duplicity and adult reserve." [13] Freud read Baldwin's book in 1898; later he included it among those which failed to deal adequately with the child's sexual development.[14]

Partly because of Hall's inspiration, partly because of growing European interest, a few Americans investigated childhood sexuality and proposed the sexual enlightenment of children. Stephen Kern has conclusively demonstrated that nearly every element of Freud's theory of childhood sexuality was anticipated by a few Europeans and Americans from the late 1850's on, chiefly in the 1890's and after.[15] With certain exceptions, the European work was bolder, concentrating more forthrightly on genital sensations, erogenous zones, and other bodily details. Yet a few Americans anticipated Freud's theory of sublimation and the role of masturbation as a preparation for adult sexual activity, the possible consequences for adult character of bodily functions such as nutritive sucking and toilet training. The European and American insights were fragmentary, and no one had worked them into a coherent system. They were by no means part of conventional wisdom. European students of degeneracy came closest to anticipating Freud, especially the psychopathologists of the French school who attempted to trace sexual perversions to associations with genital sensations in early childhood. These associative traces were lost to memory, yet reappeared in adult symptoms.[16]

The American interest in childhood sexuality grew rapidly from the 1890's on, and in 1909 James Jackson Putnam informed Freud that sex education for children was being much discussed in Boston. It had been advocated by the neurologist George Beard and was taken up by Earl Barnes, professor of education at Stanford University, who in 1892 presided over a round-table at the National Education Association on the "Development of the Ideas and Feelings of Sex in Children." He argued that "from the time that children begin to think of the world about them, they are interested in what may be called sex questions. They soon begin to inquire as to their own origin, the functions of special organs in themselves and in other animals, the relations of the family, and the meaning of words they occasionally meet." [17] Nine-tenths of them got their information from the "back alleys." Sexual questions long had interested "alienists and neurologists," and it was time parents and teachers pooled information on these subjects. Careful experimental tests were needed to determine how sexuality developed in the child.

In 1902 Sanford Bell, a teacher and former Clark student, published "A Preliminary Study of the Emotion of Love Between the Sexes," based on a syllabus sent to 1700 teachers at the Indiana Normal School and on 800 cases he had observed in the public schools. He, too, began by noting the paucity of existing information. All the authorities, from Théodule Ribot, the French psychologist, through Havelock Ellis, agreed that there had been little or no scientific study of the development of normal sexual emotions in childhood or adolescence, although there had been considerable work on adult perversions. Yet, without a knowledge of the normal, of which the abnormal was only an exaggeration, there could be no adequate sexual hygiene.

Applying the principles of evolution, Bell argued that the emotion developed from the instinct: "The emotion of sex-love, so plainly traceable to the reproductive instinct, has its evolution in each normal individual. It develops through various stages as do other instincts. It does not make its appearance for the first time at the period of adolescence, as has been thought." [18]

Sex love became manifest as early as two and a half years, but because the sexual organs were not yet mature, the child's erethism was "distributed throughout the entire body, especially the vascular and nervous system." In cases of precocity, the organs themselves became involved. Young children displayed their sexual love by caresses, by hugging, kissing, lifting, scuffling, and, sometimes, by plans of marriage. Using a genetic conceit, typical of Hall's school, Bell suggested that these activities were similar to analogous manifestations in young puppies, seals, grouse, water wagtails, goldfinches, etc. However, these early childish expressions resembled adult sexuality no more than the apple blossom resembled the fruit. Bell did not posit the existence of infantile genital sexuality or connect the perversions with the development of the sexual instinct.

Like Hall and William James, Bell defended the social progress achieved by the repressions of civilization yet also pointed out their cost: "The system of sex inhibitions which has gradually been developed . . . has been doing away with promiscuity, polygamy and polyandry; it has been establishing monogamy and postponing marriage until a period of greater physiological and psychological maturity of both sexes." This "inhibition of early sex functioning" also led to an increase in the "prevalence of such substitutes as masturbation, onanism, pederasty, etc." [19] Freud regarded Bell's study, based on the "American method" of piling up instances, as one of the few pioneering efforts in a new field.[20]

Bell, E. W. Bohannon, a Clark Fellow in Pedagogy, and several physicians, Kern has demonstrated, anticipated Freud's theory that the child's earliest experience of bodily processes could determine later character traits. In 1904 Bell observed that sucking usually disappeared in the child, yet sometimes appeared later in adult perversions. Bohannon noted that children who loved dirt often were disobedient. Control of excretions as early as ten months was related to the control of reflexes in voluntary acts, Frederick Tracy, an American physician noted in 1893. Another physician, Francis Chamberlain, in 1902 argued that in onanism the boy fell in love with himself and used his sexual organs as a "training school for the future." [21]

Since the 1890's Hall's students had been surveying the clinical and literary European studies. His students also protested what they called American "Puritanism." In 1896, in a study of "Sex and Art," Colin Scott, a Clark Fellow, called for a "modern phallicism" to replace "Puritan repression." "What we need at present is . . . a religious and artistic spirit that goes out to meet the sexual instinct, and is able to find in it the centre of evolution, the heart and soul of the world, the holy of holies of all right feeling men. . . . The purity that does nothing more than keep itself unspotted from the world is unsuited to our growing wants and larger social consciousness." All the arts, he insisted, were an outcome of and substitute for marital fulfillment.[22] Another Clark Fellow, James Leuba, noted the existence of sadism, masochism, and homosexuality; a case of inversion at Worcester Asylum was being studied under Hall's direction. Leuba also surveyed the French decadent writers from Huysmans and Baudelaire through Mallarmé and condemned their preoccupation with sensuality and death.[23]

Theories of sublimation were almost commonplace. Several psychologists, among them Scott and Leuba, were convinced that the sexual instinct provided the motive behind a wide range of cultural phenomena, including art and religion. Scott argued that all the arts were an outcome of and substitute for "sexual fulfillment." W. I. Thomas, a young sociologist at the University of Chicago, suggested that sexuality played a role in such ostensibly non-sexual activities as politics.[24]

Another Hall student, George W. Dawson, in 1899 attempted to explain sexual perversions as arrests of moral development. The sexual behavior of men was distinguished from that of animals only by intelligence and morality. If either stopped developing at any given stage, "unchastity" and aberrations would result from the discrepancy between physical maturity and moral immaturity. The fact that about 5.6 per cent of prostitutes were drawn to their work by "sexual appetite alone" proved the presence in society of women who departed "radically from the civilized type." [25]

In April 1905 Lewis Terman, who later developed the Stanford-Binet intelligence tests, contributed a study of sexual precocity. It is unlikely that he had read Freud's *Three Essays on Sexuality,* which were published that year but not listed in the American Psychological Index for 1905 or 1906. Terman emphasized Bell's earlier work, and suggested, following the Italian investigators Mantegazza and Ferriani, that sexual love "often stirred the child even more than the adult." [26] Hall argued after Mantegazza that love and hate often were felt toward the same person and might be "different degrees of the same emotive force." [27]

In 1904 Hall published the culmination of his work, *Adolescence,* a sprawling compendium of the American and European literature in neurology, psychiatry, child study, sexuality, physiology. Like Janet and most physicians, Hall believed that puberty and adolescence were the critical periods of development.

Hall assumed a "will to live," or a life force, expressing itself in hunger and love, and these proliferated into a multitude of subsidiary drives. The way these drives developed at their period of most rapid growth determined their later form and use.

Desires had to be inhibited if maturity were to be achieved, and this could encourage "irradiation" or sublimation. Yet, "faculties and impulses denied legitimate expression during their nascent periods," Hall insisted, could "break out well on in adult life—falsetto notes mingling with many bass as strange puerilities." So, too, could precociously cultivated impulses. Higher functions of inhibition and synthesis could break down, allowing earlier, more primitive traits to reappear. Thus, the personality was constructed in layers of "strangely interacting strata." Hall's frankness about the psychological and physical facts of sexuality was remarkable for the period, and was partly responsible for the book's mixed reception. Edward L. Thorndike, the educational psychologist, criticized its unscientific methods and its "unctuous" comments on sex.[28]

William James and Josiah Royce

William James (1842–1910) played an indirect but important preparatory role in the reception of psychoanalysis, and he died less than a year after the Clark Conference. He codified for a wide public America's pre-Freudian views of the unconscious, and he insisted even more strongly than Hall that psychopathology be taken seriously. To James the unconscious was both malevolent and benign. It could provide perceptions more subtle than those of consciousness. If the "grace of God miraculously operates," he wrote in *The Varieties of Religious Experience,* "it probably operates through the subliminal door. . . ." [29] Yet the subconscious also was a source of neuroses and delusions and "obscurely motivated passions." Although James did not explicitly link the subliminal with human aggression, he expressed a perception of it in 1903, provoked by lynchings in the South, that Freud first would approach in *Totem and Taboo.*

"The average church-going civilizee realizes . . . absolutely nothing of the deeper currents of human nature, or of the aboriginal capacity for murderous excitement which lies sleeping even in his own bosom. . . . The watertight compartment in which the carnivore within us is confined is artificial and not organic. The slightest diminution of external pressure . . . will make the whole system leaky and murder will again grow rampant." [30]

In his presidential address to the American Psychological Association in 1895 James argued that the discovery of the subconscious had shattered all previous notions of the self as a unity and of the mind as operating with ideas immediately available to consciousness and to common sense. "The menagerie and the madhouse," he wrote in 1901, "the nursery, the prison and the hospital have been made to deliver up their material. The world of mind is shown as something infinitely more complex than was suspected; and whatever beauties it may still possess, it has lost at any rate the beauty of academic neatness." [31]

It was in the defense of chastity, above all, that free will and effort were required. "No one," James wrote, "need be told how dependent all human social elevation is upon the prevalence of chastity. Hardly any factor measures more than this the difference between civilization and barbarism." [32]

James's extremely popular reformulation of conventional psychiatric wisdom about habit influenced a number of physicians, possibly Morton Prince, and certainly Putnam, Adolf Meyer, and Edward Cowles. James also argued that substitution was better than repression in the attempt to alter undesirable ideas.

Such insights as these made the problem of moral choice acute. James argued that evil was not fated and that moral acts could be performed through will and effort. He vehemently opposed the materialistic determinism of the early neurologists and sought to convince his friend,

Putnam, of the importance of consciousness, feeling, and will. Although James insisted that will was not a spiritual something, it did respond to ideas. Indeed, will was expressed in the effort required to keep a given idea before the mind. Such effort was especially necessary "whenever a rare or moral ideal" or a civilized inhibition was required to neutralize instinctive or habitual impulses.[33]

James's dark night of the soul, from which a belief in free will rescued him, suggests profound struggles over sexuality. He suffered from about nineteen to thirty-one from severe neurotic symptoms: mysterious backaches, an inability to use his eyes or to work for long periods, insomnia, and melancholy. He was haunted by the vision of an epileptic he had seen in an asylum, a "black-haired youth with greenish skin, entirely idiotic, who used to sit all day on one of the benches . . . with his knees drawn up against his chin, and the coarse gray undershirt, which was his only garment, drawn over him, inclosing his entire figure." [34] Perhaps the youth represented James's inner passivity, the image of the man who "let himself go time after time," and was helpless to "resist the drift." The crux of James's melancholy was a sense of "moral impotence." He began to recover when he first accepted the doctrine of free will and found he could make moral choices; he improved further when he began teaching at Harvard and finally when he married in 1878 at the age of thirty-four. His most productive years followed.

Like Freud and Hall, James protested those postponed marriages that tested the late nineteenth-century intellectual. In 1880 he wrote Josiah Royce, then twenty-five, congratulating him on his engagement: "I have found in marriage a calm and repose I never knew before, and only wish I had done the thing ten years earlier. I think the lateness of our usual marriages is a bad thing, and hope your engagement will not last very long." [35]

James's inner conflicts gave him a lively sympathy with the neurotic and an appreciation of the closeness of the normal to the abnormal. Once after showing a Harvard class through an insane asylum, he remarked: "President Eliot might not like to admit there is no sharp line to be drawn between himself and the man we have just seen, but it is true." [36]

In his explicit attitudes toward sexuality James was both conservative and rebellious. He refused to reduce love to sexuality. "The love creates the ecstasy, not the ecstasy the love," he insisted. On the surface the sexual instinct seemed above all others "blind, automatic, and untaught." [37] Yet individual taste, habit, and contrary impulses, such as shame and modesty radically could modify it. Some people, for example, never gratified their sexual desires. Elements of perversion were present in everyone; learned "inhibitions" could explain their prevalence among Orientals and the Greeks. Inhibition also could restrict the instinct to a single object, thus making monogamy possible. James was aware of other puzzling aspects of sexuality. "Why is it," he once asked a Harvard class, "that a perfectly respectable man may dream he had intercourse with his grandmother?" [38]

He also ridiculed the New England overdevelopment of modesty, which would require everyone to say "stomach instead of belly, limb instead of leg, retire instead of go to bed, forbidding us to call a female dog by name." [39] He hoped for a salutary relaxation of the New England conscience, and of its restless fear of not measuring up.

To his students and to the wider public that read his books and articles, James insisted with all the weight of his prestige that abnormal psychology be taken seriously. His tolerant eagerness to investigate the unorthodox appeared in a number of his students and friends, some of whom took up psychoanalysis.

James's Harvard colleague, Josiah Royce, the idealist philosopher, also suffered from exhaustion and severe depression, and his insatiable reading was sufficiently catholic for him to include Casanova as well as the "latest German theory of sexual insanity." He was far more than an academic philosopher, and wrote a psychology text and lectures on child raising.

Born in the Gold Rush town of Grass Valley, California, in 1856, Royce remembered himself as a "timid," "ineffective," ugly boy, red-haired and freckled. Like many American contemporaries, he developed a hate for the "Puritan" Sabbath and for dogmatic clergymen, and considered himself a born non-conformist. Later at school, he was "countrified, quaint, and unable to play boys' games." His peers promptly instilled in him a lively sense of the "majesty of the community." [40] To the lonely Royce, relations with society became an obsessive problem.

Drawing on James Mark Baldwin's genetic studies of child development, Royce suggested in 1894 that mental pathology involved a disturbance in the sense of self, especially in problems of sexuality. "Sexually tinged emotions normally have very complex social associations," he told a group of Boston psychiatrists. Consequently, "we may expect to find self-consciousness especially deranged in disorders involving the sexual functions. . . ." He referred those who doubted this to the "monumental records that fill Krafft-Ebing's too well-known and ghastly book." [41]

Royce also influenced a wide circle of students, physicians, and scientists, many of whom attended his famous seminar in logic. This was a "veritable clearing house of science" at Harvard during the early 1900's.[42] Royce promoted a sophisticated view of scientific method, partly based on an idealist critique that influenced a number of important physicians. Among those who attended his seminar was Elmer Southard, who became first director of the new Boston Psychopathic Hospital. Some of James Jackson Putnam's first psychotherapeutic conceptions came from Royce, and Putnam later consulted his philosopher friend and patient about problems that arose during a long epistolary dialogue with Freud about the nature and aims of psychoanalysis. In 1916 William Ernest Hocking, another Royce student, discussed at length the ethical implications of psychoanalysis in a *Festschrift* for his teacher.[43]

The study of dreams was encouraged by both Hall and James. In the first volume of the *American Journal of Psychology,* Julius Nelson, a psychologist at the Johns Hopkins, argued that dreams might provide clues to the nature of hallucinations, hypnotism, and insanity. Dreams fulfilled "fancies" just as daydreams did. The predominance of visual impressions in dreams had been stressed in 1892 by G. T. Ladd, Yale's first professor of psychology. Dreams closely reflected the concerns of waking life and sometimes exactly reproduced events of the day, according to Mary Calkins, a student of James and a professor of psychology at Wellesley. Freud, whose reading was remarkably wide, cited several of these American studies, notably Bell's, and Ladd's.[44] Dreams also could reproduce past experiences exactly, perhaps a recent trauma or an event of earliest childhood. In 1906 the psychologist Joseph Jastrow included a chapter on dreams in *The Subconscious.* He made a distinction close to Freud's between the primitive, unrestrained, illogical processes of dream thought and the operations of conscious logic. Jastrow briefly noted that dreams could embody the "suppressed, unacknowledged aspects of our composite temperament," and he studied the nature of the association of ideas in dreams; Francis Galton had traced a waking fantasy through associations to early half-forgotten incidents.[45] Jastrow was one of several psychologists, including Boris Sidis and Hugo Münsterberg, who experimented with hypnosis and practiced psychotherapy.

American psychologists also took an increasing interest in motivation, and Darwinism brought an emphasis on instincts and biological analogies. For William James and most other psychologists, instincts, drives, and emotions shaded into each other. Their various and standard lists included "fear, anger, shyness, curiosity, affection, sexual love, jealousy and envy, rivalry, sociability, sympathy, modesty, play, imitation, constructiveness, secretiveness, and acquisitiveness." [46] Confronting psychoanalysis, many psychologists later would refuse to reduce all drives to sexuality.

Royce, James, and many others were familiar with the old conception of internal conflict. Some of James's most brilliant passages in *The Principles* described the warring selves within. Among the several conflicts were those between habitual, instinctive behavior and envisioned ideals or standards. Sometimes, it was assumed, conflicts could occur outside awareness.

Practical application, above all, interested the growing major school of American functionalists, who recognized the importance of genetic and abnormal psychology. In 1905 Edward L. Thorndike, the educational psychologist who influenced Adolf Meyer, outlined a program for teachers that deliberately de-emphasized the classical concern with brain anatomy, perception, and sensation. Instead, Thorndike proposed that teachers concentrate on the elements of a "dynamic psychology." These were "connections between outside events and mental states, between outside events and acts, between one mental state and another, between mental states and acts. . . . Instincts and capacities, habits, the share of previous experience in perception, the association of ideas, the abstraction of elements, ideo-motor action, inhibition"—the psychology of childhood and adolescence.[47]

There was a growing tendency, shared by Freud and Janet, to ignore the traditional academic categories of will, emotion, thought, perception, sensation, etc. that had dominated the first studies of the child, as well as most psychology texts. Increasingly, thought and emotion were considered to be inseparable, and studies of memory, chiefly by the French, already had demonstrated that the emotions modified in central ways what people could recall.

Yet, there were also developments within psychology that prepared a hostile reception for Freud among some of these same Americans. Within functionalism itself an intense concern developed for scientific method. In 1903 Thorndike summed up this goal as "quantitative precision," "direct observation," "experiment," and the careful use of statistics.[48] For psychologists with this orientation the medical tradition of clinical observation within which Freud continued to work came to seem old-fashioned and unscientific. Yet experiment and quantification seemed at first largely inapplicable to those problems that most interested neurologists and psychiatrists—the messy and inexact field of the human emotions.

By 1909 American psychologists had played an important role in the crisis of the somatic style and in the work of reconstruction. Several of them—notably James, Hall, and Royce—had attacked materialistic determinism and had suggested that psychological data were as valid objects of attention as tissue. They had worked directly with physicians in the development of psychotherapy, and several had begun to practice it themselves. They also had brought to the attention of a wide public the work of the new European psychopathologists.

They had focused attention on several important new discoveries, the development of the sexual instinct and of every major psychological function, such as conscience, from childhood on. By 1909 these new aperçus had not yet been integrated into American psychotherapy.

The psychologists made one final contribution—a highly optimistic outlook, more sanguine than Freud's at his most hopeful. Because of their faith in the educability of man and their view of human character as formed by patterns of habit and association, the psychologists escaped much of the pessimism that degeneracy theory induced in some of their medical colleagues. Indeed, of all the psychotherapists, excepting the mind-cure cults, the psychologists were perhaps the most optimistic. Even so sober a functionalist as J. McKeen Cattell hoped that psychology would not only be able to prevent nervous disorders but would preside over the proper social distribution of labor wealth and power.[49]

As an outsider surveying the new psychology, Henry Adams likened the subconscious to an abyss and drew a lesson of chaos and despair that was close to Freud's final tragic view. Man, Adams wrote, was "an acrobat, with a dwarf on his back, crossing a chasm on a slack-rope and commonly breaking his neck." [50]

VI

American Psychotherapy
1885-1909

The Boston School

By 1909 a few Americans were practicing the most sophisticated psychotherapy in the English-speaking world, a psychotherapy that made acute
the first crisis of the somatic style. The new psychopathology familiarized physicians and the public with a number of general conceptions
often taken to be Freud's contributions—trauma, forgotten memories,
complexes, and the subconscious. It was through the non-Freudian version of these conceptions that psychoanalysis first was understood. By
1909 several therapies competed; no single authority dominated American theory and practice. No method was backed by a professional organization or required special methods of training. However, American psychotherapy exhibited certain typical characteristics that would select
among available European authorities and that would also mould the
first interpretations of psychoanalysis.

The first American psychoanalysts were recruited largely from those
who already were practicing other kinds of psychotherapy. Some remained satisfied with them. Others came to believe, especially as they

confronted a militant psychoanalytic alternative, that existing systems provided theoretical and practical problems that, like those of the somatic style, were becoming increasingly acute. This background made possible the persuasive impression Freud created in a few key physicians.

This chapter will describe the therapy of the "Boston school," the leader of American psychopathology, and the work of two of its major figures, James Jackson Putnam and Morton Prince. Then the unique characteristics of American psychotherapy in the years from 1885 to 1909 and the relation of these traits to incoming European influences will be described. Finally, the position of psychotherapy in American medicine will be clarified, along with some of the major problems inherent in existing non-Freudian systems.

The subtleties and the difficulties of this psychotherapy are illuminated by America's best known pre-Freudian neurotic, Miss Beauchamp, a Radcliffe girl of twenty-three whose multiple personalities and their cure were gracefully described in 1905 by the Boston neurologist, Morton Prince. She caught the attention of medical journals and ladies' magazines, Broadway and the universities. Certain overlooked aspects of this famous case bear directly on the reception of psychoanalysis and on the American norms of "civilized" morality.

The Miss Beauchamp who consulted Dr. Prince appeared to be the ideal nineteenth-century woman, the embodiment of "civilized" morality, with all the stereotyped virtues—and the stereotyped disabilities. This "saint," as Prince described her, was dependent and considerate, patient and emotional, never angry or rude. She loved prayers and church, children and her elders. Above all, she was primly reserved. She also suffered from severe neurasthenia with unusual complications— exhaustion and insomnia, fleeting pains, lapses of memory, and a weakness of will, or "aboulia," as it was called, that prevented her from doing what she intended. She experienced pervasive dread, especially of breaking promises and behaving strangely.

When conventional treatment failed, Prince hypnotized her. Either spontaneously or under hypnosis different organized personalities appeared, and when these in turn were hypnotized, still others emerged. Prince counted a total of sixteen of which only two, in addition to the first Miss Beauchamp, were important. The first of these personalities emerged early in treatment when Prince was startled to hear her begin to talk of Miss Beauchamp as "she" and of herself as "I."

This personality was the very opposite of the frail girl who had consulted him. Calling herself "Miss Devil Lady," she refused to admit that she and Miss Beauchamp (B I) occupied the same body. Prince called her "Chris" or "Sally." She was robust and tireless, loved the outdoors and champagne, and planned to travel to Europe with a male friend. She detested Miss Beauchamp's prim, scholarly life, yet envied her friends and "culture." She had thoughts, feelings, and memories, of which Miss Beauchamp knew nothing. It was Sally "who created hallucinations, paralyses, aboulia and amnesia in Miss Beauchamp. "

Finally, still another personality emerged spontaneously, a stubborn, healthy girl, "natural, tranquil in mind and body, and sociable. All nervousness and signs of fatigue ceased. She was without aboulia and chatted pleasantly." Prince designated her B IV. She remembered nothing of Miss Beauchamp's recent life. "Sally" could not read her thoughts but knew what she did. A period of war ensued among the personalities, and for a time B IV wanted to kill Sally.

"The personalities come and go in kaleidoscopic succession, many changes often being made in the course of twenty-four hours. And so it happens that Miss Beauchamp . . . at one moment says and does and plans and arranges something to which a short time before she most strongly objected, indulges tastes which a moment before would have been abhorrent to her ideals, and undoes or destroys what she had just laboriously planned and arranged." [1]

A therapeutic problem posed itself. Which was the Real Miss Beauchamp? Prince assumed there was a core personality, adapted to all situations, physically and emotionally healthy, without aboulia or amnesia. Yet the memories of Sally and Miss Beauchamp had pathological lapses. Prince decided that if B IV, because of her healthier personality, could be "synthesized" with the hypnotized Miss Beauchamp, B II, his patient would be cured.

Sally suggested that B II was the Real Miss Beauchamp asleep, and so Prince tried an experiment. He changed Miss Beauchamp into B II, then commanded, "You shall open your eyes, awake, and stay yourself, your real self. . . ." The Real Miss Beauchamp that finally appeared was "natural and self-contained . . . free from every sign of abnormality. . . . There was none of the suffering, depression, and submissive idealism of B I; none of the ill-temper, stubbornness, and reticent antagonism of B IV. Nor was there any 'rattling' of the mind, hallucination, amnesia, bewilderment, or ignorance of events. . . ." [2] "She knew me and her surroundings and everything belonging to the lives of BI and B IV. She had the memories of both. Synthesis persisted." She had achieved "freedom from amnesia, even temperament, stability, health, and absence of suggestibility and of abnormal phenomena."[3]

What, then, happened to "Sally"? She died, "squeezed out" by the fusion of B II and B IV. Prince believed she persisted, however, as a "permanent dissociated, subconsciousness" of the synthesized personality. Of Sally, the Real Miss Beauchamp remembered nothing consciously.

In comparison with the therapy of S. Weir Mitchell, Prince's methods exhibited striking innovations, and in comparison with Freud's, surprising gaps. Prince's analysis was almost entirely on a delicate and sophisticated psychological level. He devoted a chapter to dreams, arguing that they originated from the sensory impressions of conscious life, including perceptions on the fringe of awareness, such as peripheral vision. The fact that Sally, B I, and B IV had the same dreams argued that they shared a common consciousness. Yet dreams had little other significance.

Prince also sought for causes in Miss Beauchamp's daily life. What little was known of her heredity suggested "nervous instability." Her parents had been unhappily married, and she had been a "nervous, impressionable child, given to day-dreaming and living in her imagination. Her mother exhibited a great dislike to her, and for no reason, apparently excepting that the child resembled her father in looks." She idealized her mother, "bestowing upon her almost morbid affection," and believed her mother's coldness came from her own imperfections. If she could purify herself, her mother would love her. Then at thirteen her mother died after a long illness, and she felt responsible for her death. Thus heredity and childhood had provided "ample psychopathic soil" for her condition.[4] Although the emphasis on childhood was important and new, its significance was left vague, and no real connections were established between these early beginnings and her later character, apart from the exacerbation of her "nervous instability."

Yet, in an extraordinary autobiography which Prince judged to be reasonably true, Sally herself wrote that her original split from Miss Beauchamp occurred at about the time she learned to walk. Sally also emphasized more than Prince her mother's illness, feelings of unworthiness, and a traumatic memory of having a baby [a sibling?] die in her arms. Prince left these memories unexplored.

Prince insisted again and again that Miss Beauchamp was distressingly reticent. In fact all the personalities were especially reticent regarding a single important event—the trauma that Prince believed precipitated them. This was a severe fright, the result of a prank played by an adored and idealized male friend, older than Miss Beauchamp, whom she had met just after her mother's death.[5] Precisely what was said during the episode was not reported by Prince, nor was the man's relationship to Miss Beauchamp described beyond the most casual references.

Although Prince had made pioneering American studies of male sexual problems, an absence of concern with sexuality marked his analysis of Miss Beauchamp. Yet, the devilish Sally dreamed of men who passed her house late at night, whose clothes and freedom she envied; she dreamed of worms passing through her. She often made appointments with the friend who had played the prank. Indeed, Prince examined at length none of her relationships, including her relation to himself.

Yet, occasionally one of the selves would burst out in hatred for Prince, and once she imagined him dead.[6] Her deepest resistance was saved for his manipulations, especially his attempt to kill "Sally" by hypnosis. Prince constantly hypnotized his patient, sometimes by stroking or ether, as well as by verbal suggestions, and she sometimes violently reacted against all of them. Although Miss Beauchamp was a product of the traumatic prank, as were the other personalities, Prince made no links between its characteristics and those of her previous personality. A sense of development over time was conspicuously lacking. Partly this arose from the questions Prince asked of his material and the general way in which he defined the subconscious selves. "I mean simply a limited second, co-existing, extra series of 'thoughts,' feelings, sensations, etc., which are (largely) differentiated from those of the normal waking mind of the individual." Certain memories and perceptions were lost, and the psychical factors that made up the personality, such as moods and memories, became rearranged. Under abnormal conditions these rearrangements could become sufficiently organized to have a "perception of personality" and thus could be regarded as a "second self," unknown to the waking self. There was no necessary limit to the number of these selves. "At times when excited they are capable of being stirred into a fury, when they burst forth like a volcano, fermenting and boiling in crises of a pathological character." [7]

Prince's treatment rested on the assumption that the central core of personality need not share all the memories or be aware of all the traits of the dissociated co-conscious selves, as the "death" of Sally indicates. Conflicts were observed by the therapist, but need not be observed by the patient. This point became important later when Prince came to grips with psychoanalysis, and it was a logical outcome of his essentially behavioristic premises.

Two additional elements in Prince's analysis of Miss Beauchamp were significant. Prince had considerable sympathy with "Miss Devil Lady" and made it clear that he thought the priggishness of Miss Beauchamp overdone. The overwrought conscience, the "prim conventionalities" had become distasteful to several of the New England therapists.

Prince's study also strongly argued for a continuum between normal and abnormal. James Jackson Putnam, in an appreciative review of *The Dissociation of a Personality* in Prince's *Journal of Abnormal Psychology,* wrote that "A caricature, a monstrosity is not, it is true, simply a magnified image of the normal, and cannot be taken as giving more than a hint of what will there be found. But then, it is not to be forgotten that the true 'normal' exists only in name." [8]

Why did American psychotherapy, to which Prince and Putnam were important contributors, develop primarily in Boston? A lingering Unitarian and Transcendentalist tradition fostered among respectable Boston physicians a sympathy for hypnotism, suggestion, and psychic research, all major sources of the new psychopathology in the critical decade of the 1890's. Prince and James Jackson Putnam, Harvard's first professor of neurology, joined William James's Society for Psychical Research. Prince almost accepted the reality of thought transference for a time.[9] The British Society's founder, F. W. H. Myers, probably brought Breuer and Freud's work on hysteria to James's attention.

The same general atmosphere also created a formidable competition that provoked Boston physicians to create their own "scientific" psychotherapy. Christian Science and New Thought practitioners flourished, taking patients from neurologists, and, what was more galling, curing them. In 1895 Putnam argued: "In a sense, I grudge the irregulars every case that they win from us, be they few or many, because I believe that with a deeper knowledge of human nature, a better understanding of psychology, a wider range of methods and greater skill in applying them, we could cure more of these patients ourselves." [10]

New England also was reputed to foster an exceptionally large number of well-to-do nervous invalids and mental patients. Henry Adams's wife, who later committed suicide, complained that visits home from Washington were much taken up with calling on relatives in asylums. Such patients sought the neurologists and the practitioners and supported New England's public and private mental hospitals, clinics and sanitaria, which were perhaps the best in the nation, among them, the McLean Hospital and the neurological services of the Massachusetts General and the Boston City hospitals.

Partly because of an ingrained New England tradition of dissent, some Boston physicians, notably Putnam and Prince, were more willing than physicians elsewhere to try unconventional methods and champion unpopular causes. Both hewed to a Democratic tradition in politics and were active social reformers. Despite the caricature of Boston as the ark of prudery—and partly because the caricature was sometimes true—these physicians were interested in sexual problems.

Finally, medical leadership in Boston and Harvard deliberately sought to remain open to talent from every social group at a time when snobbery and exclusion were growing among the American upper classes. Openness was important because psychotherapy was developed by the co-operative efforts of American aristocratic professionals, European immigrants, Jews, and upwardly mobile physicians from middle-class families.

There is a direct link between the interest in hypnosis and the development of systematic psychotherapy. Where the former was strongest, so was the latter. Artificial separations sometimes have been drawn between a "first" dynamic psychiatry based on hypnosis and a "second" based on non-hypnotic methods. In fact the second evolved directly from the first. Every major American psychotherapist and psychoanalyst, from James Jackson Putnam and Morton Prince to A. A. Brill, began his psychotherapeutic apprenticeship with hypnotic methods.

On the whole American medical approval of hypnotism lagged behind European. In the 1880's and 1890's influential professors of neurology at major European Universities experimented with hypnotic treatment—Krafft-Ebing in Vienna; Albert Moll in Berlin; August Forel in Zurich. In the same period only a few American physicians of stature—perhaps Putnam was the only one in a major medical institution—approved of hypnotic therapy. Probably the mind-cure cults forced New England physicians to develop psychotherapy, yet discouraged physicians elsewhere. As late as 1910 the New England interest in psychotherapy, especially of a religious nature, was regarded as a regional aberration.[11]

Outside Boston interest in hypnosis was less respectable in the 1890's. New York neurologists in major teaching positions were not notably enthusiastic about it as a therapeutic tool. But they were on the whole more interested than neurologists elsewhere. For example, Charles L. Dana, professor of neurology at Cornell Medical College, believed hypnosis was valuable but extremely difficult. He was also interested in cases of multiple personality, catalogued all the known examples, and added one of his own.[12] However, he interpreted them entirely as his friend George Beard might have done, by elaborate speculations about possible derangements of cerebral function. At the Columbia College of Physicians and Surgeons, Moses Allen Starr used hypnotic suggestions as an auxiliary method, but relied primarily on rest and reassurance. A few well-known neurologists—Joseph Collins, Mary Putnam Jacobi, and Allan McLane Hamilton—regularly used hypnosis, as did Hamilton Osgood, a general physician deeply interested in psychic research. In the mid-1890's hypnosis was used at the neurological clinic of the Post Graduate Medical School and later at the Vanderbilt Clinic. By 1900 perhaps New York's best-known medical hypnotist was John Duncan Quackenbos, a popular writer and teacher of rhetoric, whose claims were as startling as those of Mary Baker Eddy.[13]

Philadelphia remained a fortress of physiological orthodoxy and scientific purism in psychology. Led by Silas Weir Mitchell and Charles K. Mills, many Philadelphia neurologists condemned hypnotism and suggestion as dangerous and ineffective. By comparison with the Bostonians, Philadelphia physicians seem narrow socially and intellectually. For a time Mitchell co-operated with William James's experiments with the medium, Mrs. Piper, but soon the enterprise struck him as "inconceivable twaddle." He reserved his psychological imagination for his patients and his novels. His distinguished father, John Kearsley Mitchell, had made experiments with hypnotism that William James considered as important as those of the English physician James Braid. The son recorded minutely a case of multiple personality but scrupulously avoided trying to explain it. Lightner Witmer, the major psychologist at the University of Pennsylvania, remained committed to scientific experimental psychology. For him, too, hypnosis and the unconscious were questionable and possibly humbug.[14]

In Chicago a few physicians, none prominent neurologists, were using hypnotism in the 1890's. The *Journal of the American Medical Association,* then in the midst of a quarrel with most of the nation's leading specialties, was sympathetic to hypnotism and suggestion and carried a number of articles about their clinical use. Occasionally physicians or obstetricians interested in sexual problems, such as William Lee

Howard, also took up hypnosis, as did a few general physicians in Memphis, San Francisco, Salem, Oregon, Cincinnati, Detroit, and other cities.[15] Psychotherapy did not become a truly respectable and relatively widespread interest among general physicians, major medical schools or neurologists until about 1904 and after.

Morton Prince (1854–1929)

The development of psychotherapy in New England arose from a combination of native and European discoveries. Prince took his first step toward creating a modern American psychotherapy in 1885 by presenting a reasoned argument for the power of thoughts and feelings. These were ultimate data, as "real" as matter itself. He wrote that a "feeling is *not* accompanied by a molecular change in the same brain; it is 'the reality itself of that change.' " If you feel pain, "the real activities in you are pain, not neural vibrations." [16] Thus Prince decisively rejected parallelism and adopted a materialistic monism.

Although a determinist, Prince, like most American therapists and psychologists, also took pains to make room for moral ideals—the good, the true, and the beautiful. Emotion, consciousness, and will could alter reflex patterns of association and direct them in new channels. Every man was free to choose in the sense that he could direct his attention as he wished. Moral principles were learned, became part of the chain of associations, and thus as "dominant and sublime as though they were the express laws of a Creator." [17] Prince's argument was favorably received by his fellow neurologists; he repeated it more briefly and cogently in an article in *Brain* in 1891, which caught the attention of Adolf Meyer, a young Swiss neurologist who, a year later, emigrated to America.[18]

Prince's interest in psychopathology was partly personal. Born in 1854 to an old wealthy New England family, he had been a frail child, but like Theodore Roosevelt, improved himself. He learned to play rugby and became an enthusiastic horseback rider and yachtsman. Like Weir Mitchell, he was nervous, and he hid during thunder storms.

His mother suffered from a severe nervous ailment, and in 1880 he took her to the great Jean Martin Charcot in his consulting room dimly lit by stained-glass windows overlooking the boulevard St. Germain in

Paris. In 1885 Prince married a nervous girl and that year was appointed neurologist to the Boston City Hospital.

Prince's little book, written as his Harvard Medical School thesis and revised between 1882 and 1883, was based almost entirely on British, and to a lesser extent, German sources. By 1890 the Americans were assimilating the remarkable new discoveries of the French psychopathologists. These provided the basis for Prince's treatment of Miss Beauchamp, and a bare outline of them requires recapitulation here. Whole sections of James's engaging *Principles of Psychology,* and chunks of his major conceptions of memory, the stream of consciousness, and the self, depended on the new French data.

These momentous discoveries opened up the whole question of what personality was and what constituted the limits and nature of human consciousness. They established experimentally the existence of the subconscious. They also demonstrated that hypnotic suggestion could induce and cure symptoms. What had seemed the undeniably physical effects of disease could be regarded as the results of an "idea" or an "emotion."

In 1882 Jean Martin Charcot announced to the French Academy of Medicine a new discovery. He could produce experimentally by hypnotism the identical contractions and paralyses that occurred spontaneously in hysteria; these symptoms resulted not from an ascertainable physical lesion, but, as Freud later wrote in his obituary essay, "from specific ideas holding sway in the brain of the patient at moments of special disposition." [19] A woman who had safely survived a grisly railroad accident later became paralyzed not from the nervous shock of the accident but from her idea of the event.

Hysterical symptoms, Charcot insisted, were real, not feigned. Hysterics were not malingerers, and men were hysterical far more frequently than usually supposed. It was from these aspects of the French clinical tradition, as well as from Charcot's studies of traumatic hysteria, that Freud began his independent investigations.[20]

From the Nancy school of A. A. Liébault and Hippolyte Bernheim came clinical evidence that an "idea" could not only cause but also cure symptoms. Freud and innumerable Americans, Morton Prince among them, studied hypnotic therapeutics at Nancy. Taking issue with Charcot, Bernheim argued that it was absurd to suppose that hypnosis was a rarity, induced only in hysterics. Hypnosis was nothing but sleep caused by suggestion, and every normal person was suggestible. Bernheim defined suggestion vaguely as the influence "exercised by an idea that has been suggested and has been accepted by the brain." He commanded patients to sleep: "Look at me fixedly, and think only about going to sleep. You will feel your eyelids grow heavy, and your eyes will become tired. Your eyelids are flickering, your eyes are watering, your vision is becoming confused. Your lids have closed, you cannot open your eyes. You no longer feel anything; your hands are motionless; you see nothing more; you are going to sleep. . . . Sleep." [21]

A verbal order during hypnosis could make a subject anaesthetic to pain, raise blisters, make women lactate, augment the ability to hear, feel, see, or remember. People could be made to commit mock crimes —murders with toy pistols, for instance—and Bernheim warned that real crimes could be induced by hypnosis.

He and his followers claimed to cure a variety of illnesses, functional and organic: loss of speech, paralyses, depressions, tremors, bedwetting, hysterical blindness. Hypnosis could make surgery or childbirth painless. Some claimed to treat children's bad habits with success; others tried to cure the insane. Freud learned from Bernheim that subjects could be made to recall after waking what had occurred during hypnosis, which they usually forgot.

Studies of memory, of automatic acts, and of the abrupt and thorough changes that sometimes occurred spontaneously in personality, especially among hysterics, led directly to the conception of a dynamic subconscious. Applying evolutionary theories, chiefly those of Herbert Spencer and Hughlings Jackson, French psychopathologists assumed that complex functions were the end product, not the elementary phases of human psychological activity. Elementary, simple acts were automatic and were exemplified chiefly in memory and habit.

Hysterics, sleepwalkers, and hypnotized persons displayed curious alterations of memory. They might recall events of years before, often of childhood, that ostensibly had been forgotten. Under hypnosis they might develop unusual capacities to remember. A cataleptic, for example, could be trained to move her arm at a bare hint from the hypnotist, which seemed to prove a cumulative memory process. Some kind of conscious awareness must be present.[22] What was its nature?

In 1886, the year Freud was studying with Charcot, Pierre Janet put forward a hypothesis to show why it was that a subject would obey at a precisely appointed date and hour a post-hypnotic suggestion as long as thirteen days after the suggestion had been made. It was as if, "subconsciously," something inside the subject were keeping careful count of time. Janet borrowed this "self" from F. W. H. Myers. Myers and Edmond Gurney in England had been investigating "automatic writing," which a subject, awake or hypnotized, produced without seeming to be aware of what he was doing. Automatic writing often recorded forgotten past emotions and events, some dating from early childhood, and could reproduce precisely passages in Greek or Latin the subject had heard or read years before. Another "self," Myers thought, dictated to the primary personality. It was this subconscious self, Janet argued, that kept track of time in post-hypnotic suggestions.[23]

If the subconscious personalities of hypnotized people or of hysterics displayed unusual abilities to remember, they also displayed unusual capacities for forgetting. Whole groups of memories become separated, and different sub-personalities with different sets of recollections seemed to merge from the same human being. The fact that these memories were not shared by a common consciousness, and that different sets of memories seemed to accompany different organizations of "volitions, sentiments, moods and points of view," formed the basis of the new studies of multiple personality. Often these personalities occurred quite spontaneously and seemed to result from a dissolution of the inhibitions that constrained the customary self. Although some important studies were French, Janet noted that the largest number of well-known cases— and there were not more than twenty-five or thirty in the world— had been observed by American physicians, Weir Mitchell, Charles L. Dana, Boris Sidis, and Morton Prince. William James suggested in 1890 that these subconscious selves seemed to possess social and cultural characteristics. In America they often were so alike they might be manifestations of the *Zeitgeist*. Often they were "Red Indians," jocular and crude. Or they might be full of "vague, optimistic philosophy—and —water." [24]

But a major issue remained unexplained. How and why did hysterics and hypnotized persons seem to forget? Why did sets of memories become detached? What caused the destruction of personal synthesis? It is important to recall that Janet viewed such a synthesis as a final result, not a rudimentary attribute of human personality. Synthesis was an achievement, one of those signs of individual development that paralleled the evolution of the race.[25]

Essentially Janet's answer was physiological, based on the tradition of hereditary degeneracy. In 1889 he argued that the "higher centers" of the brain controlling the synthesis and co-ordination of ideas became weakened and "depressed." The field of awareness narrowed, so that a single thought or emotion came to dominate the conscious personality; all others were crowded out. This explained the origin of the fixed ideas that tormented neurotics.

Between 1889 and 1892, Pierre Janet elaborated the theory that emotional shock or trauma was an important precipitating cause of dissociation. One of his most famous patients from a later period, Irene, illustrates this process. She could recall only when hypnotized how, for sixty days, she had nursed her mother who was dying of tuberculosis, suffocating, and vomiting blood. Hypnotized, Irene rehearsed every detail as though it were just happening. The subconscious seemed to contain the memory of past experiences in all the freshness of their first occurrence. Some memories seemed fraught with powerful emotions, yet the subject knew nothing of them.[26]

A year later Joseph Breuer and Sigmund Freud published "On the Psychical Mechanism of Hysterical Phenomena: Preliminary Communication." They confirmed Janet's finding that dissociation was the "basic phenomenon" of hysteria.[27] In 1896 and again in 1901 William James described with his customary felicity the range and therapeutic implications of these discoveries: "In the relief of certain hysterias by handling the buried idea, whether as in Freud or in Janet, we see a portent of the possible usefulness of these new discoveries. The awful becomes relatively trivial."[28] And in *The Varieties of Religious Experience:* "In the wonderful explorations by Binet, Janet, Breuer, Freud, Mason, Prince and others, of the subliminal consciousness of patients with hysteria, we have revealed to us whole systems of underground life, in the shape of memories of a painful sort which lead a parasitic existence buried outside of the primary fields of consciousness, and making irruptions thereinto with hallucinations, pains, convulsions, paralyses of feeling and of motion, and the whole procession of hysteric disease of body and of mind. Alter or abolish by suggestion these unconscious memories, and the patient immediately gets well."[29]

In 1890 at a meeting of the Boston Society of Medical Improvement with Josiah Royce and William James in the audience, Prince praised the "giant strides" Europeans had taken in the study of hypnotism, suggestion, and multiple personality. With his patients there for clinical demonstration, he described how he had made a normally phlegmatic woman wear her bonnet during dinner. She had put it on only after a severe struggle between her protesting, conscious self, and her subconsciousness, which carried out Prince's post-hypnotic command. He had cured many functional illnesses, including a severe traumatic neuritis of the arm and a girl who vomited whenever she was kissed. In 1890 Prince also offered a rudimentary conception of the subconscious to account for suggestion and association. These operated, he wrote, "deep down in the lower strata of consciousness," in what Dr. Oliver Wendell Holmes had called the second self or "other fellow." [30] Prince had seen no necessity for invoking this conception of Holmes's in his little book in 1885, and probably Janet's studies had made Holmes's idea seem relevant.

James cautioned Prince that it was "very easy in the ordinary hypnotic subject to suggest during a trance the appearance of a secondary personage. . . . One has, therefore, to be on one's guard in this matter against confounding naturally double persons and persons who are simply temporarily endowed with the belief that they must play the part of being double." [31] To this day how much of the phenomena of multiple personality reflects suggestion and how much is spontaneous remains in dispute.

A year later, perhaps incorporating some of James's views of habit, Prince began constructing an essentially behavioristic theory of neurosis. His point of departure was a therapeutic experiment by the Baltimore surgeon, John Holland Mackenzie, "The Production of the So-called 'Rose Cold' by Means of an Artificial Rose." Hay fever, or "rose cold" as it was familiarly known, plagued a woman of thirty-two and had first developed between her sixth and her eighth year. Mackenzie threw her into a paroxysm of sneezing by suddenly presenting her with an artificial rose so perfect it looked completely natural. He concluded that "the association of ideas sometimes plays a more important *role* in awakening the paroxysms of vasomotor coryza than the alleged property of the pollen granule." [32]

On this basis Prince elaborated an important hypothesis. Some common nervous symptoms were not to be explained by the physical condition of the nervous system or by heredity, but by "mental associations" established by previous processes. They should be called "association neuroses." Their frequent symptoms were pain, tenderness in the joints, limping, paralyses that looked deceptively real. They might begin involuntarily, as "volitional attempts to deceive." In time, passing beyond the "control of the will," they persisted as true pathological processes. Thus a mental state might become associated with a physiological function, just as some emotions increased the action of the heart; or a visual image might become associated with the terrifying experience of seeing a friend knocked down by a coach-and-four. An "automatic nervous process of considerable complexity" could become established, and afterward act as an "independent neurosis." [33]

Associations, Prince later taught, were best thought of as aspects of neurological habit. The repetition of "ideas, sensations, emotions and organic physiological processes" together caused a tendency for each element to excite all the others. Prince later cited Ivan Pavlov's neat experiments as proving that animals could be conditioned to respond to "artificial" stimuli. Anything associated with the food that excited the salivary reflex—the plate, the voice of the person who brought the food—could be substituted for the food and excite the reflex. These reflex syndromes explained how association neuroses and the effects of suggestion operated. Habit neuroses were true functional disorders.[34] By this Prince meant that the tissues functioned essentially as if they were healthy, as the psychological cure of symptoms proved.

By stressing the learned nature of neurotic symptoms Prince weakened the position that they resulted from hereditary degeneracy. Here again, criticism and reconstruction were paired. The theory of a diseased nervous system resulted in "therapeutic nihilism and social hopelessness," while his own views offered "hope and possibilities." He agreed that heredity and a "neuropathic constitution" lessened resistance to environmental influences.[35] But he allowed degeneracy far less room than many of his contemporaries. In 1898 he attacked the "dominant view" that sexual perversions occurred spontaneously without external cause, as a result of degeneracy. For physical heredity Prince substituted "psychical phenomena," learned "nervous reflexes," environment, and education.

Sexual perversions—sadism, masochism, fetishism, and homosexuality, the major subject of the paper—could be acquired like "hysteria and other neuropathic states," and therefore could be subjected to the same method of treatment. Prince argued that homosexuality could hardly be an atavism, because often only psychical, not physical, characteristics were involved. Some physicians assumed that male homosexuality indicated the presence of a female brain in a male body. This "naïve assumption" Prince likened to "old-fashioned phrenology," "a hazy sort of cerebral localization involving a different cerebral architecture for each sex." The congenital theory also involved unwarranted assumptions about the "normal development of the vita sexualis, and of the tastes, habits and modes of thought peculiar to each sex." [36]

No one, he argued, remembered the beginnings of his own sexual life. Moreover, the congenitalists overlooked the importance of "casual external circumstances," which could have been forgotten, while their effects persisted. Prince cited cases of sexual aberrations that would have appeared to be congenital had not a careful investigation of the patient's past life disclosed forgotten causes.

There was no "hard and sharp line drawn by nature between the normal personalities of the sexes," Prince insisted. "Taking a large number of people, the male personality shades into the female, and vice versa." Education, meaning "the total environment," intentional education, unconscious mimicry, external suggestion, example, all played an important role in differentiating the sexes. "I think it is extremely probable that if a boy were brought up as a girl and a girl as a boy, and absolutely freed from all counter influences, such as the unconscious influence of public criticism, etc., each would have the non-sexual tastes and manners of the other sex." [37]

Finally, he argued that in healthy people "some degree of erotic feeling or ideas may be excited by the sight or touch of the form of a person of the same sex, and, at any rate, thoughts (pertaining to anatomy) so excited may very naturally awaken secondarily associated sexual feelings. For instance, the vita sexualis in a boy is first associated with its own sexual organs." [38] In children, as Krafft-Ebing had pointed out, feeling for the opposite sex resulted from a long and complex development. Healthy people suppressed their homosexual feelings, those with tainted constitutions cultivated them, and such feelings could become exclusive and automatic. Prince insisted that sexual perversion, because it was learned, was a vice. Moreover, as Schrenck-Notzing had demonstrated, it could be cured. Treatment then would replace the laissez-faire doctrine of resignation.

All Prince's new techniques were based on his theories of habit and association. If learned associations caused hysteria and other neuroses, these associations could be un-learned. Re-training could substitute healthy reactions or "complexes," as groups of symptoms were called, for morbid ones. The therapist could suggest a different outcome to those associations that actually had been built up by past experience. This was called "side-tracking" or "re-education." For example, in a description of fear neurosis, a syndrome Prince believed he was the first to discover, he argued in 1898 that by a kind of self-fulfilling prophecy, as it might now be termed, a patient would suggest symptoms to himself and expect them to happen, which they did. A musician thus might believe himself unable to play a concert in public because he feared his hands would tremble, as indeed they would. To one such neurotic, Prince prescribed this self-healing suggestion: "I know I am the equal of others. These symptoms are only physical processes which by habit have become associated together by a previous lack of confidence in my ability." [39] This therapeutic suggestion of ability or capacity was also useful for neurasthenics who limited their activities because of a pervasive sense of fatigue.

Still another method involved educating the patient to a correct understanding of his symptoms, especially those that seemed to be somatic, but were not, a common syndrome. Often patients had fixed ideas, usually fears, about their bodies. Thus if a woman without organic illness feared she had heart trouble and would die if she left her house, she would be informed that her fears were imaginary, her heart sound. Then she would be taught to walk abroad. Prince urged his patients to suppress their trivial angers, worries and fears, lest these grow out of control. Sometimes he used hypnosis to discover trauma in the lowest strata of consciousness in order to know what suggestions to make. In the Beauchamp case, hypnosis was both an investigative and a therapeutic method.

By 1898 Prince began to experience a sense of illumination: the baffling tangle of the neuroses could be charted. After a searching inquiry into every detail of the "origin, history and character of the symptoms . . . how often what seems to be a mere chaos of unrelated mental and physical phenomena will resolve itself into a series of logical events. . . ." [40] The satisfaction of having made understandable order out of chaos came to many of the pioneering psychotherapists. They argued time and again that the order they "discovered" proved the truth of their assertions. Prince was proud of his experimental and theoretical approach. When he had visited Bernheim in 1893, he had been received as an honored colleague. It was from a secure sense of his worth as a scientific experimenter with a wide range of psychopathological phenomena that Prince would confront Freud's psychoanalysis.

James Jackson Putnam (1846–1918)

James Jackson Putnam, Harvard's first professor of neurology, would approach psychoanalysis from a significantly different outlook. More philosophical than Prince, perhaps more widely read, especially in the German literature, he, too, had absorbed the European and American psychopathology. But the most distinctive influences on him came from his philosopher-psychologist friends William James and Josiah Royce, and from Henri Bergson.

By 1906 Putnam had adopted a far more radically "psychological" position than Prince's behaviorism. Partly through an interest in philosophical idealism that grew with his interest in psychotherapy, Putnam gave a high place to autonomous laws of the mind, and later could see Freud's theories of displacement, identification, and projection as possible descriptions of how the mind actually worked. He also formulated several conceptions that somewhat resembled Freud's—of conflict between self and society, of the internalization of social forces in a sharply critical aspect of the self. Putnam strongly emphasized facing painful emotions and restoring the unity of the mind. He developed a lively interest in the child, in radical modifications of human character, and thus in the environmental and educational influences that formed them.

Some of this distinctive outlook had roots in an early Transcendentalist Unitarianism which had stressed the reality of moral conflict and of the unseen agencies of spirit that ruled the world. Writing routine essays at Harvard College in the shadow of the Civil War, Putnam expressed admiration for fanatical devotion to a great cause and for the great leader far ahead of his time, whom the world first despises, then follows. Putnam began his professional career as a militant medical materialist and rejected fuzzy conceptions of functional nervous disorder. But the arguments of his friend William James for the importance of emotion and will, his growing experience with the traumatic neuroses, and the evidence of the mind-cure cults, led him to change his mind.[41]

More strongly than Prince, Putnam argued that consciousness—including the subconscious—was the "ultimate reality."[42] Like Janet, James, and Bergson, Putnam assumed that innumerable independent processes, many of them occurring outside the "personal consciousness of the moment," made up the operations of mind and body. In health these were co-ordinated and worked independently without interference or antagonism. The result was a consistent sense of self, a harmony of action that allowed the human being to devote his consciousness "to the interest of each new question as it arises."[43] But in nervous disorder, in neurasthenia, and especially in hysteria, this harmony was broken. The patient no longer was able to focus his attention. Unrelated thoughts and emotions insistently intruded. The more the patient struggled to oust these ideas by logic or will, the deeper he carried himself into the mire. For these fears or insistent ideas were rooted in deeper, subconscious strata of the mind.

Often such ideas resulted from a "hidden" painful experience that went "even far back in the years of childhood." Drawing on Janet and Prince, Breuer and Freud, as well as his experience with traumatic neuroses, Putnam insisted that the therapist should bring this experience into the clear light of consciousness. Then, the "reorganization of the disordered forces of the mind is likely to take place of itself," a position strikingly close to Freud's.[44] Recently, Putnam noted, "minute and searching analyses of the patient's memory," chiefly carried out in studies of multiple personalities, fixed ideas, and hysteria, were leading to a modern, rational system of mental therapeutics.

Attention to human relationships was notably lacking in Prince's study of Sally Beauchamp. Putnam, on the other hand, insisted that an emphasis on social relations operating consciously or unconsciously could be supplied to psychopathology by philosophy. He argued that the "color and movement" of every man's thoughts depended on the complex entanglement of his human relationships. Insanity itself was a "social concept." [45]

The social dimensions of the self were inherent in language and symbol. Mental activity no longer could be described in terms of "brain activity" for the reason that neurological activity expressed only a limited portion of the meaning of "mental" activity. People understood each other's minds by words, and by facial and bodily expressions. What one could learn by opening up the brain and watching its action during this process of communication could only supplement what one knew by symbolic and other means. And symbolism was essentially social in its operations.

Was a human being limited by his skin?, Putnam asked. "Or, on the other hand, is he coextensive with the scope of his conscious interests and affections, and thus to be considered as complete *in himself* only when his relations to others are taken into account?" [46] Putnam suggested what amounted to a theory of the internalization of customs and traditions. This new view he derived in part from James but chiefly from Royce.

Putnam argued that the customs and traditions of social groups, families, and races, perpetuated by imitation and tradition, constituted the "social etiology" of nervous and mental disorder. "Racial" traditions created New England's "high-pitched" voice, hurried manner, and sense of strain.[47]

From Royce, Putnam derived a subtle view of the way in which social forces created the individual's "self-image," as well as his "social consciousness." The child imitated himself and others. When he learned to speak, he began to understand the meaning of his thoughts and actions, to recognize himself as a moral agent, and to view himself through the eyes of other people. At this point, disorders of social origin could begin.

First, an internal split could occur. A criticizing aspect of the self might develop which tended to judge in other people's terms. Thus one part of the self might stand over against the other part and "pass unfavorable judgments on its acts." This sense of "self-blame" could be "accentuated by the crude and foolish doctrine that disease always does arise from 'sin' or from 'infraction of nature's laws,' as if it were not well established that in plant life, in animal life, in conscious life, antagonism, destruction and loss may be the means to progress, even if not the best or only means." [48]

Divided within himself, the nervous invalid also might become separated from the community and develop an "ill defined yet grinding sense of social isolation," felt not even explicitly but as "the intangible cause" of a "nameless dread." This sense, which came primarily from within, might accompany an inability to conform to social standards, compete in occupations, take part in sports and games. Thus the patient built a wall around himself and came to see himself as a monstrosity, whom others were justified in treating as such. The resemblance of this, especially in Royce's version of it, to Freud's Ego Ideal and later to the Superego is clear. Putnam's inner critic operated unconsciously and functioned as a disapproving judge.

Because pathology began in childhood, prophylaxis also must begin then. A former patient, Susan Blow, founder of one of America's first kindergartens, interested Putnam in early childhood training. The child must be taught, Putnam insisted, to bear responsibilities, to postpone satisfactions, to carry out feelings into action, to eschew morbid self-reproach and morbid self-consciousness. The kindergarten, school hygiene, and individualized instruction could play an important part in prophylaxis. In 1898 Putnam wrote that the child should receive education about sexuality from parents and teachers.[49] Finally, educators should not force the child into a "mould" but allow the development of his own unique possibilities, yet not allow him anarchic freedom of self-expression. He objected strongly to the suggestion, associated with G. Stanley Hall, that undesirable impulses be allowed partial, therapeutic expression.[50] To mitigate the neurotic's sense of isolation was part of the physician's new task as "educator," which psychotherapeutics required. As representative of the community he was to appeal to the patient's "social motives." Putnam recalled that the aims of Weir Mitchell's rest cure were to make his patients feel the full force of their social obligations so that they might live a life of "active usefulness." The therapist worked with the patient's conscious ideas and interests as well as his subconscious.

There was another element in Putnam's theory derived partly from Royce—the reconceptualization, indeed the remaking, of the self. A "cramp-like fear of change" characterized the nervous patient, Putnam wrote. He had to be prepared to change his habits, to give up the "acting out" of his morbid fears and the indulgence of morbid thoughts that gave only momentary relief. He must be prepared to create an "image of himself freed from harassing . . . feelings." [51] Thus a neurasthenic, filled with fears of failure, would be taught to carry out acts which the patient's amateur diagnosis would have eliminated as out of the question.

Some of this treatment was frankly inspirational, designed to inculcate hope and faith in the individual's spontaneous creative possibilities. Putnam sought a philosophy to back up this therapy, and Henri Bergson supplied him with several sanguine conceptions that would also become important in his reactions to psychoanalysis. Bergson argued that the living being was not bound by iron determinism, but was a free agent of the energy or life force that lay behind evolution. Bergson also suggested that perception and character were largely determined by unconscious memories.[52]

Putnam argued that the physician himself always was a "dual or complex person" and owed it to his patient to "have a sharp lookout on the working of his own subconscious personalities," an adumbration of Freud's counter transference. As early as 1899 Putnam argued that the discovery of the subconscious required new training for the physician. Philosophy and psychology were more important for the medical student than botany and zoology. For the psychologically trained physician was "keen to see causes of disease to which the pure pathologist is often blind." [53] The physician could not treat a patient successfully without penetrating deep in imagination into his mental life. To rid a patient of a "tormenting delusion" was the equivalent of a "successful operation for a painful disease," and required equal preparation and skill, an analogy Freud also was fond of drawing.

Like Prince, Putnam granted heredity decreasing importance. Even "delusions of suspicion and persecution," usually considered results of "degeneracy and constitutional defect," in fact could be explained by a "narrow, barren, superstitious, savage . . . social envionment." One corollary of this assertion was an insistence that the physician be ready to "help change public feeling" and thus eliminate "inferior social customs and narrow social traditions," one more attitude that would become important when Putnam faced psychoanalysis.[54]

Like Prince, Putnam also argued that sexual perversions were not stigmata of degeneracy, "manifestations of an implacable destiny." Rather they were "habits growing easily on a morbid soil, but often preventable and curable, and of—so to speak—accidental origin." [55]

Sexuality played an important role in the symptoms and etiology of nervous disorder. The "emotional relations of the sexual function are so widespread that the mental disorders of sexual neurasthenics are especially severe and numerous, varying from hypochondriacal depression to serious imperative conceptions. The sexual instincts, though not consciously recognized as such, are the basis of much of the emotional instability of early puberty and middle life." [56] It was impossible to separate the physical from the emotional aspects of sexuality. Putnam had learned by 1909 to place etiological significance on social relationships, early childhood, conflict within the self and between the self and society, and on purely psychological principles of mental function.

William James, who heard Putnam's first paper on "psychical treatment" in 1895, congratulated him for taking seriously a body of phenomena James himself knew only from books. "I think there is a considerable future for this mode of treatment," he told the Boston Society for Medical Improvement. Then, in 1908 he wrote of his pleasure at Putnam's interest in Bergson and psychotherapy: "It is truly grand to see you in extreme old age renewing your mighty youth and planing yourself for flights to which those of the newest airships are as sparrows fluttering in the gutter." [57]

Partly through the work of Prince and Putnam the major foundations of a new American psychotherapy were laid between 1885 and 1900. Psychological theories of imitation, habit, and association were replacing heredity and inanation of "nervous force" as the major cause of neurasthenia. A dynamic subconscious, the locus of dissociated sets of active associations charged with emotion, caused the symptoms of hysteria. Theories of the interaction of mind and body had replaced psychophysical parallelism. Therapy had come to include hypnosis, but, more important, suggestion and re-education in the development of a new outlook and new habits.

The Psychotherapeutic Movement

Between 1904 and 1909 psychotherapy became a national medical and popular interest, and psychoanalysis became better known as one of several competing new methods. A simple chronology of events indicates the gathering momentum. In 1904 and 1906 Pierre Janet lectured in America, lending European prestige to the native medical psychotherapeutic movements. In 1905 a popular European text, Paul Dubois's *Psychic Treatment of Nervous Disorders,* translated by two neurologists who later became psychoanalysts, launched a new interest among general physicians. In 1906 a popular movement for psychotherapy was founded by physicians and by priests of the Episcopal Church in Boston and spread across the nation. By 1907 psychotherapy was hotly debated by neurologists, and in 1909 the first American medical congress on psychotherapy was held in New Haven. The *Journal of Abnormal Psychology,* founded by Morton Prince in 1906, stimulated this growing interest. Prince hoped to serve neurologists, psychiatrists, and psychologists who were experimenting with psychotherapy and psychopathology. Articles on these subjects were scattered inconveniently among the medical and psychological journals, Prince argued, and no single journal for them existed in the United States. Prince financed the venture at first himself, and he enlisted Hugo Münsterberg, James Jackson Putnam, Boris Sidis, Charles L. Dana, Adolf Meyer as associate editors. By 1908 Prince was advocating courses in psychopathology in medical schools and proposed one at Tufts College in Boston that would include all the better known new methods.[58]

American psychotherapy was marked by certain distinctive traits that selected the most congenial foreign influences and the first interpretations of psychoanalysis. Many developments associated with the "Progressive Era" had been anticipated by psychopathologists in the 1890's, notably environmentalism and optimism. The new century intensified these trends as well as another element also foreshadowed earlier, a greater emphasis on social responsibility. Like other Americans, psychotherapists absorbed the social gospel of the churches and the new emphasis on the community.

American therapists adopted a "liberal and progressive" self-image around 1907, reflecting a growing sense of polemical need. Debating with somaticists, the psychotherapists appropriated attributes of youth, zeal, and hope, as had the somaticists of the 1870's. This self-image coincided with reality only in part. Six major somaticists were only six years younger than six major psychotherapists, whose average age was forty-eight. Many somaticists, notably S. Weir Mitchell and Bernard Sachs, were as optimistic and considerate as any psychopathologist.[59]

Some psychotherapists envisioned for themselves a larger social role, and deliberately included philosophy and inspiration in therapeutics, and, like the somaticists, prescribed complete regimes for living. Since the 1870's some had observed that the physician seemed to be taking over the role of the priest. In 1907, probably following Janet, Lewellys F. Barker, son of a Quaker minister and professor of medicine at the Johns Hopkins University, argued that more people were turning to the physician "when in psychic difficulty or when in need of mental and moral direction and physicians, whether they like it or not, are thus forced into the responsibilities of 'confessor' and 'moral director.' " [60]

The New England psychotherapist Boris Sidis urged Americans to face the evils in their midst: the Moloch of Industry crushing the lives of children, the loathsome city slums, where human beings lived like worms, the poverty of the masses, the starvation, the "strikes and lockouts," the "frauds and corruption of our legislative bodies," the iniquities of corporations. Sidis had seen some of these evils when he landed in New York, poor and alone, after his expulsion from Czarist Russia for political activity. "The wonderful cures" produced by hypnotism and suggestion were no more wonderful, Tom A. Williams, a psychotherapist trained in Edinburgh and Paris, remarked in 1909, than were the changes Judge Ben Lindsey wrought in Denver's wayward boys.[61]

At the Massachusetts General Hospital came a new emphasis on the role of the family and the first employment of psychiatric social workers. Richard Cabot, an internist from one of Boston's first families, and James Jackson Putnam, chief of the neurological services, began to use social workers to mobilize the community's charitable resources and to educate families of patients in medical and mental hygiene. Unwed mothers and venereal disease were among their most pressing problems. In 1907 Putnam created a Social Service Department for neurological patients and hired social workers who acted as psychotherapists and conducted classes in occupational therapy.[62]

Although theorists of heredity placed strong etiological significance on the family, Cabot was one of the first to see the psychological implications of "family environment." An "overshadowing father or a bossing mother," he wrote, sometimes made it impossible for a neurotic child to develop independence. The family and the "hereditary conditions that formed character," the thoughts, worries, griefs, personal relationships, friendships, love affairs, domestic and family affections often could be observed more skillfully by the social worker than by the minister or physician.[63] At New York's Psychiatric Institute Adolf Meyer's wife was investigating the home-life of patients to discover the environmental causes of their illness and to estimate their chances of readjustment after they were discharged. Some physicians, Putnam and Schwab, for instance, tried deliberately to strengthen the patient's participation in social life, his sense of achievement and self respect.[64]

Inevitably psychotherapists dignified the patients they treated, whom somaticists often despised. No longer were neurotics "vampires" or "silly exaggerators," chiefly upper-class females. Most psychotherapists argued that neurosis was intensely serious and painful and could attack the happiness of individuals and families more cruelly than many organic illnesses. To be the victim of a neurosis, Putnam wrote, was like being a fly in a spider's web. If subconscious mechanisms caused hysteria, then "weak will power" or malingering were irrelevant.[65]

Physicians began to assert more strongly than they had in the 1890's that nervous disorders, including hysteria, troubled not only the rich, or the hard-driving business and professional classes. Neuroses also could plague the "poorer and middle classes," a cotton-factory operator, servants, dressmakers, milliners, nurses, the Jewish immigrant boy who worked thirteen hours a day and then attended night school. Sidney I. Schwab, a young St. Louis neurologist, concluded that 25 to 30 per cent of the nervous patients at a big-city clinic suffered from Beard's neurasthenia. Exhaustion, depression, and a sense of hopelessness were brought on by the rigors of the piece-work system and the insecurity of their jobs in the garment industry.[66]

Perhaps the most distinctive aspect of the new spirit reflected America's new prosperity and economic maturity. No longer was the painful saving of energic capital a major therapeutic goal. Many American psychotherapists, especially the New Englanders, repudiated the rest cure and the limited bank account theory of human energies. Cabot deplored the rest cure for encouraging sloth and a weakened will; like almsgiving, it pauperized. Nervous patients needed to be active and to participate in social life, rather than be coddled in passive, luxurious rest and isolation. He developed a work therapy that was adopted in several sanitaria.[67]

Moreover, human energies might be far more abundant than men had suspected. In 1906 the psychologist, Robert S. Woodworth, informed the nation's psychiatrists that the sacred doctrine that the brain was very liable to fatigue was entirely false. Experimental tests proved a surprising resistance to fatigue in prolonged mental work. The traditional "brain fatigue" was an "emotional affair," a feeling, and not a true "incapacity." [68]

In 1907, working independently, William James and his former pupil, Boris Sidis, discovered vast stores of reserve energy in everyone —a remarkable manifestation of what James might have called the *Zeitgeist.* By overcoming deadening inhibitions and routines, James suggested in his famous essay, "The Energies of Man," one could push to higher levels of energy and, thus, of accomplishment. Sidis carried the idea further. Possibly doctrines of infinite potential seemed plausible to a man, who, like Sidis, had risen in a new environment. Sidis won a scholarship to Harvard, earned his doctorate under James, and in 1909 was established by a wealthy New England lady at Maplewood Farms, his own institute for nervous and mental diseases at Portsmouth, New Hampshire. Evolution, Sidis insisted, already had selected the human beings with the greatest store of reserve energy. By increasing inhibitions and thus raising the thresholds of response, civilization added to this store. Larger amounts of reserve energy constituted the chief "superiority" of the "educated over the uneducated and the higher over the lower races." Sidis believed that by inducing what he called the hypnoidal state, in which the subject was relaxed but fully conscious, "stores of potential subconscious reserve energy could be tapped." The liberated energy caused mental synthesis and the reassociation of split-off material. Then, patients were marvelously transformed. He believed they felt like one grateful client who wrote him, "I feel that this wondrous light will never fail me." A fifty-five-year-old spinster, given to spells of tearful despondency, was changed into a radiant, energetic social worker. A dipsomaniac—and Sidis noted that the "medico-Calvinist prognosis" for this ostensibly inherited syndrome was indeed gloomy—became a paragon of American success. He had been weak, apathetic, dependent, irregular, unreliable. He became self-reliant, methodical, conscientious, firm, and trustworthy. Naturally, he became the "manager of a large concern." [69] Sidis even believed that parents could raise up a race of energetic geniuses by teaching their children not to be afraid and to avoid deadening routines. His son, raised accordingly, was graduated from Harvard at fourteen, suffered a nervous breakdown, and produced a learned study of the Boston street car transfer system.

Therapists also exploited "inspiration." Prince brought to the patient's consciousness "exalting ideas and memories." He tried to inculcate "confidence and hope" . . . and, above all, "the emotion and joy that go with success and a roseate vista of a new life." [70] The most popular of all mental therapies in America was the most eloquently inspirational, Paul Dubois's "persuasion." It made psychotherapy popular with the general medical practitioner; Jelliffe and White's translation of Dubois's *Psychic Treatment of Nervous Disorders* went through six editions from 1905 to 1909.

Dubois, a gentle, Swiss neurologist, despised suggestion and hypnotism as degrading and immoral. He preferred to exploit the patient's "rational faculties." Every emotion, Dubois argued, had a "mental representation" and therefore was open to influence by reason, excepting the "wholly animal passions"—a large omission. The therapist aimed to banish fixed ideas of fear and helplessness. Patients were to "pass a sponge" over the past and wipe out their phobias. But first they had a long, intimate conversation with the doctor, which Dubois called a "psychoanalysis." The therapist listened to the patient's life story without interrupting him, then told him exactly how he had developed in early life an exaggerated emotional sensitivity and suggested his own symptoms to himself. Like the other therapists, Dubois insisted that hysterical conditions arose by "insensible transitions from ordinary mentality." He also accepted Freud's argument that dreams revealed the most secret wishes and "aspirations." Smith Ely Jelliffe, who tried "persuasion," found the system in retrospect too empty of feeling. Others insisted that "persuasion" was no different from the suggestion they had been practicing all along. Nevertheless, Dubois augmented interest in non-hypnotic and non-suggestive therapies. Unless the patient were treated in a waking state, one popular physician argued, he could not be made "wise to himself and his troubles."

Pierre Janet astutely noted that Dubois's methods were entirely too full of the good, the true, and the beautiful, and an obligation to "behave kindly, energetically, nobly" that especially distinguished American psychotherapeutics. The extreme optimism and moralism of Dubois were missing from Janet's own worldly and realistic system.[71]

Janet's contributions after 1900 and his general approach throw unexpected, retrospective light on Freud's reception. Although both were considered by some New England physicians to be among the most hard, orthodox, and scientific of the psychotherapists, Freud still had slight influence in 1909, and his reputation was equivocal by comparison with Janet's. Their respective invitations to America indicate this difference in prestige. Freud was invited to Clark University, small and highly unorthodox, rather looked down upon by Boston and Harvard. Janet was asked by President Charles W. Eliot and Professor James Jackson Putnam to give a series of lectures at the inauguration of the sumptuous new buildings of the Harvard Medical School in October and November of 1906, and he also lectured at Columbia and the Johns Hopkins universities. In 1904 Janet had spoken at the Lowell Institute in Boston.

Janet was self-consciously conservative. Although he went through most of the important stages of the revolt against somaticism, at crucial points he appealed to physiological causes and preserved the traditional French etiological framework of heredity, including degeneration.[72] Somatic survivals can be observed in Freud's biologism and in the role he gave to the constitution in determining the quantative nature of the instinctual drives, yet Janet relied on such causes far more crucially.

Janet's rejection of somatic extravagance was vehement. He ridiculed the German penchant for "psychophysics," arguing that cortical localizations had been singularly abused to explain psychological symptoms that the physician failed to understand. Such attempts were pure phrenology and pure jargon. Whatever physical pathology ultimately might be found to underlie nervous disorders, these could be studied on a purely psychological, clinical basis. Observations, classification, and description of actual patterns were the basis of Janet's elaborate system. The results, many Americans thought, were brilliant. Janet's case studies were masterly, his language lucid, his arguments logical and careful.

Janet gave his patients long, painstaking and highly individualized treatment, which in one case, William James noted, had extended over three years. The psychiatrist Adolf Meyer commended Janet's emphasis on the individual, which coincided with his own.

At Harvard Janet spoke magisterially of the achievements begun by his masters, Charcot and Ribot, whose studies centered on hysterical patients, those experimental "frogs" of medical psychology, as Janet somewhat regretfully called them. His humanity was warm, and he indignantly inveighed against the cruel physical treatment of nervous disorders: "Do not try to count the number of arms cut off, of muscles of the neck incised for cricks, of bones broken for mere cramps, of bellies cut open for phantom tumors. . . . Not long ago I saw a patient who had had an eye excised and the optic nerve cut out for mere neuropathic pains." [73]

He devoted over three hundred pages to vivid description, an engaging *grand guignol* of hysteria, but left almost no space for etiology or treatment. He repeated that hysterics were a race apart, their nature different from human nature. "Exhaustion" of the "higher functions" of the brain caused dissociations to occur. Usually the function that became dissociated was already weak. Thus, those who had stammered would become unable to speak; the attention of those who had difficulty concentrating became seriously scattered. This function must have been one that was operating during a profound emotional crisis. And there the great Janet left matters.

He had explored etiology and therapy more fully in two eloquent volumes devoted to psychasthenia, a category he introduced in 1903. As already noted, it was one of the important contributions to the less sanguine outlook of several American neurologists. Janet had attempted to interpret psychologically and to place in a single category a number of nervous complaints that occupied the borderline between nervous disorder and insanity—weakness of will and attention, acute anxieties, obsessions, compulsions, fixed ideas. Psychasthenics displayed an unusual lack of the sense of reality, general feelings of "incompleteness" and powerlessness. Their compulsive acts were attempts to compensate for these feelings. Janet's psychasthenics suffered from nearly all the sexual symptoms Freud also observed—oral and anal fixations, impotence, infantilism, inversion. But Janet argued that Freud's insistence that these often resulted from lack of sexual satisfaction needed reinterpretation. In reality psychasthenics felt an over-all psychological inadequacy because they had difficulty carrying any act, sexual or other, to completion.[74]

The underlying cause of the disability was a lowered level of "psychological tension." Janet defined this as the ability to perform smoothly and efficiently difficult but essential psychological operations. These were chiefly concentration, decision, the capacity for an ever renewed synthesis of the new and multiform elements in experience. Such capacities made possible the adjustment to present reality, what Janet termed "la fonction du réel," and they occupied the upper levels of Janet's hierarchy of functions. Dreams, emotions, random or automatic motor acts required little psychological tension and occupied the lower levels. This reinterpretation of the classical neurological and psychological hierarchy resembled Freud's division of primary and secondary processes.

Janet argued for the interaction of predisposing and exciting causes in psychasthenia, as well as hysteria. Psychasthenia was fostered by faulty education and upbringing. Janet argued that parents encouraged psychasthenia by protecting their children from opportunities to act, to make decisions, to confront danger or to fight.[75] They gave them only lessons in prudence and withdrawal. Janet also noted that neurotics characteristically were psychologically immature, for example, disturbed men of thirty who retained all the traits of seventeen-year-olds. For all his luminous and systematic description, the major predisposing causes of hysteria and psychasthenia remained heredity and often degenerate heredity, with the addition of poor education and early training. On this "prepared" ground, the exciting cause, perhaps a shock, a difficult decision, or even a change of vocation, could cause the illness to occur. Janet reiterated this position on heredity in 1892, 1903, and 1909. Only a very small number of cases of psychasthenia were "acquired." The more marked the hereditary predisposition, the more serious the case, the less marked, the more curable. He went so far as to suggest, as had Morel, that neuropathic families, through the action of the neuroses, progressed across generations toward degeneracy and extinction, although not inevitably.[76] Janet rather avoided heredity in later years, as did the disciple who wrote the most comprehensive synthesis of his views, the Swiss neurologist and psychiatrist Leonhard Schwartz.[77] By the 1920's hereditary degeneracy had gone out of medical fashion.

Janet's treatment methods were varied and imaginative. He attempted to strengthen his patient's sense of reality and ability to concentrate. He used hypnosis and suggestion and continued to defend them against increasing attack. He also sometimes restricted his patients' expenditures of energy, so that they would live within their psychological budgets, as it were.

In retrospect it seems abundantly clear that several of Janet's characteristics ran flatly counter to recent American developments. His restrictive regimes and continued although qualified emphasis on heredity were in direct opposition to the optimism and environmentalism some American psychotherapists increasingly displayed. Equally uncongenial was Janet's closely related theory of an innate weakness in the capacity for mental synthesis of the hysteric or the drop in psychological tension of psychasthenics, whose inborn nature displayed feebleness of will, withdrawal, over-refinement, and ineffectualness. Such views as these flatly contradicted the American conviction of greater reserves of nervous energy that William James and Boris Sidis had spread with wide popular acclaim. It was possible to interpret the version of psychoanalysis Freud presented at Clark University in ways far dearer to American predilections, and the chance emphasis on sublimation added an ethical element, notably absent from Janet's cool realism.

In 1906, the second time Janet lectured in America, psychotherapy became a separate topic in the *Index Medicus.* By 1909 some ninety articles were listed, the largest total for the three decades from 1890 to 1919. About twenty-six of these were published by *Psychotherapy Magazine,* which had close connections with the Emmanuel Movement. This lay interest also increased the medical interest, and physicians from Boston to San Francisco began to discuss the subject.

Medical interest inspired popularizers, whose publicity in turn forced the attention of more physicians. In March of 1909 Hugo Münsterberg warned: "Scientific medicine should take hold of psychotherapeutics now or a most deplorable disorganization will set in, the symptoms of which no one ought to overlook today." [78] The popular interest and the influence of psychoanalysis will be described in later chapters.

By 1909 the advocates of psychotherapy included a number of strategically placed physicians. A partial list would include Putnam, professor of neurology at Harvard; Prince, who began to teach at Tufts College Medical School; Lewellys F. Barker, professor of medicine at the Johns Hopkins University Medical School; Smith Ely Jelliffe, an editor of the *Journal of Nervous and Mental Disease;* William Alanson White, director of the Government Hospital for the Insane in Washington, D.C.; Adolf Meyer, in 1910 appointed professor of psychiatry at the Johns Hopkins University Medical School; Julius Grinker, assistant professor of neurology at Northwestern University Medical School; Albert Moore Barrett, in 1906 made director of Ann Arbor Psychopathic Hospital and a former student of Adolf Meyer's at Worcester Insane Hospital; and Hugh T. Patrick, Chicago's leading neurologist.

Alternatives and Problems, 1908–1909

What was the common stock of theory and practice among American psychotherapists before the campaign for psychoanalysis began in earnest? The achievements of American psychotherapy were reviewed for the first time before a congress of general physicians in New Haven in May of 1909, five months before the Clark Conference. Prince, Sidis, Putnam, and Ernest Jones all delivered papers. No single method yet dominated American medical practice, and there were advocates of hypnosis, the hypnoid state, suggestion, and re-education.

No single view of the unconscious was yet accepted. In 1908 Prince, Hugo Münsterberg, Pierre Janet, and Alfred Binet debated six common definitions without reaching agreement. Was it the locus of ideas outside personal attention—James's fringe of consciousness? Did it consist of active ideas split off from awareness? Did it constitute a separate self or personality, a self-conscious "I"? Was it the storehouse of all memories? Or was it Meyers's subliminal self, part of a transcendental "tank of consciousness," the storage place of ideas and the source of genius? Was it a purely neural process, "unaccompanied by any mentation whatsoever"? The whole problem was discussed, again without reaching agreement, at the Sixth International Congress of Psychology in Geneva in 1909.[79]

At New Haven, Prince and others reviewed the assumptions of the new psychotherapy, happily purged of the "mystery and superstition" of the past. The basic premise was that "mental states" could induce physical malfunctioning; when mental states were altered, the physical disorder tended to clear up. The second principle governed the formation of symptom complexes. Associated ideas, feelings, emotions, sensations, movements, visceral functions, after constant repetition became linked together and were imprinted on the nervous system. The stimulation of one element excited the rest. The term symptom complex was old. By 1909, however, it had acquired two new meanings. It had become equated first with the sets of habits described by William James, and second with the dissociated states or automatisms of French psychopathology.[80]

Sometimes the experiences that became part of a complex went back to early childhood and could be remembered in minute detail. To treat obsessions, phobias, and some kinds of hysteria, the therapist needed to discover with or without hypnosis the nature of the experiences which lay buried in the past. Then the therapist could obliterate them or suggest new outcomes to the associations that had formed around them. He could strengthen the patient's general capacity for "healthy mental synthesis, or he could inculcate "healthy," hopeful attitudes. As Janet observed, all these methods and every kind of mental treatment were called "education" or "re-education" in the United States.[81]

The intensity and duration of treatment varied from a single session using hypnosis to Prince's seven years of therapy with Sally Beauchamp, which sometimes included daily interviews. Some American therapists elicited a "full and frank" confession of the patient's entire life and listened attentively to his description of his symptoms. A few, such as Barker, following Janet, were prepared to imitate the spiritual directors of Catholicism and to undertake the lifelong supervision of stubborn cases.

Putnam and Prince had developed differing theories of internal conflict: in Prince's case, among different, organized sub-personalities; in Putnam's, between the self-image and the internalized image of society. Yet, conflict did not occupy a conspicuous place in American therapeutics and was more latent than fully formulated. Putnam and Prince, indeed most New England therapists, were especially sensitive to the pathology of the oversensitive conscience, which was a notable symptom of their well-bred female patients. To all these competing theories and methods of psychotherapy, physicians usually added "rules for daily conduct," nutrition, rest, and isolation. A total regime prescribed by an authoritative physician remained, as it had for Weir Mitchell in the 1880's, the fundamental pattern of the therapy of the psychoneuroses.

Despite this achievement, some physicians were dissatisfied with the state of psychotherapy between 1906 and 1909. Many problems remained to be solved. In 1907 Prince and Isador Coriat argued that "the work of the future must be to determine the true relation between the functional disorders—physiological and psychological—and the fundamental mental fault, and thus find the rational basis for psychotherapeutic procedures." [82]

Edward Willys Taylor, assistant professor of neurology at Harvard Medical School, among others, argued that psychotherapeutics needed systematic rationalization. Therapists should publish cases that illustrated definite methods. They should "state facts simply and without recourse to over-much speculation. . . ." They should "maintain an attitude of conservatism in the interpretation of results, such as would be demanded in any problem of physical science." [83] Only by this kind of scientific scrupulousness would medical skepticism be overcome.

Morton Prince insisted that only those who had taken the trouble to obtain the requisite knowledge of the subconscious were qualified to express an opinion about psychotherapy. Those who were ignorant of this new knowledge were no more qualified "than would a person who had no knowledge of bacteriology to express an opinion on that subject." [84] Ironically, this argument would be turned against Prince himself by the psychoanalysts.

Another problem was the difficulty of defining suggestion. In 1898 Sidis tautologically defined suggestibility as "that peculiar state of mind which is favorable to suggestion. By 'suggestibility of a factor' is meant the power of the factor to induce the psycho-physiological state of suggestion in a certain degree of intensity, the suggestiveness of the factor being measured by the degree of suggestibility induced." [85]

It is hardly surprising that in 1906 the editors of the *Interstate Medical Journal* despaired. No wonder, they wrote, that the average interested physician was confused about psychotherapy. "Electricity, suggestion, hypnotism were all correlated and each used to explain the other in terms of them all." In some "mysterious way," wrote Sidney I. Schwab, who studied briefly with Freud, "the mind is supposed to be influenced by the persuasive or suggestive power of another mind . . . and the result achieved is often as much of a mystery as the manner in which it is brought about." [86]

Nearly every method produced cures of a sort—including Christian Science and New Thought. But some therapies presented serious practical problems. After 1895 many physicians in Europe and America, including Prince and Freud, grew less enthusiastic about hypnosis. Some people could not be hypnotized. It was embarrassing, Bronislaw Onuf, an American neurologist wrote, to tell a patient that his arm was "immovable as if nailed fast" and then have him move it.[87] Often patients relapsed: one symptom would be cleared up only to have others take its place. Janet complained that hypnosis was applied indiscriminately to organic as well as functional illnesses. Freud particularly came to dislike hypnosis, and remembered Bernheim's yelling in exasperation at a recalcitrant patient: "Vous-vous contra-suggestionez." Some physicians considered it impossible to divorce hypnotism from the unwelcome "trappings of occultism" with which some practitioners surrounded it. By 1909 Prince was asserting that hypnosis was necessary only in a small minority of obstinate cases. Suggestion while the patient was fully awake or in a hypnoid state was as effective.[88]

Although psychotherapists tried to foster independence and autonomy and sometimes prescribed doses of Marcus Aurelius or Emerson's "Self-Reliance," patients often became dependent. They consulted their therapists about the most trivial matters, as if unable to decide anything for themselves. "Weakness of will" seemed characteristic of some kinds of neurotics. In response to the "patient's longing for dominance," Putnam later remarked, the therapist became at once "too masterful or too intimate." [89]

Often, too, the emphasis on cheer glossed over real difficulties. Symptoms seemed to ebb and flow like the tides, despite every possible therapy, and, consequently, a few physicians assumed that they must be dealing with deep-seated constitutional defects. Yet, this kind of discouragement was seldom expressed by physicians deeply committed to psychotherapy. Many of these difficulties—confused theory, ineffectual treatment, patient dependence—became especially apparent after the psychoanalytic alternative had been encountered.

As already suggested, psychotherapy precipitated a role conflict among neurologists who were forced to take account of techniques for which their training and scientific ideals had not prepared them. For many, psychotherapy still was associated with spooks and Christian Science, and theoretical extravagance. Adolf Meyer, who directed the Psychiatric Institute of the New York State Hospitals, tried to resolve this issue by redefining the nature of "mind" and of psychotherapy between 1907 and 1909. Essentially psychotherapy was not a matter of manipulating words or abstract states of mind. Rather, it meant altering the patient's behavior by training him in habits that would fit him to meet reality.

Finally, Janet in all his serene confidence had announced at Harvard that the problems of hysteria were largely solved. But it was quite impossible to "express in formulas and in laws, what an insane person feels." [90] Meyer's attempt to create such formulas led him to introduce psychoanalysis to a growing number of young physicians.

VII

Functional Psychiatry
1890 -1907

Freud's contributions to American psychiatry, which changed more slowly than American psychopathology, can be assessed most effectively by surveying aspects of its development before his influence became important. By 1907 the American advance guard had been pursuing for almost two decades the simultaneous criticism and reconstruction that marked the crisis of the somatic style. In the words of William Alanson White, the young director of St. Elizabeth's Hospital, he and his colleagues were seeking a way to interpret the "queer and ununderstandable" behavior of the insane, especially those suffering from dementia praecox. They were discarding the conventional emphasis on "barren heredity," cellular pathology, and "hard and fast diagnosis." [1] They were replacing the paradigm of general paralysis for some kinds of insanity by a model of cumulative functional disorder, the result of habitual failure to adapt to the demands of the environment. Connections had begun to appear between a patient's symptoms and his pre-psychotic life and character. Yet, major problems remained unsolved, and a functional approach still was on the periphery of accepted psychiatric theory and practice.

Again, as in psychopathology, most of the original work was European, and mastery of it characterized the American advance guard. Nevertheless, it would be wrong to suggest that Americans were always merely catching up. Although the crisis of the somatic style had begun earlier in Europe, the issue of physical versus psychological causes in neurology and psychiatry still was being fought there between 1895 and 1907.[2] A few Americans, notably Adolf Meyer, were taking a more psychological position than many noted European authorities. Indeed Meyer's dynamic interpretation of dementia praecox was developed before the Swiss psychiatrist Eugen Bleuler published his revolutionary theories of schizophrenia.

American psychiatry was making closer ties with universities and medical schools, establishing the first small teaching clinics and hospitals on the European model. As yet there were no towering American authorities, as Meyer would one day become. There were no Emil Kraepelins, with personal systems to defend, entrenched in major university positions. Meyer still was on the way up, an impressive young psychiatrist, searching for new approaches. Fluidity and eclecticism in psychiatry, like the existence of competing psychotherapies, prepared the way for a militant psychoanalytic movement.

The first major achievement of the American advance guard was the careful study and therapy of cases on the border of insanity—obsessions, double personality, amnesias, mild mania, and depression. The influence of the psychologists and psychotherapists was especially evident in these studies, which led a number of young physicians to question the relevance of the somatic style and to take a new interest in medical psychology and, ultimately, in psychoanalysis.

Edward Cowles (1837–1919)

One of the earliest and most remarkable studies came from Edward Cowles, who anticipated several psychoanalytic conceptions. During the height of the somatic style he proposed a functional, quasi-psychological approach. Several influences had formed his outlook. He had been trained by the last exponent of the Moral Treatment of the 1840's, John Butler, at the Hartford Retreat. Cowles drew on Wilhelm Griesinger, the German psychiatrist, on French psychopathology, and on the American psychologists William James and Granville Stanley Hall. As a result of working with Hall at the Johns Hopkins University, Cowles made an analysis of a case of insistent and fixed ideas that Hall regarded as a model and published in the first volume of the *American Journal of Psychology*. Insanity was as much a "disorder of the mind" as a disease of the brain, Cowles argued. Influenced by James, he described the existence of a "habit-insanity" that could arise without hereditary taint or demonstrable organic lesion. As the power of attention and inhibition waned, the association of ideas became disturbed, and this in itself became the "very diagnosis of insanity." [3]

Cowles had a clear conception of internal conflict, the influence of childhood, the punishing conscience, and the importance of therapeutic conversation. The psychiatrist, he insisted, must "get in the midst of the patient's mind," listen, and "help him unravel his mysteries." For example, by discovering and talking over the origin of fixed ideas, a patient could dissipate their compulsive power. Cowles illustrated his argument by the case of a twenty-eight-year-old spinster, hospitalized after she had attempted suicide. "By dint of hours of talking," he wrote, "the earlier and later incidents of her mental history were made clear, and such was her intelligence that it was possible to trace to its origin the train of evolution of her morbid ideation; also to unify as parts of one process, the strange and apparently incomprehensible events of her life." [4]

Insistent ideas of doubt had begun in his patient with menstruation. At the time, she admired yet hated a beautiful girl friend and began to think harmful things would happen to her; soon she condemned herself "as if she were guilty of desiring them to happen." Then she began to imagine harm occurring to "substitutes" for her friend. One of these, a young man, died and so did a succession of other "substitutes." She saw herself as leading two lives, one within of wickedness, the other a "walking hypocrisy." Her "mental conflict" became constant between her remorse and her fear that harm would overtake her friend or the substitutes. She shot herself with a pistol, hoping to substitute physical for mental suffering and to atone for her "evil wishes." She had diagnosed herself as a case of "monomania" and wished to be confined in an asylum. "It was as if all the wrong things I have ever desperately allowed myself to do and think about now stand around me as creditors of my conscience," she remarked.[5]

She also had kept these troubles to herself. "It was a new experience," Cowles wrote, "to have a sympathizing physician, and it became a source of comfort to her." Cowles assured her that she would recover and that her delusions were absurd. She told Cowles that his words seemed "all real and true," but this feeling "goes when you go, and I am as helpless as before."

Cowles explained that her ideas resulted from repetition, habit, and fixed associations. There was no hereditary basis for them, although she had been neurasthenic as a girl, with unusual intelligence and an over-sensitive conscience. "Upon the plasticity of childhood strong impressions were made by unpleasant ideas about 'trance' and the horror of being 'buried alive' " before she was twelve years old. Her fixed ideas resembled phenomena seen in hypnosis and in hysteria, and were built by repetition into a complex system.

The relation of symptoms to childhood, the hints of masochism and self-punishment, far exceeded in subtlety the case histories of Weir Mitchell. In Cowles's theoretical work, he tried vainly to apply academic psychology to an overview of insanity. The result was largely an elaborate description of how the attention, emotion, and other faculties functioned abnormally, with almost none of the causal explanation of the earlier case history.[6]

Cowles's psychosomatic theory of insanity had similar results. Directing what Hall called the "richest and most elegant" mental hospital in the world, the McLean outside Boston, Cowles set up a laboratory and devoted it largely to demonstrating that stress and fatigue caused auto-intoxication of the nervous system, and, by a cumulative process, symptoms of insanity. The theory was never proven, and Adolf Meyer dismissed it as fanciful.[7] The laboratory director, a young Swiss psychiatrist, August Hoch, became increasingly absorbed in a psychological approach. He studied with Kraepelin and Carl Jung and succeeded Meyer as director of the New York Psychiatric Institute. Thus Cowles functioned, as much for his failures as his successes, as a link between the therapy of Moral Treatment and psychoanalytic psychiatry.

A few other alienists—still a small group within the profession —attempted psychological approaches to symptoms and treatment. In 1885 A. B. Richardson, superintendent of the asylum in Athens, Ohio, argued that William James's study of habits in *Popular Science Monthly* had opened up the prospect of restoring a patient to "lines of active energy habitual to him in health." By assiduous personal attention, it had been possible in "case after case" to "see the destructive tendencies greatly modified, if not wholly eradicated." [8]

Psychopathology, 1890–1903

Interest in dreams, the subconscious, and the psychological basis of symptoms grew among a few alienists. In 1893 Charles W. Page, who also had been trained by John Butler at the Hartford Retreat, described "The Adverse Consequences of Repression," by which he meant the cautionary negativism of New England morality and religion. Recalling Hawthorne's "Haunted Mind," he noted the ubiquity of patients who suffered from horrid, depressing dreams, filled with desires and cravings that could not be gratified. Others suffered anxiety because of objectionable thoughts. He described the case of a "modest, charming" daughter of cultured parents, a "beautiful example of angelic purity and Christian Womanhood," who, when insane, "gave utterance to a constant stream of coarse, profane and vulgar language." In concentrating on the merely negative, her training had reinforced the tendencies it was presumed to eradicate. Charles P. Bancroft, superintendent of the New Hampshire asylum, based an analysis of "Subconscious Homicide and Suicide" on Binet's *Alterations of Personality* and Boris Sidis's *Psychology of Suggestion.*[9]

The most important American application of French psychopathology to borderline patients was made by Boris Sidis, a pupil of William James. Sidis investigated the subconscious through dreams and treated amnesias by synthesizing split-off memories. His first work was undertaken from 1898 to 1902 at the Pathological Institute of the New York State Hospitals, on patients suffering from "functional psychoses"— hysterical paralyses, double personality, mild mania, and melancholia.

Psychological functions, such as feelings, ideas, and emotions, Sidis insisted, were as important as physical functions, and could be explored through the patient's introspective description of his symptoms.[10] Sidis inspired an interest in the psychological study of insanity in several young physicians, among them William Alanson White, who helped to found the *Psychoanalytic Review.*

Sidis's principle of therapeutic synthesis is best illustrated by White's account of a case on which the two collaborated at the Institute. It displayed an unusual sensitivity to details of family and sexual life and traced neurotic tendencies to childhood. His patient, a fourteen-year-old girl, was alternately elated and suicidally depressed; she was horrified by the color red and the sight of blood.

White attributed her symptoms to a forgotten memory dating from the age of eleven. Her mother and father died before she was two and a half, and thus she was "deprived of that most important of elements in the proper development of the child mind: the solicitous care and love of parents." Until her first experiences of menstruation and masturbation at the age of ten she had been a normal, cheerful, intelligent girl. Then her character changed radically. At the same time, a neighbor told her in goriest detail of a young man's suicide. She began to think of killing herself and later of killing her brother, who she believed disliked her. She entirely forgot the neighbor's story, which became a dissociated controlling subconscious memory. At the same time her wish to kill herself grew obsessive and led to several suicide attempts.

Through investigations in the hypnoid or half-waking state, White discovered the story of the suicide. Finally he synthesized this forgotten traumatic memory with her personal consciousness and all the memories he had investigated in the hypnoid states. As her consciousness was thus enlarged and synthesized, her suicidal compulsion receded. Thoughts of suicide recurred occasionally, but she could relate them to some immediate depressing event or to fatigue—they were no longer mysterious and obsessive.[11]

Sidis insisted that synthesis combined with mobilizing reserve energies could result in complete recoveries. His therapeutic zeal impressed White, who argued that Sidis's success with treatment made his methods and theories plausible. These elaborate constructions, largely attempts to base psychopathology on neurone theory, were dismissed by Adolf Meyer as fanciful "metaneurology." [12]

White's work with Sidis was more important for his own development than he himself later remembered. His case analyses gave him the same feeling of triumphant control, of solving mysteries that Prince had experienced. His patients' most incoherent, unreasonable acts could be "traced in each instance to an adequate cause, and thus what appeared as chaos on the surface was reduced to order." [13]

In 1903, the year after his experiments with Sidis were published, White unexpectedly was appointed head of the Government Hospital for the Insane in Washington, D.C. Theodore Roosevelt, when governor of New York, had met White and had chosen him over a number of candidates, including Dr. William Mabon, superintendent of Bellevue

Hospital. In 1902 Adolf Meyer, partly on the recommendation of Edward Cowles, became head of the New York Pathological Institute. Both men became leaders of the new direction in American psychiatry.

Sidis investigated dreams to determine what constituted the memories of the primary personality that in pathological cases became dissociated and subconscious. One of his most famous patients, the Reverend Thomas Carson Hanna, was healed through this technique. He had fallen on his head while getting out of his carriage and fainted. He suffered no serious organic injury, but awoke having regressed to infancy and forgotten his former life. He was taught to speak, distinguish between the sexes, and play the banjo. Sidis wished to determine whether his former life still existed as memories of which he was not aware. He had vivid "picture dreams" that turned out to be accurate recollections. By stimulating, partly in the hypnoid state, the revival of these memories, Sidis fostered a synthesis of the secondary and the primary personalities, a cure he believed to be unprecedented in psychopathology. The study drew praise in Europe as an ingenious application of dissociation theory.[14]

Adolf Meyer (1866–1950)

Without becoming a committed Freudian, Adolf Meyer became the most important early disseminator of psychoanalysis to young psychiatrists. He encountered it in his search for a medical psychology that would provide an understanding of dementia praecox. In the borderland cases, such as those investigated by Sidis and White, connections with the patient's past were relatively accessible. In dementia praecox they often were hidden in a bizarre symptomatology that contemporary physicians often interpreted as the meaningless product of brain disease. Meyer dismissed somatic orthodoxy and most current attempts, such as those of Cowles, to apply academic psychology to psychiatry.

To understand what Freud's theories would mean to Meyer, it is necessary to review his development to about 1907. By then he had become interested in childhood and sexuality, in a holistic approach uniting mind and body, and in a new view of scientific method. He had also formulated a tentative set of psychological hypotheses to explain the origins of dementia praecox.

Precise, exacting, and stubborn, Meyer became the dominating figure in American psychiatry between 1920 and 1940. Born in 1866, he was twenty-two years older than Putnam and about ten years older than Morton Prince when he left Zurich for America in 1892, lured by a vision of boundless opportunities and by a fierce desire to be independent. He had been trained for general practice, then had studied neurology with August Forel, head of the Burghölzli asylum and clinic and one of the foremost exponents of the hypnotic therapy of the Nancy school.

Two events between 1892 and 1895, Meyer's biographer, Eunice Winters, has written, aroused his commitment to psychiatry. Meyer's mother, who had stood security for loans to pay for his medical education and immigration, suffered a severe depression with paranoid complications in 1892. She was placed under Forel's care, and Meyer suggested to Forel that a possible basis for her delusions had been established by the harsh conduct of an uncle. Thus Meyer offered a common-sense explanation of aspects of her psychosis. Contrary to his and Forel's expectations, she recovered three years later. Throughout his life, Meyer later wrote, "the continual concern about my mother's illness . . . had a great deal to do with my capacity to see the setting and evolution of [mental disorder]." [15] He asked himself how it was possible that a person as apparently normal as his mother could have become mentally ill. Later this question assumed another form. Was it possible to predict how a given personality type would react to stress?

The second critical event was Meyer's confrontation with American institutional psychiatry at Illinois Eastern Hospital for the Insane at Kankakee. There he encountered political influence, complacency, a ratio of one doctor to 350 or more patients, and a doctrinaire materialism. The hospital staff assumed that clean surroundings, work, amusements, and the "latest in tonics" would lead to recovery—and the less said about prognosis the better. This might be a humanitarian outlook, but it was hardly *psychiatry* in the sense to which Meyer had become accustomed during his brief course with Forel. As hospital pathologist, Meyer discovered that it was impossible to make any connection between autopsy findings and mental disease because there had been little or no clinical study of the living patient. He began to study patients intensively himself and sought psychological conceptions to describe his findings. He combed the American and European literature and impressed his colleagues by his mastery of it.

Matching criticism with new hypotheses, Meyer moved slowly away from the relatively orthodox somaticism with which he began. He argued for the interaction of heredity and environment, yet cautiously diminished the role of the former and stressed childhood experience. In three essays published in 1895 for the Illinois Society for Child Study, he argued that heredity was a "perfect fatalism," a sword of Damocles held over people's heads. Yet it was only "to a limited extent as baneful a curse as many writers supposed." Much depended on surroundings. For example, "fundamental moral and intellectual characteristics" could be traced back to childhood years before physical development ended. Then habits and maxims formed the personality. "The child of abnormal parents," he wrote, "is apt to be exposed from birth to irrational ways of life, and with a weak constitution is even more prone to acquire unconsciously habits of a morbid character." A "precocious development" of sexual feelings often accompanied other neurotic symptoms. Unfortunately, "under the present system of making a mystery of everything connected with it," he wrote, it became a problem "almost beyond control." [16]

By 1895 Meyer had established the need for careful common-sense clinical observations closely linked to pathological findings. He had asked for a radical reform of the American psychiatric hospital to conform to European models and standards. He had seen the need for psychological tools for the description of clinical symptoms. He had questioned aspects of degeneracy theory and the physiologizing of classical somaticism. Finally, he had stressed the importance of childhood experience, and incidentally pointed to the close linkage of sexual precocity with neurotic traits, observations common in the literature of degeneracy.

In 1895 Meyer became pathologist and later clinical director of the Worcester Insane Hospital in Massachusetts, which he hoped to establish as a training center for psychiatrists. As an unpaid docent at Clark University, he gave courses in psychiatry students remembered as brilliant. He recruited a remarkable group of young staff physicians; several became interested in psychoanalysis—Isador Coriat, years later a founder of the Boston Psychoanalytic Institute; Albert M. Barrett, who headed the new Psychopathic Hospital at the University of Michigan in 1906; and George H. Kirby, a charter if temporary member of the New York Psychoanalytic Society.

Meyer made friends among psychologists and physicians in Boston and Cambridge who were pursuing psychopathology and psychotherapy —Morton Prince, James Jackson Putnam, and, above all, William

James, at whose house he occasionally stayed. He also became a friend of August Hoch, with whom he corresponded for years about psychiatric problems. James admired Meyer, and after the Clark Conference, where they had talked together at length, wrote to a friend praising Meyer's psychological gifts. James took his students to Worcester for Meyer's clinical demonstrations, having prepared them by doses of the British psychiatrists Henry Maudsley and Charles Mercier. Possibly James, who was impatient with much academic psychology, helped Meyer to formulate a broader view of what might be scientifically acceptable.[17]

The Worcester period from 1895 to 1902 is notable for three major developments in Meyer's ideas: a holistic biological concept of man, an adaptation of Kraepelin's clinical approach, and a skeptical attitude toward the subconscious. All three bore directly on the problem of medical psychology.

Meyer sought to break down the still rigid distinction between mind and body. In 1897, in "A Short Sketch of the Problems of Psychiatry," he wrote: "The body and its mechanical and chemical functions, and the mental life associated with it make out the biological unit, the person. . . . In this unit the development of the mind goes hand in hand with the anatomical and physiological development, not merely in a parallelism, but as a oneness with several aspects." [18] He opposed this "biological principle" to the materialism and the parallelism of the last half of the nineteenth century.

By 1900 Meyer became increasingly interested in human consciousness. Not every reaction reached awareness, he realized. Once an "attitude" had been formed, the mind was prepared for its repetition. This tendency to repeat was the essence of the unconscious, a position close to Morton Prince's but without Prince's "neurograms." At the same time Meyer took pains to deny the existence of "permanent ideas" below the threshold of consciousness such as those posited by the German psychologist Johann Friedrich Herbart. Meyer also opposed using the term "unconscious," because it mystified many physicians.[19]

Meyer's approach to psychoanalysis crystallized after 1902 when he taught at Cornell University Medical School and became director of the Pathological Institute of the New York State Hospitals, the nation's most important teaching and research center in psychiatry. He instituted training courses for physicians from the state hospitals and inspired a new group of physicians, some of whom became psychoanalysts.

Meyer's developing view of science combined with a temperamental dislike for hard and fast formulae to prevent him from accepting any single theory of mental disorder. Influenced by Charles Peirce and William James, he adopted Huxley's definition of science as "organized common sense." The facts of mental disorder were too various and complex to be reduced to a formula. Any theory must allow for multiple causes and for data from several different series of phenomena, somatic and psychological, for instance. Thus he rejected Kraepelin's use of general paralysis as a paradigm for mental disease.[20]

In a characteristic sentence he argued that unless "one has a chance to use, and with a feeling of justification, a free pluralistic method of dealing with things, dogmatic restrictions kill off many a possibility of seeing things for what they are worth." [21] Too often psychiatrists had indulged in "debauches" of systematization in lectures and textbooks and thus had falsified their material.

Meyer wrote his formal papers in a rambling Swiss-German English that often obscured his meaning and had little of the bite and clarity of his personal letters. Nevertheless, his essays are among the richest and most valuable in the contemporary medical literature.

In 1903 he took up the problem his mother's illness had made acute: was it possible to predict types of character or personality that might develop specific kinds of nervous and mental disorder? Often personality traits and defects appeared in childhood, or more frequently at puberty and adolescence. Meyer based his conception of type on recent work by the French psychologists, Pierre Janet and F. Paulhan. Of special interest was a "deterioration type" that presaged dementia praecox. Usually these people had been exemplary as children, overly meek and good; in adolescence they became withdrawn, morbidly conscientious, visionary, impractical, and lost energy and initiative.[22]

Two years later Meyer attempted to explain the dementia praecox type as the result of cumulative patterns of defective habits. Drawing on William James, he argued that habit should become the "unit of observation" and habit training the mode of treatment. Meyer's concept of habit deterioration was far from clear. He suggested that the deficient or precocious development of any given psychological function created a growing disequilibrium. The overly good child, who developed the habit of meekly avoiding struggle, later displayed disconnectedness of thought and deficient control. These common tendencies became dangerous when combined with poor judgment and especially with sexual defects. The latter usually did not involve "open immorality," but rather "conventional and frequently excessive observation of superficial morality." Thus patients attributed their mental illness to masturbation or to vague and ill-defined sexual misdeeds. One young woman suffered from "impure thoughts" and confessed to masturbation. She dreamed of a young man in the office where she worked, and finally gave up her job, fearing men would see her shame in her eyes.[23]

The concept of habit deterioration was developed further, without the intrusion of any major new influences, in "Fundamental Conceptions of Dementia Praecox," read at the annual meeting of the British Medical Association in Toronto, August 1906.[24] It was possible, Meyer argued, to classify patients according to a new system of "reaction types," typical ways of behaving in difficult or emergency situations.

One type of behavior gave the sense of a "full, wholesome and complete reaction in any emergency" or problem. Other types of reaction were less satisfactory—evasion, substitution, temporizing, or, more serious, fumbling, tantrums, and hysterical fits or suppression. Indeed, those who were liable to become victims of dementia praecox often developed ideas of reference, negativism, fault-finding, suspicion, and hallucinatory, dreamlike episodes. Above all, they were dependent rather than aggressively managing their adjustment to the world.

Meyer believed that this typological approach could aid prevention, and he drew an analogy to tuberculosis: warning signs could be found early and habits changed to prevent it. Similarly, signs of dementia praecox, although more complex, also could be recognized and altered by mental and physical hygiene.

When this paper was presented on October 2, 1906, to the New York Neurological Society, Smith Ely Jelliffe, editor of the *Journal of Nervous and Mental Disease,* asked a question. Meyer's analogy to tuberculosis was crucial. Would he present "some psychological foundations, if such had been observed, whereby the pre-dementia praecox stages might be recognized? Were there mental types which reacted disastrously to their environment, types that might be classified by any of the newer modes of investigation of mental character? Were there certain memory types, certain reaction types, certain association types, which in line with Dr. Meyer's idea of a habit psychosis might offer a clue as to the very early stages of a deteriorating process?" [25]

Jelliffe's question led to the heart of what Meyer was trying to do. In fact, Meyer could not answer Jelliffe's question except in the very general terms already discussed—mysticism, lack of coherent habits and thoughts, dreaminess, etc. Two case descriptions summarized from the "Remarks on Habit Disorganization" will indicate the kinds of systematic observation that then were lacking. At adolescence a healthy boy who avoided other boys and cruel sports became cranky and jealous of his sister, whom he threatened with a razor. He was worried about masturbation and nocturnal emissions. In the hospital he was diagnosed at first as a sexual neurasthenic. Then he began to smell a bad smell and finally drifted into a paranoid dementia. A young girl, whose father was an alcoholic and had been married three times, developed religious delusions and compulsions, and "precocious sexual instincts." She became inefficient at school and later in factory work; she experienced neurasthenic complexes with "head and intestinal symptoms, growing seclusiveness, amenorrhea, and finally peculiar tantrums."

In restrospect, there were glaring lacunae in Meyer's conceptualizations and his elaborate, detailed case histories. Although seeking "causal" chains, he was in fact describing largely the early appearance of the symptoms of dementia praecox. Although his "unsatisfactory reactions" represented attempts to deal with difficult situations, he did not examine what the latter might involve. Sexual symptoms were frequent, yet beyond noting their precocious appearance, and his patients' feelings about them, they were largely unexplored. Although a patient might say, "I am a good girl—my mother is dead—it's all my father's fault," family relationships were seldom examined on an emotional level. Meyer's deliberate rejection of system led to case histories that were full and minute, yet often disorganized and miscellaneous. What he notably lacked was a theory of development, of sexuality, and of the meaning of symptoms, which required some method of penetrating beyond the outward behavior to its significance for the patient.

The new directions Meyer represented were beginning to have cumulative effects. One symbol of these changes was the growing acceptance of the term psychiatrist. Hack Tuke's standard *Dictionary of Psychological Medicine* in 1892 had listed only "psychiater" as a "mental physician" and "psychiatrie" as German for "psychological medicine." James Mark Baldwin's *Dictionary of Philosophy and Psychology* in 1902 defined a "psychiatrist" as a synonym for "alienist." In 1906 G. Alder Blumer, superintendent of the Danvers Insane Hospital in New York State, noted a gradual *rapprochement* between "psychologists" and "psychiatrists," which the French always had emphasized and "the importance of which we in America have but begun to realize." [26]

Yet, the *rapprochement* of these two disciplines and the attempt to attract a more able group of young physicians to psychiatry provided problems. There are scraps of evidence to suggest that the prospects for some of the young men Meyer trained were not entirely rosy. In 1905 Isador Coriat found advance at the Worcester Insane Hospital closed to him.[27] Salaries still were low, the chance to do important research still slender. Finally he resigned to begin private practice in Boston and work at the Boston City Hospital. Other young physicians found the state hospital atmosphere depressing, and like Frederick Peterson or Francis X. Dercum before them, turned to private practice. A new technique of therapy performed at the office, for instance, would enhance the opportunities outside public institutions.

American Psychopathology and Psychiatry in 1907

On the eve of Freud's visit to America, what point had been reached by the advance guard in their search for new knowledge and in their reconstruction of neurology and psychiatry? How did their achievements compare with those of their European colleagues? What did Americans know about the subconscious, the role of sexuality and childhood in nervous and mental disorder, about dreams and hallucinations, the relation of normal to abnormal? What position did the new knowledge occupy within American neurology and psychiatry? What in retrospect were major unsolved problems? Finally, what did Freud contribute that was essentially new? What elements of his Clark lectures might have surprised his audience?

Except for a few studies of childhood and psychopathology, the originality, quantity, and richness of the European work exceeded the American and more closely anticipated Freud's. Nearly every element of his system had been foreshadowed in a partial way. Yet, Freud's formulation of the insight was the most consistently integrated and radical, and he often worked out connections among different fields in highly original ways.

Americans had pioneered studies of childhood, and the importance of the early years was a commonplace. Then, through imitation and habit, the elements of adult character were formed. Neurologists and psychiatrists, beginning with the degeneracy theorists, long had observed that pathological traits often appeared in the very young. European physicians, more often than American, had traced adult nervous states and sexual aberrations to early experiences.

Yet childhood was by no means all determining. Puberty and adolescence still were the critical periods of development when the unruly sexual instincts and the major outbreaks of pathology first were likely to appear. Moreover, the human character continued to develop and change into the late twenties when true biological and social maturity was reached and when professional habits finally were formed. Thus, Freud's insistence on the supreme importance of infancy and childhood was novel. Evolutionary child study had familiarized Americans with the child's aggressive, cruel impulses, and his shameless love of nakedness. Shorn of the innate conscience of the Scottish Common Sense philosopher, the child, like the savage, was instinctive and willful and needed to be tamed. No one, American or European, had stripped family life of pastoral fancy, to individualize in all children murderous hostilities.

Freud's sexual theories were anticipated in many scattered insights.[28] Europeans pushed more explicitly than Americans, a recognition of the normality of sexual emotions and activities in childhood. No American, like Wilhelm Stekel, studied "coitus in childhood" or pushed the sexual implications of nutritive sucking as far as did the Hungarian pediatrician S. Lindner. Americans were aware of some of the European work that most closely anticipated Freud's, although by no means all of it. A comparison of advanced European and American views of sexual impulses in the child with Freud's will illustrate the latter's systematic radicalism.

In 1895 Jules Dallemagne, a Belgian physician, in the closest European anticipation, observed: "It may seem daring to speak of genital impulses in childhood; however they exist. Perhaps more than we believe they form the basis of later sexual life." [29] In 1904 Stanley Hall observed that if we knew enough about genetic psychology, the "whole symphony of adult sexual feelings and acts" could be traced to infantile genital sensations. At Clark University, Freud insisted: "A child has its sexual instincts and activities from the first; it comes into the world with them; and, after an important course of development passing through many stages, they lead to what is known as the normal sexuality of the adult. There is even no difficulty in observing the manisfestations of these sexual activities in children; on the contrary, it calls for some skill to overlook them or explain them away." [30]

The Oedipus complex, that mixture of sexuality and aggression, remained, for the most part, embedded in late nineteenth-century European belles lettres, although earlier American novels had hinted at maternal incest and the American heroine of Henry James's *The Golden Bowl* displayed an attachment for her father so intense that it drove her husband to adultery. American physicians and educators warned against too much love or too much harshness, and assumed the child's character would largely be formed by his mother. They taught that the hysterical child must be separated from his parents, whose unhealthy training had helped to bring on his illness. But the sexually-toned attachment for the parent of the opposite sex and concomitant murderous rage toward the rival parent or toward siblings was not a part of American or European medical insight. Finally, despite many piecemeal anticipations, Freud's assertion stands: the chapter on sexual development was as a rule omitted from the literature on childhood, and from texts in neurology and psychiatry. Although Americans recognized that children might experience diffuse genital sensations in infancy, overt manifestations of childhood sexuality still were considered abnormal, and Stanley Hall observed that children reared normally paid little attention to their sexual parts. For the most part American views of childhood were closer to Wordsworth than to Zola's description of children's attempts at coitus or to European anticipations of Freud's oral and anal erogenous zones. Except for the direct influence of early experience in the formation of adult sexual perversions or the more general view that childhood training might combine with heredity to foster neuroses, the relation of the early years to adult symptoms was left vague. Childhood experience was given increasing weight by the psychopathologists, but was far from playing the role Freud attributed to it.

The operations of America's still powerful "civilized" morality can be detected clearly in ambivalent attitudes toward the new European studies of adult sexuality. Eminent neurologists and psychologists such as Granville Stanley Hall regarded Krafft-Ebing's *Psychopathia Sexualis* as prurient. The work of Havelock Ellis, whose influence will be surveyed in detail later, was increasingly welcomed by American physicians and psychologists, perhaps because of his essentially optimistic view of the beneficent and self-regulating nature of the sexual instinct. A few Americans had studied sexual problems frankly, but on the whole less profoundly and extensively than the Europeans. Americans had observed the existence of sadism, masochism and homosexuality, and some believed these were widespread, especially in cities. They had seen the ubiquity of sexual symptoms in nervous and mental disorders, as well as overscrupulous obsessions with the minutest observations of sexual morality. They had protested against excesses of prudery and reserve and favored the sexual education of children. No American psychologist or neurologist had gone as far as Freud in suggesting that the civilized deprivation of sexual satisfaction, like the oats withheld from the village horse, might prove fatal.

Although it was a commonplace that the abnormal was only an exaggeration of the normal, the latter still provided the standard by which the pathological was to be judged. Thus the incomplete fragmentary associations within dreams had been dismissed as meaningless because they did not conform to the standards of normal adult logic.

In Europe and America popular literary and scientific interest in dreams was intense and growing. Again, although the closest anticipations of Freud's views were European, Americans already were aware of a number of insights that paralleled his. The dreams of normal people resembled the hallucinations of the insane and the phenomena of hysteria and suggestion. Dreams often preceded the outbreak of symptoms. Dreams reflected waking life, often of the day before, were closely integrated with it, and could recapture the past with unusual vividness, especially forgotten traumatic events or early childhood memories. Reason and inhibition slept during dreams, permitting immoral behavior. Dreams could reflect undercurrents of emotional life, perhaps the opposite to those that moved the waking self. Beginning with William Hammond and especially with Boris Sidis and Morton Prince, physicians studied their patients' dreams, yet this occupied only a small place in the neurology and psychiatry of the advance guard.

Nearly all studies of dreams, as Freud observed, reflected the somatic orientation of psychiatry and psychology. Somatic stimuli were the fundamental cause of dreams, including their symbolism. No American and only one or two Europeans saw in dreams any special psychological function or any dynamic purpose within a system of mental economy. No American physician or psychologist sought to interpret them as symbolic ways of making coherent statements about the dreamer's emotional life. No one had developed a set of common psychological mechanisms, such as displacement, condensation, symbolism, and substitution, or psychological motives, such as repressed wishes, to connect dreams, neuroses, insanity, and normal life.

American work in psychotherapy and the unconscious compares favorably in depth and sophistication with the European. Except for William James, Americans created fewer original conceptions, but they did provide important extensions and applications of European discoveries. For the American advance guard, as for the European, the unconscious, no matter how defined, was an essential part of the new psychopathology. Americans knew that habits, emotions, and associations could develop outside the awareness of the personality to form complex autonomous systems. They adopted Janet's clinical conception that subconscious, dissociated memories were a major cause of nervous symptoms. Morton Prince's co-conscious states approached Freud's unconscious more closely in that they were emotive psychological complexes that functioned exactly like conscious ideas, although outside awareness. Boris Sidis came close to Freud's description of the primary function when he regarded the subconscious as a primitive, illogical, and eminently suggestible aspect of the self. The Europeans who most closely anticipated Freud's view of the unconscious were the philosophers Schopenhauer, Nietzsche, Bergson, and Herbart. Americans were familiar with all of them. Educators in the 1890's and early 1900's had revived interest in Herbart, who probably influenced Freud and who believed that concepts persisted outside awareness and were inhibited from entering it by the force of other concepts. But this view played no role in American medicine. No American or European physician combined in a single conception the unconscious as the locus of instinctual drives striving for discharge, primitive thought processes, repressed memories, and repressing forces. No physician or psychologist insisted as strongly as Freud on the range of the determining power of the unconscious.

The central role Freud attributed to repression had no real counterpart in American or European therapy. It was a commonplace that the confession of troubles was helpful. Some physicians, especially New Englanders, were aware that the "repression" or "suppression" of emotions and instincts could be harmful. Some, like Putnam, discussed the conscience as an agency representative of internalized social judgments. Yet Putnam did not envisage the conscience as an agent censoring the content of memories and dreams. For Janet, it was not an unconscious ethical force but a traumatic experience in a constitutionally weak human being that caused the narrowing of consciousness, the equivalent of Freudian repression. Repression was unique to Freud's psychopathology, although, once more, Nietzsche and Schopenhauer had closely approached his general conception.

The analysis of conflicts also was unique to Freud. Theories of conflict within the self were, of course, ancient, and James and a few others had strikingly reformulated these traditional insights. Prince had observed his patients' warring traits, but had not made awareness of them central to treatment. Free association, the analysis of transference and resistance also were Freud's contributions. So, too, was his standardization of a detailed therapy, applicable alike to selected cases of hysteria, compulsions, and fixed ideas, without the conscious admixture of suggestion, inspiration, or detailed regimes for living.

In psychiatry Adolf Meyer took a more radically psychological position than most European psychiatrists, excepting the Swiss, who by 1905 were becoming strongly influenced by Freud. Meyer's major contribution had been to minimize the role of heredity, as the psychotherapists had done, and to interpret mental illness as a set of accreting, disorganized habits, faulty modes of reacting to environmental stress. A few other physicians, applying the model of the borderline disorders, notably hysteria, had begun to elicit new connections between mental symptoms and the patient's previous life. At the Clark Conference Meyer's position on the psychogenic origin of dementia praecox was more radical than Jung's. Jung assumed that the stress of emotional complexes might set up a metabolic auto-intoxication causing the psychosis, a view Meyer dismissed as biological mythology.

The new knowledge of psychotherapy and the unconscious still played, on the whole, a peripheral if growing role in European and American neurology and psychiatry. Both were still dominated by the somatic style, and many of the new developments since 1885 were not part of textbook wisdom. A comparison of American and European views of hysteria and dementia praecox by the major authorities illuminates the role psychoanalysis would play.

Clearly Janet's was the most elaborately psychological view of hysteria and psychotherapy next to Freud's. Yet, Janet's views and therapeutic methods found little place in major German texts, even those of H. Oppenheim, professor of neurology at Berlin, who had a reputation as a psychotherapist. Oppenheim's famous letters of advice to neurotic patients mainly attempted to persuade them that their notions of a physical basis for their symptoms were false and that these were mental in origin. The German psychiatrist whom Americans regarded as the most interested in applying academic psychology to psychiatry, Theodore Ziehen, by 1904 was little influenced by Janet's psychological theories or therapies. Although he suggested that latent, that is, unconscious memory pictures were unusually active in hysteria and that symptoms often appeared in childhood, Ziehen still relied on sanitaria, elaborate daily schedules and "purposeful neglect." The Germans were more aware of Freud by 1907 than the French or Americans, but often judged his methods to be harmful. On the whole German psychotherapy as expounded in the major texts was less subtle and developed than the French.[31]

Several major French authorities, notably Joseph Babinski and Gilbert Ballet, disputed Janet's interpretation of hysteria as a narrowing of consciousness, and it played a role chiefly in the work of Emmanuel Régis, who also was the first major French psychiatrist to become interested in Freud.[32] On the whole, American authorities were more accepting of Janet than the Germans, reflecting the strong American interest in French psychopathology. Janet's theories and his method of finding and destroying the fixed ideas that caused symptoms were included in several major texts between 1898 and 1909. Yet, the work of Morton Prince and Boris Sidis was largely ignored.[33]

To most American and French neurologists, psychotherapy was the most important treatment for the neuroses, along with rest, diet, electricity, and isolation. But psychotherapy meant chiefly the general moral influence of the physician, changing the patient's attitude by "kindness, patience, firmness, interest and sympathy." Many included suggestion, often on the level of "This pill will make you well," and a few, such as Janet, still used hypnosis. But the French and American method of discovering the past traumatic causes of symptoms and changing systematically the patient's mental associations still was peripheral to neurological psychotherapy. The highly developed therapies of Janet and Prince or Sidis were only on the fringe of European and American neurology. It was not until 1909 that August Forel organized the International Society for Medical Psychology and Psychotherapy. Its members included the advance guard of European psychotherapists—yet only a few occupied major teaching positions.[34]

In most texts, European and American, heredity remained the major cause in 50 to 75 per cent of all cases of hysteria. The prognosis for a complete and permanent recovery was unfavorable, although individual symptoms might clear up. Only a few (notably Sollier, Dubois, Babinski, and Prince) claimed to cure hysteria. Most authorities, American and French, agreed that hysteria was a mental or emotional disorder, a few still speculated that it reflected a disturbance of cortical function.[35] Surprisingly, most texts continued to include the classic illustrations from the iconography of the Salpétrière depicting women in the various stages of the grand paroxysms of hystero-epilepsy.

Freud's view of hysteria, forcibly expressed at the Clark Conference, was more radical than Janet's or than that of most authorities, except perhaps Babinski, who regarded hysteria as a result of suggestion. Informed by Stanley Hall of Janet's influence in the United States, Freud flatly contradicted his view—the prevailing one—that hysterics suffered from inferior constitutions. Unlike most neurologists, Freud offered not the palliation of symptoms, but radical cure of the disease in selected cases.

In European and American psychiatry a more exclusively somatic orientation prevailed than in neurology, despite a growing interest in psychological approaches to insanity. There was rich European work on associations, alterations in memory, perception, will, and, to a growing extent, the emotions. Although the role of degeneration was increasingly questioned, Kraepelin and most authorities assumed dementia praecox to be a cortical disease, with a constitutional basis. In most American and European texts degeneration still played an important etiological role.

Adolf Meyer's functional psychiatry, still highly tentative, had made little headway between 1903 and 1907, although his influence was growing. Symptomatic of the state of American psychiatry was an important new text published in 1905 by Stewart Paton, a student of Kraepelin. Paton mentioned Janet's new discoveries and the French emphasis on tracing symptoms back to childhood. He took up the clinical analysis of associations and affects by Theodore Ziehen and Carl Wernicke, and he cited the contributions of Edward Cowles and of psychologists, such as James Mark Baldwin. Yet none of this affected Paton's views of treatment. Fascinated by the neurone theory and by a primarily physiological outlook, Paton described a regime that relied on rest, isolation, baths, diet, and drugs. He opened his text by arguing that "actual cure" of cases was less important than prevention, and that "changes in consciousness, anomalies in the emotional life, impairment of volition, are merely expressions of a disturbance in equilibrium of the functions of the brain." Meyer denounced precisely these aspects of Paton's work. As August Hoch observed, psychological factors still occupied a perfunctory place in most psychiatric texts.[36]

In retrospect there were significant omissions in the new work. Except for some of Janet's observations, there were few if any systematic attempts to describe emotional relationships among family members, parents and children or siblings. Yet children became hysterical when parents or siblings died, or when they nursed a parent through a distressing illness. Or the hatred of a brother might drive a sister to attempt suicide. Yet these data of family attachments still were largely reserved for novels. In those breaking with the importance of heredity, no new modes of assessing pathogenic family relationships had been developed. It was Freud's achievement to bring such literary and common-sense insights, which the somatic style largely had made irrelevant, within a psychological model.

There were few means of interpreting hallucinations and delusions in terms of the patient's motives or experiences, although important beginnings had been made notably by those experimenting with the borderline disorders. Sexuality was a frequent symptom in nervous and mental disorder, yet no coherent theory of sexuality informed these observations, except, as a very few were beginning to believe, that the sexual instincts and emotions and their associations developed in stages beginning in the first years of childhood and became important at puberty and adolescence. Already, a few American physicians realized, detailed hypotheses had been offered by Freud to account for some of these phenomena.

Psychoanalysis was injected into the fluid and critical first stage of the crisis of the somatic style beginning about 1905. Freud's probably was the most systematic and radical of all the rich parallel developments in neurology and psychiatry. Yet the advocates of all the new systematic psychological approaches, including Janet's, still were a minority. As they pushed their views, they encountered a growing defensiveness among the orthodox. There was little unity or organization among the psychotherapists. Prince, for example, deliberately refrained from creating a school in order not to antagonize the medical profession, and his views sometimes met with frank contempt. In 1906, for example, he spoke of curing cases of hysteria by psychotherapy. A colleague retorted, "We can cure symptoms, but the hysteric is a biological type, born that way." [37] The mind-body controversy was as lively in

Europe as in America. The first major psychoanalytic studies in psychiatry by the Swiss were regarded by Emil Kraepelin with an incredulity that increased with the years: dementia praecox probably was an organic disease, not a psychogenic product. In 1913 the French psychiatrist Emanuel Régis dismissed Freud's theories of etiology because "classical objective" studies had determined that the psychoneuroses closely resembled organic brain diseases. Thus, although important beginnings had been made, only a few were willing to redefine the role of neurologist and psychiatrist to encompass as major techniques the new systematic medical psychologies and psychotherapies.

PART 3

*The First Impact
of Psychoanalysis
1908 - 1918*

VIII

Acquaintance and Conversion
1885-1911

The status of psychoanalysis changed radically between the spring of 1908 and the spring of 1911. In 1908 it was merely one of a number of competing medical psychologies. Most neurologists and psychiatrists then viewed Freud's theories skeptically; none considered himself a "psychoanalyst" in Freud's sense. By the spring of 1911 two psychoanalytic associations had been founded; James Jackson Putnam, professor of neurology at Harvard, was president of one of them. These organizations formally marked the advent of a new sense of identity and the beginning of psychoanalysis as a professional movement in the United States.

This swift, if limited, success resulted from the commitment of a few physicians, some of them influential, who devoted unusual zeal and energy to the spread of their new beliefs. Such a degree of conviction was not an abstract process of coolly appraising a new scientific theory. Rather, it involved a deeply felt conversion to what Freud and his disciples called "the cause." [1] Those who came to consider themselves "psychoanalysts" experienced the heat of a new faith, generated by the unconventionality of Freud's theories and the contumely of the opposition.

For those who became converts and for some who did not,

psychoanalysis acted as a catalyst, forcing them to take more extreme positions than before in the reconstruction of neurology and psychiatry in which they already were engaged. There were several reasons for this. Freud's theories were radical and systematic, filling gaps in existing knowledge, providing new and fruitful explanatory models. Because Freud's views were seen as revolutionary, others with more moderate positions had to take them into account. From the beginning—well before professional organization was contemplated—opposition on moral and scientific grounds in both Europe and America reinforced the pugnacity of Freud's defenders, and this in turn heightened the resistance of opponents. On a social level, arguing for the superiority of psychoanalysis, converts mounted a campaign that reflected the intensity of their newly won convictions. Finally, the professional organization of psychoanalysis further inflamed the opposition and made the psychoanalysts more belligerent still.

Conversion usually resulted from a long and complex process. Physicians became committed to psychoanalysis only after prolonged argument and after being impressed by the strong personalities who advocated this new therapy. Only then, as a rule, were the meaning and the significance of Freud's views more clearly grasped.

Despite parallel developments, many of Freud's theories seemed to those who seriously confronted them difficult and obscure. First might come a casual interest and then a period of testing; one aspect of Freud's changing views might be accepted, another rejected. Usually even this kind of casual concern was not an individual but a social act. Most physicians found Freud's therapy difficult to apply; clarification and confirmation were essential and usually were undertaken with colleagues or friends. Often the results of this kind of knowledge were ambivalent: many found psychoanalysis stimulating but not convincing.

At this point, personal contact with an already committed and better informed physician decided the issue. Psychoanalysis would be tested, perhaps on one's own problems and dreams, as well as those of one's patients with a greater suspension of disbelief. The advantages of psychoanalysis and the failure of other methods would become apparent, perhaps for the first time. Then, with a new loyalty to the theories and to those who advocated them would come the sense of conversion. The reception of psychoanalysis in the United States illustrates this pattern. This chapter will explore first the period of casual acquaintance, then the period of personal confrontation and conversion that began in 1908, when a few men already passionately committed to psychoanalysis came to America. Abraham Arden Brill returned to New York after a year's study in Zurich. Ernest Jones moved to Toronto from London; in September of 1909 Freud visited Clark University.

Freud carried furthest the simultaneous critique of the somatic style and the construction of a systematic, psychological alternative, and this constituted his fundamental radicalism. A few examples will indicate this development. In 1893 Freud rejected the classical analogy between organic and functional disorders. Traditionally neurologists speculated that some localized "dynamic" lesion like "oedema" or "anaemia" caused hysterical paralyses. The lesion presumably was "located" in a part of the cortex controlling the paralyzed area. This analogy was absurd, Freud argued, because hysterical paralyses did not obey the laws of anatomy, but rather, as Janet suggested, a purely psychological system, i.e. popular, lay conceptions about the body. Thus hysteria never simulated the kind of paralysis that would be caused by brain damage. If a "leg" were paralyzed in hysteria, the whole leg would be inert in a way that no brain lesion could create. In place of a "lesion" Freud substituted a purely psychological "mechanism": paralyses were caused by an undischarged subconscious impression.[2]

Next, Freud rejected, between 1893 and 1896, the classical role of heredity in the etiology of hysteria, obsessions, compulsions, and anxiety. Freud attacked degeneracy theory as at once too vague and too inclusive. The hereditary diathesis included every nervous disease no matter how slight. How could one explain "dissimilar heredity," by which one nervous disease became substituted for another? Instead Freud offered the hypothesis that "specific" determining causes could be discovered. Disturbances in the sexual life played for Freud the etiological role traditionally assigned to heredity. Although after 1898 Freud gave a larger place to "constitution," he defined it in terms of the interaction of heredity and very early experience.[3]

As heredity assumed less etiological importance for Freud, psychological hypotheses at once began to play a major role. Freud's became the most self-consciously psychological of all the psychopathologies. Defence, the warding off of painful feelings, later interpreted as repression, became the cornerstone of psychoanalytic theory beginning in 1894. Morton Prince still relied on speculation about "neurograms." Janet, self-consciously conservative, still insisted that "exhaustion of the higher functions of the encephalon" was the starting point of hysteria. Freud retained somatic views chiefly in the sense that the models on which he based his psychological "mechanisms" were derived from German neurological doctrines, such as the tendency of the organism to maintain a constant level of excitation and other biological analogies.[4]

Finally, within cautious limits, Freud's therapeutic claims were the most daring. His cures did not envisage careful curtailment of the patient's activities but rather a restoration to normal capacity for work and enjoyment. This could only have become a possible therapeutic goal with the rejection of the orthodox theory, which held that neurosis recurred at the slightest "exciting" cause in those with inherited taint.

From the beginning the complexity of Freud's theories and the language barrier made an accurate assessment of them difficult. Some Americans learned about psychoanalysis only from available abstracts and articles in English. Until 1909 none of his work had been translated, and only one systematic summary of the theories of hysteria had been published in English.[5] A few read Freud in German with varying degrees of skill, by 1907 few had studied Freud with reasonable completeness.

Freud, Neurologist and Psychopathologist, 1895–1908

Freud's initial American reception was marked by irony and anachronism. His early research on "Living Nerves and Nerve Cells" in fresh-water crawfish was praised in the *Journal of Nervous and Mental Disease* in 1882. Yet his major work, *The Interpretation of Dreams,* was largely ignored when it was published in 1899.

Parallel developments in America did not lead to an appreciation of similar elements in Freud when these took a truly radical turn. For example, Putnam's interest in childhood and the emotional ramifications of sexuality did not lead at first to an understanding of Freud's sexual theories or his emphasis on early experience. Nor did Putnam's understanding of internal conflict lead him to grasp this element in Freud. Similarly Morton Prince's rejection of the overwhelming role of heredity or his investigation of the conflicts of multiple personalities did not help him to understand Freud's stand on these issues. Only during the period of conversion and after psychoanalysis became reasonably familiar were Freud's departures more clearly understood. The psychiatrist Adolf Meyer noted in 1913 that Freud's essay on paranoia published in 1896, a "perfectly lucid presentation," was ignored "and only partly grasped until . . . the mode of thinking had become more common property. . . ." [6]

Freud first gained a reputation in the United States as a promising neuro-anatomist, and this aura of solid achievement may have prepared the serious reception of his first psychopathological work. Adolf Meyer recalled that he first knew Freud as the author of a lucid study of aphasia and as an authority on the cerebral palsies of children. Freud's two reputations, as neuro-anatomist and as psychopathologist, overlapped between 1893 and 1898. From the American perspective Freud's contributions to neuro-anatomy were minor, but significant. His research on crawfish furthered a more accurate knowledge of living tissue and was "painstaking and objective," made "with as little teasing as possible in the transparent blood of the freshly killed animal." [7]

Two students of Theodore Meynert, both of whom later fought psychoanalysis bitterly, were chiefly responsible for bringing Freud's neurological work to American attention. The New York neurologist Moses Allen Starr abstracted Freud's two papers on research related to the cerebellum in the *Journal of Nervous and Mental Disease* in 1885 and 1886. Another New Yorker, Bernard Sachs, who claimed that his interest in psychology was aroused by William James at Harvard, but who clung to a somatic viewpoint, praised Freud's work on the cerebral palsies of children.[8] The aphasia monograph, as already noted, received attention because of its psychological and holistic emphasis. Freud himself in 1884 described his method of staining tissues with chloride of gold in the British neurological journal *Brain,* which was widely read in America, and the technique also was noted in the American *Journal of Nervous and Mental Disease.*[9]

Freud's reputation as a psychopathologist began in 1894 and 1895 in Boston, Chicago, and New York and developed in three phases. Until about 1904 psychologists and neurologists, for the most part, were interested in his theories of hysteria, anxiety, and obsessions and in his cathartic treatment. Beginning around 1904 Freud became better known in both Europe and America on a wave of medical enthusiasm for psychotherapy. Interest in psychoanalysis was aroused for the first time among American psychiatrists in 1905 by Carl Jung's association tests and his studies of dementia praecox in 1907. By 1908 psychoanalysis was known with greater or less accuracy to the major American psychopathologists and had influenced the work of several of them.

The development of an American psychotherapeutic movement was responsible not only for the attention paid to Freud but for what initial influence he exercised. J. Mitchell Clarke, a British neurologist, for example, published an account of Freud's views of hysteria in 1896 more accurate and systematic than anything printed in the United States.[10] Yet, Clarke was entirely uninfluenced by Freud, and in 1906 still was prescribing the classical somatic remedies for hysteria. In 1908 another British neurologist, Charles Beevor, informed the American Neurological Association that psychotherapy had not yet "invaded" England's shores. No British neurologist used it; few had "studied the French books." [11] So although the British account was more accurate, because there was no strong psychotherapeutic movement, Freud's influence in Great Britain was negligible.

Americans persisted in understanding Freud in terms of theories he formulated from 1892 to 1898, his years of "sturm und drang." Partly because his views became far more abstruse, his later work at first received less attention. His contributions were interpreted chiefly from the standpoint of the changing styles of American psychotherapy and of conceptions already familiar. At first his methods were seen as variants of French hypnotic and suggestive therapy; then, after 1904 as a mode of education. For some time Freud's theories were understood as contributions to the new school of psychopathologists led by Pierre Janet and with the Americans who elaborated the French approach. For example, Freud's repressed contents were interpreted as about identical with Janet's dissociated states.

In assessing Freud's role in American medical psychology it will be useful to separate, as the Americans usually did, Freud's therapeutic technique, his theories of sexuality, dreams, and the psychopathology of everyday life, and Jung's contributions. Because American attention was selective, some account must be given of Freud's own development to point up the American choices. Freud remained a minor figure, yet because of his sexual theories he had a reputation by word of mouth that exceeded what the relatively modest printed record suggests.[12] He was not considered sufficiently important to receive the kind of thorough, systematic attention that Charcot had been paid. Nevertheless a few Americans saw Freud as a significant psychopathologist. They judged his and Breuer's work to be "important" and "suggestive." [13] Freud's modifications of cathartic therapy, his theory of a common sexual etiology for neuroses usually believed to be different in origin, such as hysteria and the morbid fears of neurasthenia, set him apart.

Perhaps the first American notice of Breuer and Freud's theories of trauma and catharsis was published by William James in 1894.[14] His attention may have been drawn to them by his friend, F. W. H. Myers, who had written a long description of their "Preliminary Communication" on hysteria in the *Proceedings* of the British Society for Psychical Research in 1893. James's one paragraph notice of "this important paper" followed a four-page review of Janet's *L'État Mental des Hysteriques.* In the past ten years Janet had "revolutionized" views of hysteria, and had destroyed Charcot's theory of the hysterical attack as a "natural entity." Now Breuer and Freud independently had confirmed Janet's theory that hysteria resulted from a shock and a consequent splitting of consciousness. Reminiscences of the original shock, James wrote, "fell into the subliminal consciousness." Unless discovered in the hypnoid state, they acted as "permanent 'psychic traumata,' thorns in the spirit, so to speak. The cure is to draw them out in hypnotism, let them produce all their emotional effects, however violent, and *work themselves off.* They make then (apparently) a new connection with the principal consciousness, whose breach is thus restored, and the sufferer gets well." The method of treatment they had stumbled on, James wrote, had supplied them with their theory of hysteria.

It is true that Breuer and Freud in fact had announced their confirmation of Janet's theory of dissociation. But James omitted one of the most significant features of this first essay: the insistence that the symptoms of hysteria were not "protean" and random but were determined by the precipitating trauma, either directly or symbolically, just as vomiting might follow moral disgust. As Ola Andersson has emphasized, they had extended Charcot's interpretation of traumatic neuroses to cover ordinary hysteria. They had centered their investigations on the third phase of Charcot's classic attack, the "attitudes passionelles." For James, the significance of the paper had been its evidence for a subliminal consciousness and its unique therapy, not its neurological departures.

In May 1895 James's friend Putnam argued that a new therapy, common to Breuer and "Freund" (perhaps the first of many confusions of Sigmund Freud with the German neurologist, C. S. Freund) as well as to Janet, went beyond a mere improvement in the patient's hopefulness or the temporary relief of his symptoms. This technique traced back the patient's "fixed and morbid thoughts" to their original source, then "neutralized" them by hypnosis.[15] What Putnam described was not Breuer and Freud's catharsis but Janet's suggestive therapy.

About a month after Putnam's brief notice, on June 11, 1895, came the first American appreciation of the theory of repression. Significantly, repression was seen as an aspect of social manners and customs and as the creator of new symptoms of nervous disorder. In the annual Shattuck Lecture to the Massachusetts Medical Society, Robert T. Edes described "The New England Invalid" in all "her" varieties. A friend of Putnam's, Edes had been president of the American Neurological Association in 1883, directed the Adams Nervine Asylum, a private sanitarium for nervous patients near Boston, and had just published a novel about a case of amnesia during the Civil War.

Probably basing his arguments on George Beard, Edes suggested that the inhibitions and repressions of modern civilization might be causing new symptoms of nervous disease. The theories of Breuer and Freud explained how this could occur. Hysterical symptoms developed in the absence of "appropriate motor reaction" to a mental or moral shock. Normal functioning required a "proper balance between an inflow of irritations and the outflow of energy." Inhibition destroyed this balance, required as much "nervous expenditure" as action, and was "peculiarly unremitting and wearing."

Edes interpreted "repression" and "inhibition" as voluntary acts practiced by the upper classes with increasing assiduity. His attitude reveals a faint contempt for female hysterics and for the "refinements" of late nineteenth-century civilization: "It might perhaps be questioned whether the more modern habit or manners of repression, of keeping the feelings concealed, a habit which increases with civilization and fashion, with higher social position, and is especially strongly marked in our Anglo-Saxon race, has not a good deal to do with the diminished prevalance of the more outspoken and striking forms [of hysteria], and the substitution therefor of the quiet, insidious, obstinate paralyses which so closely counterfeit organic disease, and are in reality so much more serious than a good old-fashioned hysterical 'fit' which comes on slight provocation and is soon over. Would it not be better if our customs and 'good form' permitted a patient to scream, as she so often says she wants to, instead of restraining her feelings for propriety's sake, and developing a neuralgia or paralysis or an attack of 'nervous prostration'?"

The correctness of the theory seemed proven by the effectiveness of the therapy derived from it. For when the "history of the affair was completely elaborated under hypnosis and fully talked over, the hysterical symptoms disappeared." [16] Probably Edes based his remarks on the "Preliminary Communication" rather than on the *Studies in Hysteria,* which was published only about a month before his lecture.

Edes interpreted repression as a deliberate, apparently conscious act, and this remained the American view. Freud's "verdraengen" usually was translated "suppression." At first Freud himself had suggested that repression was an intentional, although not necessarily a conscious act; by 1896 he was placing great emphasis on the fact that it was unconscious and that the forces within the personality that caused repression also tended to be unconscious. The American version in this period never made the unconscious nature of repression clear. Memories were "suppressed" because they were painful, much as Janet's patient Irene "forgot" her mother's painful death. Occasionally, it was observed that such memories might conflict with "moral convictions" or "social environments." But for the most part patients simply "pushed" painful ideas "out of their heads."

The *Studies in Hysteria,* published in 1895, provided the basis for most early American impressions of Freud's technique. At the first Clark Decennial Celebration of 1899 the Swiss psychiatrist August Forel noted that Breuer and Freud's essentially hypnotic and cathartic method often worked, but that their "doctrine of arrested emotions" had been developed into a "one-sided system." [17] Freud, of course, already had given up hypnosis and had discovered the existence of "resistances" that blocked the patient's ability to remember. The patient lay relaxed, with eyes closed, on a sofa in front of him. As Freud placed his hand on the patient's forehead he insisted that the patient tell everything that occurred to him, whether it seemed irrelevant, unimportant, or disagreeable. In 1896 Freud first called this technique "psycho-analysis." Getting the patient to relax, with or without hypnosis, questioning him about his symptoms, urging him to say whatever came to mind became standard in most American descriptions of "Freud's cathartic method." Freud modified the technique in *The Interpretation of Dreams* and in essays published in 1904 and 1905. The patient no longer closed his eyes, his forehead was no longer touched, he was no longer questioned. He spontaneously brought up material; instead of pursuing the origin of each symptom systematically, this was uncovered piecemeal. Attention now was focused on resistances, and on filling in the gaps in the patient's memory. The cathartic model persisted, however, probably because it was simple and because it was easy to understand in terms of the dominant French views of trauma. For example, Lewellys F. Barker, professor of neurology at Johns Hopkins, described psychoanalysis in 1907, which he had just begun to study, as an elaborate form of "confession." As late as 1910 a Boston alienist, Isador Coriat, who soon became a psychoanalyst, argued, directly echoing William James, that talking out mental thorns made them prick the less.[18]

Putnam's first report of the experimental use of Freud's method in an English-speaking country also described it as narrowly cathartic. In 1906 Putnam tried psychoanalysis with patients of the poorer and middle classes on a ward at Massachusetts General Hospital.[19] Nearly all of them suffered from the classical symptoms of hysteria—paralyses, anaesthesias, spasms, epileptoid attacks, inexplicable bitter tastes in the mouth, etc. In addition to "psychoanalysis" they were also treated with high frequency electricity, Zander exercises, baths, "occupation of 'pleasant and artistic sorts,'" "careful explanation," and "friendly encouragement."

Recall and abreaction, almost automatically, were expected to eradicate symptoms. Cure still was affected by allowing the emotion to work itself out in some *"adequate* and *natural* expression." Repression, in Putnam's version, still was a voluntary act. The patient "studiously" turned his attention from the emotions connected with a painful sexual experience. Or, Putnam added vaguely, the affect was repressed because it was "incompatible with his prime interests and characteristics, or from some analogous cause."

Putnam's reason for undertaking the experiment is uncertain. Possibly Boris Sidis, who had moved to Brookline from New York around 1904, encouraged Putnam's interest in Freud as he had William James's. Possibly Putnam had been influenced by James himself or by Adolf Meyer or Stanley Hall. Freud's "fluent style," Putnam wrote, his "abundance of illustration," "general cultivation," and "imaginative ability" had "secured for him an attentive audience as well among professional psychologists and among neurologists of his own stamp." [20]

Possibly the new emphasis on therapeutics that did not involve suggestion or hypnosis was an important consideration. Putnam observed that Freud had applied the paradigm of hysteria to obsessions and the morbid fears of neurasthenia. His method's distinction was that it did not employ suggestion. Putnam mentioned all Freud's later major work, omitting the essays on psychotherapy, but he did not grasp its significance. According to Putnam, Freud "urged" patients to search their memories, "as a house-keeper dusts the remotest corners of her rooms, in order to bring to life anything and everything, no matter how disagreeable, how offensively sexual, which may be related to the condition which is at stake or may even come into the mind, without at first seeming to have any relationship to this condition." [21]

Putnam carefully questioned his patients, sometimes after placing them in a semi-hypnotized state. Often they recalled in great detail memories of past experiences, unhappy love affairs, or accidents that had occurred long before. Yet recall did not always clear up their symptoms. Putnam attributed some of the improvements that did occur to the adjunctive therapeutic measures. Moreover, he seriously doubted Freud's explanation of why catharsis was successful. It could not possibly "work off" and annul repressed emotions, for the "uncompleted emotion" was a fiction that did not exist. Essentially, Putnam was denying what he took to be Freud's supporting assumptions, first that an experience was conserved intact and that the origin of a given act of itself determined the act's later significance. Past experiences were not recorded in the brain like "dents in a piece of brass." Rather, the brain was as "fluid and changing as our memories," and "each succeeding mental 'structural' event" modified all past as well as all present and future adjustments. Darwin's theory that the bodily expression of emotions represented "inherited survivals of purposive acts" did not define necessarily the present meaning of a similar expression. "The fact that the carnivorous animal once curled his lip to show his teeth may account for the curling of the lip in anger or scorn, even suppressed and unconscious scorn," Putnam wrote, "but it is yet to be proved that the 'habit' of lip-curling may not outlive this meaning." [22]

"In fact, however, no mental process ever really repeats a previous one," Putnam wrote. Conscious or unconscious memories and experiences were not "possessions" like books in a library or bricks that could be picked up and used over again. "When we say that one memory 'calls up another,' under the law of association, all that we have a right to mean is that a new mental state comes into existence which contains as one of its features something which *makes us think* of a past mental state." [23] Therefore it was not true that the old, unfinished experience worked itself out and thus led to a cure.

Putnam especially objected to the continuing effect of earlier sexual traumas: "If any one goes beyond the range of hospital cases of hysteria, and interrogates private patients of cultivation and intelligence, accustomed to rational self-analysis, he will find instances enough of persons of fine character, who, in their youth, have had emotional experiences of sexual character, serious enough to have made a deep impression on their thoughts for many years. Yet these experiences may have been not alone outgrown but utilized for the broadening and strengthening of the character and the will." [24]

It was objectionable to force patients to dwell on "repugnant details" when these might not have the etiological significance for the present state that Freud attributed to them. It was preferable to use "side-tracking" or even suggestion, and to discuss *the principles* rather than the details involved in "morbid" associations. Freud's method was very difficult "and implied a degree of skill which few physicians can attain, if not the possession of personal characteristics of an unusual sort."

Edes, revising his earlier opinion, also noted that "ignorance has its advantages as well as its dangers." For "tedious questioning" might even vivify ideas the physician wanted to rid the patient of. Besides, few physicians could listen "sympathetically to a daily rehearsal of imaginary, self-developed woes, or the maudlin details of a self-accuser." [25]

At the New Haven Congress in 1909 Morton Prince argued that Freud was wrong in asserting that the patient's awareness of the origins of his symptoms accomplished the cure. If patients kept ideas out of consciousness because they were painful, making them conscious surely would not make them any less unpleasant. Something else, then, had to occur. Without being entirely lucid, Prince seemed to be arguing that the patient's attitude toward the repressed ideas changed: he learned to tolerate them. Then, the conflict and dissociation ended, and the ideas could be reintegrated into the conscious personality. But this could be achieved by suggestion or hypnosis without the patient's having to become aware of the ideas at all. Prince insisted that the patient chiefly learned a "new point of view." Psychoanalysis, then, was "nothing more" than a special form of education and had precisely the same "therapeutic value." [26]

The term "psychological analysis" came from Janet. Prince, however, used psychoanalysis to describe his own "painstaking and exhaustive" diagnostic and therapeutic methods. By this he meant tracing the origin of a symptom or a fixed idea, possibly informing the patient about this, suggesting new alternatives, and reintegrating the split-off elements of the personality.[27]

A therapeutic gloss on Freud's analytic method that probably reflected the new emphasis on Dubois's "persuasion" was described in 1907 by E. W. Taylor, a neurologist who worked with Putnam at the Massachusetts General Hospital and taught neurology at Harvard Medical School. After uncovering the "suppressed mental state," Taylor would inform the patient of it and at the same time attempt to modify his attitude. A girl who had suppressed information about painful scenes with her insane brother was taught that she need not feel ashamed about having a case of insanity in the family.[28] This emphasis on "informing" the patient about the source of his troubles was carried over by those who considered it a "psychoanalysis" to find the patient's "complexes," then to point these out to him.

In 1908 two reasonably accurate descriptions of Freud's later technique, one hostile and another friendly, were published. The latter was written by Bronislaw Onuf, a New York neurologist, born in Russia and trained in Zurich. The former, by Francis Xavier Dercum, a conservative Philadelphia neurologist, was based with considerable care on Freud's own account of his method in Loewenfeld's text on compulsion neuroses published in 1904.[29]

Thus by 1908 Freud's cathartic method was well known. The version of 1895 was included in a description of psychotherapy by a member of the graduating class of 1908 at the University of Pennsylvania Medical School, and it was the subject of a paper delivered at a forum on psychotherapeutics at the New York Neurological Association. The next year psychoanalysis was being discussed by general physicians in San Jose, California, in Alabama, and in Illinois.[30] It was an elaborate, radical, and experimental method. The best known and most respected authorities, such as Putnam or Barker, found it troublesome and usually unnecessary. Others, like Dercum, judged it an abomination. Its more widespread use undoubtedly had been prevented not only by inherent complexity but because of the notorious "etiological factor." Putnam acknowledged in 1906 that Freud pressed his claims for the role of sexuality with increasing insistence, with a "keenness and force" and a "richness of detail that demanded attention. . . ." On this momentous sexual issue Putnam had written that he was reserving judgment.[31]

The early reception of Freud's sexual theories was complicated by several factors—the complex evolution of the theories themselves, the anachronistic American view of them, and the operations of "civilized" morality. The confusion caused by Freud's own development is nowhere clearer than in the reception of these hypotheses, which to Sidney I. Schwab, the first American to study with Freud in the 1890's, remained among the most perplexing in neurological literature.[32] This sympathetic estimate is worth taking at face value. Some of the most widespread and confused interpretations of Freud derived logically from the way his theories in fact did develop. Like his therapeutic methods, Freud's sexual theories changed drastically between 1897 and 1905, yet were understood in terms of the first formulations.

The first objections to Freud's sexual hypotheses were not based on moral grounds, probably because his views in many respects were quite conventional. Rather, they were made on scientific grounds. The editor of *Alienist and Neurologist,* C. M. Hughes, condemned Freud's theories in 1898 as "wildly conjectural" and "unprovable." They were an example of the "absurd lengths to which medical men will go in their conclusion, either when seeking medical notoriety or when they take leave of their season [sic]." [33] To a considerable extent, the medical version of "civilized" morality—chiefly the absence of sexual desire in women and the horror of sexual perversions—made the descriptions of American physicians euphemistic and cautious. Sometimes key positions in Freud's hypotheses were omitted; once more, the later theories were largely ignored.

To some physicians Freud's views represented regression to an earlier period when masturbation or sexual excesses were thought to cause neurasthenia or insanity directly. The dangers of masturbation and of premature awakening of the sexual instincts were widely recognized.

Reticence may have led the New Englanders Edes, Taylor, and Putnam to ignore at first the sexual aspect of Freud's theories. Reticence is also apparent in Onuf's brief abstract of Freud's theory of the neuroses of defense published in the *Journal of Nervous and Mental Disease* in 1895. A "painful or disagreeable impression in the sexual sphere" caused the neuroses Freud had grouped together in this classification— "acquired", i.e. non-congenital hysterias, phobias, compulsive ideas, some kinds of hallucinations. Onuf cited several of Freud's cases, a girl who accused herself of crimes because she was ashamed of masturbation, a girl who developed the hallucination that a man she loved returned her affection, which he did not. But he omitted other cases in which women expressed sexual desires more directly. Onuf also treated less than adequately Freud's insistence that these sexual feelings or experiences were intolerable morally, and therefore productive of conflict.[34] Freud's theory of anxiety neuroses also was described periphrastically by Hugh T. Patrick, a Chicago neurologist, in the same journal in 1895. "Some abuse or deep impression in connection with the sexual function" caused the anxiety that Freud wished to detach as a separate syndrome from neurasthenia.[35] The anxiety neurosis was characterized by irritability, anxious expectation, and hyper-conscientiousness. Patrick listed the etiological situations Freud described—sudden presentation of the sexual problem to maturing young girls; continence in young men; frigidity in young wives; lack of gratification for husbands; coitus interruptus; masturbation; impotence. These were complaints neurologists saw often enough. Yet Patrick omitted Freud's claim that essentially anxiety resulted from the inadequate psychical discharge of accumulated sexual excitation.

Freud's first hypothesis about the role of childhood sexual traumas in the etiology of the neuroses was briefly noted by the *Journal of the American Medical Association* in 1896. According to Freud's "grave assertion," "conscious or unconscious memories" of sexual occurrences instigated by adults in earliest childhood caused hysteria, paranoia, and compulsory ideas. Yet, the *Journal* omitted Freud's insistence that the child was capable of "delicate sexual sensations" and that the "occurrences" often involved perverse acts.[36]

Havelock Ellis devoted a long, perceptive article to Freud's views of "The Sexual Emotions and Hysteria" in the St. Louis journal *Alienist and Neurologist* in 1898. Freud placed the sexual factor in proper perspective once more, Ellis argued, after the recent denial by Charcot and most other modern authorities that it played any etiological role at all in hysteria. Unlike Charcot's patients, Freud's belonged to the educated classes, and for this reason he had been able to undertake delicate "psychic investigations." These disclosed that his patients suffered from conflicts between their ideas of right and the "bent of their inclinations." [37]

In Freud's first formulations published through 1898 the traumatic sexual experiences usually were either masturbation or "coitus-like" acts initiated in childhood by adults. In 1897 these views began to change drastically. He concluded that his patients' memories of sexual experiences in childhood were fantasies, the expression of wishes. The destruction of his earlier hypothesis of actual traumatic outrages, which he had announced in 1898 as the "caput Nili" of neuropathology, left him with a feeling of triumph. He had demonstrated, at least to himself, that his hypotheses came not from mere speculation but from "honest and effective intellectual labor." Subjective psychological feelings became as important as "actual reality."

He moved sharply away from the earlier description of "coitus-like" sexual activity in childhood. The earlier perverse acts became united with the child's capacity for "delicate sensations" to become his polymorphous sexuality. The agents of seduction became the parents, who through caresses and affection first awakened the child's sexually toned gratifications and became the objects of his ambivalent feelings. The theory of defense was applied now chiefly to the memory not of actual "coitus-like" behavior but, above all, to wishes and fantasies and to auto-erotic activity.

Some of these radical departures, notably the male child's murderous hatred of his father and desire for his mother, enacted in the legend of Oedipus, were first published in *The Interpretation of Dreams.* The development of sexual maturity out of the component instincts and their relation to adult sexual perversions were described in the *Three Essays on Sexuality* in 1905.

The impression that remained, however, seems to have been the initial one of sexual acts performed by adults, masturbation, or outrages against children. The Oedipus complex was ignored entirely. In his 1906 evaluation of psychoanalysis Putnam took from the *Three Essays on Sexuality* neither the Oedipus legend nor the sexual development of the child but the etiological role of sexual inversion. Although the patient did not know it, Putnam wrote, "the ordinary symptomatology of the neuropsychoses contains elements that hark back to the subconscious feelings which represent, in embryo, the grosser manifestations of the most abandoned sexual perverts." [38] This summary almost reversed Freud's developmental theory, for Putnam's focus was on adult acts, not on the universal childhood components from which they emerged. In 1908 Onuf argued that deprivation of sexual intercourse was the "nucleus" causing a young wife's psychosis.

Adolf Meyer's understanding was clearer in what probably was the only notice of the *Three Essays* published in the United States. He argued in 1906 that Freud "opened the eyes of the physician to an extension of human biology which differs very favorably from the sensational curiosity shop of the literature on perversions" and was "especially illuminating on account of the pedagogically important study of the infantile period." Such material, he wrote, was "of the kind concerning which the great social compact of ethics has created an anomalous sensitiveness." "Freud's work is an absolutely essential, though less documentary, supplement to such presentations as Stanley Hall's in his work on *Adolescence,* and is to the psychopathologists as important as the study of dietetics to the general physician." [39]

The indignation of a conservative defender of "civilized" morality showed clearly in Dercum's essay and his remarks to the American Neurological Association in 1908. It must be remembered that to most physicians, and certainly to Dercum, the neurotic was primarily a "she." To attribute to delicate and refined women who became neurotic overt sexual experiences in which they participated "actively" in childhood was undoubtedly shocking. Dercum's appraisal was based almost entirely on Loewenfeld's criticism of Freud's work up to 1899. Like Putnam, Dercum failed to understand the theory of infantile sexuality, for he argued that the "sexual immaturity of children" made Freud's theory of etiology biologically improbable. He was clearly working from a view of sexuality that involved adult acts and not "fantasies" or wishes.[40]

Not until the period when certain prominent Americans were involved with Freud's ideas did Freud's later sexual theories receive a full and frank discussion in print. In 1908 Brill, just back from Europe, contributed a careful abstract of Freud's essay on the relation of bisexual fantasies to the neuroses. In 1910, as a result of the Clark Conference, Adolf Meyer finally published a summary of Freud's views of sexuality. They were important, Meyer insisted, because, as he and others had observed, sexuality played an important role in neuroses as well as in the symptoms of insanity and the habit conflicts that seemed to lead to them.[41]

Freud's method of interpreting dreams and his theories of determinism in the slips and errors of everyday life charmed when they did not convince. Such determinism was too sweeping, Boris Sidis wrote to William James in a letter recommending *Die Traumdeutung* and *Zur Psychopathologie des Altagelebens* in 1905. In a review of the latter for Prince's journal Sidis concluded that "although perhaps not quite convincing this little volume is interesting and timely." Frederick Peterson, professor of psychiatry at Columbia University, wrote that anyone who mastered the book became a "sort of Sherlock Holmes," and spent "much of his time in detecting and unraveling the purposes of the sub-conscious in his friends." [42]

The Role of Carl Jung, 1905–1909

In 1904, without indicating acquaintance with Freud's psychological work, including his early essay on paranoia, Adolf Meyer wrote that he hoped for a medical psychology that would illuminate the patient's "mental attitudes and habits." In 1905 and 1906 he welcomed Carl Jung's new association tests as offering a way to study these factors. The tests made it clear that psychiatrists no longer could dismiss the content of symptoms as merely anomalous or absurd. Kraepelin's pessimistic aversion to psychological analysis was now groundless, for the tests promised to go beyond description to "cause." They showed that associations were not merely the result of chance, but were determined. Physicians could draw inferences from the tests about a patient's adjustments, prevailing motives, and the causes of his "stream of mentation." The studies were the year's best contribution to psychopathology and pointed to a possibly important role for the theories of Breuer and Freud.[43] Meyer now began to use the term "habit conflicts" in place of "disharmonies," and argued that it had become possible to "demonstrate chains of mental happenings which fulfill all the conditions of an experiment."

In March 1907 Meyer introduced a course in the "most recent developments of psychopathology" at the New York Psychopathological Institute, renamed the Psychiatric Institute in 1908. "There is a growing realization of the importance of the knowledge of the fundamental factors of mental life, in harmony with the present growth of a dynamic psychology," he wrote. He suggested that the subconscious, until then a "semi-popular" subject scorned by "cut and dried scholastic psychology," had become of scientific interest and was being stripped of mystery. That year he reviewed Boris Sidis's recent experiments in psychopathology at length. They were useful, but the work of Janet, Freud, and Jung was less "artificial," "more life like," and closer to the facts of observation. Above all, Meyer noted, Sidis did not explain *why* dissociation took place.[44]

Jung believed his association tests provided a "good means of fathoming and of analyzing the personality." [45] A slow response or no response to a set of stimulus words indicated that the word touched something that was repressed, an emotional complex. For example, a woman hesitated over the test words "water, ship, lake, swim." It turned out that she had thought of drowning herself shortly before, and her "complex" had betrayed itself by pauses over these words.

The tests quickly were taken up by a few psychologists and psychotherapists. Hugo Münsterberg mentioned their use in crime detection, and his student, Robert M. Yerkes, a pioneer of psychological testing, worked on them during a Harvard summer school course in 1908. Some physicians, including Ernest Jones, used them as a preliminary way of finding out what was troubling nervous patients. William Alanson White argued that complex indicators would provide reliable guides to the patient's emotional life, and would "open the flood gates so that all there is left for us to do is listen." [46]

What, then, was a complex? Jung had taken the term from Theodor Ziehen, and assumed, as did his teacher, Eugen Bleuler, that emotion or "affect" determined every thought and every action. Jung made a second assumption: sensations, ideas, and the affects connected with these were grouped in consciousness as entities—"molecules," Jung called them. Every memory, with its accompanying emotion, made up such a unit, or "affective complex."

Jung realized that his views might overstep the "limits of Freud's." By this he probably meant that his definition of a complex went beyond Freud's description of repressed contents. Jung sometimes treated the complex as unconscious and as equivalent to one of Janet's "dissociated states." Indeed, William Alanson White saw the former as German terminology for the latter. At other times, Jung gave examples of complexes that were conscious emotions. In fact, he used "complex," "impression," and "emotion" almost interchangeably. Being in love was a "possession-complex"; there could be complexes of unsatisfied sexual desire or of hopeless love. Education was nothing more than the inculcation of "complexes." [47]

Jung's vague definition of "complex" readily lent itself to assimilation by medical psychologists as an elastic convenience. Prince used the word to mean any organized set of associations that tended to become autonomous. By 1909 he asserted that to discover "subconscious complexes" was one of the goals of his preliminary psychoanalysis. Later he explained that he did *not* use the word in Jung's sense, because he did not mean something of which the subject need be unconscious. But then, neither, invariably, did Jung.[48]

Jung aroused interest in *Die Traumdeutung*. The initial lack of attention to it is odd, because for almost two decades a few American psychologists, medical and lay, had been studying dreams and their possible relation to nervous and mental disorder. Psychiatrists long had noted the resemblance of insanity to dream states, and Janet had observed that hysterical paralyses sometimes followed vivid dreams. Freud himself cited some of the American work. He used a statistical study by Mary Whiton Calkins, professor of psychology at Wellesley and a pupil of William James, to support his argument that dreams could not possibly be explained in all their richness of detail solely on the basis of somatic stimuli. By 1906 Putnam and E. W. Taylor were aware of the outlines of Freud's dream theory.[49]

Jung had read *Die Traumdeutung* in 1900, failed to understand it, and read it again in 1903.[50] This time he became interested in the mechanism of repression in dreams, for he already had noticed a similar phenomenon in experiments with word associations. He also was impressed by Freud's analysis of a case of paranoia published in 1896. Jung began to apply Freud's principles to the psychoses. Jung's *Psychology of Dementia Praecox,* published in 1907, attempted an ordered explanation of the contents of delusions and hallucinations. Jung's translators, A. A. Brill and Frederick Peterson, described the basic problem: "We recall a patient whose auditory hallucinations were attributed to a child, and another who heard the voice of God. The mannerisms of one were characterized by a continuous rubbing on the top of his head, while another for hours described certain figures in the air. Are these diversities accidental or have they a reason? "[51]

Jung's solution was elucidated by August Hoch. Son of the director of the Basel University Hospital, Hoch had emigrated to America in 1887 and been trained by William Osler at the Johns Hopkins. He worked as a pathologist at the McLean Hospital under Edward Cowles and met Adolf Meyer in 1895. Hoch moved to the Bloomingdale Hospital in 1902, and in 1909, after studying with Jung in Zurich, at Meyer's suggestion succeeded the latter at Ward's Island. He had a keen and sensitive appreciation of the elements of normal life in insane patients.

In borderline types of insanity, Hoch argued, especially the "anxious melancholia" of older people, the "exciting cause" usually was quite apparent—a death, a loss of fortune, an event that would have made anyone seriously depressed. In dementia praecox, however, such causal links were usually obscure or seemed, on the surface, non-existent.[52]

Freud's dream theories provided two important clues. First, Freud provided examples of the speech and the dreams of normal people that resembled closely those of patients suffering from dementia praecox. Second, symbols played a crucial part in dream construction. Like the symptoms in hysteria, the symbols in dreams represented underlying conflicts. Just as hysterical symptoms were constructed by condensation, displacement, substitutions, logical transformation of complexes, Jung argued, so too, were the delusions, hallucinations, the bizarre behavior of dementia praecox.

Jung applied symbolic interpretation and the association tests to dementia praecox patients. Their apparent apathy and lack of feeling was illusory; they showed an abundance of delayed responses to stimulus words. When associations to these words touched on complexes, bursts of activity were set off. In dementia praecox the complexes acted like a "cancer" and overwhelmed the conscious ego or sense of self. Internal conflicts, longings, and wish fulfillments determined the nature of these complexes.

Hoch then integrated these discoveries with what Adolf Meyer had been emphasizing independently. The symbols and "complexes" represented what Meyer might have called "inadequate reactions" or adaptations to the difficulties of life. Hoch provided an example. A woman became infatuated with a man not her husband. She blamed herself bitterly and finally developed dementia praecox. She talked constantly about the "two in one," a mystical fusion of her husband and the man, and insisted that her mission was to "show the world how to lead a pure life." Hoch argued that all the defective adjustments that showed up in the insanities had been present to a much smaller degree in the patient's ordinary life. Thus it was possible to conclude that the insane really were in many ways like normal people, and that insanity represented an exaggeration of reactions present in normal living. Hoch elaborated a suggestion of Karl Abraham's and Adolf Meyer's that a "shut-in" personality, withdrawn, given to fantasy, unadapted to the real world, was most likely to develop dementia praecox. He also argued that sexual abnormalities and conflicts were characteristic of it.[53]

The *Psychology of Dementia Praecox* became widely influential. In it Jung, foreshadowing his break with Freud, expressed his disagreement with Freud's "overemphasis" on sexuality.[54] Jung also concluded that hysteria and dementia praecox, despite great similarities, must have different basic causes: hysterics recovered, and patients with dementia praecox did not. Dementia praecox, therefore, might be the result of some unknown physiological process, possibly a toxin caused by the complexes.

To Adolf Meyer the search for a "disease entity" behind insanity was the fetishistic search for a foolish *Ding an sich*. He thought Jung argued that the complex not only determined the nature of the psychological symptoms but also exacerbated acute attacks of insanity and in other ways actually played an etiological role in the disease process itself. Around 1907 he introduced psychoanalysis, and especially Jung's association tests, theories of complexes and of dementia praecox to his staff at Ward's Island. Medical residents there soon were joking that they had better wear rubber-soled shoes so as to be able to sneak up and catch a patient's hidden complexes. At the Clark Conference he rejected Jung's toxic explanation, which seemed to him a regression to the orthodox, physiological viewpoint he was trying to modify. Habit conflicts and life experience sufficed to cause the disorder. He urged psychiatrists to think in terms of a "dynamic psychology" of actions and reactions, "irrepressible instincts and habits," a "metabolism of conduct," a "psychobiology" as he called his own theories. Throughout his own long teaching life he acquainted scores of Americans with his detached and changing attitudes toward psychoanalysis, which had a strong admixture of ambivalence. Until about 1913 he believed that he and the psychoanalysts were exploring the common ground of a functional interpretation of nervous and mental disorder. Years later he wrote that Freud's great contribution was to have seen these disorders in truly "human and vital terms." Yet at the Clark Conference he insisted that the acceptance of a functional approach by physicians would be prejudiced not only by somaticism and routine, but by an aversion to any extreme dogma, such as sexual etiology or complexes.[55] At Ward's Island and after he became professor of psychiatry at the Johns Hopkins University, he introduced a number of physicians to psychoanalysis, and some took it up, if only for a time— among them Brill, John T. MacCurdy, Morris J. Karpas, Trigant Burrow, C. P. Oberndorf, Charles Macfie Campbell, George H. Kirby, Laurence Kubie, and others. It was chiefly through Freud's theories, A. A. Brill insisted, that a functional approach replaced the "soulless" description and cellular pathology of the old psychiatry. Psychoanalysis had made possible a fuller understanding of the individual case.[56]

Until 1908 Freud's reputation was modest. Two major articles by Ellis and Putnam, a few abstracts and brief reviews, a dozen or two references, and the discussions of Jung's work sum up the attention psychoanalysis had received. The impression of psychoanalysis constructed in this early period remained the version of less well informed physicians for many years and supplied the basis for the first popularizations. Psychoanalysis meant catharsis, the discovery of suppressed complexes, and a not very explicit view of the importance of sexuality in the neuroses.

The Clark Conference also tended to reinforce some of these first impressions of psychoanalysis. Freud himself devoted most of his first and part of his second lecture to Breuer's discovery of the cathartic method. He commended Jung's association tests as useful for getting a preliminary impression of a patient's complexes and for studying the psychoses.

Although Freud's most important specific theories, such as the nature of childhood sexuality, were neither fully understood nor accepted, there is evidence that Freud strengthened certain tendencies already in operation. Indeed, Freud helped give shape to some of them. Morton Prince, as already noted, probably was influenced by Freud to make "psychoanalysis" the preferred treatment in certain kinds of neuroses. To Putnam, Freud was one among several important contributors to the new psychopathological tradition. In 1904 he had argued that the "minute and searching analytic case histories" of Breuer and Freud, Janet, Prince, Sidis, Sturgis, and Osgood were rationalizing mental therapeutics. Their methods of "mental analysis" were useful for both diagnosis and treatment. Even more significantly, Putnam argued that these "psychologists" had demonstrated that a man's acts were determined by his past history and his habits and that no one was as free or as integrated as he might wish. Disorganization of the personality had roots in childhood.[57] In this eclectic assessment, Freud probably contributed to the emphasis on childhood and mental determinism, while the elements of "disorganization" and habit came from Janet and Prince respectively.

An example either of parallel development or of borrowing without acknowledgement can be seen in a series of articles by Boris Sidis in the *Boston Medical and Surgical Journal* in 1907. Early childhood experience, the role of overt sexuality, and a cathartic treatment in which symptoms were cured simultaneously with the analysis of their origin characterized these essays. Sidis at once denied charges of plagiarism, which apparently already had been levelled against him. He conspicuously disclaimed any similarity in his own work to that of Janet or Freud. Employing the hypnoid state, the original experience that caused the symptoms could be uncovered, sometimes an experience of early childhood.[58] A clothing fetish and an ideal of feminine beauty were traced back to a nurse a patient had had up to his fifth year who had fondled and embraced him. A homosexual idea was traced to the age of eight when the patient barely escaped abuse by older boys at school. None of Freud's theories of infantile sexuality were employed. Nevertheless, the importance of traumatic sexual experiences in early childhood was emphasized far more frankly than in any other case history by an American psychopathologist. Moreover, Sidis defended the continuity of present mental states with those of the past against what he took to be Putnam's "metaphysical" objections in his article on Freud in 1906.

Psychoanalysis already had aroused antagonism, and in 1908 was denounced during the neurologists' debates over psychotherapy. Dercum condemned the length of psychoanalytic treatment, its psychological emphasis, and its sexual elements. To carry out that "newest of all fads," he told the American Neurological Association, the patient was placed upon a "couch on her back" and induced to talk from one to three hours at a time about her case. Patients were expected to recall "memories of sexual aggression and sexual assault" in which they had been "active participants in childhood." "No one need dwell upon the wholesomeness of endeavoring to elicit such histories from patients, no one need dwell upon the harm that may be done, nor upon the unpleasant and unethical attitude which the physician must assume to elicit such histories."[59]

No one present seriously defended psychoanalysis. Putnam announced that he disbelieved in the Freud method, although it had been a "distinct contribution," and he insisted that neurasthenia was a "bodily disease" accompanied by distinct "stigmata such as a slender build." Adolf Meyer argued that psychotherapy did not mean "psychoanalysis" or some "special" formula, but rather training in proper "mental activity, conduct, and attitudes" and in "the meeting of difficulties."

Conversion, 1908–1911

Meyer's influence illustrates the role of personal contact in creating a clearer understanding and a more sympathetic attitude toward psychoanalysis. To some physicians the combination of determinism and sexuality made Freud's theories seem fantastic, possibly warped by a personal obsession. For this reason Freud's integrity and that of his immediate disciples was of crucial importance; so, too, was their ability to expound and defend their views.

Perhaps the first American to become convinced of Freud's integrity by personal contact was Sidney I. Schwab, a young neurologist from St. Louis. In 1896, just out of Harvard Medical School, he studied for a year and a half in London, Berlin, and Paris, where he knew Pierre Janet. In Vienna, he joined a small group, then known, he recalled, as "die Anhaenger von Freud." [60] In 1906 and 1907 he argued that only Freud had begun to construct a "systematized" technique of psychotherapy. During the wars over psychoanalysis he defended Freud's honesty and scientific acumen. Schwab himself did not become a psychoanalyst, and he argued that Freud's theories were often insufficiently supported by facts.

Nevertheless, as Bernard Hart observed, Freud's ideas "astonished by the intensity of their illumination" and "inevitably aroused an answering thrill of conviction." [61] Conversion to the psychoanalytic cause was like an intensification of this "answering thrill of conviction." It usually resulted from personal confrontation with a deeply committed advocate as well as the satisfaction of personal needs, seldom fully known.

Jung's conversion and role as a proselyte were archetypal, and became important for the American movement because of his many students. Opposition to Freud first provoked Jung's own declaration of support. After a moment of what he recalled as devilish hesitation, he decided to publish the experiments that corroborated some of Freud's conclusions. In 1907 he met Freud in Vienna, talked with him for thirteen hours, and developed a "father transference." Jung also remembered that he found Freud enigmatic and puzzling from the first, and suspected a personal bias behind his sexual theories. From contemporary accounts, however, Jung became an enthusiastic Freudian, and his ardor in turn infected others.

In some physicans it seems likely that a commitment to psychoanalysis filled important personal needs. For example, its iconoclastic as well as therapeutic possibilities may have appealed to Ernest Jones, whose fiery, autocratic temperament kept him from a post at London's best hospitals despite his obvious brilliance. Psychoanalysis fulfilled for Trigant Burrow a long painful search for a life's work. In 1909, then thirty-four, he wrote to his mother from Zurich, where he had gone to study with Jung:

You must have felt these many years of your loving hope for me, that there has been a talent, an interest, an aptitude—what you will —a tendency to penetrate into certain types of character which however wide of the strictly medical concern, with its knife and stethescope, has not been without signs of significance for the handling of suffering humanity. You must have felt how apart this strange, keen interest has been from conventional medical standards. You must have felt my own pain and disappointment and embarrassment when at every turn, no matter what heart and enthusiasm I brought to each new direction of endeavor, I was ever confronted with the same old uninspiring, unimaginative mechanical physical tools and physical problems. . . . For to my unspeakable misery I had all along *not* found my work. But a new day has dawned and I *have* found my work. . . . So you may judge the comfort and encouragement I hug to my heart at this revelation! . . . Judge of my happiness.[62]

To A. A. Brill, by 1909, Freud had become a hero, a man who like himself had labored on "despite all discouragement." Freud may have substituted for the father Brill had left Europe partly to avoid, and his sense of Freud's greatness may have helped to reconcile him to his own Jewishness.

Brill had emigrated to New York from Austria-Hungary at the age of fifteen, had been defrauded of his few dollars on ship, and had landed penniless and alone. He had lived on the Lower East Side and never lost its accent. He was ambitious, energetic, eager to learn, to read the classics, and to become a doctor. He supported himself by sweeping out bars, giving mandolin lessons, and teaching. After a long and bitter struggle he worked his way through New York University and in 1904, at the age of twenty-nine, received his medical degree from the Columbia University College of Physicians and Surgeons, where he met Frederick Peterson, professor of neurology, who had been president of the New York State Commission in Lunacy. Like two other poor boys, Smith Ely Jelliffe and William Alanson White, Brill got a job in the New York state mental hospital system and came under the influence of Adolf Meyer and August Hoch. He worked as a pathologist for two years, then changed to clinical psychiatry. Meyer taught him to write detailed case histories, and presented abstracts of the German psychiatrists Emil Kraepelin, Theodor Ziehen, and Carl Wernecke at Ward's Island. Like the other young men, Brill was eager to follow Meyer, while the older men seemed either lazy or upset by him. Brill organized the pathological laboratory at the Central Islip State Hospital and performed autopsies. At the same time he began working at the Vanderbilt outpatient clinic of Columbia University. He treated neurotic patients with tonics, sedatives, hypnosis, and Dubois's persuasion.

In 1907 he scraped together enough money for a year in Europe, postponing marriage to do so. He went first to Paris and found French neurology and psychiatry old-fashioned by comparison with Meyer's. Babinski's famous isolation treatment that had so much impressed Lewellys Barker seemed another version of Weir Mitchell's rest cure, Brill recalled many years later. Then Frederick Peterson came to Paris. "Why don't you go to Zurich—to Bleuler and Jung," Peterson urged. "They are doing that Freud stuff over there. I think you would like it," Brill recalled. In early August he went to Bleuler's clinic. The methods were impressive, although the buildings were ordinary: "But after attending the first staff meeting, I felt inspired. The way they looked at the patient, the way they examined him, was almost like a revelation. They did not simply classify the patient. They took his hallucinations, one by one, and tried to determine what each meant, and just why the patient had these particular delusions. In other words, instead of registering phenomena, they went into the dynamic elements which produce those phenomena. To me, that was altogether new and revealing." [63] Karl Abraham, one of Freud's early followers had just resigned his post at the clinic, and Brill took his place.

Jung was a "very ardent and pugnacious Freudian," Brill remembered. He was brilliant, mercurial, and ardent. He would sail across the lake, tirelessly discoursing on his theories of insanity and his endless lore about mysticism, alchemy, occultism, insanity, psychoanalysis.

During Brill's year at the Burghölzli, for the first time in his life he discussed sexuality with medical colleagues even at the dinner table, analyzed his own dreams, investigated a case of schizophrenia with Jung's association techniques. At first he was skeptical about psychoanalysis, but experience with his own parapraxes began to convince him. In 1908 Brill returned to New York eager to begin psychoanalytic practice. In 1909 he and Peterson published their translation of Jung's *Psychology of Dementia Praecox* in the *Nervous and Mental Disease* monograph series, edited by Jelliffe and White.

The conversion of James Jackson Putnam, the most important American, and probably the most distinguished physician, with the exception of Bleuler, to become interested in psychoanalysis, began in

1908. Putnam was a founder of the American Neurological Association and one of its most respected members.

Ernest Jones, who had left London and was teaching psychiatry at the University of Toronto, was invited to Boston by Morton Prince in the winter of 1908 to discuss psychoanalysis with his circle of physicians and psychologists interested in medical psychology and psychotherapy. Jones spoke at Prince's house before a dozen or more physicians and psychologists, including Hugo Münsterberg, E. W. Taylor, George A. Waterman, Isador Coriat, and Putnam.

Jones had a prodigal capacity for working hard and cultivating both friends and enemies. He became a friend of Prince, Brill, and Meyer and spoke in 1913 at the opening of the Phipps Psychiatric Clinic at the Johns Hopkins. Before his return to London in 1913, he had lectured to physicians in Niagara Falls, Detroit, Chicago, Boston, Baltimore, Washington. Many of the twenty papers he wrote during these four years entered the psychoanalytic canon. Freud read the manuscript of Jones's Hamlet essay in Worcester.

Jones was short and had a round, pleasant face that seemed to indicate unusual good nature. But he has left rather acerb descriptions of most of the American therapists; perhaps he felt bitterness in proportion to his failure to convert them. Prince, who remained skeptical and whom he liked, was handsome, debonnaire, and hospitable, but stupid, with "no great depth of mind or patience for difficult problems." Sidis, who became an indignant opponent, was "conceited, self-sufficing . . . not open to ideas from the outside. . . ." Coriat, who found Jones's expositions of psychoanalysis incomprehensible, was painstaking but uninspired. Adolf Meyer was "too narcissistic" for more than a nodding acquaintance with the unconscious.[64]

Jones judged harshly; he was impatient with fools and lesser talents. He was born and reared in Wales and occasionally some combination of accident and temperament seemed to work against him. He attended Fabian lectures, read Nietzsche, and questioned the "stupidities and irrationalities" of society. In his teens he had been fascinated by religion and philosophy, then rejected both. These early concerns, he later analyzed, reflected attempts at "atonement with the Father." A more lasting solution to his doubts, he briskly concluded, was found "once the unreal elements were abolished with their supernatural beings and their imagery of another world." [65] The solution, of course, was psychoanalysis. At age thirty-eight, in 1907, Jones developed severe rheumatoid arthritis, and fought the ailment for the rest of his life. Once, suffering from overwork—that classical complaint—he toured the continent as therapy.

He was puzzled by the functional illnesses he found in his neurological practice. Hunting for solutions, he combed the French medical psychologists, the English hypnotists, the Swiss, Austrian, and German neurologists and psychiatrists, and William James. Of all the authorities, only Janet, Sidis, Prince, and James were "thinkers"; the rest "mere practitioners," he decided. He corresponded with Prince and wrote for his journal. In 1903 his close friend Wilfred Trotter, with whom he carried on wide-ranging intellectual explorations, brought Freud to his attention. He began improving his German, read the Dora analysis, and was struck by the fact that Freud listened to his patients. This, oddly enough, Jones "had not heard of any one else doing." By 1906 he was trying out psychoanalysis.

The sudden sense of illumination that psychological analysis seemed to bring was as intense for Jones as it had been for Prince or White. Freud's "free association," he wrote years later, "with its idea of flawless continuity . . . at once appeals to us as an intelligent example of law and order in apparent chaos. . . . That apparently disconnected remarks should from the mere fact of their contiguity prove to be bound together by often invisible (i.e., unconscious) links was a brilliant illustration of determinism reigning in the sphere where it was most often denied: it was a most impressive extension of scientific law." [66]

It is possible that for Jones, too, this sense of illumination came partly through personal contact. In 1907 he published a paper based entirely on Janet's theory of dissociation.[67] Then Jones met both Jung and Freud, and endured a severe professional crisis. In 1907 he worked for a time at Kraepelin's clinic in Munich, and returned to London via Zurich. He had encountered Jung at the international Congress of Neurology in Amsterdam three months before and was impressed by his quick, active mind and his attractive breeziness. Later he recalled in Jung a "lack of clarity and stability," just as Jung recalled his doubts about Freud. Jones met Brill at the Burghölzli and listened to a discussion of symbolism in dreams at the Zurich Psycho-Analytical Society. In February 1908 Jones resigned his hospital post in London after a child had accused him of discussing sexual topics. In fact he had been investigating the application of Freud's theories to the child's hysterical paralysis. With enthusiastic recommendations from William Osler and other distinguished physicians Jones secured a post as assistant in psychiatry at the University of Toronto.[68] Jones embarked first on six months of work on the continent before leaving for Canada. With Jung and Brill he attended the Psychoanalytic Congress at Salzburg and heard Freud, whose modesty and acumen deeply impressed him. Sensing something almost feminine in Freud's nature, Jones wished to protect him. Jones and Brill traveled to Vienna, where they dined with Freud and attended Wednesday evening meetings of the Vienna Psychoanalytic Society. With Brill he also visited Ferenczi in Budapest. Brill returned to New York with Freud's permission to translate his work into English. Jones later set out for Toronto, taking with him, ostensibly as his wife, a charming neurotic Dutch girl, Loe.

Jones convinced only one member of the Boston circle—Putnam. On the surface, two men could scarcely have been less alike. Jones was an atheist, an unconventional outsider; Putnam, a Puritan aristocrat of impeccable morals, an idealist who believed in the innate purposes of the universe. To Jones, then twenty-nine, Putnam's deferential modesty seemed absurd.[69] Modesty never was one of Jones's distinguishing virtues.

Putnam may have looked mild and shy, with his pale blue eyes and short stature. But to Jones's astonishment Putnam had no trace of that "puritanical intolerance" Jones associated with the strictest morals. Putnam's "charming amiability" was extended to those who adopted unconventional opinions and behavior. Both men had a lively sense of the starkness of inner conflict, which Jones thought of as St. Paul's war of the members against the law of the mind. Jones was moved by sympathy for neurotic suffering, a hatred of injustice and of the crass attitude which asserted that success was proof of virtue in the social struggle for survival. Putnam was noted for his kindness, his endless patience with nervous cases. Both men attacked the established "materialism" in medicine that precluded a psychological point of view. Both hoped to help patients lead more socially useful lives and both hoped to reform, if only slowly, the social world. The two became good friends, and Putnam traveled to Toronto to support Jones at a meeting where two American neurologists planned to discredit him. The wife of one of them, Jones's first psychoanalytic patient, had been treated successfully for a conversion hysteria in 1905–1906, and had subsequently divorced her husband.

Jones was touched by the fact that Putnam was the only man he knew who ever admitted publicly that he had been mistaken. Putnam changed his mind about psychoanalysis, but his loyalty was not easily won. The persuasive powers of both Jones and Freud were required to achieve it. Fluent in German, keeping up with the latest literature, American and European, Putnam had dismissed Freud. He had been repelled by an article in the *Neurologisches Centralblatt:* "I was impressed by the boldness and confidence of the statements, but rashly attributed these qualities to eccentricity and perhaps notoriety seeking on the part of the writer, and laid the paper down with a distinct feeling of disgust; the reasoning, I thought, could not be correct." [70]

Yet, much in his own development and temperament had prepared him to give psychoanalysis a sympathetic hearing. Like Freud he had moved from a physiological to a psychological and functional approach to the neuroses, though to a less thoroughgoing one. He had found hypnotism and suggestion useful and had come to believe in the role of the subconscious in symptom formation. Bergson, as well as the idealism of Josiah Royce and his friend Susan Blow, had aroused his interest in symbolism, in the role of society in creating the image of the self and in memory as a process of adjustment to the present. He had seen Christian Science and other mind cures regenerate the moral life of patients as well as relieve them of their neuroses. Yet, all this preparation had not given him a clear view of Freud's theories nor led him to accept them. Nevertheless, Putnam was singularly open-minded and considered himself a student to the end of his life. He was painstaking and scholarly, and, with his classical education, admired Freud's style even when rejecting his conclusions. In December of 1908 Jones led Putnam to reconsider his objections.

On December 15 Granville Stanley Hall wrote to Freud inviting him to lecture at the twentieth anniversary of the founding of Clark University:

> Although I have not the honor of your personal acquaintance, I have for many years been profoundly interested in your work which I have studied with diligence, and also in that of your followers. I venture herewith to send one or two documents that may serve to identify myself.
>
> The purpose of this letter is to ask if you can come to this country and to this University the first week in next July, and give perhaps four or six lectures, either in German or English, setting forth your own views, either the substance of those already printed or something new —whatever shall seem best to you.

The occasion is the Twentieth Anniversary of the founding of this institution, and we hope to attract a select audience of the best American professors and students of psychology and psychiatry.

Janet, who has visited this country and given a similar course of public lectures, has had a profound influence in turning the attention of our leading and especially our younger students of abnormal psychology from the exclusively somatic and neurological to a more psychological basis. We believe that a concise statement of your own results and point of view would now be exceedingly opportune, and perhaps in some sense mark an epoch in the history of these studies in this country.

We are able to attach to this proposition an honorarium of four hundred dollars, or sixteen hundred Marks, to cover expenses. You will, of course, be free, after the week's engagement here, to make any others in this country.

Hoping that you may be so situated as to give us a favorable response, I am, with great respect. . . .[71]

Freud refused at first because he could not afford to take the time from his practice. Hall wrote again February 16, 1909. The fee had been raised to $750 and the date changed from June to the week of September 6. Freud accepted, and Hall informed him on April 13, 1909: "We have given out no notices as yet; nevertheless, in some way news of your coming has reached a number of people in this country who have been profoundly interested in your work and have written us expressing their pleasure and their desire to hear whatever you may have to say to us." [72]

On August 9 Hall wrote:

I am writing you now merely to say that I hope you will be my guest as long as you can stay in this city. It would be well to take a carriage from the Worcester station to my house, 94 Woodland Street, where you will find your rooms in readiness. Should Madame Freud accompany you, it would give Mrs. Hall and myself great additional pleasure. We not long since entertained M. & Madame Raymond Cajal of Madrid.

Some guests prefer the liberty of a hotel to being entertained at a private house. Should that be your choice, we will provide the best rooms we are able at the Hotel Standish on Main Street, which is the best in the neighborhood of the University.

If you have already determined on what the general or the special topics of your lectures are to be or how many lectures, or anything else you desire to send concerning them in advance, we can use it to advantage in our announcements. There is a wide and deep interest in your coming to this country and you will have the very best experts within a wide radius.[73]

Hall had judged Freud to be significant later than had William James. In 1895 Hall had reviewed Robert T. Edes's "New England Invalid" without mentioning Breuer and Freud, whose work Edes had used. In *Adolescence,* published in 1904, Hall indicated his acquaintance with Freud's essays through 1896. Breuer and Freud's research confirmed the "cathartic usefulness of play, as a purgative of irritability which otherwise might be drained or vented in wrong directions," just as psychic traumata might result in "hysterical convulsions." They had indicated the "peculiar vulnerability of early adolescence" in girls. Freud had been the "first to insist that all forms of morbid anxiety were loosely associated with the *vita sexualis,* and always arose in cases of retention of the *libido.*" Freud's assertion that early sexual experience was the "caput Nili" of neuropathology might be only partially correct, but it indicated the "wide psychic and somatic resonance" of the "sexual function." In 1908 in an article on sex education Hall argued that Freud, Jung, and Janet had indicated the profound and "often all conditioning psychic ramifications of sex." Hall's psychology course in 1908 for the first time included the "psychology of sex" as a major division and covered the theories of Freud, Moll, and Ellis. But Hall had indicated no real grasp of the scope of Freud's system.[74]

In May 1909 the American Therapeutic Society, an organization of general physicians, held its conference on psychotherapy in New Haven. Hall and Jones both attended. Jones delivered a provocative, somewhat haughty address on psychoanalysis. Putnam, in the course of a paper on character formation, announced that he was becoming more and more convinced of the worth of psychoanalysis as a new and radical treatment. He repeated several of Jones's best arguments: that the patient's own insight into the meaning of his symptoms was a crucial element of therapy, that these symptoms could be traced to early childhood conflicts. He added moral arguments to these technical considerations. Psychoanalysis treated patients as if they had intelligence and will of their own; it provided a basis for the radical improvement of character, a reversal of traits caused by hidden survivals of old environments and misinterpreted emotions.[75] The symposium went through three printings, and the Congress probably helped to stimulate the already considerable curiosity about Freud, who arrived in September 1909.

For Hall and for the "grand old man of Harvard," as Ferenczi described Putnam, the meetings with Freud were decisive. Putnam satisfied his New England conscience as to the integrity and high-mindedness of Freud and his disciples. They were "so kindly, unassuming, tolerant, earnest and sincere," he wrote.[76] He discussed psychoanalysis with Freud at Worcester, at his camp in the Adirondacks, and through a correspondence that continued until the fall of 1916. He and Freud conceived a respect and liking for each other, and this personal loyalty was important. Freud treated Putnam with much the same affectionate candor he showed to the Swiss pastor Oskar Pfister, despite serious and growing disagreement.

"Your visit to America," Putnam wrote to Freud on November 17, "was of deep significance to me, and I now work and read with constantly growing interest on your lines. . . . I can truly say that I do succeed in obtaining insights into my patients' minds and thoughts of a far deeper sort than I ever got before. Some of your conclusions I cannot as yet verify but there is so much I can verify that I am in no mood for hostile criticism."

"Be assured that I shall pay respectful attention to everything that you write," Putnam concluded, "even if I cannot at once agree with all you say, and believe that I shall enjoy and shall incline to accept your views much the more for having had the pleasure of making your acquaintance." [77]

Freud promptly replied: " . . . the exchange of ideas with you, in spite of its brief duration, particularly strengthened my hopes that there might be a future for psychoanalysis in your country. Although you are a decade older than I am, I found in you a high degree of general open-mindedness and unprejudiced perceptiveness to which I really am not accustomed in Europe. These qualities, which I value, lay the foundation for our relationship. It is not at all important that you agree with me in every particular. My work demands from the reader only this: that he seek to undergo the experiences on which it is based. . . . I neither demand nor expect that others accept anything I say without having first gone down and explored the sources of my observation." [78] Putnam wrote that he was reading all of Freud's work, as well as Jung's, and was getting a clearer grasp of the technique. "Psychoanalysis begins where the primary confession ends," he observed. Before Freud had left the Adirondacks, Putnam had promised him to write an article, probably to dispel misunderstandings about psychoanalysis. "Personal Impressions of Sigmund Freud and His Work. With Special Reference to His Recent Lectures at Clark University," was published in Prince's journal in the issues of December 1909-January 1910, and February-March 1910, and it pleased Freud greatly. Until the Conference, Putnam argued, psychoanalysis had been known in America largely "through the gossip of prejudice and misconception. . . ." He took pains to defend Freud's emphasis on sexuality as a natural result of what his patients themselves revealed. To ignore facts for the sake of prudery was absurd; it was "the cry of the church against Darwin . . . the scientific student of human acts and motives is considered a disseminator of morbid tendencies. . . ."

"All this is wrong. A fool's paradise is a poor paradise. If our spiritual life is good for anything it can afford to see the truth. No investigation is wrong if it is in earnest. Knowledge knows nothing as essentially and invariably dirty." [79]

Freud was grateful for this defense. "I owe you special thanks for the seriousness with which you came to my aid in the matter of sexuality. . . . I hope your words will make a strong impression in America and will secure for psychoanalysis the lasting interest of your countrymen. I found your words . . . beginning with 'All this is wrong,' particularly fine. This does credit not only to your qualities as a physician and a scientist but as a man. . . ." [80]

Putnam devoted all his energy and eloquence to what he and Freud called "the cause." To William James, Putnam wrote that he was reading Freud's *Der Witz,* and was impressed by Freud's "keenness and tireless readiness to penetrate and penctrate." [81]

By the spring of 1910 Putnam had tried psychoanalysis on twenty patients suffering from a variety of disorders—hysteria, anxiety neuroses, neurasthenia, fears, impulsions, impotence, stammering. Psychoanalysis proved unusually effective with these patients, whom he had been treating off and on for several years with all the other methods. He described the results to the American Neurological Association in May 1910. Putnam's paper appalled some neurologists but favorably impressed others. William James wrote on June 4 from Bad Nauheim, where he had gone for treatment of heart disease, that he had heard of Putnam's "ovation" in Washington.

"My 'ovation' in Washington to which you refer," Putnam replied, "was of course a Freudian victory alone, if victory it was. I do think as much interest was manifest as one could fairly look for. Of course there was much opposition also. A Dr. Walton brought down the house by saying that we would rather have an ounce of Muldoon than a pound of Freud. But I countered on him by asking about his book *Why Worry* and saying that it seemed cynical to ask 'why worry' without ever enquiring 'Why do you worry?' For certainly no conceivable inquiry of that sort can be made except à la Freud." [82]

James died two months later, and Putnam wrote a long appreciation for the *Atlantic Monthly* that incidentally revealed much about his own reasons for taking up so doubtful a cause as psychoanalysis. He praised James's life-long battle with depression and invalidism and the fact that he had "never lost his courage and his buoyance." He also praised James's talent for leaps in the dark, acts of the will that added "something new to the forces of the world" and made for the world's progress. James had been able to take a position, even if not absolutely certain of it, if he believed it would help to "create" truth. Putnam quoted a letter James had written him after defending the right of Christian Science and New Thought healers:

"If you think I like this sort of thing you are mistaken. It cost me more effort than anything I have ever done in my life. But if Zola and Col. Picquart can face the whole French army, cannot I face their disapproval? Far more easily than the reproach of my own conscience." [83] Since his Harvard College days, Putnam had championed unpopular causes as part, perhaps, of that personal contribution to progress by which he believed a man's life would be judged. The very disadvantages of psychoanalysis were, then, all the more reason for defending it. An unsympathetic colleague, Phillipps Coombs Knapp, remarked that Putnam was an "invincible" optimist with a "propensity to discover the precious jewel in the head of the ugliest and most venomous toad." [84]

To Sidney I. Schwab, Putnam's speech inaugurated a "new era" in American neurology. [85] The speech was part of an aggressive campaign by neurologists and psychiatrists already committed or who were beginning to think of themselves as psychoanalysts—among them, Putnam, Brill, White, and Ernest Jones. For the first time Freud's views were presented with greater pains and accuracy. Psychoanalysis was not yet faced by internal schism, and was, as Freud had assured Putnam, still experimental and "incomplete." Only as internal divisions developed did Freud and his immediate followers create criteria for what constituted "correct" psychoanalytic doctrine. Thus in 1909 it was still possible for Ernest Jones to place considerably more importance than Freud or Putnam on Jung's association tests. The experimental nature of psychoanalysis in these years reinforced the American tendency to eclecticism, so that Freud's theories continued to have a selective interpretation.

The psychoanalysts advanced a number of common arguments. Psychoanalysis achieved better results than any other method. It was the most scientific and systematic of all the medical psychologies and extended furthest the great revolution begun by Beard and Charcot, Janet and Prince. It advanced the progress of neurology and psychiatry. It was the most widely applicable of all the medical psychologies, with important implications for philosophy, sociology, and even for social reform.

To some of those who became seriously interested, psychoanalysis acted as a catalyst. Their search for a functional approach to nervous and mental disease now became more clear-cut and uncompromising. This process can be seen in Putnam, who had the widest background and experience with classical somaticism as well as with rival methods of psychotherapy. All the points in Freud's theories he and other Americans had largely ignored now stood out clearly: the role of infantile sexual conflicts and infantile sexuality; the mechanisms of projection, substitution, introjection, identification; the importance of dreams.

Putnam now saw psychoanalysis not as "catharsis" but as a minute exploration through free association of the patient's symptoms and dreams that invariably led back to childhood conflicts about sexuality. Not only was this the most efficient therapy for the less serious neuroses, it also could help conditions previously considered hereditary and therefore "beyond relief," notably the compulsions, phobias, and obsessions of psychasthenia. Here Freud opened a "distinctly new door for hope." [86]

This "most radical mode of treatment" was forcing physicians to redefine their conceptions of what was hereditary, degenerate, and incurable. For the first time Putnam formally attacked the degeneracy theories of the late nineteenth century. Perhaps borrowing an argument from William James, he wrote that they exalted mediocrity at the expense of genius and innovation. Worst of all, doctrines of degeneracy fostered pessimistic fatalism and materialism. Janet and others had argued that the signs of inherited tendencies to neurasthenia and especially psychasthenia appeared in very early childhood. But psychoanalysis, Putnam insisted, proved that these very signs in young children—nervous weakness, shyness, self-consciousness, timidity, excitability, sensitiveness—themselves resulted from or were reinforced by early experiences. Even where nervous disorder ran through several generations in the same family, the parent's neurotic attitude toward the child and the child's imitation of the parent—not a mysterious physical heredity—were the true causes. Education in the widest sense, and experience, were more important than inheritance.

Although Putnam's views on sexuality always had been moderate and tolerant, psychoanalysis inspired in him a new and surprising rebellion against prudery and harsh moral judgments. Just after the New Haven Congress he had written to the *Boston Medical and Surgical Journal* protesting an editorial condemning psychoanalysis. The "fierce yet often needless conflict between natural instincts" and a cruel, blind, and artificial social organization caused a "vast number and an immense variety of neuroses and psychoses." Also for the first time he accepted here the basic similarity of patient, physician, and "degenerate." "In their grosser sorrows and misdeeds our milder mischances and delinquencies and our patient's depressions and obsessions are often to be seen in a new and truer light," he insisted.[87] In the paper of 1910 he condemned "blind obedience to an artificial code of social rules. . . ." This as well as laziness and self-interest had prevented men from looking candidly at the repressed and rejected element of sexuality. Attempts to conceal facts that wounded pride and convention led only to "mental conflicts and to illness and to false standards of civilization." [88]

Also for the first time Putnam assessed Freud's place in the neurological tradition. Briquet and Janet, Putnam wrote, had been observers and describers; Charcot, Weir Mitchell, Bernheim, and Dubois had been therapeutists; Freud was one of the few "analysts, inductive reasoners and broad thinkers." He had extended the scientific revolution in psychology and had "set on foot an equal revolution in character-study, educational psychology, race-psychology, child study, and folk-lore." [89]

Freud was so pleased by this paper, "On the Etiology and Treatment of the Psycho-Neuroses," that he translated it himself for the *Zentralblatt.* "You convince me," he wrote to Putnam, "that I have not lived and worked in vain, for men such as you will see to it that the ideas I have arrived at in so much pain and anguish will not be lost to humanity. What more could one desire? . . . You express my convictions more clearly than I do, in the broad cultural context in which they belong, and have imbued them with the all-inclusive viewpoint of a friend of humanity." [90]

Putnam and other analysts coupled their claims for the superiority of psychoanalysis with an attack on rival methods. Once nettled by opponents at a Boston medical meeting, Putnam retorted: "I should not have ventured to present these histories tonight, except that all of the patients have been treated unsuccessfully by members of this audience." [91] Suggestion, especially, glossed over symptoms and perpetuated the patient's neurotic dependence. Its results were temporary and often ineffectual, as William Alanson White demonstrated in an example. A woman whom he treated for a phobia for the color red lost this fear but developed an obsession with suicide. White tried to make her think of the harmless idea of a cat whenever this obsession occurred. But the substitution failed to work; whenever she thought of a cat, she became weak and depressed. Thus one symptom was replaced by another in an endless chain. Suggestion did not reach the underlying conditions and was a cloak for ignorance.[92]

Ernest Jones took pains to analyze on a theoretical level why hypnosis and suggestion failed. His argument is worth retracing because it constituted a psychoanalytic departure from Charcot and Bernheim. Following Ferenczi, Jones argued that hypnosis and suggestion perpetuated neurotic dependence in order to work. Hypnosis was not effective through any special power or talent of the hypnotist, but rather because of the subject's inner psychological disposition. As Charcot had observed, not a single state could be induced by hypnosis that did not occur spontaneously in hysteria—paralyses, anaesthesias, parasthesias, trances, ecstasies. But then, in an unexpected way, Bernheim's insistence that hypnosis was normal and that most people were suggestible also had been confirmed, for everyone was a bit neurotic and hysterical. This Freud had proven.

Authorities on hypnotism, including Janet, had noted certain peculiarities of hypnotic suggestibility. For suggestion to cure persistent symptoms, the patient had to feel a special "rapport" with the hypnotist. This included affection, dread, jealousy, veneration. Symptoms would clear up, then recur, and the patient would experience a passionate desire to be hypnotized again. After hypnosis the patient felt a sense of temporary fatigue and blissful well-being. Jones pointed out erotic aspects of this sequence.

He explained the patient's dependence on the hypnotist by a psychoanalytic paradigm: the relation of a child to his parents, especially his eagerness to please them. Hypnotists, Jones noted, usually assumed either a paternal or maternal role. Some, like Svengali, commanded; others worked with protective devices, a darkened room, rhythmic stroking, etc.

Even the things used to induce trance, monotonous sounds or bright objects, commonly were employed to divert or soothe a child. As a further confirmation of this interpretation, he noted Freud's discovery that neurotics sometimes unconsciously carried out commands issued years before by their parents. Hypnotism, then, was nothing more than the fixation of infantile incestuous feelings on the operator. The patient's unconscious infantile cravings found an outlet in symptoms. By hypnosis this emotion or craving was withdrawn from the symptom and directed toward the operator. So the symptom ceased—but only at the price of maintaining this transference.[93]

In Freud's system, the patient's own insight cured. Freud's watchword, Putnam remarked, was "You can do better if you *know;*" that of most other therapists, "You can do better if you *try.*" [94] Psychoanalysis was a radical re-education, which worked because it was systematic and scientific.

Freudians pointed to Freud's superiority as a theorist: he was the first to construct a truly conceptual medical psychology. Over and over the psychoanalysts insisted that Freud's method was radical in the literal sense: it eradicated the roots of symptoms. It was truly causal and carried determinism farther than anyone yet had attempted. This accounted for its superior results.

On what basis did Freudians advance the claim that psychoanalysis dealt in causes? Some of the reasons were explicit, others must be inferred. First, Freud placed primary importance on what Putnam called the "great springs of emotion and of motive." In most other systems motivation occupied a secondary place. To Prince, for example, a neurotic compulsion was the result of an acquired habit. To the Freudians the compulsion might be caused by a wish to ward off a forbidden impulse. Second, Freud located motivation in the unconscious, making it an active, purposive agent. Third, Freud united the miscellaneous collection of traditional drives in sexuality, interpreted as desire or craving. Sexuality worked beneath the play of other emotions such as fear or pain.

Then, psychoanalysis offered a comprehensive and compelling sense of continuity. By being able to trace a symptom back through past manifestations to one or more events that had been forgotten and were therefore in the "unconscious," patient and therapist achieved a sense of coherent pattern. For example, Putnam and one of his first twenty analysands traced a tiresome, apparently senseless hypochondriasis back to the age of four, to "infantile passions and repressions, parental domination . . . and early training." In another case a very early craving for protection remained unchanged, so that later, in acute anxiety attacks, a patient would call, unexpected, at a friend's house to be reassured. Or the patient would use illness as an excuse to legitimize the same protective need. Or in adult life a patient might be unwilling to part with a nickel for carfare or mail a letter because of emotions associated with an early unwillingness to part with feces. These kinds of clinical discoveries had overridden Putnam's earlier objection that experiences could not possibly repeat each other. It was as if one recognized at a single glance, "in a transparent microscopic section construction of the brain, the whole course of a great neurone tract. Childhood and age are seen as if now present and coalescing, in effective form." [95] In the unconscious time was annulled.

The unconscious also operated according to definite mechanisms that controlled the processes of symptom formation—forgetting or repression, condensation, distortion, secondary elaboration, etc. All these conceptions were psychological constructs. This was their great virtue, according to a pupil of Ernest Jones's, the British psychiatrist Bernard Hart.

Freud's system disposed of the disputes over mind and body. This battle, Hart insisted, was absurd and unreal because both physical and psychological approaches were equally valid—in their separate and distinct ways. Both were equally scientific. For the aim of science was to organize and predict that William James had called the "flux of sensible reality," not to deal with some "fabulous entity called matter" which existed outside the mind that perceived it. In fact, both psychology and physiology were different ways of organizing perceived experience. For this reason Freud's unconscious was as scientifically valid as the "atomic theory, the wave theory of light, the law of gravity and the modern theory of Mendelian heredity," Hart argued.[96] Freud's unconscious was a construct, designed to do what all theories of the unconscious since Leibnitz had been designed to do: establish continuity between the seemingly discontinuous flashes of psychic life. This, it may be added, was precisely what Janet had done when he borrowed F. W. H. Myers's "secondary self" to explain post-hypnotic timekeeping. Physicists and psychologists, then, were dealing with exactly the same entities, sense impressions, but using different conceptual models. The physicists used those of space-time, the psychologists those of personality and consciousness. Within each scheme there could and must be rigorous determinism. Above all, however, neither scheme should be confused with the other. Thus, Prince's favorite scraps of "science," "memory traces in the brain," "nervous energy," were as artificial as Freud's "unconscious," and meaningless because they were constructs borrowed from another discipline. A physician would hardly introduce "ideas" to explain the causation of physical phenomena. Nor should a psychologist consider the unconscious as a "brain fact," or "fill up the gaps in his reasoning with mythical 'nerve currents.' "

Hart sneaked in all the comforting fixity of the old "objective world," with its eternal natural laws, which he thought he had gotten rid of. He reasserted Berkeley's argument that all we immediately know are our sense impressions. But these impressions—this "flux of sensible reality"—was itself orderly. Human experience, Hart wrote, "does not take place in an entirely haphazard and chaotic manner . . . but . . . events follow one another with more or less regularity. . . ." [97] Freud considered Hart's the first "clever word" about his theory of the unconscious, the "best on the damned topic . . . and enormously superior to Morton Prince's trash." [98] (Freud had taken Prince to task for never explaining why repression occurred.) Hart's interpretations of Freud were influential. His essay was included in Prince's symposium on the subconscious and later reprinted; his *Psychology of Insanity,* which interpreted the psychoses in terms of conflict, was both a medical and lay success.

To some physicians Freud's apparent psychological sophistication was perhaps less important than the analogies between psychoanalysis and physical cures of disease and between Freud's description of mental mechanisms and the causal laws of chemistry and physics. Putnam came to believe that the psychoanalysts were too impressed by the magic word "cause," which in science meant "invariable sequence." Diseases were caused by specific entities, and Freud once explained that just as no one caught tuberculosis from any source but the tubercile bacillus so no one developed a neurosis except on the basis of specific sexual experiences. These, as Jones was fond of arguing, were the "pathogenic material." A neurosis, then, could be drained like a bad sinus—by psychoanalytic surgery.[99]

In placing a central importance on instinctual motivation, Freud seemed to be penetrating to the evolutionary roots of human conduct. By uncovering this instinctual substructure, psychoanalysts thought they were dissecting "down" to a "deeper" and therefore more significant level.

The sense of understanding and control that psychoanalysis gave its practitioners needs further elaboration. Every psychological approach seemed to elicit this sense of discovery; but for many physicians the insights gained from psychoanalysis were more illuminating and explanatory than those from other methods. Schwab, who was as well acquainted with Janet as with Freud, remarked that laying bare the chain of associations from the dream or obsession to its origin revealed the "fabric of a patient's intimate psychical life." Freud's method, he wrote in 1911, lifted the fog "from what would otherwise have been a darkened chaos of disconnected events often separated by such long periods of time that their relationship seemed utterly out of the question." [100] To nearly every neurologist and psychiatrist the symptoms of neurosis and insanity at first appeared to be baffling, irrational, and bizarre. They seemed to have no order or understandable logic. Above all, psychoanalysis supplied a pattern and a sense of order. White insisted that it allowed him to see the mind, even in insanity, not as made up of scraps in a "wastebasket" but of ideas built into a "coherent and harmonious structure." [101]

The ecumenical sweep of the psychoanalytic synthesis was attractive, especially to men like Putnam, Brill, and Jelliffe. Jelliffe, for example, had been using Henry James's *Turning of the Screw* [sic] and Weir Mitchell's novel, *Constance Trescott,* to illustrate the formation of delusions in paranoia.[102] By comparison the narrower medical psychologies of Janet and Prince seemed thin and, the analysts charged, largely descriptive. Janet's work, Putnam argued, did not lend itself to "great philosophical and sociological generalizations." But psychoanalysis was "revolutionizing" child and character study, educational psychology, mythology, anthropology, the critique of art and literature. Amid all these diverse phenomena the analysts thought they discovered unity as well as corroboration for their theories. Putnam believed that Freud provided a "master-key to many of the mysteries of life." [103]

This search for unity probably fulfilled an emotional craving, seen at its strongest, perhaps, in Putnam, Jelliffe, Hall, and White. These men had been brought up on the vast systems of the nineteenth-century builders, Spencer and Hegel, for instance, and tended to look for a synthesis that would span all chasms. Putnam interpreted such psychoanalytic theories as the infant's belief in his own omnipotence as evidence for a Hegelian view of the Mind's functions.[104]

One aspect of the sense of illumination from psychoanalysis was important in treating insanity and neurosis. By insisting that both were merely degrees of exaggeration from the normal, physicians not only could feel a closer rapport with their patients but they could also interpret these illnesses in terms of their own feelings and experiences. Freud once remarked that he did not hope to meet a human being with whom he could not feel some bond of understanding. For Brill the psychotic patients who had seemed so alien and strange now aroused echoes within himself. Brill remarked of his experiences in Zurich: "I was captivated by the case histories because the patient no longer represented something entirely foreign to me, something insane, as I had hitherto regarded him when I merely described his strange behavior. Now even his most peculiar expressions as I traced them back to his former normal life struck familiar chords in me." [105] Putnam told his colleagues that before taking up psychoanalysis he had never really understood his patients. His previous knowledge of their "lives, characters, capabilities and needs" as well as of the seriousness of their sufferings had been "utterly superficial." [106]

The superior practical results, the sweep of theory, the sense of understanding helped to form and intensify progressive currents, for psychoanalysis became influential almost from their beginning. The new psychological approach to insanity dated from about 1895; the popularity of psychotherapy was as recent as 1905. Bancroft's rejection of pessimistic fatalism occurred in 1908, based not on Freud but on American psychology and sociology. Thus psychoanalysis not only "fitted in" with a strong movement of reform but also helped to define that movement.

Contemporary neurologists and psychiatrists used the term "progressive" in two senses. The first and chief meaning was "modern," "up-to-date," reforming of old ways. Jones called psychoanalysis a "progressive movement" in 1910, probably without reference to the American social context.[107] Yet, the word "progressive" had another meaning closer to that social context. This meaning included hope, efficiency, and social usefulness. Psychoanalysis was "progressive" in all these senses. It also supplied a polemic at once moralistic and levelling.

The time and effort expended on each case seemed to the psychoanalysts to be both scientific and humanitarian, while to their opponents it seemed unnecessary nonsense. Ernest Jones told the New Haven Congress in 1909 that Freud spent as much as three years of almost daily sessions with a single patient. The physicians shivered, Jones recalled. In

1910 a neurologist, commenting on Putnam's brief for psychoanalysis, argued against dignifying the patient's obsessions by listening to them. Such "parasitic ideas" must be ruthlessly eradicated. In two treatments he had cured a patient by making him say to himself every time an obsessive thought obtruded: "What a damn fool I am." [108]

Because the patient's own insight cured in psychoanalysis, the role of the therapist changed substantially. Putnam made fun of charismatic guidance. No longer was the therapist to be an "inspiring leader" who urged the patient to "project his vision beyond the obvious darkness into a region of imagined light." [109] Nor should the therapist prescribe solutions for a patient's problems. After sorry experience most analysts discovered that direct advice often had disastrous results. The patient worked out his own solutions best. Moreover, both therapist and patient were democratically engaged as equals in exploring the unconscious. Putnam was delighted that Freud's dream theories were confirming some of the naïve hunches of the common man.[110] Putnam also believed that psychoanalysis provided a bracing corrective to prevailing ethical snobberies. In the unconscious, as he emphasized in defending Freud's sexual theories, all men were alike. Pretensions to virtue might be inspired by the worst possible motives.

Yet this levelling appeal was combined with highly selective demands on both practitioners and patients. The analyst should combine an aptitude for detective work and an ability to listen with precision, tact, and impeccable morals. To these personal qualifications were added those of special training to master the method, a training available to very few. The analysts issued this professional challenge summarily. Their new scientific hierarchy was not Anglo-Saxon and nativist. It was open to Jews and foreigners and poor men who could work their way up. Psychoanalysis appealed, then, to a kind of qualified expert professionalism relatively new in American life.

Demands on patients also were both new and high. They must be intelligent and educated, of sound moral character, those members of society, Ernest Jones remarked, "whose value to the world is greatest." [111] This requirement was partly inherent in psychoanalytic theory. That people developed neuroses because of conflict testified to the strength of their moral nature. This consolation Freud had offered to one of his most famous early patients, the paralyzed Elizabeth von R. But it was more than a consolation. High morals were essential for the process of judgment to take place after repressions had been lifted. Patients also had to be young, preferably under fifty; otherwise they were too set in their ways to be re-educated.

The age limitation was new. But the appeal to the worthwhile citizen, as patient or physician, could not have better fitted the very people American neurologists chiefly had been treating all along in private practice—the professional and business classes. Unlike institutional work, psychoanalysis, Burrow insisted, was "only adapted to the educated classes." [112] According to common belief, these classes suffered most from nerves. Certainly they were best prepared to pay for treatment. Yet, the snobbery of psychoanalysis must not be overestimated. Analysts chose their patients—partly from necessity—from all classes in American life. Neither Freud nor the Americans were quite as lofty as Ernest Jones.

For most professional and business groups the efficient use of human resources and energies was an important goal, and this psychoanalysis advanced in two ways. It claimed to cure more kinds of patients. It also purported to reduce the amount of hospitalization as well as to obviate the highly restrictive artificial regimes imposed on neurotic patients. As much as Richard Cabot's "work treatment," psychoanalysis was meant to bring the eclipse of the sanitarium, the spa, and the Weir Mitchell rest cure. Freud had discovered—as had alienists long before him—that patients who recovered in sanitaria tended to relapse when they returned to the environments in which they had fallen ill. It was in these daily, ordinary settings that the analysis of neurotics should be conducted.[113]

Janet's psychasthenics were among the first to benefit from this psychoanalytic approach. Lewellys Barker and Charles Dana, for example, usually hospitalized psychasthenics, who, they believed, suffered from a brain disease that only eugenics could eradicate in some distant future. Sometimes, Putnam remarked, patients who were obsessed by the fear that they might kill some member of their family lived "practically exiled from their homes." [114]

This environmentalism of the psychoanalysts must be qualified. The Americans, although more environmentalist than Freud, were less so perhaps than Meyer and Sidis. Putnam, in the debate that followed his discussion of psychasthenia in 1910, conceded to Barker that in this illness a "hereditary influence" might be at work.[115] Jelliffe and White worked out a compromise with heredity in their popular textbooks on neurosis and insanity. Just as the analysts sometimes continued to use rest, drugs, and even suggestion as adjunctive therapies, so, too, their position on heredity represented a new direction of emphasis, not an absolute shift from previous beliefs.

At the same time, the analysts discarded that other fashionable "exciting" cause of nervous and mental disorder—"the strain of modern living." This strain, Putnam remarked, was a very obvious problem, whereas in nervous disorders the truly insidious worries were those which seemed to result "without external provocation." [116] Men really set the pace of life to suit themselves; their inner conflicts were the most troublesome, he argued. This position had important consequences in determining the objects of psychoanalytic attention.

The analysts also couched their appeal in terms of will and social usefulness. Therapy would strengthen will power and self-mastery. Psychoanalysis stood for reality as against pleasure. Putnam insisted that it was no mere wallowing in morbid introspection. Rather, it was a relentless self-scrutiny, a ceaseless exposure of primitive and childish cravings. Here Putnam's rhetoric had something in common with that of others who sought to eradicate social evils in progressive America. Psychoanalysis was a kind of ethical realism, analogous to the muckraking exposure of social abuses.

The personalities and tactics of the Freudians had much to do with the way their arguments were received. Because of his distinguished career and the high regard in which he was held, Putnam was listened to with careful attention. He was considered "broad minded," "just and kindly," even by those who disagreed with him. Brill, too, for all his outspokenness was a persistent but modest advocate. Jones, however, because of his belligerence and his brilliance, was a formidable, sometimes a nasty adversary. He was equalled in this by Smith Ely Jelliffe, who once had the temerity to suggest that God alone had cured a patient Morton Prince claimed to have treated successfully. Some Freudians struck critics as too keenly antagonistic and too dogmatic. Prince remarked that they were enthusiastic and ingenious and included minds as able as any in medicine, yet, their zeal and some of their tactics were not calculated to endear them, even to their friends.[117]

Early in 1911 Morton Prince noted that the Freudians were beginning to "style themselves psychoanalysts." Psychoanalysis as a synonym for "mental analysis" was familiar enough. But a psychoanalyst was unique and new. As Prince implied, the term signified a self-consciousness and cohesion exhibited by no other group of psychotherapists.

Psychoanalysts thought of themselves as a younger generation ready to revolutionize neurology and psychiatry. In 1909 Brill was thirty-five, Jones thirty, White thirty-six, Putnam sixty-three, and their average age was about ten years less than that of their principal opponents among the neurologists. On the whole the Americans exceeded Freud in *hubris*. Invariably the psychoanalysts interpreted their cases as corroborating Freud's theories. They welcomed new recruits as if they had seen the light. Brill greeted Jelliffe's first psychoanalytic paper in 1913 by pointing out Jelliffe's new found "earnestness and conviction." They tended to dismiss their critics as neurotic, ignorant, and unwilling or unable to master the psychoanalytic method.

At the American Neurological Association meeting in 1910 Jones had argued that physicians would oppose this "progressive movement" only through blind prejudice and at the peril of looking absurd in the future. In all America, Jones asserted, only "three or four men" were qualified to criticize psychoanalysis at all. But opposition could not then crush psychoanalysis, although it might have done so fifteen years before. For, he said, in prideful exaggeration, some two hundred "trained workers had confirmed Freud's conclusions in most of the countries of Europe." [118]

Freud, whom Jones kept informed of the American campaign—if not of his own tactical exaggerations—was pleased with his "superb letters, full of victories and fights." [119] Although the Freudians did win a hearing for psychoanalysis, they also aroused a formidable and angry opposition. Inevitably psychoanalysis, because of its medical techniques and its criticism of the prevailing customs of "civilized" morality, became involved in wider struggles within American culture. The popular interest in mind cures and the slow repeal of reticence in matters of sexuality provided the context in which the battles over psychoanalysis were fought.

IX

Mind Cures and the Mystical Wave:
Popular Preparation
for Psychoanalysis
1904-1910

Emmanuel

Freud lectured at Clark University during the high point of the greatest popular interest in mental healing, hypnosis, and suggestion in American history up to the Great War. This movement aroused the first public enthusiasm for psychoanalysis. A few important journalists and co-operating physicians began to turn from the older methods to Freud's. The popular movement crystallized a special style of journalism about psychotherapy as well as stereotypes about treatment and the physician. Some of these tendencies are apparent in the two-column interview with Freud printed in the *Boston Evening Transcript* on September 11, 1909.

The enthusiasm of Freud himself and of Freudian converts, as well as the hyperbole of journalists, laid the groundwork for Freud's legendary public reputation. Like every "latest" development, Freud's psychoanalysis was presented as perhaps the most significant event in the history of psychotherapy. The *Transcript*'s correspondent, Adelbert

Albrecht, who in 1913 served as associate editor of the *Journal of Criminal Law and Criminology,* declared that Freud's *Interpretation of Dreams* marked a "considerable step forward on a way along which we have wandered for centuries in darkness in the fullest sense of the word, and his studies on the sick soul-life of children are nothing short of epoch-making. The 'Story of Little Hans' will probably ever remain a unique and model study of a child's soul." Journalistic reticence cloaked Little Hans's major complaint—severe castration anxiety.

Albrecht went on to describe Freud as a charismatic healer with appropriate phrenological stigmata: "One sees at a glance that he is a man of great refinement, of intellect and of many-sided education. His sharp, yet kind, clear eyes suggest at once the doctor. His high forehead with the large bumps of observation and his beautiful, energetic hands are very striking. He speaks clearly, weighing his words carefully, but unfortunately never of himself. Again and again he emphasizes the merits of his colleagues, particularly of his friend, Dr. Jung of Zurich, who is staying with him at President Hall's."

Probably as a result of Albrecht's questioning, Freud predicted that the Emmanuel Movement, a popular experiment in religious psychotherapy then at the height of its vogue, would soon die down. Admitting that he knew little about it, Freud promptly condemned it:

> When I think that there are many physicians who have been studying modern methods of psychotherapy for decades and who yet practise it only with the greatest caution, this undertaking of a few men without medical, or with very superficial medical training, seems to me at the very least of questionable good. I can easily understand that this combination of church and psychotherapy appeals to the public, for the public has always had a certain weakness for everything that savors of mysteries and the mysterious, and these it probably suspects behind psychotherapy, which, in reality has nothing, absolutely nothing mysterious about it.

Freud at once constructed a counter-image that became in turn an important psychoanalytic stereotype—psychoanalysis was austere and difficult, requiring extraordinary expertise but promising radical cure. The "analytic method" gave the greatest insight into the "mechanism of the sick soul"; it translated what was unconscious in the patient's mind into consciousness. Thus it ended compulsions, "for the conscious will reaches as far as the conscious psychic processes, and every psychic compulsion is caused by unconsciousness."

The method worked against the constant resistances of the patient, especially "in cases where the sexual life plays a part. It is difficult enough to bring patients to a complete confession of these things when they are entirely conscious, how much more difficult then if they are unconscious of them!"

When Albrecht noted that "several psychotherapeutists in America claimed to have cured hundreds of cases of alcoholism and its consequences by hypnotism," Freud argued that psychoanalysis was superior:

> The suggestive technique does not concern itself with the origin, extent, and significance of the symptoms of the disease, but simply applies a plaster—suggestion—which it expects to be strong enough to prevent the expression of the diseased idea. The analytical therapy on the contrary, does not wish to apply a plaster, does not wish to inject any thing; but to take away, to get rid of, and for this purpose it concerns itself with the origin and progress of the symptoms of the disease, with the psychic connection of the diseased idea, which it aims to destroy.
>
> I have given up the suggestive technique and with it hypnotism because I despaired of making the suggestion strong and durable enough to effect a permanent cure. In all severe cases I saw the suggestion crumble away and the disease again made its appearance.
>
> Further there is another important point for my method. Particularly in the most complicated cases suggestion and particularly hypnotism do not help at all, because as Bernheim has pointed out long ago, most people cannot be hypnotized. Do not believe the wonder-doctors who try to tell you that every human being is subject to their hypnotic art.

Hypnotism also was a morally doubtful kind of trickery that resembled the "dances and pills of feather-decorated, painted medicine men."

"In what my technique consists I cannot explain so easily. The instrument of the soul is not so easy to play, and my technique is very painstaking and tedious. Any amateur attempt may have the most evil consequence."

"The psycho-analytical cure, as I have implied, makes difficult demands on the patient and on the physician. Also it is not yet complete. Of the patient it requires absolute frankness, occupies a great deal of time, and is therefore expensive." But it was also radical. Freud had used it successfully with patients other physicians had given up as hopeless. Although suited only to a limited number of cases, in which symptoms were not acute, "for all forms of hysteria and in the wide field of delusions it is, in my opinion indispensable." [1]

Freud was not alone in his distrust of the Emmanuel Movement; yet it gave psychoanalysis some of the first publicity it received. Because the Emmanuel Movement claimed to combine religion and medicine, some American physicians were more angered by it than by Mrs. Eddy's Christian Science.

In 1908 the Movement's founder, the Reverend Elwood Worcester, who liked to paddle canoes and climb the Rockies, deplored Theodore Roosevelt's "loud roar for the Strenuous Life." The Strenuous Life, Worcester argued, was likely to make Americans more nervous than they were already.

He preached that America faced a crisis of nerves and national character more subtle and trying than the Civil War. To prove his point he cited the early deaths and suicides of famous citizens; the "alarming increases" of insomnia and that "most prevalent disease of modern times" —nervousness. He deplored the widespread consumption of too many cocktails and dangerous drugs. Venereal diseases and prostitution flourished. Illicit sexual relations and aberrations were increasing; the divorce rate was rising. Finally, Americans were distraught by the erosion of religion and the "vastest doubts which have ever dismayed the minds of men." [2] To combat these evils Worcester inaugurated a cheerful crusade combining liberal Christianity, the powers of the Subliminal Self, and the latest in medical psychotherapy.

Worcester consulted his neurologist friends—Silas Weir Mitchell, Isador Coriat, James Jackson Putnam, and the internist Richard Cabot —who suggested he organize a class for neurasthenics. Putnam and Worcester worked out the details. On November 11, 1906 in the vestry of Emmanuel Church the first class was held for the "moral and psychological treatment of nervous and psychic disorders." Worcester had intended it to be a modest addition to his parish's medical work, which already included a successful program for the tubercular poor. However, on the first morning, instead of the handful of people he had expected, one hundred and ninety-eight appeared for treatment. This was more than Putnam and the other physicians could examine. The newspapers became interested, and for the next four years the Emmanuel Movement, as the press christened it, spread across the country among a number of Protestant denominations, Baptist, Congregationalist, Unitarian, Presbyterian.[3]

Freud's strictures against the Movement indicated his distance from an important strain of American belief—faith in the ordinary man's inner light. This survival of Transcendentalism was typified at its best in William James, and the newspaper account of Freud's scientific pretensions angered him. James wrote to Mary Calkins, professor of psychology at Wellesley, that Freud was a "regular halluciné" with his dream theories. To the Swiss psychologist Theodore Fluornoy, James complained that Freud had "condemned the American religious therapy (which has such extensive results) as very 'dangerous because so unscientific.' Bah!" [4]

James's own wide interests had helped to stimulate the curiosity of Americans about hypnosis, psychic research, automatic writing, multiple personalities, mental healing. Richard Cabot in 1908 congratulated European psychotherapists for not having to put up with an organized, lay enthusiasm for mind cures. But James, with characteristic iconoclasm had observed that the public was practical enough to accept mental healing while doctors and clergymen were doubtful.[5]

In March 1890 James stimulated the popular vogue by a long article, "The Hidden Self," in *Scribner's Magazine.* The untidy data of mysticism, the occult, spiritualism, faith healing, and hypnotism largely had been ignored by science, James argued, because they could not be neatly ordered or completely understood. But now, for the first time, a French scientist, Pierre Janet, by the most careful experiments had demonstrated the co-existence within a single human being of several personalities, which differed in traits and in memories. This was the great achievement of Janet's *L'Automatisme Psychologique,* and James called for comparative studies of hypnotic and hysterical phenomena to exploit this rich, new vein of information.

By 1895 the "psychotherapeutic movement" that had begun in Europe in the middle 1860's with Liébault and developed rapidly in the next decade was making headway in the United States, principally through the efforts of psychologists such as James, G. Stanley Hall, Joseph Jastrow, and James Mark Baldwin, and of a few physicians.

The more "liberal spirit of the age," the weakening of a narrowly materialistic conception of science, was permitting people to take seriously what only a few years before they would have dismissed as quackery, according to the neurologist Allan McLane Hamilton.[6] The vogue of medical psychology and mental healing and their relation to psychoanalysis can be charted with relative completeness from American magazines.

Both medical and popular interest in hypnosis, suggestion, mental healing, and multiple personality peaked in the early 1890's, declined slightly after 1895, then waxed rapidly after 1900. Articles devoted to these subjects doubled between 1905 and 1909, chiefly because of the Emmanuel Movement, which also aroused a new interest in the mysterious subconscious. This subliminal self of James and F. W. H. Myers, with its hidden powers and possible connections with the Great Beyond, was the first and most important public image of the Unconscious. After the Great War began, enthusiasm for the helpful subliminal self declined as suddenly as it had arisen. While the interest in psychoanalysis grew rapidly from 1915 to 1918, the subliminal self remained the possession of a few prophets, such as Edward Carpenter, or of literary figures, such as Maurice Maeterlinck, and finally, in the early 1920's, of the French auto-suggestion theapist Emile Coué.

The vogue of psychotherapy and the mysterious subliminal were important aspects of the American progressive mood in the years up to 1914, and occurred during a time of prosperity and confidence and the rise of a new style of mass journalism. Journalists exploited new developments in neurology and psychiatry, and physicians collaborated with them or else wrote for magazines and newspapers themselves. Some of the journalists were young muckrakers, such as Ray Stannard Baker, for whom nervous disorder and mind cures provided a novel and inspiring subject. A Canadian who joined the staff of the *Outlook,* H. Addington Bruce, made a specialty of articles about nervous and mental disease.

For some journalists psychotherapy represented personal salvation. James Oppenheim, Floyd Dell, Lucian Cary, and Max Eastman all suffered from neurotic symptoms, mysterious backaches, depression, writer's block. They wrote about psychotherapy and psychoanalysis after experiencing its benefits.

These journalists employed the new techniques of the mass magazines. Sometimes their accounts of "the latest in science" were sober and accurate. More often, every development was couched in the hyperbole of doom or salvation. When doctors themselves were dramatic or incautious, journalists exaggerated their worst faults. Journalistic simplicities destroyed important nuances and qualifications. Prince, who wrote arresting case histories, was horrified to discover that an article on hypnosis had been sensationalized by a newspaper headline. Worcester was even more distressed when, after he had aroused a woman in a faint, a newspaper reported, "At Auto-Suggestion Meeting, Dr. Worcester Claims to Have Brought Dead Woman to Life Again." Journalists played up human interest and chose cases with which readers could sympathize.[7]

Physicians, too, sometimes borrowed these journalistic devices. The neurologist Frederick Peterson conducted a personal column, "The Nerve Specialist to His Patients," in *Colliers,* a brash, progressive magazine. He advised readers how to cultivate constructive habits, overcome hereditary weaknesses, and minutely regulate the daily life of a nervous child. Peterson recommended to parents all the traditional prescriptions: cold baths, no schooling before the age of seven, no novels to read, an emphasis on physical, not intellectual pursuits. At maturity, the neuropathic child should pursue a career of manual labor, preferably in the country. Peterson's popular articles reflected his growing interest in psychotherapy. In 1907, the year he and Brill were in Zurich, he published an article in *Harper's* on "The New Divination of Dreams." Without mentioning Freud's name, it was a popularization of *Die Traumdeutung.* Dreams, Peterson wrote, often referred to childhood, to matters of "deep significance" hidden under trivial disguises. They fulfilled wishes and desires of an unspecified nature. Peterson closed on a note typical of the American view of the subconscious: dreams and wishes were products of the "One Great Will, lying at the foundation of all consciousness and subconsciousness." [8]

The popular vogue aroused the interest of a few people who later took up psychoanalysis. In New York young A. A. Brill had become fascinated with hypnosis by reading du Maurier's *Trilby.* In Topeka, Kansas, Karl Menninger followed the story of Morton Prince's Sally Beauchamp in the *Ladies' Home Journal.* From 1890 to 1910 hypnotism and suggestion enjoyed their greatest vogue.[9] In 1905 a dozen popular or semi-popular books on mental healing were published, ranging from Dubois's *Psychic Treatment of Nervous Disorders* to Ralph Waldo Trine's *In Tune with the Infinite.* Mental treatment was discussed in genteel, expensive old magazines like the *Atlantic* and in middle-brow digests such as *Current Literature,* but most often in new magazines like *McClure's* or the *American* for which the muckrakers wrote.

In 1908 appeared a new, elegantly printed, lavishly illustrated large magazine devoted exclusively to medical psychotherapy and the Emmanuel Movement, *Psychotherapy: a Course of Reading in Sound Psychology, Sound Medicine and Sound Religion.* It was edited by William B. Parker, a Congregationalist, a member of the New York Harvard Club, a former lecturer in English at Columbia University, and literary editor of the uplifting *World's Work*. The new magazine, which lasted for three volumes in 1908 and 1909, served up easy, popular articles and comments by distinguished authorities: the psychologists Robert S. Woodworth, Joseph Jastrow, and James R. Angell; the philosopher Josiah Royce; clergymen from the Emmanuel Movement; the psychotherapists James Jackson Putnam, Richard Cabot, Paul Dubois, and Frederick Peterson. The first issue announced that Freud would contribute an article. But nothing by him appeared, and A. A. Brill finally furnished a polite, two-part series about therapy and the psychopathology of everyday life. Woodworth described the physiology of the nervous system; Putnam argued the need for a grand new philosophic synthesis which would combine the insights of Bergson and Wordsworth. Dubois wrote about his persuasion therapy. The tone of the magazine was impeccable, the cases inspiring.

By far the most attention to nerves and mind cures was paid by women's magazines. These carried slightly less than half of all the articles published on nervous disorder between 1905 and 1909. *Good Housekeeping* devoted a "Happiness and Health" section to psychotherapy and the Emmanuel Movement, and printed short articles by most of the writers for *Psychotherapy*. The rival *Ladies' Home Journal* persuaded Worcester to write eight articles for $8000. He accepted, explaining, "I knew that I should never earn that much money again in my natural life." [10] Some physicians peevishly complained that the Emmanuel Movement was being spread entirely by women. Most neurasthenics were women, and most women were deeply religious, Richard Cabot insisted in *Psychotherapy*. In 1908 the *Ladies' Home Journal* urged readers to try the relaxation and breathing exercises of Annie Payson Call, a New England healer recommended by William James. She claimed to follow the James-Lange theory that an emotion and its physical expression were identical. If one were angry, one should relax one's muscles completely; the anger would vanish. Using her infallible methods, women could become poised, calm, and active. *Good Housekeeping* even carried a recipe by Hereward Carrington, a psychic researcher, on how "To Become Beautiful by Thought." [11]

The Crisis of "Nerves"

After 1905 the interest in mental healing was accompanied by a wave of publicity about a growing "menace" of nervous and mental disorder.[12]

Alarms and salvation, that characteristic combination of progressive rhetoric, were offered together, especially by the Emmanuel Movement, which inspired perhaps the largest number of articles about the nervous crisis. Worcester exploited its dangers vividly in the Movement's Bible, *Religion and Medicine,* which enjoyed a phenomenal success. The book was written with a witty Irish Episcopal clergyman, Samuel McComb, and their official consulting neurologist, Isador Coriat. It went through eight editions within seven months, and was included in a list of three or four "great" books on psychiatry by the Surgeon General. Every important newspaper across the country reviewed it.

Worcester's statistics were alarming. He estimated that there were about 500,000 neurotics in a nation of eighty million people. In 1907 H. Addington Bruce insisted that mental disease was increasing faster than the population, although it was still less common than in England or Wales. He cited an asylum census of 1903 which showed 150,151 asylum inmates, a 100 per cent increase over 1890, although the population had risen only 30 per cent. Rates in the Northeastern United States, from Maine to New York, were especially high; the Midwest's were appreciably lower. California, however, had a larger proportion of insane than any other state, and San Francisco the highest suicide rate in the nation—72 per 100,000. In Cincinnati alarm over the number of suicides had led to the founding of a municipal commission to prevent self-destruction.[13]

The "alarmist proclamations" were derided by a number of physicians. A Baltimore psychiatrist wrote that Worcester's were on the level of "nostrum advertisements," and his brand of psychotherapy was just "one more gospel from New England," that land of "witchcraft and transcendentalism."[14]

The New York neurologist Joseph Collins argued that nervous disorder was no more frequent in 1909 than it had been three hundred years before. The *Nation* suggested that the apparent increase might reflect better hospitals and treatment and a greater public willingness to seek them because insanity no longer was considered a sign of "moral obliquity."[15] The reality or unreality of the increase is less significant than the worry about it and what this worry signified. A number of Americans prominent in the Progressive Era had suffered nervous crises: William James; Jane Addams; Woodrow Wilson, who had taken a therapeutic hiking trip in the English lake country; Frederick Taylor, the apostle of scientific management; Senator William Beveridge. John B. Watson, the behaviorist, suffered severe "angst" at the University of Chicago Graduate School, and so, he acknowledged, was prepared for Freud's message. Neurasthenia and insanity touched in one way or another hundreds of thousands of less distinguished Americans—sufferers and their families, philanthropists, social workers, family doctors, legislators who passed appropriations for public hospitals. The concern over nervousness and mental disorder also expressed a part of what contemporaries thought of as a widespread "unrest," a sense of rapid change and of grievances against American civilization.

Journalism reflected the crisis in neurology and psychiatry. Among a small minority of doctors the old optimism of the founders of American psychiatry lingered on. The majority of physicians still were battling these hopeful claims, while a growing minority were expressing the new, more hopeful outlook. Those with novel methods of treatment invariably tried to arouse a sense of need to stimulate interest in what they had to offer. Some prominent philanthropists and psychiatrists preached that insanity and tendencies to suicide were inherited and that only about one person in six sent to an asylum recovered.

One of the best measures of informed opinion about insanity was the reception in 1908 of a crusading book by Clifford Beers, a young Yale graduate. Adolf Meyer, H. Addington Bruce, and William James had approved the manuscript before publication, and James, with characteristic generosity, had given Beers $1000 for his cause. *A Mind That Found Itself* was an eloquent example of personal muckraking. "It reads like fiction," James remarked, "but it is not fiction." [16] It was an account of Beers's two years as a mental patient and a plea for humane treatment. Reviewing the book in the *Journal of Abnormal Psychology,* Ernest Jones asserted that many of Beers's grievances were correct. Nurses were ill paid, and doctors ignorant. Jones urged that the state maintain institutions for the treatment of "both rich and poor, though, not, of course, alike." Laymen greeted the book with condescension and incredulity. Many reviewers indicated that a nearly unbridgeable chasm separated the insane from the normal. "We cannot easily look upon it [insanity] as an illness, but rather as a dethronement of all that is vitally human," one wrote. Insanity implied violence, and it seemed impractical daydreaming to assume that restraint was not necessary. It was surprising that a man who had once been insane could reason correctly or even remember what had occurred during his illness. The *Nation*'s reviewer argued that in the majority of cases, "short of Utopia," insanity was an "unpreventable and incurable disease." [17] Because Beers had recovered, he could have been suffering only from some kind of acute, functional melancholia or psychasthenia caused by worry and fear, and not from true insanity caused by inherited taint.

Beers personally did as much as anyone to change this attitude. In 1908 he launched a typical progressive crusade, the Mental Hygiene Movement. Its purpose was to stimulate research for the causes of insanity and to arouse public demands for improved mental hospitals. The Movement became an important intermediary between specialists and the public, and, later, after 1918, a transmitter of diluted psychoanalytic ideas.

Progressive journalists had begun to write glowing reports of hopeful new developments, and in 1909 social workers were warned against the "fairy tales" in the press about easy cures for mental disease.[18] These reinforced the views of a minority of laymen that insanity could be cured and prevented relatively easily. All along, an important ingredient of this attitude was liberal Christian optimism. In 1900 a sermon at the National Conference of Charities and Corrections asserted a sanguine, psychological approach: "There is probably not a mental delusion, not an incoherent or foolish expression [of the insane] which is not pregnant with meaning, could we but interpret it aright; and by the comparative study of the mental manifestations deduce from them the physical and mental history of the sufferer, as a comparative anatomist can reconstruct the entire body of an extinct animal from a single bone." [19] Magazines eagerly printed almost any hopeful medical news: the speculation that insanity might be caused by a toxin and cured by an anti-toxin, that insanity was not a "disease" but an inability to adapt to environment. Some physicians urged that the old words "chronic and incurable" be abandoned. Psychiatry, they insisted, despite lack of improvement in recovery rates, was on the threshold of a golden era.

Popular Psychotherapy

Worcester and the popular psychotherapy movement did much to spread a more sanguine outlook. *Psychotherapy* magazine printed Paul Dubois's argument that, although psychotherapy was no panacea for insanity, still it was an important adjunctive treatment if applied early and skillfully enough. Worcester asserted that half of all cases of nervous disorder were the result of hereditary tendencies. Still, he argued, men were not "prisoners of fate," and insanity was transmitted far less frequently than nervousness. Moreover, all children of nervous parents did *not* "inherit a diseased nervous system." Even when "dangerous tendencies" appeared very early, these could be overcome by a "wide training and wholesome mode of life." [20]

Worcester also revived George Beard's environmentalism. It is no accident that this believer in the nefarious effects of modern civilization should have been disinterred in the Progressive Era. Worcester based almost an entire chapter concerning "environmental causes" of neurosis on Beard's *American Nervousness*. Beard's arguments allowed Worcester to combine moralizations about the evils of modern life with alarms about the spread of neurasthenia. Alcohol, syphilis, sexual misconduct—all these purely "environmental" causes weakened the nervous system. The point, Worcester emphasized, was that these factors could be controlled.

Journalists soon also stressed the new environmentalism. They were attracted by the theory that poverty and modern industrial conditions were increasing nervousness and insanity. Disease was a social problem. Slums and tenements, the frantic, unwholesome life of great cities, the "struggle for existence" were worse than ever. When they could, they gave psychotherapy a coloring of progressive social reform. Morton Prince, it was emphasized, had fought the gas and railroad interests. Boris Sidis was a "Russian revolutionary," and one of his most spectacular cases demonstrated the relation between bad social conditions and nervous disorder. A young man from the slums suffered from fits of uncontrollable shaking. Sidis discovered that as a child the patient had been afraid of being gnawed by rats while he slept in a filthy tenement. In 1910 *Living Age* took issue with Hugo Münsterberg's argument that modern civilization actually was easing rather than exacerbating tensions. The *Nation* reported without endorsing a widespread and wholesale condemnation of the age: "There is work without recreation, excitement without rest, gayety without pleasure—in short, nervous expenditure, without corresponding satisfaction. . . . There is a struggle and stress in social life unknown in other times. Men, women and children are overworked. There are sweat shops, unwholesome factories, and long hours." [21]

The evils of business competition received special attention. In *Good Housekeeping* Dr. Putnam urged Americans to reconcile themselves to the possibility of failure and warned that spectacular financial success did not sufficiently compensate for the strains it entailed. A "hard-working lawyer," described by the editors of *American* magazine as a "big, husky fellow," blamed American business competition for his breakdown in the "Autobiography of a Neurasthenic." For three years he had been haunted by the fear that he might jump off the Brooklyn Bridge. He was shocked to discover that at times he hated his wife and children. He left his practice, was cured in a sanitarium, and concluded, "Never again will I listen to the promptings of the miserable fever of emulation that drives so many American men to sickness, despair and madness." [22]

In important ways the "nervous breakdown" marked the failure— and the protest—of upper-middle-class professionals in what Robert Herrick, a popular novelist, called the "via dura" of America. The architect, the physician, the professor, the office worker, whoever labored "with tightening nerves" could be broken by the "great machine" of civilization. Herrick's popular short novel *The Master of the Inn,* published in 1909, described a nerve sanitarium that offered an antidote to the strenuous American pace. A fictional portrait of Dr. Gehring, a famous psychotherapist of Bethel, Maine, the novel described his ideal therapeutic community. It was a little like the utopias of the early nineteenth-century socialists except that it was entirely male. Everyone worked for the common good; money was never mentioned. "Habit and tradition" ruled. In the pure country air, far away from the city, the sanitarium occupied a New England farmhouse with a "dainty Italian" pool and colonnade—the quintessence of the genteel retreat, laid out by an architect patient. Men were cured not by medicine, but by hard manual work, rest, and above all, by a cathartic confession reminiscent of Dubois, and administered by the mysterious Master.[23]

Some of the more subtle grievances against civilization also received attention. In 1907 social workers were told that "natural man" was being crushed by "ill fitting social conventions" as well as by poverty and misfortune.[24] Signs of growing public discontent with life-long monogamous marriage distressed Worcester. Democratizing transgression, he noted that sexual problems were not "peculiar to the working classes": "One of the revelations of our work has been the large number of men and women, frequently of the highest station, who are suffering from disturbances originating in the sexual life." [25]

The period's sexual preoccupations were apparent in some of the popular treatments of psychopathology. *The White Cat,* a novel of 1907 that expressed the current interest in dual personalities, appeared on the surface to be impeccably moral. The hero was an upstanding young

American, vigorous, idealistic, devoted; the heroine, a gentle girl with a private income who could live quietly in a country cottage and hide her dual personality from the world. Her second self, cruel, childish, and indelicate, it was hinted, had been indulging in sexual lapses with a foreign hypnotist who wanted her money. At the novel's climax the hero decided to frighten to death this second self. He reasoned that, because multiple personalities were caused by shocks, they could be cured by the same means. In a pistol duel between hero and heroine, her bosom heaved, her "embroidered waist" and her skirt were "half torn off," revealing her "brown and white breast." Her evil self was destroyed; the pure heroine emerged from the battle to marry the hero. The point is that within the most moralistic conventions, sinful conduct could be hinted at and perhaps made to seem even worse by being left inexplicit. This book appeared after Janet's lectures at Harvard and was connected by one reviewer with the vogue for his theories of hysteria.[26]

The psychotherapy movement helped to change the image of the female neurotic from absurd to serious. The older stereotype had been conveyed by S. Weir Mitchell in a series of popular novels. The neurotic female was a "pallid, feeble creature" who reclined on "silken down-lined coverlets"[27] in darkened rooms, where every sound was muffled by "heavy-piled" Turkish carpets. She exhausted friends and relatives by demands for sympathy, kisses, and compliance with her slightest wish. In 1910 Weir Mitchell's neighbor Agnes Repplier scolded female neurotics. Isabella d'Este had witnessed the sack of Rome without suffering nervous prostration. Why, then, couldn't American women at least exhibit the fortitude of their grandmothers?[28] Judging from pictures, Miss Repplier was thin and hawk-like, the epitome of Weir Mitchell's nervous female who lacked "fat and blood."

Mitchell's novels also had portrayed tragic nervous and mental disorder in men—paranoia, depression, and dipsomania. He described in detail the plight of a brilliant socially ambitious alcoholic, a self-made success who was powerless to restrain his periodic craving for liquor.

The Emmanuel Movement and the psychotherapists insisted that nervous disorders be taken seriously. They emphasized that hysteria, for example, affected as many men as it did women. A nervous breakdown could force a physician to give up surgery because his hand trembled beyond control, or a businessman to leave Wall Street because he could not stand the strain. Worcester asserted that he would rather break his thigh or have tuberculosis than "endure for thirty days even sub-acute melancholia and insomnia." [29]

The therapist's realm broadened to include every aspect of social and emotional life. Neuroses, Worcester emphasized, included important "moral symptoms": weakness of will, inability to concentrate, lack of self-control, apathy. Transient moods, or crises of identity, Putnam argued in *Psychotherapy,* were really minor nervous breakdowns. The overwhelming importance of the emotional life was a special contribution of Freud and psychoanalysis, Ernest Jones asserted in *Psychotherapy.* Cabot's stress on the reactions of family members to each other placed this intimate area squarely in the province of the therapist.

The neurologist's reputation as a rich man's expert, a specialist who soothed the nerves of the idle, also was changing. In the 1890's Weir Mitchell's office had been likened to the library of an elegant country gentleman. But in Ray Stannard Baker's stories on medical social service in 1908 Putnam and Cabot were described as busily at work in the bare, clinical cells of a great metropolitan hospital. They tried to help nervous immigrants, Jews, Italians, Russians, poor students, people in the "lower walks of life." Sometimes help was difficult to give because, Baker reported, the poor gave the neurotics among themselves "black looks" and little sympathy. At Dr. Putnam's neurological clinic at the Massachusetts General Hospital, social workers used the new psychotherapeutic techniques.[30]

First, they gained the patient's confidence, so that he could unburden himself of his troubles. They prescribed auto-suggestive exercises, and patients were to repeat to themselves: "I will be calm tomorrow. I will be patient and fearless and trusting." The teamwork of doctor, psychologist, social worker, and clergyman Dr. Cabot called the "American type of psychotherapy." [31]

The Emmanuel Movement also hoped to bring psychotherapy to the city's poor, who had left or never joined the Protestant churches. Psychotherapy spoke directly to the soul, and met one of the most practical needs and crying sorrows of the day—the menace of neurasthenia. The Emmanuel nerve clinic provided a unique medical social service, and between October 1907 and April 1909 125 patients out of 5000 were accepted. They were treated by psychotherapy only after each patient had been examined by a qualified physician to make certain his symptoms were not physical in origin. At first the medical consultants did the examining, later, each patient was required to be examined by his own physician.[32]

Worcester combined psychotherapy with weekly services, prayers, hymn singing, and meditation. He claimed to be able to treat melancholia, insomnia, nervous exhaustion, alcoholism, sexual perversions, fixed ideas and phobias, indecision, and hysteria. Those who most often sought help were spinster teachers or married women, mostly mothers of "moderate or restricted means." Loneliness and fear of failure were the most common complaints, especially among teachers, and financial worry among businessmen.

Worcester seated his patients in an old Morris chair in a quiet, darkened room. He would ask them to relax and would try to induce a "hypnoid" state. Then in a quiet voice he would repeat his healing suggestions. He claimed to have cured a sufferer from pseudo-angina pectoris after five months of telling him that his heart was beating quietly and regularly and that he would feel no pain. Versions of the Emmanuel Movement spread to Chicago, New York, Jersey City, Detroit, the New England towns, San Francisco, and Berkeley. Jews and Roman Catholics, with their clergy's approval, participated in Emmanuel services.

Worcester had studied psychology in Germany with Wundt and Fechner, and had heard Charcot, Janet, Bernheim, and Kraepelin. Abnormal psychology, he wrote, gave him a lively sense of the "reality of the soul as a living organism," as well as of the power of ideas.[33]

Years later, looking back on the state of psychotherapy in 1909, Worcester paid high tribute to Freud. By emphasizing the importance of instinct Freud and William McDougall had made psychology dynamic. Yet Worcester's appreciation of Freud, like that of so many other Americans, was divided. He thought Freud magnificent but often mistaken; he had an "unearthly power" to "describe the unseen." He was fearless in proclaiming "truths most humiliating to man." His logic was tenacious, his clinical single-mindedness admirable; he had an "uncanny art of coining names and phrases which stick." He had invented a precision instrument for the investigation of the mind, one that revealed the "hidden springs" of human conduct. Surprisingly, Worcester accepted Freud's wide application of sexual symbolism and the importance of sexuality in childhood. He praised Freud's "shrewd" discovery of rationalization. Granting all this, Worcester argued that sex was not the sole motive of human life; the Oedipus complex was "rare," especially in the United

States. Still, Freud, for all his drawbacks, was one of those geniuses who appear once or twice in a century.[34]

The unconscious of Worcester and McComb was derived not from Freud but from Von Hartmann, Schopenhauer, and American traditions of mental healing and Transcendentalism. The subliminal self was the source of instincts in men and animals that regulated action precisely and automatically: the trout's change in color in a new environment, the rhythmic beat of the human heart, the delicate responses of the nervous system. The subconscious also was the source of memory and of all those peculiarly accurate time-keeping functions that the hypnotists had studied. With Schopenhauer, Worcester asserted that the subconscious was the source of the disorderly, coercive powers of love and sexual attraction; in dreams the subconscious temporarily subdued reason and morality. Worcester also fervently believed that the subconscious was powerfully helpful. It was the locus of emotion and of will, which Worcester took to mean a force partly rational but largely instinctual and affective. The subconscious also was the source of man's reserve energies. As Sidis and James had argued, these could be tapped by lifting repressions, unclamping oneself from the grip of dead routines. Then the subconscious could pour forth its healing powers. Worcester argued that the subconscious was not all-powerful, nor did its existence obviate human will and effort. Yet, somehow the subconscious was "purer and more sensitive to good and evil than our conscious mind." [35] Its roots were the Infinite, it was closer to the Universal Spirit. The subconscious then was uncanny: it healed; it remembered everything; it solved problems; it could impart glorious, undreamed-of resources.

Bergson, the Prophet

Other popular enthusiasms besides the Emmanuel Movement strengthened this exaltation of the subconscious and closely identified it with instinct, intuition, and the freer expression of emotions. In 1909 William James inaugurated a vogue for Henri Bergson, the French "philosopher-scientist," and four years later Bergson lectured in America in overflowing halls. Several American psychoanalysts would interpret Freud through Bergsonian assumptions. What Bergson actually said is less important than what Americans thought he said. At fifty-two he was a miracle of success, according to the American rules, a shy professor, read and quoted all over the world. He appealed to a remarkably wide group—liberal Congregationalists, socialists and syndicalists, Theodore Roosevelt, society ladies, feminists, and rebellious young intellectuals. Only Catholics and cultural conservatives found him sinister. Irving Babbitt judged him to be a clever, insidious latter-day Rousseau.[36]

Bergson's appeal came from the fact that he was regarded as a competent scientist who yet rejected scientific positivism and materialism. Lyman Abbott, editor of *Outlook,* and Henry Ward Beecher's successor at Plymouth Congregational Church in Brooklyn, thought Bergson's doctrines of evolution described the workings of God. Bergson taught that evolution proceeded by unpredictable, stunning leaps, sudden mutations that need have no ascertainable connection with the past. The evolutionary process—and process, not fixity, was Bergson's central conception—seemed to be the expression of a Life Force, an *élan vital.*

Bergson also appealed to the younger generation of the non-churched, which considered itself "restless, searching, groping, questioning. . . ." Just out of Harvard, Walter Lippmann in 1912 welcomed Bergson as the "most dangerous man in the world" to stand-patters. Bergson, he wrote, was to philosophy what Roosevelt was to politics— "a fountain of energy, brilliant, terrifying, and important." Bergson offered "instead of a world once and for all fixed, with a morality finished and sealed . . . poverty accepted, learning taken from schools like a pill . . . a world bursting with new ideas, new plans and new hopes." Bergson spoke for all those who wanted the "whole of life in its tantalizing abundance." [37] This abundance, however, was to be appreciated less by reason than by instinct, feeling, and intuition. Feminists, too, found Bergson congenial. Because of his emphasis on human solidarity, on the instinctive and intuitive perceptions of the ordinary man as against the rational exercises of philosophers, some Americans could consider Bergson a democratic levelling influence. It was easy to identify instinct and intuition in everybody with the beneficent subconscious and the Life Force.

Bergson took a lively interest in psychopathology and in dreams. He had read *Die Traumdeutung* and ransacked the literature on aphasia, including Freud's study. His views of perception and memory formed the basis of his dream theories. In that oceanic imagery which contemporaries enjoyed, he once described life as an "immense wave" in which past and present formed an organic whole. Perception, he insisted, was chiefly memory. Memories, like "steam in a boiler" were "under tension" and constantly pressed toward consciousness.

Dreams chiefly arose from physiological sensations, much like the perceptions of waking consciousness. In an article specially adapted for the *Independent* in 1913, and in a section his editors subtitled "Freudian Theory," Bergson wrote that dreams could recapture youth and infancy in infinitesimal detail. In a vague and exhilarating passage he asserted that the memories in dreams were those "which can assimilate themselves with the color dust that we perceive, the external and internal sensations that we catch, etc., and which, besides, respond to the effective tone of our general sensibility." To this he appended a footnote: "This would be the place where especially will intervene those 'repressed desires' which Freud and certain other psychologists, especially in America, have studied with such penetration and ingenuity. (See in particular the recent volumes of the *Journal of Abnormal Psychology,* published in Boston by Dr. Morton Prince.)" [38]

Dr. Prince no doubt would have been displeased to have been linked so closely to Dr. Freud. But he would have approved another twist of Bergson's amateur psychopatholgy. This kind of research, Bergson daringly hinted, might even prove the reality of telepathy. He concluded with a stirring challenge that the psychoanalysts and their followers enjoyed quoting: "To explore the most secret depths of the unconscious, to labor in what I have just called the subsoil of consciousness, that will be the principal task of psychology in the century which is opening." [39]

The *élan vital* became a pleasant way to define the libido for some psychoanalysts. Putnam cited Bergson's views on perception as precedent for Freud's theories of memory. Bergson had gone beyond the "inadequate" theory of evolution that depended solely on "adaptation" and had restored the importance to human progress of intuition and faith. The *Independent* found his view of dreams truer and more inspiring than Freud's sordid speculations.

The enthusiasm for Bergson swept together several currents: Freud and abnormal psychology; a sanction for radical change in morals and society; a benediction for instinct and intuition. It was not hard to identify an overthrow of unhealthy repressions with Bergsonian liberation. For the Life Force surely would dispose progressively. This ebullient irrationalism was the polar opposite of Freud's reclamation of the unconscious.

Bergson also helped to inflate some of the popular balloons of the psychotherapy movement. He was invoked by H. Addington Bruce as one of those "animists" who along with William James and Sir Oliver Lodge made a belief in the soul respectably "scientific." Some of Bruce's chief "evidence" came from very loose interpretations of psychopathological research. Like Bergson he denied that the brain and the "mind" were identical. Bruce cited Boris Sidis's famous case of the Reverend Hanna. Bruce argued that his alternating memories proved that the "ego" which remembered past experience was not the same as the brain, which presumably had been damaged. He concluded that the brain was the "instrument" of a "purely psychical entity" that "uses the brain as a medium of expression and fashions it to serve the needs of expression." The brain was merely a "telegraphic exchange" facilitating or blocking communications from the soul.[40]

Bruce considered Sidis's provocative discovery of "latent reserve energies" to be one of his greatest contributions to psychopathology. Indeed, Sidis and James's doctrine of "vital reserves" became a widespread and exhilarating conviction, common to New Thought, the Emmanuel Movement, and some medical psychotherapists.

James himself wrote a popular article on the subject for the *American*. One had merely to break through the first layers of fatigue to discover "amounts of ease and power that we never dreamed ourselves to own. . . ."[41] As his prize example of a man energized by challenge, James singled out Theodore Roosevelt. George Santayana observed that in this essay James almost suggested that the "resources of our minds and bodies are infinite, or can be infinitely enlarged by divine grace."[42]

James's pragmatism and his belief in reserve energies were used to justify the most sanguine gospels of healing. Medical therapists such as Sidis had advanced claims sufficiently grandiose. But his were as nothing compared to those of practitioners on the fringe of respectability, for example, the nattily dressed, gorgeously moustached John Duncan Quackenbos of New York. He hypnotized patients with a red carnelian or a diamond mounted on the end of a gold pencil. A physician and teacher of rhetoric at Columbia, Quackenbos carried on a brisk practice among children and adults. He claimed to cure "neurasthenic insanity," alcoholism, drug addiction, moral diseases, kleptomania, erotomania, cigarette smoking, tea and coffee drinking, amnesia, neurasthenia. He also could alleviate physical diseases, including sclerosis and diabetes, or psychological quirks such as morbid self-consciousness, stammering, gluttony, delusions and obsessions, even pre-natal shocks.

It was awe-inspiring, he wrote, that a suggestionist like himself could "transfigure" human character in a single hour. Suggestion appealed to the super-sensible ethical forces that existed in the transliminal or "higher spiritual" sphere of life. But this was not all:

> Not only may dull minds be polished, unbalanced minds adjusted, gifted minds empowered to develop their talents, but the educating mind of the school child may tread the royal road to learning which ancient philosophers sought for in vain; the matured mind of the scholar may be clothed with perceptive faculty, with keenest insight, tireless capacity for application, unerring taste, and the imaginative mind of the painter, poet, musician, discovered, may be crowned with creative efficiency in the line of ideals that are high and true. And, finally . . . the state of hypnosis would seem to prove that we have within us an immaterial principle entirely independent of sense organs and sense acquisitions.[43]

A. A. Brill asked Dr. Quackenbos how he hypnotized his patients so efficiently: "If they don't succumb to suggestive sleep, I just give them a stiff dose of paraldehyde and they soon 'go under.' " Suggestions given then were "just as efficacious." [44]

The most popular mental treatments were offered by Christian Science and New Thought. Since the 1890's both had grown spectacularly: in 1882 there were not one hundred "Scientists"; in 1908 there were 85,000, a solid phalanx of upper-middle-class citizens. New Thought and Christian Science claimed to obviate every illness. They gave to the popular interest in psychotherapy a "moon-struck optimism," as James remarked.

"Despised by select and superior people," Santayana observed, Christian Science and New Thought flourished outside the pale of medicine. James, with his sympathy for the unorthodox, regarded them as examples of what he ironically called healthy-mindedness, a "wave of religious activity, analogous in some respects to the spread of early Christianity, Buddhism and Mohammedanism." New Thought, as one of its leaders, Horatio W. Dresser, explained, had grown up in an atmosphere of revolt against Calvinism: "The 'old thought' was undeniably pessimistic; it dwelt on sin, emphasized the darkness and misery of the world, the distress and the suffering. The new dwelt on life and light pointing the way to the mastery of all sorrow and suffering." [45] It assumed the goodness of the natural man and his capacity to transcend human limitations. In 1919 Dresser praised Freud as a profound psychologist, who, although not "spiritual," had thrown light on "desire, the will and the love nature." Psychoanalysis was closer to New Thought, Dresser argued, than were hypnotism or suggestion.

At a New Thought sanitarium on the Hudson young Max Eastman, who later took up psychoanalysis, sought treatment in 1906 for mysterious fatigue and backache. The sanitarium's décor included reproductions of Greek sculpture, plaster casts of American Indians, bouquets of roses, a real skeleton, a "draped and canopied couch," and pictures of Darwin and Thomas Huxley. Treatment combined psychic experiments, "non-church religion," admonitions to "conquer the world with sheer sentiments of optimism," electric shocks delivered through a serrated gold crown, hypnotism, and suggestion.

Eastman asked one patient, "All Well?" "Not perfectly well," she replied, "But *perfectly able to get well.*" Her emphasis, he recalled, "was evangelical," and there "ensued a conversation about psychic power, self-culture, the road to poise, the therapy of love, the accessibility of the universal life force. . . ." Another sufferer attributed his cure to "Christian Science . . . and pure will power backed up by whiskey." At the sanitarium Eastman met a charming young physician who was investigating mind cures, Beatrice Hinkle, who later became a Jungian analyst.

But the New Thought treatment failed, so Eastman went to the famous Dr. Gehring, who told him that his backache was only an "idea in the subliminal." He was set to chopping wood and to writing out a complete history of himself, his parents, and his grandparents. Gehring used suggestion, and this time Eastman seemed to be cured. He set about an article on the blessings of psychotherapy. Since his backache had gone, he was determined to demonstrate that there was no real distinction between functional and organic illness. The backache had been conquered by his own will power. The essay, published in the *Atlantic Monthly* in 1908, predicted a "momentous reform in the practice of medicine." [46]

Journalists tended to exaggerate, as did Eastman, the rosy consequences of medical psychotherapy. Much illness was merely functional, involved no lasting changes in the body, and therefore could be cured by purely psychological means. Character defects, too, could be transmuted into socially useful virtues. Bruce made these claims for the achievements of Janet, Prince, Sidis, and other psychopathologists:

"Lost personalities have been restored, deep-seated delusions have been uprooted, neurasthenics and hystericals . . . defective children, vicious boys and girls . . . have been remade into useful members of society; victims of liquor and drugs have been saved." [47] Racial powers of resistance could be strengthened by psychotherapy and this strength transmitted to future generations. While Bruce was insisting that "much so-called insanity" was curable, William Alanson White told social workers in 1908 that there had been no advances at all in the rates of recovery. Bruce translated the interaction of mind and body into the "power of mind over matter." The editors of the *Ladies' Home Journal* held out dazzling hopes to the insane. They prefaced a popular account of Dr. Prince's Miss Beauchamp with this notice: "Many people who for untold centuries have been shut up to die as hopelessly insane might have been cured as Miss Beauchamp was cured if the true nature of their malady and the methods Dr. Prince used—those of hypnotism—had been understood." [48]

The latest discoveries of abnormal psychology seemed to be shoring up lay beliefs in the power of "mind." Dr. Peterson remarked that it was the "age of rehabilitation and verification of old popular beliefs and superstitions." Journalists hailed Bergson's popularity, the mind cure cults and psychotherapy as proof that in America, at least, the twentieth century was opening in a great surge of "idealistic philosophy" and "spirituality" after the crass materialism of the preceding generation. It was as if the age of Transcendentalism were being reborn and its claims endorsed by modern science. America's "greatest modern discovery" was that mind was not drably bound to inevitable laws of nature. Rather, mind was itself "creative," and evolution could now take a spiritual direction.[49]

In a story about the Emmanuel Movement, Ray Stannard Baker wrote that a new living faith in free will had superseded the old fatalism: "The basis of the whole system is a vital belief based partly on religion, partly on the applications of the new psychological knowledge that a man is, indeed, largely the master of his fate; that there is new hope for the weakest and the lowest; that if a man will place himself where he is in the current of good and high thoughts, if he says, 'I do,' 'I will,' instead of saying weakly and hopelessly 'I cannot, I do not,' his life will become a new thing." [50]

The mind cure cults offered some very crass assurances of financial and social success. *American* magazine was especially eager to help readers over the "rough" parts of the upward road. Often "progress" meant simply getting ahead. Much psychotherapy was oriented toward binding up the wounds of those who failed or assuring the ambitious of new personal powers. Even the poor, it was implied, could uplift themselves, and the insistence that disease involved a social problem meant no lessening of the emphasis on individual responsibility. Some of the books that taught mental poise also taught the frank business inspiration of Orison Swett Marden's studies in "poise and power"—meaning ways to get rich and stay happy.

After 1908 the Emmanuel Movement drew increasing criticism from physicians. There were rumors that patients with cancer were being treated by suggestion. Cabot withdrew his support because his medical reputation and advancement at Harvard were being threatened. Putnam also withdrew after writing a polite but firm article condemning clergymen who treated illnesses, especially the neuroses, which demanded the highest level of medical knowledge and skill. Let clergymen comfort the sick, let doctors heal them, he concluded. In February 1909 the psychotherapy clinic was closed. Informally, however, Worcester continued the work, later through an independent foundation, and Isador Coriat con- remained as consultant.[51]

The Emmanuel Movement functioned as a transition from the supernaturalism of the mind cure cults to scientific psychotherapy. When Worcester hinted that the subliminal self disposed of the infinite resources of the universe, he retained a strain of Transcendentalism and of the hopes of psychic researchers, of whom he was one. Yet, the Movement also endorsed and publicized the medical therapies of Sidis, Prince, Janet, and Freud. Indeed, Emmanuel purported to be solidly grounded in the "latest" scientific advances, although religion still was central, as it was to the New Thought. The Movement's therapists were priests; the physician still was an adjunct, not yet the "scientific medicine man" of a new cultural order. Medical psychology always had operated to reinforce social norms. But the Emmanuel Movement enlarged the psychotherapist's role by making him explicitly responsible for moral problems. Worcester and McComb probably helped to make medical psychotherapy acceptable to thousands of members of the Protestant churches, and they inspired pastoral psychiatry.

The endorsements of psychoanalysis by Henri Bergson and the Emmanuel Movement exemplify the ambiguity of Freud's appeal to Americans. Samuel McComb expressed this precisely when he insisted that Freud was an ally against the "transcendental mystery mongers" as well as against the contemptuous scientific materialists. It was against this accent on mystery as well as medicine that Freud protested in the *Boston Evening Transcript* interview.

The Emmanuel Movement not only helped to prepare the way for psychoanalysis but, together with the enthusiasm for the mysterious subliminal self, also helped to compromise it. To cultural conservatives the insistence that the subconscious contained the "best" elements of personality undermined the older faith that "character" rested squarely on fully conscious moral choice and that the battle against evil and temptation lay in the power of the will to control the passions. To psychologists who prided themselves on their scientific Wundtian heritage the subconscious of the Boston school of psychotherapy, and of Josiah Royce and William James, represented a regression to the "occultism" of the Middle Ages. In this atmosphere of popular enthusiasm and medical and academic skepticism the professional debates about psychoanalysis were carried on.

X

The Repeal of Reticence
1911 - 1914

Crusades

During the formative years of the psychoanalytic movement, from about 1911 to 1914, the crisis of "civilized" morality became acute, embittering debate about Freud's theories. On the level of national discussion came a repeal of reticence about venereal disease, prostitution, and the sexual enlightenment of children. A very few medical authorities revised "civilized" sexual hygiene, a revision that at first affected the outlook only of a few. Nevertheless the revisionist position, formulated chiefly by Havelock Ellis, became closely identified with psychoanalysis by 1914 and became increasingly influential in the 1920's.

Throughout the nineteenth century social radicals had rejected every aspect of "civilized" morality. The Owenites and the Fourierists repudiated not only monogamy but the family, industrial capitalism, and Christianity. This rejection of an entire civilization horrified English Victorians and respectable Americans. Anarchists, Emma Goldman among them, continued this tradition in the twentieth century.

But the most influential criticism came not from a tiny group of radicals but from representatives of the dominant minority, who were deeply committed to the central values of "civilized" morality yet found some of its provisions onerous to live with. These critics included physicians, educators, psychologists, feminists, clergymen. Some of them complained, for instance, that late marriage made the preservation of chastity a heroic achievement during youth when the sexual passions were strongest. Chastity created painful conflicts in some professional and intellectual Americans who as a group married latest. G. Stanley Hall, a poor farmer's son who married at thirty-five preserved the language of revivalism to describe the battle with lust: "The chief sin of the world is in the sphere of sex, and the youthful struggle with temptation here is the only field where the hackneyed expressions of being corrupt, polluted, lost and then rejuvenated, of being in the hands of a power stronger than human will become literally true." [1]

Critics also agreed that the trials of chastity were made more difficult by ignorance and misinformation, for which reticence as well as medical quackery were partly responsible. Physicians who dealt with sexual problems and women reformers had been pleading for the enlightenment of children since at least the 1870's. Parents exhorted children to lead pure lives, yet seldom enlightened them about sexual functions. They learned about these in garbled and distorted ways. Often laymen absorbed outdated or extreme medical views.[2] Popular tracts on sexual hygiene, even those written by physicians, were often inaccurate and alarming. G. Stanley Hall assumed for years that nocturnal emissions were abnormal and were a sign of waning manhood. But that was exactly what Dr. Cowan's widely read *Science of a New Life* and thousands of tracts and advertisements taught. Girls rarely were informed about sexuality at all.

Ignorance was so widespread, especially among cultivated women, that Sir James Paget, Queen Victoria's personal physician who was widely respected in the United States, insisted that "the method of copulation" needed to be taught. Physicians described situations of nuptial ignorance extreme enough to be implausible, without some knowledge of the pervasiveness of "civilized" reticence. One exasperated physician from Red Lodge, Montana, pleaded in 1896 for instruction in the physiology of sex and bitterly protested the identification of sexual desire with uncleanness: "Great God! the Creator of all Nature, is it a crime to possess a penis or a vagina? Or possessing one, is it a crime to feel the natural desire to put it to its intended uses?" [3]

The "civilized" code had other self-contradictory results. The ascription of sexual passion almost exclusively to the male had unfortunate results; fear of male lust sometimes kept girls from marrying. A Boston Unitarian novelist, Mrs. Kate Gannett Wells, reported in 1891 that girls were learning from such literature as Tolstoy's "Kreutzer Sonata" and the novels of Zola that male sexual behavior was bestial.[4] Girls also heard their mothers complain that their husbands' marital demands made life unbearable for them. Physicians lamented that wives were sexually anaesthetic and submitted without pleasure to their husbands' embraces.

The male counterpart, horror at displaying sexual passion before a pure woman, sometimes caused impotence. William Hammond, the neurologist, successfully treated a patient who had been "profligate" with prostitutes but could not bear to subject his "intelligent, refined, beautiful" wife to "such an animal relation as sexual intercourse." "I ought," he told Hammond, "to have married a woman used to this sort of thing." A. D. Rockwell, a gynecologist who worked with George Beard, noted that sexual disturbances caused a "vast sum of misery which is either silently borne, or finds expression in some sort of crime or in suicide." [5]

The "taboo of silence" was broken by crusades against prostitution initiated by reformers in the name of purity. These began in the 1880's, gathered momentum in the 1890's, and became associated with progressivism, especially after the turn of the century. Like other American crusades, the purity drives relied on exposure and shock, sometimes on Quixotic legislation, but chiefly on the universal panacea of education.

The first crusades were directed against any government regulation —and thus official tolerance—of prostitution. Like many other reforms, this was based on English precedent. Prostitution had been an urban institution since colonial times, but there were alarms that it was increasing dangerously toward the turn of the century.

The first purity campaigns enlisted old anti-slavery crusaders such as William Lloyd Garrison, Julia Ward Howe, and Aaron Powell. Feminists, clergymen, temperance leaders, medical extremists, and purists who detested the ballet and the theater met at the first National Purity Congress in 1895. Theodore Roosevelt, then president of the New York Board of Police Commissioners, sent a note denouncing the double standard: "I will not have one law for men and one for women; they shall be treated exactly alike, so far as I am concerned." [6]

The crusaders' first demand was for frank and open discussion. "Purity through knowledge," Benjamin O. Flower told the Congress, was the "supreme demand of the day." So strict had convention become, he indignantly reported, that the clergyman who edited the *Christian Life* was arrested on charges of obscenity for printing a candid sermon calling for "higher moral concepts in the marital relation." Reticence, Flower charged, in fact had protected vice. He attacked the immorality of his own social class:

> Generations after generations of men [have indulged] in licentiousness as a recognized prerogative of their sex, forgetting that they tincture their offsprings with the poison which colors their thought world and courses through their veins. . . . This false though popular dictum that is known as the double standard of morals . . . furnishes the cases upon which the vicious 'wild oat theory' rests, and we find our young men poisoning their imaginations, polluting their souls, enervating their will power and corrupting their blood by wild debauchery and unrestrained licentiousness, after which, with the sanction of conventional society, they marry virtuous girls and curse civilization with morally tainted progeny.[7]

The Congress heard the Reverend S. S. Seward, a New York minister, advise parents to inculcate in their children a horror of masturbation, the "first step" toward later impurity. Dr. J. H. Kellogg praised the invigorating effects of continence. Frances Willard, president of the National Woman's Christian Temperance Union, congratulated the Congress on its frankness. "Ten years ago we dared not talk about [purity] nor even tell each other what we thought." A Philadelphia clergyman complained that those who tried to speak out were reprobated as "poor zealots," and he urged "God's prophets" to "lift up their voices like trumpets." [8]

The Purity Congress and especially the muckraking revelations of prostitution received considerable newspaper publicity. Newspapers traditionally exploited divorce trials, vice, and sexual scandal. Indeed, some purity groups hoped to purify the press as well as eradicate prostitution. Yet, the Congress and purity agitation did not receive much publicity from respectable national magazines, still the arbiters of taste. Organs of reform, such as Flower's *Arena,* the publications of the Women's Christian Temperance Union, and church periodicals, sometimes discussed these issues. On the whole, however, there had been no major, national destruction of reticence. As late as 1912 Jane Addams declared that clergymen, physicians, newspapers, and decent people still preserved hypocritical silence.[9]

During the 1890's and the early 1900's, the purity movement, David Pivar has suggested, developed a new leadership.[10] The old moralistic tone gave way to a more pragmatic and "scientific" approach. Perhaps the chief new allies of purity were distinguished physicians, whose support represented an important change in attitude. In the 1870's the president of the American Medical Association had appealed for the regulation of prostitution. By the late 1880's medical organizations had dropped this demand, largely because of the failure of regulation in Europe and pressure from purity crusades and women's groups. In the 1890's and especially after the turn of the century physicians began to insist on the prophylaxis of chastity. The ranks of the purity crusaders were joined by sociologists such as E. A. Ross, social workers such as Jane Addams, philanthropists like John D. Rockefeller and Julius Rosenwald, former president Charles W. Eliot of Harvard, G. Stanley Hall, prominent businessmen, and highly respected physicians such as the New York urologist Prince Morrow.

The new campaigns, directed against the "great black plague" of syphilis, were as much medical as moral, and proclaimed the saving power of scientific knowledge. It was the progressive crusades that repealed reticence on a national scale, in respected magazines such as the *Ladies' Home Journal* as well as in reform organs and the daily press. Prince Morrow, perhaps the most important organizer of the movement, opened the campaign among sociologists, physicians, and women's clubs around 1905. He insisted that the public be informed of the seriousness of venereal diseases, how widespread they were, and how readily innocent women and children contracted them. "Every moral reform," he wrote, "comes from the exposure of human suffering." [11] He estimated that 75 per cent of all adult males had or had had gonorrhea, and from 18 to 20 per cent syphilis. Far worse, large numbers of "pure" wives contracted these diseases from their husbands. Ignorance and the double standard were the root of the evil. The Wassermann test and the Salvarsan treatment for syphilis, discovered in 1906 and 1910 respectively, gave new hope for practical results from an educational campaign.

Between 1905 and 1917 purity campaigns were carried on by literally dozens of state and local groups, and by 1913 voluminous reports publicized the work of vice commissions in twenty-one major cities from New York to San Francisco, and these received considerable attention in the press. In 1911 Edward Bok opened the *Ladies' Home Journal* to a crusade against prostitution. No topic should be banned if it were in the interests of true morality to discuss it.[12] President Eliot, who gave up an ambassador's appointment to crusade for purity, voiced the reformers' demand that the importance of sexuality be openly recognized. The swirls and tumults of sex, he wrote, provide much of the normal satisfactions of life, as well as life's worst anxieties and afflictions.[13] Yet the repeal of reticence troubled him. Among the most conservative of the crusaders, Eliot thought that publicized medical knowledge might be a mixed blessing. It must spread the idea that the innocent could become infected. Inevitably it also would advertise the fact that these diseases, once believed to be the wages of vice and sin, could be cured, and thus immorality actually might be fostered. The best hope, Eliot held, was a campaign against male lust and against the weakness of complaisant women who supplied the "demands of men." Brothels should be closed, and the profit taken out of vice. Schools, churches, parents, the YMCA, and the press should unite in a campaign conducted with sufficient delicacy not to arouse prurient and morbid suggestions. Pure thoughts as well as pure deeds must be encouraged. Other proponents of sex education, more radical than Dr. Eliot, also were troubled by the social effects of their efforts. For in practice a distinction between prurient and educational publicity was hard to maintain. The movies, the theater, and the novel all took up the theme of prostitution with a new boldness.

Ironically, the anti-prostitution campaigns publicized how popular immorality in fact was. One educator feared that young men might gain the impression that vice was excusable because it was so widely indulged. The investigations revealed shocking discrepancies between "civilized" morality and actual behavior, clashes between upper-middle-class attitudes with those of other social groups, and antipathy between urban and rural values.

Vice commissions testified to the ubiquity of the "social evil." It was absent, the New York Commission reported, only from "white rural districts." In Chicago possibly one quarter of the city's males visited prostitutes, and paid them fifteen million dollars a year. More than half of Chicago's prostitutes were recruited between the ages of sixteen and eighteen and came from farms and small towns in the Midwest.[14]

The fact that many young immigrants came alone, without families, provided inevitable problems. Possibly each group, Catholic farmers from Ireland or Italy, orthodox or free-thinking Jews from Austria-Hungary or Russia brought their own slightly differing sexual customs and attitudes. Some observers thought that the Jews, regarded as an especially chaste race, quickly degenerated in the urban American environment. Jane Addams believed that loosened family ties, loneliness, and despair contributed not only to immorality but to actual mental derangement among the new immigrants.[15] Analyses of the "social evil" often associated prostitution with such "foreign" amusements as drinking, dancing, and music halls.

Not only immigrants but Negroes and the very rich also were considered especially immoral. The rich, one reformer was convinced, flouted convention because they could afford to. Theodore Roosevelt was certain that Newport society was "rotten" with "frivolity and vice." High venereal disease rates among Negroes were cited as evidence of immorality.

It was in the cities that new patterns of moral behavior were emerging from the clash of standards. The city was hated by many rural newcomers, because it seemed to violate every sacred precept. One American writer recalled "the sense of shocked and shamed decency I felt when first I came to the city, a boy almost, and fresh from the country; how I tossed in my bed trying to see as right things that everyone in the city accepted as a matter of course, but that, from earliest boyhood I had been taught to regard as wicked. I could not for many months become accustomed to seeing immodestly dressed women or to hearing half-veiled indecency flaunted from the stage, blazoned in the newspapers, or used even in ordinary conversation. I could not get used to . . . scenes and actions directly forbidden as unforgivable at home." [16] Brand Whitlock, the reforming mayor of Toledo, Ohio, detested the harsh judgments of rural preachers who believed that all prostitutes deliberately had chosen a life of shame. The vice investigations did much to muffle the customary thunder directed at the fallen.

Most students of prostitution, including the municipal vice investigators, blamed economic and social injustices as fundamental causes. Starvation wages for women failed to keep pace with the rising cost of living; keen competition kept women workers unorganized and unprotected. The Minneapolis vice report suggested that the "demoralizing and extravagant display" of the "newly and suddenly rich" led poor girls to take up an "easy way" to remove the contrast. Poverty had been singled out as "the hotbed of impurity" at the Purity Congress by William Lloyd Garrison: "In every city the tenement house, with its crowded occupants, and unhealthy ventilation, with the impossibility of individual privacy and delicate reserve, produces its natural fruit." [17]

Because the poor were often immigrants and formed the mass of unskilled workers in many cities, the association grew between vice and the foreign born. In 1913 a New York vice report suggested that "foreigners by birth" furnished the majority of the "exploiters of prostitution." [18]

The vice campaign publicized devastating attacks on "middle-class" prudery and hypocrisy. With her customary indignant clarity, Jane Addams noted that the mistresses of servant girls who had been "ruined" usually fired them as "disgraced forever and too polluted to remain for another hour in a good home." Actually confronted with prostitutes, she noted, social workers experienced such "distaste and distress" that sometimes they suffered an "actual nervous collapse." [19]

When John D. Rockefeller, Jr. supported a performance of Brieux's *Damaged Goods* in New York in 1913, he was abetting a criticism of middle-class prudishness that was in fact hostile to much of the American moral outlook. Richard Bennett, whose best known role was Sir James Barrie's *Little Minister,* played to a house packed with "physicians, settlement workers, eugenists, philanthropists, authors, suffragists, ministers and university professors." [20] The performance had been organized, by coincidence, by Freud's nephew, Edward Bernays, then working on the staff of the *Medical Review of Reviews.* It was edited by Ira S. Wile, a New York pediatrician who became interested in Adlerian psychoanalysis. Fearing that *Damaged Goods* would be closed by Anthony Comstock, as Bernard Shaw's *Mrs. Warren's Profession* had been in 1905, Bernays organized a Sociological Fund Committee of the *Review* made up of distinguished men and women to sponsor the performance. It would be a blow against the reticence that prevented frank discussion of venereal diseases, Bernays thought. The denunciation of prudery was conventional enough. The physician berated an indignant politician whose daughter had been infected with syphilis by her husband: "I should like to know how many of these rigid moralists, who are so choked with their middle class prudery that they dare not mention the name of syphilis or when they bring themselves to speak of it do so with expressions of every sort of disgust, and treat its victims as criminals, have never run the risk of contracting it themselves." [21]

The doctor went on to suggest that of course young men kept mistresses before they married. So the only way to prevent the spread of syphilis was this: "To live with only one woman, to be her first lover, and to love her so well that she will never be false to you." Brieux's play treated sexual passion with a casualness and a French practicality entirely alien to the spirit of the purity crusade in whose name it was performed.

The campaigns against prostitution initiated the public movement for sex education that some educators, psychologists, and physicians long had been urging. Instruction in sexual hygiene would prevent vice by publicizing the dangers of venereal disease and would encourage purity by a decent knowledge of the facts of sexuality. Sex education became the social hygiene movement's essential panacea for the destruction of vice. Because of the prevailing reticence, almost no suitable literature was available. Beginning around 1909 a number of texts and pamphlets were prepared. On the whole these made few changes in the older doctrines of sexual hygiene, except perhaps in tone.

Every hygienist attacked a male folk code that prescribed the opposite of chastity. Endemic among men of all classes, according to its scandalized opponents, this code taught that purity was feminine, a "priestly contrivance" fostered by "old men and home-keeping wits." Incontinence was a sign of manhood, and men should exercise their sexual organs to ensure good health and virility. Repeatedly American physicians and the American Medical Association announced that continence was entirely compatible with health and was an absolute necessity for the unmarried.[22] Many of the older doctrines were retained almost intact: the belief that women were purer and less "sexual" than men, that true masculinity and feminity were expressed by "right living," that will power must control the dangerous sexual passions.

The chief departure was an emphasis on affection and an insistence that sexuality was sacred not base. Sex hygienists rejected "monasticism," "asceticism," and "celibacy," as well as the notion that sexuality was dirty and shameful. "We must teach our sons," one reformer wrote, "that the union of man and woman is a sacrament . . . that this passion is good, not bad; that it can . . . and must be put to the noblest uses." [23] Sexual intercourse was legitimate only when sanctified by love.

In the instruction of boys, warnings about masturbation were less dire. Nevertheless, as eminent a hygienist as Dr. Winfield S. Scott, professor of physiology at Northwestern University and author of pamphlets distributed by the YMCA, argued that masturbation would lead to loss of virility. His theory was that nocturnal emissions did not evacuate the product of the interstitial glands, as did masturbation, which was therefore debilitating and un-manning. Scott continued to teach young men that the only justification for sexual intercourse was reproduction.[24] G. Stanley Hall, who played an active role in the social hygiene movement, like the others preached an end to reticence, and yet a conventional sexual hygiene.

In 1912 Dr. Ralph Reed, a Cleveland psychiatrist and member of the American Psychoanalytic Association, placed Freud, as did Putnam, among those who were for the first time seriously investigating sexuality. Freud was more profound than Krafft-Ebing or Havelock Ellis. Reed noted that his own generation had been spoiled by their upbringing, and that Americans did not have a "popular yet proper language" of sexuality. He advocated the sexual enlightenment of children, not by puerile and foolish books in school, but by parents, especially mothers. The impact of Freud and psychoanalysis on the movement was largely indirect before the war.[25]

Havelock Ellis and the New Sexual Hygiene

However, the social hygiene movement gave limited publicity to a far more radical authority, Havelock Ellis, the British physician and writer who reversed nearly every doctrine of traditional sexual hygiene. Because of his close connection with psychoanalysis, the enthusiasm he aroused in America, and his scientific reputation among many physicians, Ellis played an important role in reformulating theories of hygienic sexual conduct. His prestige among American and European intellectuals and physicians before and after the war was high. A *Festschrift,* issued in 1929, included tributes from Ellen Key, H. L. Mencken, Judge Ben Lindsay, Waldo Frank, Franklin H. Giddings, Bertrand Russell, Margaret Sanger, Clarence Darrow, and Bronislaw Malinowski.[26] Ellis stood for the very opposite of the common American view that sex was dangerous and must be painfully disciplined, for he believed in the positive beneficence of the sexual instinct.

Ellis enjoyed both a public and an esoteric reputation in the United States. From 1890 to 1910, while Freud was still relatively obscure, Ellis gained praise and popularity as a writer of scientific distinction and literary charm. He began as an admirer of Galton and Lombroso; and in books, magazines, and journals he discussed eugenics, criminology, dreams, hysteria, psychology, and sex. He brought this miscellaneous knowledge to bear on the characteristics of *Man and Woman,* which went through four editions from 1894 to 1904. He considered it an introduction to his life's work, *Studies in the Psychology of Sex.* The first six volumes were written between 1897 and 1910 and published in Philadelphia because of legal difficulties in England. In 1898 a London bookseller was convicted of selling "a certain lewd, wicked, bawdy, scandalous and obscene libel," Ellis's *Sexual Inversion.* The author, whose hair turned white during the trial, at once arranged to have the *Studies* brought out by an American medical publisher, F. A. Davis Co., with a "high reputation and an army of travellers." Ellis agreed that the *Studies* be sold "only to physicians and lawyers." [27] An exasperated young intellectual protested: "It is worth your life to get Havelock Ellis's six volumes from a bookstore or a library. You can do it only with a doctor's certificate or something of the sort." [28] Nevertheless, they sold briskly, and went through several printings. The *Journal of the American Medical Association* predicted that because of "morbid demand" *Sexual Inversion* might reach a dozen editions, unless Anthony Comstock interfered. *The Evolution of Modesty,* which appeared in 1900, was widely acclaimed, and an excerpt appeared in the *Psychological Review.* Physicians, on the whole, considered the book "painstaking, careful, thorough, scientific," a sign of an "awaking public interest in the settlement of sexual problems" in all civilized countries. Even the *Nation* was enthusiastic, although noting that the book could hardly be recommended to Mrs. Grundy: "If ever Professor William James's motto, 'Be not afraid of Life!' takes on a profounder significance than elsewhere in the world of human phenomena, it is in the presence of these facts to the elucidation of which Havelock Ellis is devoting the best years of his life . . . nowhere else can one find a resume of the best scientific knowledge about sexual life in its narrowest and in its broadest implications. . . ." [29]

This praise was followed by discreet silence. The *Nation* and other American magazines ignored most of Ellis's subsequent volumes in the *Studies.* For they contained much more frank detail, often of practices considered perverse, full and unexpurgated case histories, and a thorough attack on conventional norms. Even the psychology journals treated the *Studies* judiciously. The *Review* ran two notices of *Sexual Inversion,* an unfavorable one by a little known psychiatrist and high praise by James

Mark Baldwin. Hall's *American Journal of Psychology* printed a part of volume six, *Sex and Society,* but contented itself with the barest summaries of the others. Probably Hall felt ambivalent about Ellis, whom he considered a courageous authority who was, nevertheless, too concerned with perversities.[30] Physicians, however, generally continued to praise the *Studies.* The most enthusiastic medical reviewers were Smith Ely Jelliffe and William Alanson White, who welcomed Ellis's work before their conversion to psychoanalysis. White pronounced the fourth volume, *Sexual Selection,* "careful, painstaking," quite the best modern work in English on this "most obscure domain of psychology." [31]

Ellis was one of the first writers in English to appreciate Freud's importance. Nearly every volume of the *Studies* and most of Ellis's other work praised him and cited the findings of psychoanalysis. A casual reader hardly could have read Ellis without learning about Freud. In 1895 in *Man and Woman,* Ellis had suggested, as had a few other British physicians, that sexuality might play a role in hysteria. The next year he came across a fourteen-page review of the Breuer and Freud *Studies* in *Brain,* the British neurological journal. He began a correspondence with Freud that continued sporadically for years. Then he sent Freud a copy of his article in the St. Louis *Alienist and Neurologist* on hysteria, which, Freud noted with pleasure, began with Plato and ended with himself. Freud, Ellis contended, had restored the sexual emotions to a role in the etiology of hysteria with a series of brilliant, laborious, and insightful studies. This was one of the few lights of praise from the outside world that in these years mitigated Freud's sense of gloom and isolation. Freud considered Ellis "highly intelligent" and later expressed mild envy at his "richer and happier nature," his "many-sided and harmonious life." [32]

It was a totally wrong interpretation, for Ellis was tortured much of his life by impotence and by a tangled marriage. His own volumes on sex were perhaps as much personal therapy as scientific investigation. Ellis's St. Louis essay was incorporated into *The Evolution of Modesty,* which together with *Sexual Inversion* and the *Analysis of the Sexual Impulse* may have had some effect on Freud's sexual theories. In 1905 Freud acknowledged that his *Three Contributions* had been influenced by case histories similar to those of Ellis's studies. These had concerned adults who remained normal despite having been seduced in childhood. So Freud had revised his views to take more account of the "sexual constitution" as affected by both heredity and by very early experience.

Freud later adopted the term "Narcissism," which Ellis had introduced.

Ellis in turn continued to use Freud's rapidly growing contributions. Ellis claimed that his own studies and Sanford Bell's led Freud to push back "the sexual origins of neuroses to an even earlier age, and especially to extend this early origin so as to cover not only neurotic but ordinary individuals. . . ." [33] Certainly his most provocative application of psychoanalysis came in the sixth volume of the *Studies, Sex in Relation to Society,* published in 1910. In five out of ten citations, Ellis used Freud to prove the baneful effects of "civilized" sexual morality and especially sexual abstinence. He included the most radical of Freud's assertions in the 1908 essay on "Modern Nervous Disorder" without the cautionary reservations. Women tolerated abstinence less easily than men; abstinence caused anxiety and fostered weaklings; most people's capacity for sublimation was strictly limited, and a certain amount of "direct satisfaction" was necessary, lest "morbid manifestations" result. The apparent intellectual inferiority of women came from "the inhibition of thought imposed upon them for the purpose of sexual repression." Ellis went on to quote the popular poet Ella Wheeler Wilcox that "sorrowing virtue is more ashamed of its woes than unhappy sin, because the world has tears for the latter and only ridicule for the former." [34]

Jelliffe greeted *Sex and Society* ecstatically in the *Journal of Nervous and Mental Disease* in 1912. Ellis's previous volumes, he wrote, "are well known to our readers." He interpreted exactly the spirit of Ellis's liberating message: "Society is encrusted with the living and dead barnacles of former times, and few individuals are aware how formal, borrowed, and imitative are their speech symbols and their attitude of mind towards all questions." [35]

Ellis approached sexuality from a relativistic, historical, and anthropological standpoint in order to destroy nearly every tenet of "civilized" morality and nineteenth-century sexual hygiene. A true understanding of sexuality could come only by studying its infinite variety in animals as well as in men, primitive and civilized. With all the voracious international scholarship of the pre-war period, Ellis soaked up the existing European and American literature. The *Studies* summarized the lively and extensive American research in sexuality from Sanford Bell's study of children to the neurologist Smith Baker's analysis of "Conjugal Aversion" and Judge Ben Lindsay's "Why Girls Go Wrong."

With William Graham Sumner, Ellis insisted that sexual customs were humanly ordained: "Immoral, never means anything but contrary to the *mores* of the time and place." Therefore religious grounds for determining desirable behavior were irrelevant. Where Acton had cited the church fathers for their wisdom, Ellis often used them to demonstrate the blindness of Christian asceticism. He chose this from Saint Bernard: "Man is nothing else than fetid sperm, a sack of dung, the food of worms. . . ." [36] Ellis enjoyed arguing that marriage did not become a Christian sacrament until after some thousand years of church history.

Where William James had argued that chastity was the hallmark of high civilization, Ellis insisted that savages, although not necessarily chaste, regulated sexuality more carefully and were less promiscuous than civilized men and women. It was only among the "lazy classes of human society" that sexuality was both "unnaturally stimulated and unnaturally repressed" so that "the instinct of detumescence" was "ever craving to be satisfied." [37] Perhaps, judging from his own case, he argued that rather than needing to be crushed, the "instinct of tumescence" in men and women required the most elaborate and prolonged stimulation.

Ellis rejected the nineteenth-century physicians' identification of the sexual and the reproductive instincts. Like Freud he argued that they were by no means the same. Those who referred to sexuality euphemistically as "the reproductive instinct" usually were "unconsciously dominated by a superstitious repugnance to sex." Reproduction was the "natural end and object" yet was not part of the "contents" of the impulse.[38]

On both psychological and medical grounds Ellis abandoned the virtues of continence. Physicians had rejected Acton's hypothesis that semen was reabsorbed and thus vitalized the body and the secondary sexual characteristics. This function now was reserved for the interstitial glands. Thus Acton's physiological grounds for continence as conducive to health, energy, and success were no longer valid. Instead of laying down rules for the desirable frequency of sexual intercourse, as most physicians had done, Ellis, like George M. Beard earlier, insisted that individual variations within the limits of normality were immense. Even the perversions, he argued, were germinally present in normal people. He was far more tolerant of masturbation and homosexuality than most American physicians. The former was nearly universal and comparatively harmless, except in excess. Homosexuality usually was the result of "congenital abnormality," as well as environment, but was not among the "stigmata" of degeneracy, as many authorities, among them Krafft-Ebing and Morel, had assumed. Homosexuality was frequent in the United States and England, and he included a number of American cases and studies. Ellis was skeptical of "cures" for homosexuality and argued that if an invert were made "healthy, self-restrained and self-respecting, we have often done better than to convert him into a feeble simulacrum of a normal man." He was especially critical of the German psychiatrist Schrenck-Notzing, who claimed successful cures by a combination of hypnosis and the brothel.[39]

Ellis also rejected the later nineteenth-century's insistence that women were without sexual passion. The strength of this attitude is clear from a review in 1907 in the *New York Medical Journal,* which noted that Ellis had overthrown the "very common view" that women were "naturally frigid and liable to sexual anaesthesia. . . ."[40] Rather, Ellis argued that they were as capable of sexual feelings as men, although for the most part they had not been trained to experience them. Male ineptness and ignorance of the arts of love, Ellis argued, contributed to this mistaken stereotype.

His zealous investigations turned up numerous case histories of middle- and upper-class English and American men and women who violated every canon of "civilized" morality. Often they reflected nearly all the nineteenth-century's sexual preachments: disgust, remorse, religious fears, crises of conscience. Yet, they did not become severely neurotic and gradually had ceased to feel these compunctions. Perhaps it was these object lessons in conduct that gave Margaret Sanger, the pioneer of birth-control clinics in America, "psychic indigestion" when she read Ellis's six volumes all at once.[41]

Ellis took pains to demonstrate that Americans, both physicians and laymen, had exaggeratedly "Puritanical" attitudes toward sexuality. American authorities dilated on the evils of masturbation; a distinguished gynecologist solemnly announced, "I do not believe mutual pleasure in the sexual act has any particular bearing on the happiness of life. . . ."[42] Ellis ridiculed those who believed that sexual lapses could be prevented by laws. American legalism was "clearly a legacy of the early Puritans." Like Calvin's Geneva, America had a passion for useless legislation. In 1907, for example, New York had made adultery punishable by prison or fine as the result of a campaign by the National Christian League for the Promotion of Purity. But the new law had become at once an unenforceable dead letter. The Chicago Vice Commission advocated searchlights for the public parks. Since sexuality was in no way "impure," nor the loss of sexual secretions debilitating, the older strictures against lascivious thoughts vanished from Ellis's sexual hygiene.

Considering the extravagances of "civilized" morality, Ellis probably performed a real and courageous service. Yet his positive solutions had their share of paradox and naïveté.

He had a transcendent faith in what he called the "gracious equilibrium of Nature" or "Nature's rhythmic harmony." Once the conventional and religious basis of sexual morality was destroyed (he once likened legal marriage to a corset), the beauties of body and spirit would flower. With his far more radical friend, Edward Carpenter, anarchist and defender of the intermediate sex, Ellis asked, "Why . . . should people be afraid of rousing passions which, after all, are the great driving forces of human life?" [43]

He by no means advocated that promiscuous coupling replace Christian marriage. Rather he desired a new chastity, emptied of "unnatural" forms, that steered a course between abstinence and license. "Asceticism and chastity are not rigid categorical imperatives; they are useful means to desirable ends; they are wise and beautiful arts." "An absolute license is bad," he continued, "an absolute abstinence . . . is also bad. They are both alike away from the gracious equilibrium of Nature. And the force, we see, which naturally holds this balance even is the biological fact that the act of sexual union is the satisfaction of the erotic needs, not of one person, but of two persons." In a "sane natural order," he insisted, "all the impulses are centred in the fulfillment of needs and not in their denial." Such fulfillment could best be attained by restoring the "art of love" to that honored place in society from which Christianity had removed it. Indeed art, he wrote in 1914, which involved the gentle training of Nature, was the best governor of morals. Morality should be neither entirely rooted in instinct and tradition or in abstraction. As the Greeks and Walt Whitman had preached, ethics was an aesthetic achievement.[44]

Because of the importance he placed on sexual fulfillment, Ellis argued that intercourse was an entirely private matter. He advocated trial union as a way of assuring greater sexual compatibility in marriage. Only if children were born did intercourse become the concern of the state, which then could insist on the couple's obligation to the child. This obligation, however, did not restrict either partner's sexual freedom. Lifelong fidelity, like lifelong sexual attraction, simply could not be guaranteed and therefore should not be promised. How this would affect the raising of children Ellis did not rigorously explore. He merely maintained that liberty would foster not license but responsibility. Again, he invoked the magic of Nature. Consulting the habits of apes, wasps, and primitive peoples, Ellis concluded that monogamy was clearly ordained. In "fluid form" it was a natural impulse that had existed "in human nature from the first." Ellis also believed in progress. Civilization was moving toward ever greater social order and cohesion and ever greater individual freedom.[45] The new ideal of sexual hygiene, then, was an experienced couple, who filled each other's sexual needs, and for whom, it is probably fair to say, children were a secondary consideration.

In the years that Bergson and the popular psychotherapists were preaching trust of the unconscious, Ellis was rehabilitating the sexual instinct. Their message, founded on a belief in human goodness, was primarily one of liberation from the old restraints.

In any case, Bergson and Ellis were important, because it was often through their eyes that many partisans of psychoanalysis interpreted Freud, whose assumptions were quite different. He was dubious of the beneficence of the sexual instinct and by 1913 argued that the attainment of normality and sexual love were complex, often never completed processes. Yet, Ellis was blind to these sterner aspects of Freud, for they clashed so completely with his own optimism. That Freud agreed with him he assumed in most of his work. Finally, in a later essay, at once shrewd and mistaken, he made the identification explicit. Like himself, Freud was rehabilitating the flesh. Ellis attributed to him the spirit of one of his own heroes, James Hinton, a Victorian surgeon, philosopher, and marriage reformer who preached a gospel of sexual innocence. "How utterly all feeling of impurity . . . is gone from the sexual passion in my mind! . . . It has taken its place . . . as one with all that is most simple and natural and pure and good." [46]

Ellis continued: "It was in this spirit that Freud formulated his theory of 'libido,' with its infantile manifestations and marvellous transformations, serenely pursuing his life, while the conventional world was shocked. . . ." Pushing the misunderstanding to its limits, Ellis delivered two compliments that Freud must have taken as deadly insults. First, Freud was an artist, not a scientist. Next, this Columbus of the unconscious had proved that "Spirit is as indestructible as matter . . . Freud's work is the revelation in the spiritual world of that transformation and conservation of energy which half a century earlier had been demonstrated in the physical world." [47]

What Ellis had achieved was a union of his faith in Nature, which he derived in part explicitly from Rousseau, with Freud's clinical testimony to the malevolent effects of repression and "civilized" morality. This explosive combination was exhibited in much of Ellis's writing; indeed he was far more widely read up to 1920 in the United States than was Freud. Ellis's *Task of Social Hygiene* was commended to teachers of the subject by the two best American social hygiene texts. One of them also listed Ellis's six-volume *Studies in the Psychology of Sex.* Another writer suggested that, although these were "beyond the needs of the ordinary teacher" who would have to steel himself against the pathological material, they were still an "exhaustive study of the entire problem from the sources." [48] Who could resist such a recommendation? The prevailing censorship made Ellis's *Studies* seem all the more indispensable. Probably they were widely read by physicians and psychologists, and to some extent by social workers and teachers.

Cultural Wars.

Ellis's revision of sexual hygiene and his positions on marriage, divorce, and sexual perversion represented a radical, if not the most radical, outlook in the growing battles over "civilized" morality. To the left of Ellis were a few laymen committed to free love and anarchism. On the right were adherents of conservative culture and conservative religion, Protestant and Roman Catholic, who opposed any alteration in "civilized" morality, including the repeal of reticence. In the large amorphous center were moderate religionists and non-religious adherents of the nineteenth-century compromise that combined faith in morality and progress. Some members on the left of this group sought moderate "reforms" of marriage, chiefly easier divorce, and a few advocated birth control.

Inevitably psychoanalysis was drawn into these cultural battles. Psychoanalysts were forced to take stands on some of these issues, as the remainder of this study will indicate. Perhaps more important for the reputation of psychoanalysis, however, were the attitudes of the extremists. Psychoanalysis was invoked by some radicals to justify their demands; it was rejected by some conservatives as subversive of the moral order. Struggles centering on "civilized" sexual morality continued into the 1920's, and indeed into our own time. Perhaps the deepest issue was the extent to which human instincts could be trusted to act without formal authoritative restraints. Would the pursuit of personal happiness have salutary results, or lead inevitably to license and destruction?

Up to 1914 the sexual issues that were most often discussed in national magazines were prostitution and divorce. These debates illustrate the general lines of cultural cleavage. To conservatives who accepted immutable standards of right and wrong, religious or secular, the American divorce laws, the loosest in the world, and the divorce rate, the world's highest, were tragic errors. Lifelong marriage, Theodore Roosevelt insisted, was a "realizable ideal" and a social duty; personal happiness in marriage was an irrelevant consideration. This was the crux of the issue. Cultural conservatives such as Roosevelt believed that human nature required strict, absolute standards of conduct. Duty was more important than pleasure. The reproduction of the race transcended all other obligations; it was the primary, perhaps the only, justification for sexual intercourse.

Liberal reformers and radicals took a more benign view of human nature; greater freedom would bring purer morals. They were eager to destroy what they considered a rigid "Puritanism" associated with the Protestantism they had absorbed as children. The opposing sentiments were expressed during a notable debate in 1909 on the topic, "Is the Granting of Freer Divorce an Evil?" J. P. Lichtenberger, professor of sociology at the University of Pennsylvania, put the liberal view: "Two generations have witnessed the passing of the dogmatic age in Protestant theology. . . . The stern morality of Puritanism, based on theoretical standards, is being replaced by the practical morality arising out of our changed social conditions." He was answered by Pennsylvania's representative in Congress, Walter George Smith, who believed that the liberal position destroyed all morality and embodied a "darkness that is visible." [49]

Sexual satisfaction for both partners—to an unprecedented degree for women—was a central, if often covert issue in some of the debates. Reformers and radicals assumed that personal fulfillment counted as never before and could be achieved by trusting the wisdom of the body. Ellen Key, a Swedish feminist whom Ellis often quoted approvingly and who was widely read in America, uncompromisingly took this position: "Either we believe that the sensual instincts are pitfalls and obstacles, or we regard them as guides in the upward movement of life on a par with reason and conscience." [50]

A moderate proponent of easier divorce, Lichtenberger insisted that "mutual attraction and preference" should provide the basis for the new morality. Love, based on such preference, was the ideal by which all conventions and institutions were to be judged. Most marriage reformers agreed that marriage was a human institution fashioned for human happiness. Above all, it was not a sacrament divinely ordained by "outside providence."

Some advocated divorce by mutual consent, for as Ellen Key and Havelock Ellis insisted, lifelong fidelity, based on sexual attraction, could not be guaranteed. Another and more extreme proposal was to replace premarital chastity by trial marriage, i. e. unions without formal social ties. Ellis argued that trial marriage would foster more satisfactory sexual adjustments within marriage and hence greater marital faithfulness. Nearly all marriage reformers agreed in advocating eugenics, the breeding of a nobler race by the sexual union of its best specimens. However, Ellis and Ellen Key admitted that people might require considerable education before they learned to be sexually attracted to eugenically suitable mates.

Perhaps the most famous and temperate book about divorce by an American, William Carson's *The Marriage Revolt,* published in 1915, carried a long argument by Carl Jung that originally had appeared in the *New York Times.* Americans, Jung insisted, were the world's most tragic people because the most neurotic. They tried to control themselves too rigidly and consequently suffered from devastating conflicts. American neuroses caused the failure of American marriages. The American male spent too much libido on business and too little on his wife, whom he treated as if she were a mother. Jung had argued in the *Times* that, if prudery were eliminated, America would become the "greatest country the world had ever known."

Psychoanalysis was invoked by one of America's most famous anarchists, Emma Goldman, to justify the complete sexual liberation of women. Possibly it had been Miss Goldman who at the Clark Conference had passed Ernest Jones a note asking that Freud discuss sexuality. Freud had drily replied that he could no more be driven *to* the subject than *away* from it. The *Boston Evening Transcript* reporter at the Conference noted that Miss Goldman was "plump, demure, chastely garbed in white," and a trifle fidgety.[51] She wore a rose at her waist and rimless glasses and could have passed for a visiting schoolteacher. She defended free love and political assassination with equally high-minded, pedagogic seriousness. To her free love meant not promiscuity but sincere emotion unshackled by convention; and she insisted that free lovers stay together to raise their children. In 1910 in an essay on "The Hypocrisy of Puritanism," she cited arguments Ellis had taken from Freud that sexual repression caused the intellectual inferiority of women. Puritanism imposed "absolute sexual continence" upon "the unmarried woman, under pain of being considered immoral or fallen, with the result of producing neurasthenia, impotence, depression, and a great variety of nervous complaints involving diminished power of work, limited enjoyment of life, sleeplessness, and preoccupation with sexual desires and imaginings. The arbitrary and pernicious dictum of total continence probably also explains the mental inequality of the sexes. Thus Freud believes that the intellectual inferiority of so many women is due to the inhibition of thought imposed upon them for the purpose of sexual repression." Miss Goldman's magazine, *Mother Earth,* carried almost nothing about psychoanalysis but a great deal about Havelock Ellis and sexual topics. From 1910 to 1915 she was crossing the country on annual lecture tours, sometimes speaking to groups of college students. She discussed such subjects as "The Intermediate Sex (A Study of Homosexuality)," "The Limitation of Offspring," "Is Man a Varietist or a Monogamist?," specific methods of birth control. She was harassed from time to time for her political and sexual opinions. Anthony Comstock held up an issue of *Mother Earth* in which she denounced the moral crusades against "the white slave traffic," which she considered to be the result of capitalist exploitation.[52]

Because of the prevailing censorship and the conservative distrust of radical politics and radical sexual views, the right to advocate these causes became a burning issue. One of the stoutest defenders of Emma Goldman's right to be heard was Theodore Schroeder, a lawyer and the mainspring of the Free Speech League founded in 1911 by Lincoln Steffens, Brand Whitlock, and Hutchins Hapgood, among others. Schroeder added another element to the image of psychoanalysis, a shrill critique of religion and censorship. Schroeder, whose parents were members of the 1848 generation of German revolutionists and close friends of Carl Schurz, had moved to New York after a few stormy years practicing law in Salt Lake City. His father was a strict Lutheran disciplinarian, and at eighteen Schroeder began reading the atheist pleas of Ingersoll, Feuerbach, and Schopenhauer. After becoming embroiled with the Mormons, he turned to a study of the sexual origins of religion, which led him to read the European sexologists, especially Krafft-Ebing and Havelock

Ellis. His expertise in sexual lore commended him to Benjamin Flower, who, like Schroeder, was campaigning for divorce reform, as well as for purity.

Schroeder's "wild swooning desire to talk, in season and out, about the obscenity of religion. . . ." led Lincoln Steffens to remark, "I believe in Free Speech for everybody except Schroeder." [53] In 1905, in Flower's *Arena,* Schroeder, in a typical tirade, denounced Christianity for fostering perversions and erotophobia and for making woman a "sex slave" in bourgeois, Christian marriage. The titles of two subsequent articles indicate their tone and content: "The Erotogenesis of Religion," "Revivals, Sex and the Holy Ghost." In "Our Prudish Censorship" he argued that filth and lewdness existed entirely in the mind of the observer; moreover, everyone was slightly lewd. He regarded himself as an "ethical scientist" fighting the "dogmatic morals of ascetic theologians." Schroeder advocated Free Speech, birth control, voluntary sterilization, and the widest possible dissemination of information about the prophylaxis and treatment of venereal diseases. Schroeder contributed a long article to the first volume of the *Psychoanalytic Review,* "The Wildisbuch Crucified Saint, a Study in the Erotogenesis of Religion," and his essays appeared there for the next two decades. Schroeder's studies of religion were cited in an article in the *New York Times,* "Is Sex the Basis of Religion?" by James Van Teslaar, a Roumanian follower of the heretical psychoanalyst Wilhelm Stekel. Van Teslaar, who had settled in Boston, edited a Modern Library volume on psychoanalysis published in 1924 that contained the Clark lectures. [54]

Another defender of Emma Goldman and a militant advocate of birth control, Dr. William J. Robinson, published the first English translation of Freud's essay on "Civilized Sexual Morality and Modern Nervousness" in his *American Journal of Urology and Sexology* in 1915. Earlier he had invoked Freud's authority to support his argument that continence was harmful in an editorial in the *Medical Critic and Guide.* [55]

Although the radical position was held by only a few and probably not widely known, even liberal reformers began to fear that the repeal of reticence was going too far. Maurice Bigelow worried that young people were beginning to discuss sex for "pleasure" instead of for sober instruction in right living. Another sex educator, William Foster, president of Reed College, Oregon, lamented: "Subjects formerly tabooed are now thrust before the public . . . films portray white slavers, prostitutes, and restricted districts, and show exactly how an innocent girl may be seduced, betrayed, and sold. . . ." The new frankness constituted literally a "social emergency." [56]

Conservatives were even more vehemently outspoken. Former President Taft believed that sex education abetted "lubricity in literature, on the stage and indirectly in education. . . ." As the repeal of reticence continued, conservative opposition grew fiercer. In 1912 Joyce Kilmer, the author of "Trees," attacked Havelock Ellis's *Task of Social Hygiene* in the *New York Times*.[57] The book was likely to teach young people, too many of whom already had photos of Walt Whitman on the mantle, to be "socialists, individualists, Anarchists, Eugenicists, Pacifists, Internationalists, Suffragists, Free Lovers, Neo-Malthusians." In 1914 Agnes Repplier, Weir Mitchell's neighbor, a spinster and a Roman Catholic, condemned the entire repeal of reticence as the work of "uninstructed missionaries" who had "lightly undertaken the rebuilding of the social world." Too much reliance was being placed on crude, undigested knowledge, too little upon "religion and discipline." [58] People already knew quite enough about the deadly sins. She was especially indignant at forgiveness being extended to prostitutes and at the theatrical exploitation of brothels. Academic neo-humanists such as Paul Elmer More and Irving Babbitt attacked the "liberation" of human impulses. Not Bergson's *élan vital* but only the *frein vital* of self-restraint could save civilization.[59] These critics for a time were an influential force on the *Nation,* and the magazine's reviewers began to attack some of the new frankness about sexuality. In 1911 G. Stanley Hall's *Educational Problems* was denounced for containing pages of nasty and nude detail worthy of Krafft-Ebing or the Italian sexologist Mantegazza.[60] Even the *Bookman* condemned the book for stimulating impure thoughts and an interest in "forbidden themes" in the "average mind." Hall himself for the first time maintained that purity of thought probably was impossible. The tone of the *Nation*'s reviews of psychoanalytic work grew more hostile, although they were by no means entirely unfavorable. What was especially objectionable in Freud's *The Interpretation of Dreams,* for instance, was the "morbid tendency to over-emphasize the potency of erotic influences in all of experience," which in turn led Freud to "improbable and revolting explanations." [61] Yet this stricture was coupled with an admission that sexual forces were "powerful in infantile life." After the Great War began, the *Nation* published letters and reviews linking Freud with decadent German pansexualism.

For the prevailing mores, psychoanalytic works, especially those not explicitly written for lay audiences, represented a degree of frankness and sexual detail exceeded only by Ellis and Krafft-Ebing. Psychoanalytic theory made it perfectly clear that maintaining purity of thought, one of the important tenets of late nineteenth-century "civilized" morality, probably was a hypocritical impossibility. Even Freud's Clark lectures probably would not have been printed by a national magazine without expurgation. Perhaps the *Outlook*'s comments on Brieux's *Damaged Goods* best indicates the tone that still prevailed in 1914: "the subject is openly and without circumlocution the terrible character of the disease which attends sexual immorality." [62]

XI

Opposition and Debate:
Science and Sexuality
1909 - 1917

By 1910 psychoanalysis had become one of the most fiercely debated subjects in American neurology and psychiatry.[1] Members of the American Neurological Association at their Washington meeting in May discussed James Jackson Putnam's new Freudian convictions for seven recorded pages.

Their reactions reflected the crisis within neurology and medical fears of the popular craze for psychotherapy. Thanking his colleagues for their courtesy, Putnam remarked that it had taken "a little courage in a person not particularly courageous" to read a paper on this "unpopular subject." [2] He had done so for one reason: to open a full, friendly discussion of the functional or psychological approach to neurology. This issue, he explained, had been represented from the Association's founding by George M. Beard and now divided neurologists by antagonism and misunderstanding.

Of the eleven physicians who spoke their minds, seven, nearly all classical somaticists, had listened respectfully out of their high regard for Putnam. But psychoanalysis, they remarked, was "tainted" with sex and such baffling and valueless "metaphysical" elements as dream interpretation.

Four neurologists, however, already using psychotherapy, pointed to the knowledge of the psychoneuroses gained by psychoanalysis, quite apart from "the sexual sphere." They were pleased that so fine a man, in whose integrity they had absolute confidence, had tested so suspect a method. Yet, Ernest Jones wrote to Freud that Joseph Collins, an influential New Yorker, had told the annual banquet: "It was time the association took a stand against transcendentalism and supernaturalism and definitely crushed out Christian Science, Freudism and all that bosh, rot and nonsense." [3]

Thus began a pattern of recurring debates—before the Philadelphia Neurological Society in 1911, the New York Academy of Medicine in 1912, the International Congress of Medicine in 1913, the American Medico-Psychological Association in 1914, and at most meetings of the American Psychopathological Association. These disputes were characterized by violent opposition, considerable partial acceptance, and a hardening of position among both psychoanalysts and their opponents. The more successful psychoanalysis became, the more widely it spread, the more adamant became the arguments. There were three chief issues at stake: the scientific status of psychoanalysis, the validity of Freud's sexual theories, the values of "civilized" morality and American culture.

On the whole, opposition in America was more moralistic than in Europe. Nevertheless, before psychoanalysis became an organized movement, European opposition to Freud's sexual theories was serious, based on their "exaggeration," "implausibility," and in some measure on the "disgust" they aroused.[4] From a sense of persecution partly caused by inner tensions, partly by reactions to real events, the psychoanalysts exaggerated the degree of rejection on these grounds.

The psychoanalytic battles pitted against each other not only different professional generations with different values but men of different social and professional prestige. The psychoanalysts usually were ten or more years younger than their opponents among the neurologists. The latter came, as a rule, from more exalted social origins and held positions of greater prestige. Five had been presidents of the American Neurological Association, while among the psychoanalysts only Putnam had comparable professional status. Opposing psychologists were the youngest of all, and were imbued with the new ideals of experimentation and quantification that issued in behaviorism and the new educational psychology.

The disputes aroused animosities among colleagues and strained old friendships. Putnam, who bore the brunt of the first attacks, became alienated, at least for a time, from some of his Boston colleagues, and his neurological practice diminished. A. A. Brill, who assumed leadership of the movement around 1914, saw his friendship with Frederick Peterson temporarily ruptured. Ernest Jones chafed under his juniority in years and status; he was thirty-two in 1911 and for a time at the bottom of the academic ladder in Toronto. Hoping to manage the tactical polemics of the movement, he counseled that dignity was the best defense. Yet he seriously underestimated his own capacity for withering sarcasm, and he damaged for a time his relationship with Morton Prince.

By 1914 major points in the brief against psychoanalysis had been formulated; opponents often borrowed and elaborated each other's arguments. The psychoanalysts' version of these disputes, especially Ernest Jones's, has become the best known. To Jones and his colleagues, their opponents were ossified and blindly prejudiced. None was well informed about psychoanalysis.

"Genuine criticism" of Freud's work was badly needed, the sympathetic British psychiatrist Bernard Hart argued in 1910; most criticism had been vitiated either by somatic or moral prejudices. "Derision and indignation" from the anti-Freudians and "repartee rather than evidence" from the psychoanalysts characterized the early exchanges, according to the less sympathetic American psychologist Robert S. Woodworth.[5] Both he and Hart were only partly correct. Exchanges of contumely were by no means the entire story; some of the early criticism repays careful attention for, occasionally, it raised enduring problems.

Psychoanalysis was debated roughly, but often searchingly when it was a fresh, new heresy. Some of the arguments, especially those from friendly psychopathologists, and even those from moderate somaticists, still are advanced in principle today. Only recently has psychoanalysis become sufficiently secure to confront some of these objections. Most opponents, except for extremists, combined their assaults with limited praise of Freud's achievements. Yet, the nature of psychoanalytic theory, the polemical tactics of both sides, and the paramount issue—what constituted a correct application of the psychoanalytic method—made fruitful discussion difficult.

By 1917, in contrast to the period up to 1908, most of Freud's major works were available in English, chiefly in Brill's translations, whose quality still arouses dispute. Putnam thought them graceless but conscientious. Recently they have been criticized more severely on grounds of obscuring Freud's meaning and of illustrating his text with inappropriate examples of Brill's own. Freud's prolific followers published their own expositions, and thus the opponents of psychoanalysis had a wide variety of sources to choose from.

Many critics had a sketchy acquaintance with Freud's work, derived chiefly from secondary sources. Pierre Janet, for example, relied largely on French and American secondary works, such as the summary by Régis and Hesnard or the criticism by Ladame. Charles W. Burr claimed to use Freud but in fact referred largely to Brill's essays and Bernard Hart's *Psychology of Insanity* which omitted Freud's sexual theories because Hart judged them to be doubtful.[6]

A few opponents, even one or two of the most hostile somaticists, were well informed and presented Freud's theories carefully. In 1911, for example, Francis X. Dercum criticized Freud's conception of dreams after giving an account of it that followed Freud's own presentation in "Ueber den Traum." [7]

Despite all the new sources of information, much of the irony and even the anachronism that had attended the first American acquaintance with Freud persisted.[8] Many critics accepted Freud's work in the 1890's, purged, however, of the "over-emphasis" on sexuality. Still other critics accepted the joint achievements of Breuer and Freud yet rejected Freud's later contributions. A number of physicians were willing to accept the "mechanisms" of condensation, displacement, and so forth that were developed in *The Interpretation of Dreams*. Thus what many critics praised helped to perpetuate the focus on Freud's earlier conceptions.

Now, however, Freud was also criticized for the views of his followers. Academic psychologists tended to rely on Jung's association tests for the model of what Freud meant by association. Psychoanalysis also was criticized for weaknesses that Freud later attempted to correct. For example, repression and resistance were often misunderstood, partly because Freud's theories about them were not fully worked out until 1914 and after.

Science: Somaticists

To conservative American somaticists, especially the Philadelphia neurologists, psychoanalysis represented the most psychogenetic position of all the new psychopathologies. For a time they were goaded into taking a strong stand against all theories of psychological etiology in the insanities. It was a stand so extravagant that a European critic of Freud regarded it as a return to views typical of the nineteenth century.[9]

Psychoanalysis flouted existing somatic knowledge. According to Francis X. Dercum, the more physicians learned about nervous and especially mental diseases, the more "psychic" causes disappeared from "our etiology." Psychoanalysis was a regrettable return to the moral theories of the "older writers" who foolishly had included "worry, care, sorrow, remorse, reverses, disappointments, misfortunes" among the causes of "mental disease." Manic depressive insanity was no more the result of emotional perturbation than paresis was, and everyone knew that science had demonstrated that paresis was caused by syphilis. Reversing Bernard Hart's argument that the virtue of Freud's theory was its exclusively psychological nature, Dercum's colleague, Charles W. Burr, argued that precisely this aroused antipathy and skepticism. To those inclined to believe that "there is always a physical cause for a mental act, and a perversion of physical function whenever there was a perversion of mental acts," Freud's psychological hypotheses were "very fanciful." Such "dogmas" were not susceptible to proof, and Freud offered none; they were "harebrained jargon." Psychoanalysis marked a "return to darkest Africa"; it was unscientific because it insulted common sense. A cause of chronic paranoia, Dercum proclaimed, was unmasked by Freud's followers as an "irritation of the anal erogenous zone." [10]

Freud's unconscious seemed peculiarly unverifiable in traditional somatic terms. What was the nature of this mysterious entity? Where was it located? How could one prove its existence, let alone the attributes Freud blessed it with? Burr protested: "I cannot understand how a mental thing of which we are by definition unconscious can influence conscious life." [11] He questioned the role psychoanalysts attributed to the unconscious in causing neuroses. Acute, painful conflict usually was conscious, and most nervous trouble came from a conscious, remembering mind. Painful situations, far from being forgotten, were precisely what stuck in most people's heads.

Dercum granted that the construction of a "modern psychology of insanity" was a respectable, if "largely speculative pursuit." But the applications of psychoanalysis to dementia praecox by Bleuler and Jung endangered accepted theories of both etiology and classification.

Dercum countered the theories of psychoanalysis with an extreme somatic position that included lightly disguised degeneracy theory. The fundamental "intrinsic" cause of insanity was "neuropathy," which he defined as a nervous system "fundamentally aberrant and defective in development." Neuropathy weakened resistance to nervous and mental diseases. Only those who were weak became ill, not those subjected to strain. This was a more extreme position than that taken by most neurologists in the 1890's, who usually assumed the interaction of both stress and constitution. In support of the role of neuropathy in insanity, Dercum cited the liability of dementia praecox patients to die "in large numbers" from tuberculosis, and the recent findings about paresis. "At most," he concluded, a psychic factor gave "to a pre-existing morbid state a special coloring, a special detail to a delusion or an obsession," but could not "primarily bring it about." [12]

The most effective treatment for mental disease—including dementia praecox—was "full feeding and simple physiologic measures" with some adjunctive "rational psychotherapeutic procedure," in the sense of reassurance and a long talk with the doctor. Dercum was especially angered by Bleuler's suggestion that psychoanalysis was the treatment of choice in dementia praecox.

The only hope for prevention and cure lay in studies of heredity, biochemistry, and physiology rather than in the "fanciful interpretation of hysterics' dreams. . . . In actual material solution of the great biochemical problems of the day lies the hope of a real advance in psychiatry, not in closet-born theories and sterile speculations."

E. Stanley Abbott, an assistant physician at the McLean Hospital, attacked Adolf Meyer's theories of dementia praecox for "leaning toward a Freudian manner of interpretation" and tending to "overlook the physical elements." It is true that in 1911 Meyer argued that there was no contradiction between his own theories of habit deterioration and the complexes of the psychoanalysts. But he argued that in some cases of dementia praecox predisposition seemed to play a major role. Yet, Abbott insisted that Meyer was reversing the "general tendency of medical science to take morbid conditions out of the functional class and put them into the organic." Instead, Abbott proposed a "parallelogram of forces" in which physical and psychological factors would play interacting roles. Smith Ely Jelliffe, who had not yet become a psychoanalyst, protested in 1911 that he could not accept an exclusively psychogenetic theory of dementia praecox because a "fairly definite pathology" seemed to be building up. In a typical flight of neurological speculation, he suggested a somatic explanation for Meyer's theory: "Our pathological findings may represent atrophies of unused association tracts which have resulted from the so-to-speak petrification of bad habits of mental adjustment." [13]

Science: Psychopathologists and Psychologists

Psychoanalysis divided the psychotherapists and psychopathologists who were forced to take Freud's theories into account. Although this group made the most astute criticisms of psychoanalysis, their debates with Freud's followers seldom resulted in enlightened reconsideration on either side. The analysts made systematic assumptions that were different from those of their critics, and the result frequently was talk at cross purposes. Proposals in 1910 to found psychoanalytic organizations further embittered the early debates. The critics attacked Freud's conceptual apparatus, his standards of evidence and logic, and above all, the problem of the psychoanalytic method: the validation of psychoanalytic theory depended almost entirely on evidence derived from psychoanalysis. Friendly opponents asked who was to judge what was in fact meant by psychoanalysis. When had it been validly carried out? Why could not other methods in the hands of skilled psychopathologists or psychologists achieve the same results? At this period in the history of the psychoanalytic movement, teaching of the technique had not yet been institutionalized. Practitioners learned informally from other analysts, from reading, and from correspondence with colleagues.

The analysts' debating tactics confused this crucial issue. At first Ernest Jones insisted that psychoanalysis was an elaborate procedure and required exacting technical knowledge. Later, faced with criticism based on the esoteric nature of psychoanalysis, he urged opponents to try it because it was relatively easy: ". . . if you take the trouble to learn Italian you can read Dante in the original. Any one can learn either Italian or psycho-analysis. Neither is very difficult. . . ." [14]

An acrimonious exchange over Freud's theory of dreams illustrates how Morton Prince was forced to reformulate his own views. Yet his assumptions prevented him from granting to the psychoanalytic method the status Freud's followers insisted upon. In this debate bitterness on grounds theoretical and personal finally resulted in charges of cultism and stupidity.

On May 2, 1910 the American Psychopathological Association was founded in Washington, D.C., to embrace physicians and psychologists interested in abnormal psychology. Prince delivered a paper he had given twice before, praising Freud's theory of dreams yet outlining his objections and his own views of their "mechanism and interpretation."

It had been Freud's "brilliant stroke of genius," he wrote, to discover that dreams were not mere "fantastic imagery" or "haphazard vagaries" but "orderly determined phenomena, capable of logical interpretation." The remembered dream, made up of raw material symbolically transformed, expressed underlying thoughts. This raw material consisted of "complexes," which Prince considered equivalent to "brain residua," or "neurograms" of past experience, of the day before, and especially of thoughts of the period just before falling asleep. Using this material the dream itself was determined by a "motive," a strongly organized system of ideas, usually deeply rooted and highly emotional. This motive was expressed symbolically, sometimes but not always in distorted form, and usually concerned thoughts and feelings that had dominated the subject's psychological life over a long period.

All these assumptions were new departures in Prince's approach to dreams and directly represented Freud's influence. But on fundamental points—the underlying motive of the dream, the role of repression, and the method to be used in obtaining memories and associations— Prince radically disagreed.

Prince's patient, a widowed social worker, had suffered from dissociated states and was keenly interested in abnormal psychology, which she pursued as a therapeutic hobby. However, she could not remember her dreams. Prince hypnotized her to obtain them. To discover her associations he tried voluntary recall and what he thought to be Freud's method, free association to elements of the dream with the patient in a state of "abstraction" in which her critical judgment was suspended. He used both methods while she was in her normal state and also while she was hypnotized. He claimed that the associations under hypnosis were fuller and more detailed than those recovered in the waking state.

He also encountered not the slightest sign of "represession or resistance" while she was bringing up associations. Although some of her dreams fulfilled wishes, they did not invariably do so, especially censored and disguised wishes. Some of her dreams in fact represented fear or anxiety, emotional aspirations or dominant previous attitudes.

Actually Prince's interpretations were highly allegorical. For example, she dreamed of toiling up a hill, and said, "I must not show that I am frightened, or this thing will catch me." She saw two clouds or shadows, one black, one red, and said, "My God, it is A and B! If I don't have help, I am lost," and she called Dr. Prince.[15] This dream expressed her fear of relapsing into her dissociated states. Toiling up the hill represented her view that life was a "constant struggle and toil against difficulties." The rough path was associated with a climb in the White Mountains, a painting she had seen, and despairing thoughts about life that had recurred just before she fell asleep.

He dealt cursorily with several striking dream elements. She dreamed she was in danger of being smothered by countless cats unless she kept silent. This he interpreted as her own insistence that she not complain. She dreamed of a host of wild men, frightful creatures dressed in skins, with bare limbs. This symbolized her feeling that life itself was "wild." She dreamed of Dr. Prince, bound and helpless, clubbed by Lilliputians. This symbolized her son and her own sense of helplessness and heavy-hearted despair.

These intriguing items may have prompted Jones to suggest a critique of Prince's paper. Prince wrote to Putnam on November 21, 1910 that he welcomed criticism, with a proviso about sexuality that will be described later. At this point the discussion was embittered by a lengthy denunciation by Jones of the methods Prince relied on—suggestion and hypnosis. They had no scientific basis and were merely transference phenomena tinged with Oedipal eroticism, Jones insisted in the December-January issue of the *Journal of Abnormal Psychology*.[16]

Prince wrote Putnam a long, anguished letter. He borrowed the pen name "Fiona McLeod" of the Welsh poet, William Sharp, to express to Putnam what lay outside the bounds of polite convention—his affection for him and the intensity of his anxiety for the future of psychopathology. He urged Putnam not to get too "mixed up with Jones and his crowd," whom he considered to be fanatics. They were threatening the common bond that ought to unite those interested in psychopathology against the "indifference and antagonism of the profession." Jones he dismissed as nervous, highstrung, self-centered; and Prince, Jones dismissed, in letters to Putnam, as careless, irresponsible and boyishly belligerent: "When I first saw him—after this paper was written he had not read the *Traumdeutung*." [17]

Putnam suggested to Jones a set of principles on which all psychopathologists might agree and urged that they refrain from divisive criticism. Those who accepted Freud's work in general terms were "not compelled to give their adherence to every statement or interpretation that he has made. I mention this because Prince seemed to think we were so obliged. . . . Freud's doctrines are not a 'theory' but a series of provisional conclusions based on observations. . . . Some of the special facts that cause so much animadversion, so far from being suggested, come out spontaneously in statements and reports of dreams, even dreams of young children."

"Finally, I think from my experience at our local meeting a few weeks ago, we ought to guard ourselves as carefully as possible from criticizing other methods or minimizing their value." [18]

Worse followed. The February-March number of the *Journal of Abnormal Psychology* contained a courteous yet sarcastic critique by Jones of Prince's theory of dreams, a riposte in the same spirit by Prince, and two articles severely criticizing psychoanalysis, one by Boris Sidis and another by the Frankfurt psychiatrist A. Friedländer.

By then the issue had become Freud's "esoteric" method, the psychoanalytic technique. This lay at the heart of Jones's criticism. He argued that Prince had not really used psychoanalysis, because he had not analyzed resistances. Only psychoanalysis, not hypnotism, penetrated deeply enough to reach the real meaning of a dream. If Prince could only recover his patients' dreams by hypnosis, this in itself indicated that there was a difficulty in their reaching consciousness in a waking state, in short the very resistance Prince denied. Any psychoanalyst would recognize the sort of "content" Prince had reached as "belonging to the first stages of any investigation into the sources of dreams, but [as] quite unlike the latent content . . . revealed by psychoanalysis."

Prince could find no "disguise" in the dreams, yet admitted that the manifest dream thoughts were symbols. "Well, all that one means by disguised thoughts is the allegorical representation in consciousness of thoughts not accessible to introspection, in other words just what Dr. Prince records." [19]

Prince argued that "it seems as if the followers of Freud care more for the acceptance of their method—for being baptized in the faith, than for the determination of truth." Freudian literature was sprinkled with such expressions as "proved," "established," "well known," "accepted." "Such expressions take the place of 'theory,' 'possibility,' 'probability' to which we are accustomed in progressive science, and make it all the more difficult for unbelievers to embrace the faith. This cocksureness of youth brings only the smile of amusement to the older and more philosophical reader, but to the younger it seems undoubtedly to smack of arrogance." This attitude hampered the adherents of psychoanalysis, who included "minds as able as any in medicine." [20]

Prince argued that if the antecedent experiences he uncovered adequately explained the meaning of the dream and his conclusions were logical, whether one used Freud's method or not was irrelevant. The problem to be solved was the interpretation of dreams, not some special method of doing this. Hypnosis, Prince insisted once more, revealed more memories, especially in strongly dissociated states, than did Freud's psychoanalysis.

"If the Freudian hypothesis be true that every dream is the fulfillment of a repressed wish, and the repressed wish continues to meet with a repressing force which prevents its coming into consciousness, then, as Jones says, it would be inevitable that the patient would come sooner or later to an obstacle. But suppose the hypothesis is not true, what then? To assume an inevitable resistance is to assume the question at issue." "But surely Dr. Jones cannot hold that inability to remember in the waking state episodes that can be remembered in the hypnotic state is necessarily due to resistance. . . . Such a view reduces the whole matter of resistance to an absurdity, as it disregards a vast number of data collected by hypnotic investigation." [21]

Jones concluded tartly: "Dr. Prince's paper, interesting and welcome as it is, in no way invalidates my previous statement that up to the present no one who has taken the trouble to acquire the psycho-analytic method has failed to confirm Freud's theory in all essential particulars." Prince replied: "No one who has shown by his writings that he is thoroughly trained in and conversant from first-hand knowledge with all the phenomena of abnormal, experimental and functional psychology has accepted Freud's theory." [22]

Psychopathologists and psychologists failed to practice psychoanalysis to test Freud's theories; psychoanalysts, including Freud himself, were for the most part untrained in any psychology except their own. Jones was so angered by Prince's reply and the other unfavorable articles that he offered to resign as an editor of the *Journal* and wrote to Putnam that the psychoanalysts should found their own publication. Prince had not informed him of the forthcoming rejoinder, and the dismissal of psychoanalysis as a "cult" was inexcusable.[23] Disappointed by this bitter exchange, Putnam wrote to Jones on February 14, 1911: "I have read Friedländer's article and also yours and Prince's with much sorrow and dismay. . . . I think it would be utterly unfortunate if those of us who really care about psychopathology in the large sense should drift apart, no matter what the provocation." Jones replied on February 27 that Putnam had his "unity complex," while he himself had a "truth for truth's sake complex:"

> With regard to unity, that is in the lap of the gods. No one desires it more warmly than I do, except at the price of self-respect and honest convictions. I suppose no big movement can proceed harmoniously, without involving some alienation. Pour faire une omelette il faut casser les oeufs. But it is very evident that here the disruption is coming entirely from the other side. Such language as the current number of the *Journal* contains makes it very difficult to keep the peace. . . . One thing above all else I hope, and that is that in America abusive language (Dercum, Collins, Prince, Sidis, etc.) will remain the monopoly of our opponents. Then it is bound to react on them in time, while if any of us were to reply in the same strain it would lead to a never-ending muddle. So long as we maintain a dignified attitude we are safe, and the onlooker will soon see which side has the fanaticism and high feeling.

Prince wrote to Putnam once more on March 3, 1911: "Nobody cares a rap what theories a man holds. What scientists do care about —all men care about—is the mental methods by which theories are arrived at." Freud's was the method of philosophy, not science. Prince remarked that he could "only accept conclusions based on a consideration of *all* related data and drawn by logical processes of thought which do not admit any other possible conclusion. Or if the data are insufficient for such conclusions then only tentative hypotheses." But Freud "piled concept upon concept,—until finally we have a structure which falls to the ground with the first crumbling of the foundation."

"When he comes across something in Freud which to him seems to be true, to be supported by facts of observation and logic, a thrill of real delight goes through [him]. . . . Then presently he comes across a statement of fact which is contradicted by experience; then an interpretation which defies every law of logic, and disregards other further interpretations, sets every logical process at defiance. At this point and not until this point he puts the book down in disdain. . . . You are raising a cult not a a science." [24]

Putnam acted as a buffer between Jones and Prince and urged Jones not to make an additional reply. Jones stayed on as editor, and Prince continued to print articles by both the psychoanalysts and their opponents. Jones, however, was to have the last word. Abstracting the *Jahr-buch für Psychoanalytische und Psychopathologische Forschungen* for 1911, he included a demonstration by Jung of the "incompleteness" of Prince's dream analysis, which Jones commended to all interested persons.[25]

These issues remained to divide the psychoanalysts from psychopathologists and psychologists: the "esoteric" nature of the psychoanalytic method, Freud's standards of evidence and logic, the ignoring of relevant data gathered by other workers with other methods. Psychoanalysis discarded Prince's life's work, his observations of multiple personalities, his hypnotic investigations and experiments, his case histories.

Academic psychologists, too, were put off by the apparent psychoanalytic disdain for their discipline and for their laborious attempts to be scientific. Woodworth noted that he was prejudiced against the psychoanalysts because they were "contemptuous of our modest efforts," dismissing them as "barren and superficial," while the analysts' methods were excessively "rough and ready." The psychiatrist Bernard Hart once called Freud more poet than scientist.[26]

Friendly critics asked for more experimental testing of psychoanalytic conclusions. Adolf Meyer wanted "simple and accurate methods of reporting" and the observation of a case by several different people. Woodworth argued that because the Freudians rested much of their system on theories of infantilism they should carry out more direct studies of children. Psychoanalysis carried determinism to regrettably naïve lengths, argued Frederick Lyman Wells, one of the first clinical psychologists, who worked at the McLean Hospital and had studied with William James.[27] The attack on determinism centered on the logic of free associations and their interpretation.

By what criteria could a correct interpretation be established? The fact that one association evoked another and that this called up still another proved no causal relation among them. The mind, academic research had demonstrated, tended as a general rule to "get away from the context of the starting point in free association" and displayed "multiple possibilities of reaction." Associations were conditioned by time, circumstances, and individual differences. The fact that an association was made in the present was no criterion for assuming the same set of associations in the past. How was it possible, psychologists argued, to assume that current associations had been factors in the antecedent production of a dream?

If the Freudian mechanisms, such as condensation and displacement, were required to demonstrate that a dream fulfilled a wish, this was highly equivocal logic. How could one select a relevant association from a train of ideas, without an a priori notion of what that association should be? "Freud somewhere says," Woodworth wrote, "that it would be impossible for A to lead to B, unless B had been operative in the production of A." [28]

But Freud had never described free association so simply. At Worcester he had said that if the patient were allowed to speak with complete freedom, nothing could occur to him that did not bear on the element both analyst and patient were seeking. Woodworth probably derived his model of associations chiefly from Jung's tests, for which Freud also had given sanction at Clark. To Freud, associations and chains of memory were intricate processes and always involved a subtle interweaving of multiple factors. The "correct" association—and this was seldom a single item—could be determined by two general criteria. First, the association bore a distinct, ascertainable relation to the symptom or starting point. This relationship had to satisfy both analyst and patient, and its discovery could be expected to have therapeutic results. Some of these arguments were advanced in a spirited riposte to Woodworth by the New York analyst Samuel A. Tannenbaum. The real difficulty, he argued, was that the psychologists lacked the clinical experience of conducting a psychoanalysis. For in that process both the relevance and the precise "fit" of the associations could be observed. [29]

Somaticists, psychologists, and psychotherapists—sometimes crudely, sometimes subtly—raised another objection, that patients fabricated their free associations. Burr insisted that if you relied on the evidence of dreams, you were dependent on the veracity of the dreamer whose statements you could not check. Neurotics were notoriously unreliable when describing their own symptoms. Possibly patients suggested the material to themselves in a state of auto-hypnosis, or the analyst fostered the associations.

Prince and Dercum both argued that there were close analogies between hypnosis and free association. The state of abstraction in which patients said whatever came into their minds was "in principle, hypnosis." Dercum, at a symposium in Philadelphia in 1911, added sinister moral overtones. The patient was completely relaxed; "her" thoughts drifted unrestrained by judgment or criticism, as if these very functions became "dissociated." In such an "abnormal" state, Dercum argued, the patient would pick up a suggestion from the analyst, even if it were inadvertent on his part. Sidney I. Schwab tried to answer this objection by arguing that as of 1904 Freud never questioned patients but let them bring up subjects for analysis themselves: "There is no surrounding of the patient by mysterious lights and shadows nor any clap-trap of that nature. . . . He has not told these patients, 'Now, I am endeavoring to find out whether such and such a thing happened to you sexually, now be on the lookout for things like that,' as has been suggested . . . this evening." [30]

Prince and others pointed out, however, that among their patients —social workers, clergymen and professional people—there were some who had read much about psychoanalysis and were eager to test its theories. Unwittingly exhibiting the powerful identification of shame and sexuality, Woodworth argued that the very caution not to censor associations established a "mental set": "The subject is warned time and again that he must keep back nothing if he wishes the treatment to succeed. It is easy to see that such instructions tend to arouse a definite set of mind toward that which is private and embarrassing; and this easily suggests the sexual." [31]

Occasionally attacks were leveled against points in Freud's system that he himself later changed, for example, his hypothesis that anxiety resulted from repressed libido. Prince asserted that the cause of some anxiety states was a fear of being helpless and alone without the possibility of aid. Some years later, Freud developed a new theory of anxiety. The feeling of helplessness, he wrote, could arise from threatening inner impulses or punishments of conscience, both repressed. The repression itself was a defense against the anxiety the inner stimulus had aroused. What made the repressing tendency so powerful that it could overcome unconscious drives? Psychoanalysts unsatisfactorily attributed this power to personal aspirations, or social compulsions. Freud only addressed himself to this problem beginning in 1914.[32]

On the basis of learning theory, Woodworth attacked the Freudian assumption of the persistence of drives and affects. Freud insisted that drives that did not "work themselves out in action" became suppressed into the unconscious where they remained damned up, likely to break out in unexpected ways. But this was false, except possibly for a very few tendencies, such as anger or sex. In fact, every day a person selected from among a plurality of stimuli that called for action. Tendencies that were rejected, "nipped in the bud," simply disappeared. Thus the sexual impulse was not "sublimated," rather, interest was withdrawn from sexuality and directed elsewhere. Moreover, drives were not merely the product of instincts, but of learning. New motives developed with a force of their own and could become as coercive as an original native tendency. Thus the persistence of a drive and its transmutation, displacement, and sublimation were myths.[33]

As in the case of the extreme somaticists, psychoanalysis induced a hardening of doctrines among some psychologists. For example, Knight Dunlap, professor of psychology at the Johns Hopkins University, was led to define scientific psychology more narrowly than before. In 1912 he opened a campaign against the Freudian unconscious, as well as Morton Prince's co-conscious, as "fantastical" and unnecessary hypotheses. Like the most orthodox nineteenth-century somaticist of the 1870's, Dunlap argued that science must drive out the "soul" and "psychic" phenomena. The more psychological principles agreed with the "structure and function" of the brain and nerves, the faster psychology would progress. He insisted that the "anecdotal method" practiced by the psychoanalysts was unscientific. By this criterion he would have had to exclude much medical and far more psychiatric observation.

Yet, by 1913, the pscychoanalytic issue had forced him to take a more sympathetic view of Morton Prince, whose methods he now regarded as truly causal, while the Freudians merely led patients, with lengthy subtlety, to substitute for an old, sick association, a new one as if the new one were the patient's own discovery.

In 1920 Dunlap argued that Freudians were "tender minded" mystics who could not stand the tough, boring demands of experimental science. Indeed, psychoanalysts were trying to "strangle" science from within. Dunlap reduced Freud's topographical view of the mind to a caricature in which repressed contents lived in the basement of a house, whose upper floors represented consciousness. He also argued that on Freudian principles the whole universe could be defined as sexual. Analysts seized on behavior in the child that only became part of sexual activity later, then cited that behavior as evidence of childhood sexuality.

In John Broadus Watson, psychoanalysis helped to provoke a denial of consciousness altogether. What the psychoanalysts described as the result of unconscious processes, Watson argued, could be formulated in more common-sense terms as the result of habits or habit systems. This critique was directed as much against Morton Prince as it was against Freud.[34] As behaviorism became influential after 1913, many psychologists and a few psychoanalysts hoped to reduce the unconscious to an aspect of neurological functioning.

Freud's royal road to the unconscious, dream interpretation, also was severely criticized. Critics were willing to study dreams seriously but objected to interpreting them by "fixed symbolism," especially sexual. Alfred Reginald Allen, a former assistant of S. Weir Mitchell's and one of the few Philadelphia neurologists favorable to psychoanalysis, charged that some analysts made arbitrary interpretations of manifest content. Even Freud, following the lead of his pupils, had begun to elucidate symbols according to a kind of "dream-book formulary." Sexual symbolism was ridiculed. Toes, toadstools, umbrellas, daggers, etc., signified the male organ; caves, fireplaces, circles, the female. Why should a snake, Woodworth asked, not stand for "sinuousness, or slyness, or danger or wisdom? Why did it always have to represent the male organ?" [35]

The Freudians retorted that absurdity was no criterion of falsity. Brill remarked that, after psychoanalyzing patients twelve hours a day for several years, even the "wildest symbolism" no longer seemed strange to him. He protested that the sexual symbols some of his patients presented he himself could not possibly have thought of.[36] Moreover, Freudians argued, literature, mythology, anthropology, and art were confirming the almost universal meaning of some symbols.

Somaticists, psychologists, and psychopathologists attacked the pragmatic argument that psychoanalysis worked better than other methods. Evidence from cures was no evidence at all. There had been successful cures from Christian Science, Lourdes, osteopathy, and yoga. Moreover, neurotic illness often cleared up spontaneously, certainly in less time than the three years often required for psychoanalysis. In 1911 Schwab posed a question that he insisted Freud and his students had not asked themselves: "By what standards are we to measure the curativeness of a method as intricate as this, and further by what standards are we to measure a patient in respect to his state of being cured." [37]

Other critics argued that psychoanalysis failed, or was unnecessary and even dangerous. The most elaborate analysis had to be supplemented by deliberate "re-education and training in control." Patients also recovered without any deep-going analysis or with only the most superficial discussion of immediate causes. E. W. Taylor proposed a common-sense therapy based on simple explanations of repression and conflict, without any use of free association, the couch, or transference.

Citing a survey by the German psychiatrist A. Hoche, Dercum charged that psychoanalysis was misused on insane patients and had made some neurotics distinctly worse.[38]

Finally, the inclusiveness of psychoanalytic theory reinforced the impression that psychoanalysis was not only a school but a "cult." Morton Prince argued that its concepts "have a peculiar fascination for many minds. From the fact that they offer an explanation of human thought and conduct, and the methods seek to penetrate within the hidden depths of human nature, to unravel its mysteries, and to lay bare the unsuspected motives which lie behind the apparent motives that determine conduct and thought, the psychology has for some a resistless attraction. It is difficult for me, too, not to feel it." Dercum charged that the psychoanalysts purported to reform "manners, aesthetics, art, mythology, pedgogy, criminology, and moral psychology," which was true. It was not difficult to ridicule these far-reaching and seemingly cock-sure "incursions," as G. Stanley Hall called them admiringly, or to dismiss them as "arm-chair anthropology," as Woodworth did. Prince predicted that whether true or not, psychoanalysis would triumph.[39]

Many of the objections raised in principle, then, still trouble psychoanalysts and their critics: the criteria for a correct interpretation, the nature of recovery, the effectiveness or ineffectiveness of analytic treatment, shortened methods, experimental corroboration, the systematic formulation of analytic hypotheses, methods of observing analytic procedures, extrapolations in related disciplines.[40]

Sexuality

The issues of science and sexuality were inextricably connected because Freud's findings concerning the latter were derived almost entirely from the psychoanalytic method. The early debates embalmed the values of American "civilized" morality during its period of crisis. Then whenever psychoanalysis was attacked, so, too, were Freud's sexual theories. On this difficult, never dispassionate ground, most battles were fought.

Freud's conceptions about the role of sexuality in the neuroses changed drastically from 1896 to 1905, and this complex development inevitably affected their reception. Opponents seldom criticized these conceptions systematically, and once again indiscriminately relied on the earlier material or stressed the incompatibility of the earlier and the later definitions.

Alter one of Freud's favorite metaphors: that from existing ruins an ancient building can be reconstructed. Note, rather, the old foundation stones still supporting Freud's newer structure. As already suggested, Freud's earliest essays defined sexuality in terms of adult acts, coitus, masturbation, and perverse variants. Childhood sexual traumas were outrages initially perpetrated by adults. In these essays Freud used analogies from physics concerning the nature of electrical force to describe somatic sexual excitation, its accumulation, displacement, and discharge. In replacing the seduction theory of trauma with fantasy, infantilism, and the Oedipus complex, Freud was forced to give a wider definition to sexuality.

Nevertheless, he preserved some elements associated with the earlier theories. Libido, for example was treated as if it were a measurable quantity of energy, and the metaphors of excitation, accumulation, and discharge were kept, helping to preserve the earlier focus on adult sexual acts. Nor did Freud greatly alter his view that neurasthenia was caused by current "disturbances of the sexual metabolism," i.e. by unsatisfactory psychic and physical discharge of accumulated sexual tension through masturbation or coitus interruptus.

Although Freud gave up the single traumatic experience as an etiological factor and replaced it with a multiplicity of causes, trauma continued to play an important, if disguised role. Freud gave greater importance to the child's sexual constitution, i.e. his individual combination of partial sexual drives, determined by his physical nature and his education and culture. Infantile sexual activity, fantasy, and repression became the three major factors in the etiology of the psychoneuroses. Yet, it was as if analyst and patient still were searching, not for a single shocking moment but for a decisive set of interacting traumatic early experiences.

In both the earlier and later essays, a provocative contempt for prudishness appeared. The Clark lectures and "Civilized Sexual Morality and Modern Nervousness" condemned the extravagances of repression and advanced, to a limited extent, the claims of individual happiness. In 1910, aware that he was being misconstrued, Freud issued a warning against "wild analysis"—the prescription of sexual intercourse for the relief of neurosis. Freud insisted that he did not define sexuality in terms of "coitus or its analogy, the processes causing the orgasm or the ejaculation of the sexual product." He was using sex in the same broad sense as the word "love." No amount of coitus or other sexual acts could cure neuroses. These could exist "where there was no lack of normal sexual intercourse. . . ."[41] Sexual satisfaction, then, was in large proportion a psychological gratification.

Freud believed that his explanations, which often aroused distressing "skepticism and laughter," even in patients, were based solely on his own creation, psychoanalysis. In 1896 in an essay on sexuality and etiology read before an audience of Viennese colleagues, he ironically and obliquely pleaded for others to corroborate his work. "For the method is so difficult that serious study of it is indispensable," he said, "and I cannot remember that any one of my critics has expressed any wish to learn it from me." Later, in *Three Essays on Sexuality,* he urged that "one single observer could not fill the gap in our knowledge."[42]

A realistic need for corroboration dictated in part the importance he gave to followers who would use his method and test his findings. It is hardly surprising that he was deeply disturbed in 1912 by Jung's desertion of both the psychoanalytic method and the theories of sexuality. A few months earlier, perhaps sensing the coming break, Freud had fainted in Jung's presence after a harsh dispute. He worried, too, lest Putnam, with his deep idealism, also desert the cause.[43]

The international psychoanalytic movement split in 1911 and 1912 over the sexual issue, including Freud's definition of libido. A tiny minority, what might be called the left wing, insisted that sexual intercourse was the only adequate outlet, the only preventive of neurosis. Jung's adherents denied infantile sexuality and the Oedipus complex, and defined libido as a generalized "life force." Neurosis, Jung believed, could be prevented by the fulfillment not of sexuality but of "life tasks."[44] Both these directions would have important consequences in America.

Pronouncements on the sexual issue by Freud's followers, orthodox or schismatic, complicated the issue. Stekel, for example, continued to present Freud's earlier analogy between the physical symptoms of anxiety and the sexual act. Brill, especially in *Psychoanalyis: Its Theories and Practical Application,* tended to discuss sexuality in flat, crude terms, although insisting on Freud's broad definitions. Brill questioned his patients directly and sometimes relentlessly about their sexual lives. Putnam, on the other hand, had insisted on defining sexuality broadly, although keeping Freud's views on infantile sexuality.[45]

Opponents criticized Freud's sexual theories even less systematically than other aspects of his work. Certain topics were especially likely to be misconstrued: the nature of sexuality in childhood, the Oedipus complex, the universality of the component instincts, the relation of early component drives to later perversions and to the neuroses.

Customary usage made Freud's definition of sexuality difficult to grasp. Physicians and psychologists were accustomed to defining sexuality as the distinguishing anatomical and psychological differences between men and women or as sexual intercourse. Frederick Lyman Wells, the clinical psychologist, who accepted the traditional definition, condemned Freud for eviscerating this meaning and for making sexuality identical with undifferentiated sensations of pleasure. At a meeting of the Psychopathological Association in 1913 he insisted: "Few phases of these doctrines can have done more harm to their own cause or to the cause of truth. Where the function of science should be to delimit our concepts and give them clearer meanings, psychoanalysis has reduced this term to the level of an affective expression, deprived of every connotation that gives it a distinctive place in the language of realistic thinking." [46] Wells apparently failed to understand the heart of Freud's genetic definition: the connection between infantile sexuality on the one hand and object choice, perversions, and neuroses on the other. Charles W. Burr in 1914 cited in some bewilderment an extreme example of extending the meaning of sexuality by an American Freudian, C. C. Wholey: sexuality "covers a broad and comprehensive field of experience and activity, whether bodily desires or mental longings. It embraces all desires, instincts, wishes, ambitions, like hunger, sex, acquisition, aspiration, the social sense, love of art etc." "Some of us would like to know," Burr added, "specifically what the 'etc' includes." [47]

In 1913 at the International Medical Congress in London, Janet tried to make psychoanalysis appear at once solemn and banal, and much of his criticism centered on the definition of sexuality. He contrasted ideas from the early essays with later ones and with those of Freud's followers. He cited Putnam's assertion that sexuality included all the "affectionate and uplifting sentiments" and Ernest Jones's surprising admission that the libido was equivalent to Bergson's *élan vital*, Bernard Shaw's life force, or Schopenhauer's will to power: "Thus the psychoanalysts, while admirably sublimating the word 'love' continually speak of the 'Oedipus complex,' the masturbation of Narcissus, the little children who observe dogs during the sexual act, and of the railway station which represents the coming and going in coitus. Such confusion is not favorable either to the study of the *élan vital* or to that of sexual phenomena in humanity."

He included a not entirely implausible caricature of the genesis of homosexuality derived from Freud and from the anthropology of some of Freud's disciples: "Little boys always imagine that their mother has a male organ, identical to their own; the ancient hermaphrodite divinities were represented as female with male organs super-added, exactly as little children imagine their mothers to be; in this way the children only repeat the old beliefs of the race." [48]

It was impossible, Janet insisted, "to know what a traumatic memory is, or a subconscious memory and, above all, just what the Freudians mean by sexuality and sexual disturbances." Janet argued that Freud had illegitimately extended the traumatic paradigm to cover every type of nervous illness and even dementia praecox. The older physicians, including Charcot and Briquet, Janet wrote, had considered the role of sexuality, chiefly in its physical aspects. They assumed that puberty, the menopause, disease involving the sexual organs, excessive masturbation, sexual perversions, violations of morality, concealed pregnancies, excessive indulgence, and illicit relations played an important role in the neuroses. But Freud had entirely rejected the role of biologic sexual development. "The sexual phenomena with which he is occupied are those which have moral consequences and which affect the neuroses by the intermediary of psychological phenomena."

Yet, Janet recognized the problems Freud's elastic definition attempted to encompass, and admitted that "emotions of a sexual order" are the most "frequent, strongest and the most fertile. . . ." Yet, "true sexual excitation" was not always involved, because love was a "complex sentiment which can clothe itself in many forms." [49]

In 1914 Charles W. Burr charged that the Freudians wanted their sex two ways at once: "physical and spiritual." They claimed to define the libido symbolically, to write of sublimation and "higher" things. But in actual case histories or in interpreting dreams they meant sexuality "in terms of ordinary sexual desire, sometimes normal, sometimes perverted." According to Alfred Reginald Allen, an allegation that by sex the Freudians meant the animal thing the man in the street did was expunged from the records of the American Psychopathological Association.[50]

The way in which sexuality was defined was important for many psychoanalytic conceptions, including childhood sexuality, the perversions, and the nature of repression. If the customary definition were accepted—and many critics assumed that was the only possible definition—then the psychoanalytic strictures against "repression" seemed to clash with all the traditions of "civilized" morality and its accompanying sexual hygiene. For as Janet had implied, the prevailing view held that "yielding to the yoke of animalism" contributed to nervous and mental disorder. He and most physicians believed that a moderate exercise of normal adult sexual functions was healthy. But they all insisted that control of the sexual passions, like control of daydreams and emotions, was essential in the preservation of nervous and mental health. In 1895 Dercum had written that sexual abstinence led to neurasthenia in only a "very small" number of cases. Paul Dubois, the favorite psychotherapist of American general practitioners, argued that the lives of priests demonstrated that continence held no dangers.[51]

The universalization of the perverse was more shocking than hypotheses about the sexual etiology of the neuroses. The partial drives of childhood, sadism and masochism, and homosexuality received less discussion than perhaps any other facet of Freud's theories. For the most part they were ridiculed as preposterous; anal eroticism was especially horrifying. These aspects of psychoanalytic theory probably clashed most markedly with previous tradition. Usually physicians assumed that the perversions existed only in patients, not in everyone, and were unmistakable signs of degeneracy. There had been a growing interest in the sexual perversions in the medical journals for nearly two decades, and much of it exhibited exactly that "prurience and prudery" Freud detested.[52] Along with this "clinical" interest went unequivocal moral condemnation. The perversions must be condemned as morbid, otherwise the patient's will to be cured of them would be undermined. The distrust of Freudian psychology, G. Stanley Hall told the American Psychopathological Association in 1915, represented a "complicated protest of normality," a healthy turning away from the morbid. Most people believed that only in "perverts and erotomania or other abnormal cases" was sexuality so overwhelmingly important as the Freudians believed.[53]

Physicians divided on what role should be assigned to sexuality in the etiology of nervous and mental disorder. A small group believed this role was minor. Reversing his position of 1907, Boris Sidis argued in 1911 that "where present the early sexual experiences can be shown to be ineffective and inessential." Such experiences had happened to many healthy people, and had not happened to many psychoneurotics. Psycho-analysis, Sidis urged, was "but another aspect of the pious quack litera-ture on sexual subjects." Dercum insisted that "sexual psychogenic fac-tors" were "exceedingly infrequent" in asylum experience. Disturbances of the "complexes" relating to self preservation and community relations were far more common, especially in paranoid states.[54]

Nevertheless, most physicians were willing to concede a considerable role in nervous and mental disorder to sexuality, and most thought of themselves as unprejudiced and frank. In 1909 Frederick Peterson wrote: "Freud . . . is perhaps, extreme in his opinion, but we might grant him half, since, roughly speaking, fifty percent of the trends, wishes, desires that inspire our activities are for the perpetuation of the species and fifty percent for self-preservation." At the London Congress, Janet conceded that in three-fourths of all nervous cases sexual distur-bances occurred. But most physicians balked at the conclusion that dis-turbed sexuality was an etiological factor in *every* case.[55]

Psychologists, especially, argued that the psychoanalytic emphasis on sex was narrow, one-sided, and exaggerated. They assumed a plurality of drives or instincts, an assumption strengthened by the growing popular-ity of William McDougall's social psychology. McDougall divorced the love and parental instincts from the sexual, distinctions Prince thought sound. William James had taught that habit and other drives such as modesty could profoundly alter the sexual instinct. Some people never gratified their normal sexual desires, yet showed no adverse effects. When Jung attempted to explain libido as generalized psychic energy and Adler began emphasizing the "will to power," psychologists felt that their conclusions were borne out by the psychoanalysts themselves. Woodworth wrote: "When one lays everything to 'libido' and another everything to the 'masculine protest,' it may soon be recognized that both of these factors, and probably several others, are operative in pro-ducing abnormal, and also normal, mental results." [56] G. Stanley Hall in-sisted to psychopathologists that "anger" was as primary a motive as sex, and functioned in accordance with the Freudian mechanisms of repres-sion and displacement.[57]

The psychoanalytic version of the theory that valued aspects of civilization were rooted in sexuality seemed especially scandalous, a viru-lent and denigrating reductionism. At the London Congress Janet re-marked that the doctrine of sublimation confused "the highest tenden-cies of the human mind with instincts which are common to all the animals." An explanation for the psychoanalysts' attempt to resolve all instincts and emotions into one primary category was offered by Elmer Southard. The Freudians had a special predisposition: they were "emotional monists." This, he thought, made possible Putnam's adherence to both Hegel and Freud.[58]

All critics, friendly and hostile, agreed that psychoanalysis flouted the customary reticence of "civilized" morality. To Dercum the psychoanalytic technique inevitably involved a "lesion of the proprieties." Freud's own sense of reticence was such that he could not bring himself for years to publish the full sexual details of his case histories, lest he embarrass his patients, Adolf Meyer noted this scruple in his brief review of *Bruchstück einer Hysterie-Analyse* and *Drei Abhandlungen.* The norm of reticence seems to have been more stringent in America than in Europe, both natives and foreigners agreed. Meyer, a Swiss, and Barker, a Canadian, neither of whom were psychoanalysts, believed that prejudice and convention made it difficult to discuss and to try Freud's methods in America. Meyer wrote in 1909: "No experience or part of our life is as much disfigured by convention as the sex feelings and ambitions. Not to speak of them at all or only under the cover of symbol is the pedagogical and social ideal of our civilization. . . ." [59] Many natives prided themselves on the wholesome moral superiority of America over decadent Europe. Ernest Jones was amused when an indignant lady after one of his lectures declared that Americans had nothing to repress; their dreams were always pure and unselfish.[60]

Yet many critics were sympathetic to the repeal of reticence. A growing appreciation of the importance of sexuality occurred among physicians in both Europe and America beginning in the 1890's. The curiosity about perversions was one sign of this, the studies of childhood sexuality another. Janet, after his lengthy and bitter criticism in 1913, concluded: "But I am well aware that underneath these exaggerations, and perhaps because of them, there has been developed a quantity of valuable studies on the neuroses, on the evolution of thought during childhood, and on the various forms of sexual sentiments. These studies have called attention to facts which were not well known and which, owing to a traditional reserve, we have been too much disposed to neglect." [61]

Yet, psychoanalysis seemed an insult to women and children, precisely those objects the American code was designed to protect. Children were the agents of future progress. Mothers guarded the nation's moral purity. The Oedipus complex, then, aroused special indignation. Dercum fulminated in 1914: "What must we think of the wounding of the feelings of a sensitive and innocent nature when to a loving son or daughter is suggested an incestuous love for the mother? What injury can be greater than to give to one of the most beautiful relations in human life the most shocking and the vilest of interpretations?" [62] A. Friedländer, the German psychiatrist, wrote, quoting Sadgaer: "As we learn from psychoanalysis all boys have the fancy of putting themselves in their fathers' places and impregnate their mothers." A. Moll, author of *The Sexual Life of the Child,* with an introduction by Edward L. Thorndike, argued that Freud saw "the sexual in the life of the child where it cannot possibly exist." The Philadelphia neurologist James Hendrie Lloyd, who had suggested a sexual origin for hysteria in 1873, described Ernest Jones's essay on Hamlet as an "unholy effusion": "Hamlet as a boy has an intense passion for his mother. This was not what the world in its ignorance calls the natural affection of a child for his mother, but it was distinctly erotic; in short, the first manifestation of sexual passion. We are assured that this is not uncommon, and that women of a passionate nature frequently inspire this feeling in their young sons. Jones even quotes with approval Freud's dictum that the 'mother is the first seductress of her boy.' " [63]

To Charles W. Burr mentioning sexuality to women was in itself shocking. The "average woman," Burr remarked in 1911, "is not going to confess [to a sexual matter] unless it is put to her." It was wrong to be "catechising young girls and probably put into their heads ideas of sexual insult they had as children which never really existed." Making explicit the relation between class and "civilized" morality, Charles L. Dana argued that psychoanalysis had proved disastrous with "educated, refined and delicate minded" women and that dwelling on "a mass of old memories" and the details of early emotional experiences threw such patients "into almost delirium of distress." Dana had stopped using the method with the "educated and intelligent type of patient." In "dispensary practice," where some of his assistants already were trying psychoanalysis, "there had been rather more scope for the method." "Dwelling" on past "sexual mistakes" contradicted not only reticence but much of the previous psychotherapeutic teaching that "new associations" must replace the old.[64]

Some charged that psychoanalytic treatment fostered overt immorality. Sensitive youths came to fear that they were "congenital perverts."

Some patients committed suicide. It was scandalous that analysts carried on intimate conversations, sometimes for years, with recumbent females. To Boris Sidis psychoanalysis was a "worship of Venus and Priapus," a direct invitation to masturbation, perversion, illegitimate births, unions out of wedlock.[65]

Opponents used the sexual issue to discredit psychoanalysis. Yet, the psychoanalysts themselves also used the issue to make their opponents seem hidebound, obscene, and squeamish. Instead of answering cogent criticism, they often counterattacked on the sexual issue. This tactic is apparent in a letter of September 5, 1911 from Ernest Jones to the Philadelphia neurologist Alfred Reginald Allen:

> Yes, I am of course familiar with the penetrating criticism of Freud's work that you so well describe. It really is only another way of saying that these gentry in question know that *they* are incapable of dealing with sexual topics in any shape that would not be indeed obscene, and they very naturally concluded that no one else can; to them sexuality is synonymous with obscenity. . . . Such fanaticism is of course at bottom a reaction against repressed desires; they cannot allow themselves to "wallow in talk" with lady patients, and they won't allow anyone else to— just like a reformed drunkard.[66]

To Putnam and most other analysts, prudery was the paramount, almost the only recognized cause of opposition. They advanced four principal counterarguments. They tried to correct "misunderstandings" of the theories themselves; they insisted on a need for frankness; they asserted Freud's unimpeachable integrity; they argued that everyone was preoccupied with sexuality; they sometimes leveled their own vitriolic, *ad hominem* attacks. Freud had set the precedent for this kind of polemic by arguing that those who opposed his sexual theories suffered from the same resistances he encountered in his patients.

Just after the Clark Conference, Adolf Meyer tried to define these theories for the New York State *Hospital Bulletin*. His discussion of them was not very lucid, but he stoutly defended their worth. He wrote that he himself had noticed that dementia praecox victims showed special sexual compulsions and preoccupations. Freud was not "wallowing in dirt and smut." He was facing human nature as it was, not as "we might wish it to be." Dealing with sexual details, Meyer and Schwab argued, was no more "immoral or unclean" than examining urine or feces.[67] This argument revealed something of the same contemporary equation of sex and scatology that patients also made. Putnam was at pains to show that Freud, Ferenczi, and every other psychoanalyst had had to overcome initial disgust in the pursuit of psychoanalytic truth. Novels, myths, tragedies, terrible sufferings, and terrible crimes daily thrust sexuality to public attention. There was a growing group of serious students of sexuality—Moll, Krafft-Ebing, August Forel, Havelock Ellis. Yet these serious investigators of the subject were rejected as "disturbers of the peace": "The cause of formal modesty and reticence has indeed had many noble martyrs, both before the days of Paul and Virginia and since. But there is such a thing as paying too dear for this niceness, especially when, through the opposite course, we can have all that we should gain by this, and more besides." [68]

Frankness and truth were essential prerequisites to moral progress. Perhaps unwittingly exploiting feelings of guilt, Putnam asserted that everyone was at heart troubled by childish and sensual cravings. To be ignorant of them, to fail to face them, merely held men back on the Bunyanesque "Hill of Difficulty." The best virtue, he wrote, came not from inheritance, but from struggle. Confronting our cravings would result in "liberality, tolerance and purity" and an end of the "tyranny of ignorance and prejudice." Putnam even asserted that the tensions created by sexuality were the very motives of progress.

Some of Putnam's younger colleagues charged their opponents with being old and incompetent. White once insisted that they were the "established authorities who have passed through the formative period of their lives. . . ." Burr was moved to complain: "We are considered men of no standing . . . 'back numbers' . . . unprogressive and ignorant. . . mere describers of symptoms . . . not intelligent enough to understand such profound matters of philosophy." [69]

Sometimes the psychoanalysts' counterattacks brought reprisals. Ernest Jones noted that the denunciation of Freud by Moses Allen Starr before the neurological section of the New York Academy of Medicine in 1912 was unprovoked. In fact, Putnam had provoked it. He had denounced medical prudery at the meeting in uncompromising terms, and asserted that most physicians shared the ignorant prejudices of the public. Prefacing his remarks with a quotation from Emerson, "When half gods go, the gods arrive," he implied that doctors resisted psychoanalysis because unconsciously they were obsessed with sex. Most of the important American Freudians were present—White, Jelliffe, Brill,

Hoch. After Putnam's speech, Starr attacked Freud bitterly. Starr claimed to have worked with Freud in Brücke's laboratory in Vienna. Starr then advanced an argument that Janet and others had used to explain the sexual emphases of psychoanalysis: Freud's theories reflected the immorality of Vienna and of Freud's own "peculiar life." "Freud," he asserted, "was not a man who lived on a particularly high-minded plane." [70]

Starr was speaking as a physician sympathetic to psychotherapy who had asked Janet to lecture at Columbia in 1906. Psychotherapy, Starr insisted, was an instrument as useful as the scalpel or the microscope. But the psychoanalysts were perverting psychotherapy by frivolously exalting sex and by ignoring the vital instinct of self-preservation. This instinct, he noted, manifested itself in America in the struggle for wealth.

Putnam, who had been so shocked he could not speak, wrote to Freud: "I felt very, very badly about Dr. Starr's most undignified and discourteous and absurd attack after my paper and several others. I went to the meeting determined to take all criticism quietly and good-naturedly, and was so taken aback by what he said that I could not trust my words in answer.

"Since then I have indulged in a great deal of *esprit d'escalier* to no purpose, but have also had a talk with him personally and made him feel ashamed." [71] Freud replied reassuringly on June 25, 1912: "I am sorry that you were upset by Dr. Allen Starr's remarks. I was able to remain cool and untouched by them because I had never known Starr. Nor did the issue, no matter what its precise origin could have been, seem important enough. His information about my early years amused me mightily. Would that it had been true!" [72]

Opponents sometimes insinuated that psychoanalysts were perverts or sexually obsessed. After denouncing psychiatrists who explained perversion as caused by environment, not heredity, Burr remarked, "there is something in degenerates which attracts the affection of other degenerates."

But Freudian explanations of neurosis were provocative. For example, Isador Coriat had explained stammering as a reaction to unconscious sexual complexes. A stammering psychiatrist protested that in his case at least this explanation was irrelevant.

The Freudians defended their own integrity and Freud's as well. In 1911 Schwab, who considered himself a moderate, attacked the "prejudice and ignorance" of the Philadelphia organicists he had just addressed: "It seems the strangest thing to hear the objections raised here tonight to Freud's method on the grounds that it is immoral and that the discussion pertaining to it is unpleasant.

"Those who have ever known Freud personally know that he is not a man to arrive at conclusions in a hurry or without careful consideration of the facts. He is in no sense an advocate of any morbid sexual tendencies nor one that delights or is unduly interested in things that are not seemly. He has succeeded in removing the purely personal attitude toward these subjects in his work." [73]

The suggestion that a personal bias or a unique sociological environment gave rise to psychoanalytic theory became a standard objection. Even Schwab remarked that it was difficult to find "suitable"cases in America. Here was a very real issue that requires explanation. Men sympathetic to psychoanalysis, such as Morton Prince, who insisted they had no prejudices, failed to find in their patients what the psychoanalysts did. Once more the discussions among members of the American Psychopathological Association are revealing. Putnam and Prince, for experimental purposes, both studied the same patient in 1912 and came to quite different conclusions.

The issue had come to a head a year earlier when Prince's assistant, John Donley, attacked the sexual basis of the anxiety neuroses. Putnam's former assistant, George Waterman, cited two "asexual" cases, an eighty-year-old Civil War veteran whose symptoms seemed to be the result of serious cardio-vascular damage and a "fearless, reckless" Boston fireman who, after driving his engine into an electric car, developed severe anxiety attacks. Then Ernest Jones had challenged: "We have heard nothing about the sexual life of these patients." [74] He urged the opposition to bring forward fully analyzed cases in which sexual factors played no role. Perhaps the joint Prince-Putnam experiment resulted from this dispute.

Their patient was a forty-one-year-old spinster, probably a social worker, physically robust and intelligent. She had been treated unsuccessfully by many doctors for eight years for severe phobias, chiefly of insanity and death, and for terrifying anxiety attacks in which she believed that the air was "unreal" and that she might suffocate. Prince described her as a consciously "asexual" type. He traced the beginning of her phobias to the age of sixteen. She had been ill then, and a friend of her father's vividly described to her Michelangelo's *Last Judgment,* which he had just seen in Italy. She began to brood about death and thought that if she died she would go to Hell. Some years later and independently of the first phobia, the milder fears about the unreality of air began. Then, having entered a period of depression in which the earlier thoughts of death and judgment obsessed her once more, she concluded she was going insane. Later still, when she fell ill with jaundice, she associated the obsession with the sense of unreality. From then on, when alone on the street or in her room, she would have attacks in which she thought she was suffocating. Prince had traced the origin of what he considered the "objects" of her fears and the "settings" in which they occurred. Thus he considered physical illness, depression, and the thoughts of death as causes in themselves.

Putnam made a far more complex set of assumptions, and turned up different and much earlier material. He explored her childhood and her relations with her parents. Her hearty frankness, he believed, covered a rich and abundant life of sexual fantasy. She had recalled childhood masturbation and later dreams of taking a bath in a tub in the gutter. Freud had interpreted this dream for Putnam as undoubtedly a "prostitute-dream" in which her "cravings were so strong that she was, as it were, whispering to herself that she would go to any length to gratify them." Her fears of death derived partly from her identification with her weak and vacillating father, who had died slowly and painfully. Her chief difficulty, he concluded, had been her inability to face her inner cravings, her desire for love, marriage, and a sexual life. Putnam insisted he based his interpretations on the large body of knowledge psychoanalysts had gathered from other cases. Neither he nor Prince had relieved her symptoms more than temporarily.

In the ensuing debate, Prince argued that the Freudian interpretations displayed naïve *a priori* reasoning. He was incensed when Brill suggested that in his experience spinsters were difficult to treat successfully because of their unsatisfied libido. Prince retorted that he had cured many unmarried women; Brill was reading his own assumptions into the case.

The Freudians present praised Putnam for his more "profound" interpretation. Prince, they said, had only described the surface, although in photographic detail. L. Pierce Clark remarked that Putnam had "gone deeper into the sexual constitution and its make-up; his analysis is therefore more preferable." Adolf Meyer tried to make peace by asserting that two honest and competent observers simply had come to differing conclusions.[75]

A year later, in 1913, Prince presented still another case of phobia, again carefully observed, again offering no sexual material. While he was discussing the patient's fear of bell towers, the Freudians in the room nodded knowingly. Prince insisted that she was *not* afraid of towers, which he knew were a phallic symbol to the Freudians, but of the sound of bells, which she associated with her mother's death. Ernest Jones argued that the towers *were* phallic and that the entire symbolism was ingeniously neat: the ringing of the bells signified not only her mother's death but her joyful wedding to her father.

Prince had examined the patient minutely by means of hypnosis and automatic writing. Her phobias were "singularly free" of sexual associations. He attributed most of her problems to a complex of "self-blame" and added that many instincts besides the sexual could cause traumas. In her case, McDougall's instinct of self-abasement was appropriately explanatory. If his diagnosis were wrong, he challenged, why had she recovered from her phobia? Jelliffe ironically suggested, "God took care of it." Prince remarked: "I will tell you the reason why she ceased to fear: I changed the settings which gave meaning to her conserved and harmful ideas and thus changed her point of view." He taunted Jelliffe to search deeply enough to find other primitive instincts besides the sexual. "Why do you stop at the sexual instinct?" he asked. "I don't stop anywhere," Jelliffe answered. "But you do, that is just what you do," Prince concluded.[76]

This bad-tempered dialogue involved important theoretical considerations. Prince insisted that the neuroses were perversions of normal memory processes and that buried memories were best reached by hypnosis. It was regrettable, he said, that a man of Freud's genius had given up hypnosis at the very moment when he might have made important discoveries by its use. If hypnosis efficiently revealed forgotten material, why didn't it turn up sexual associations? [77]

A speculative answer may illuminate the role of "civilized" morality, the relation of the psychoanalytic technique to the resistances structured by this norm, and aspects of Prince's personality. Two issues are involved here, the recovery by hypnosis of memories of sexual experiences and fantasies and the nature of "resistance." Originally, Freud had used hypnosis to enlarge the patient's memory and this resulted partly in the discovery of sexual trauma. When he gave up hypnosis he encountered what he termed "resistances," forces apparently blocking the patient's memory. The analysis of resistances revealed conflicts between internalized standards and instinctual drives. Resistances, Freud concluded, provided the most significant material for analysis. Ernest Jones, as already noted, pointed out that Prince had failed to analyze these blocks to memory, and Prince had countered by arguing that in his patient they had not appeared. How in fact could hypnosis have been expected to affect "resistances"?

At the time, students of hypnosis disagreed about the degree to which the technique could overcome a subject's moral conscience. Bernheim's pupils, for example, thought that subjects could be induced to commit crimes by hypnotic suggestion, and staged experiments in which subjects attacked doors with paper knives. Physicians who took the opposite view pointed out that these were not real, but mock crimes. Janet, for example, insisted that patients would not perform acts contrary to their own moral sense: a hypnotized girl would not undress before a group of medical students.[78] "Civilized" morality, it must be remembered, placed thoughts and discussions about sexuality, at least for women, on almost the same level as sexual acts. For this reason, possibly hypnosis did not overcome this internalized norm in Prince's patient, a well-bred spinster.

The matter had been quite different with the patient whose dreams Prince analyzed in his study, "The Mechanism and Interpretation of Dreams." She was a widow with a son. Prince wrote to Putnam asking that her dreams not be interpreted sexually for several reasons. She already had "yielded a great deal in allowing her dreams to be published at all, and if they were made the subject of public discussion over her sexuality she would feel it horribly as I think anyone would." In fact some of her dreams were frankly sexual; she was not squeamish, but "quite frank" in discussing these aspects with Prince.[79] To interpret her dreams publicly in sexual terms would have breached medical confidence. From the point of view of "civilized" morality, such a discussion would have violated the norm of public silence about the sexuality of a refined woman. One aspect of Prince's analysis is noteworthy: the total absence of any discussion of the patient's relationship to himself, which, judging from the dreams themselves, was highly charged with emotion.

Possibly Prince did not wish to discover sexual material in dreams that were not overtly sexual; perhaps his patient subtly perceived this feeling. Moreover, Prince's personality may have affected his therapeutic emphasis. His mother and his wife suffered from nervous disorders. He had killed hypnotically precisely the troublesome, sexually disruptive aspect of Sally Beauchamp's multiple personality. In 1912 he indignantly rejected the Freudian subconscious, which he likened to a murderer: "I conceive the unconscious not as a wild, unbridled conscienceless subconscious mind, as do some Freudians, ready to take advantage of an unguarded moment to strike down, to drown, to kill, after the manner of an evil genii, but as a great mental mechanism which takes part in an orderly, logical way in all the processes of daily life, but which under certain conditions involving particularly the emotion-instincts becomes disordered or perverted." [80] Thus, a personal orientation, a technique that did not focus on resistances but used a method that the Freudians claimed obliterated them, and the operations of the "civilized" norm had affected Prince's results.

The Freudians, on the other hand, insisted that they discovered resistances and sexual material in their patients' associations. One analyst argued that sexuality was deeply disturbing to "carefully nurtured girls" from "homes of refinement, where reference to sexual matters is synonymous with indecency." [81] It is important to recall that psychoanalytic techniques focused precisely on overcoming the reticence that "civilized" morality so strongly emphasized. Thus, the technique of the Freudians and the personal orientation that a commitment to psychoanalysis entailed operated to affect *their* results. What they claimed to find in patients and the relation of these findings to the civilized norm will be described in the conclusion of this study.

The description of the unconscious Prince attributed to the Freudians would have been endorsed by few, if any, American psychoanalysts in 1912. Freud was just beginning to consider the unconscious to be full of destructive potential, a view that grew stronger with the years.

Cultural Values

Prince may have felt that psychoanalysis was among the evil genii that menaced important values of American culture. On cultural decay, Prince, the Boston psychotherapist, and Dercum, the Philadelphia organicist, saw eye to eye. In 1885 Prince had taken care to see that his own psychological determinism did not undermine the value of the "true, the beautiful and the good." By 1916 he sensed a growing disintegration of the common meaning of national "ideals" which he considered the heart of social consciousness and social organization. At a garden party in

Tokyo he said that the social conscience, or censor, as he called it, which regulated individual conduct, was weakening. He praised the severity of Japanese education; for education's purpose was to repress the "barbaric instincts with which every child is born." Repression, Prince also believed, could inhibit effectively women's sexual instincts for long periods of time. America's once great social consciousness, devoted to high ideals of freedom and morality, was being fragmented by polyglot races, selfish individualism, heterogeneous class divisions, demoralizing social philosophy, and conflicting interests.[82] Dercum in 1914 also warned against cultural collapse, of which the "epidemic" of psychoanalysis was one symptom: "There is an abandonment of all previous standards. . . . The real gives place to the unreal, the beautiful to the unbeautiful, the wholesome facts of life to the morbid untruths of disease . . . the evidence of the senses are replaced by the phantasms of exhaustion." [83] Attitudes toward moral and cultural issues inextricably were bound up with psychoanalysis. The lines were not always sharply drawn. Nevertheless, those who expressed themselves most vehemently against psychoanalysis generally disliked the growing frankness about sexuality, as well as the new theater of Ibsen and Shaw and the paintings of Gauguin and Duchamp. There were exceptions. Elmer Southard, the brilliant young Boston psychiatrist who enjoyed twitting the psychoanalysts, appreciated modern painting, while some analysts thought it deplorable.[84] But this does not alter the general pattern of extremes on either side.

Attacks on psychoanalysis grew more vitriolic from 1911 to 1917, reflecting cultural conflicts and increasing Freudian successes. Opponents argued, correctly, that most of the material printed in America favored psychoanalysis and that they merely were redressing the balance. They deplored the feverish zeal of the psychoanalysts, who struck one unsympathetic observer as possibly "idealistic" sexual neurasthenics with a tendency to scribble. To Freudians American civilization was not deteriorating, but progressing, and creating for them a more influential place. As Dercum wrote, they were convinced the future was theirs.[85] It was partly this growing prestige, partly the confident and brash advertising put out by the publishers, W. B. Saunders, for Brill's translation of *Die Traumdeutung* that incited one of the most cruel and brutal of all the American polemics. The publishers claimed that "the main facts of *The Interpretation of Dreams* have never been refuted . . . the book has led to a revolution in the science and methods of psychotherapeutics." "Flimflam and fustian," commented Victor J. Haberman, a Berlin trained neurology assistant at Columbia University: "If you look through a dozen articles of these dream seers you will find each one, in spite of his lullabying about Freud's meaning of *sexual,* making out his patients to desire 'cunnilingus' from her father, or as wishing to perform fellatio on him, or having her pockets picked etc. You get the opinion that almost all daughters desire this. (What an orgy such an 'evening' at the Psychoanalytic Society must be with all these brethren munching their themes!)" [86] The fact that this diatribe was printed in Prince's journal may have indicated his growing impatience. In 1912 Adolf Meyer noted that the trustees of one new hospital had forbidden the use of either hypnosis or psychoanalysis. He attributed this rule to misinterpretations of psychoanalysis, and possibly these involved sexuality, a subject that quite independently of psychoanalysis was arousing public discussion. The new degree of literary, dramatic, and journalistic interest prompted a Catholic psychiatrist from Fordham University to assert in 1912 that America had gone "sex mad." He did not attribute this to psychoanalysis, which he dismissed as a variety of suggestion therapy, but to wealth and luxury and over-attention to sexuality which were making America more and more like decadent Europe.[87]

Precisely the interpretation Freud most feared was made in 1911 by Samuel A. Tannenbaum, a young psychoanalyst, at a symposium in New York, which Brill also addressed. Using Freud's 1894 essay on the anxiety neuroses, Tannenbaum dilated on the grave dangers of sexual abstinence. Abstinence could only be recommended to those with "no great sexual needs." Before marriage it could not be required "of all men because a large number of those who heeded such advice would undoubtedly develop apprehension neurosis." He suggested that young men engaged to chaste women should frequent prostitutes to avoid "frustrated excitement," a course he also advised to onanists. In the tradition of sexology manuals, he counseled prudent stimulation. "Well advised instruction," he wrote, "will rarely fail to convert an unhappy, quarrelsome home into a peaceful and happy one, and change a thin, pale, sleepless, irritable and morose wife into a healthy, beautiful and contented mate."

Brill, whose own speech probably followed directly, took pains to insist that by sexuality Freud meant not only the "coarse sexual, but also everything appertaining to sex in the psychic and physical spheres." [88] Obviously, Brill was alarmed. For in the second edition of *Selected Papers on Hysteria,* published in 1912, he included Freud's essay on "wild analysis." In the preface Brill wrote: "We can well afford to disregard our uninformed opponents, but heaven protect us from our friends who accept everything without knowing what they do." [89] Tannenbaum's speech aroused the editors of the *Medical Review of Reviews* to denounce him as vicious and disgusting. "It is doubtful if Freud would sanction such a heinous offense," they wrote. "Passing by all the moral considerations of these atrocious principles, the physical hazards involved stamp them as hostile to the welfare of the race." [90]

However, Tannenbaum disregarded criticism. In 1913 at the height of the public furor over "white slavery," when John D. Rockefeller was publicly endorsing the Broadway production of Brieux's *Damaged Goods,* Tannenbaum urged that prostitutes be licensed. He thought prostitution a regrettable institution, but, nevertheless, he suggested that the dangers of abstinence might be worse than those of venereal disease. He also linked the cause of psychoanalysis with birth control, easy divorce, and early marriages—based on Ellen Key's *Love and Marriage* which advocated something close to free unions.[91] Tannenbaum was an able polemicist against the critics of psychoanalysis, and addressed himself to Woodworth, Meyer Solomon, and the Jungians. In one curious essay, he identified the source of libido as the sun and, ultimately, God. He also wrote an indignant muckraking article on medical fee-splitting. He was a friend and a favorite psychoanalyst of the rebellious young intellectuals, who about 1912 made Greenwich Village their headquarters. They made a common cause of psychoanalysis, sexual liberation, and a sweeping, popular mysticism that emphasized the joys of the Bergsonian Life Force. This proved to be a new and important combination. One of its major elements was anti-Puritanism. In 1916 the New York *Medical Record* argued that survivals of the "strait-laced fanaticism of our Puritan forefathers" partly motivated the opposition to Freud's theories. Just as Puritanism dictated that Americans judge works of art from a moral rather than an aesthetic viewpoint, so, too, the psychoanalyst had been described as a "peddler of pornography, a sink of salaciousness and in general about three shades worse than the mayor of Gomorrah." The psychoanalytic theory of the child's sexual and hostile feelings was especially upsetting. "A common cry against psychoanalysts for intance, is their conception of the sexual life of the child and the various loves and hates he may develop within the home." Convention held that a child loves its parents "naturally and inevitably" because it was right that it should. "We should advise the psychoanalyst then who wishes to promulgate his method to avoid mentioning the intimate method of taking anamneses, not to discuss children at all if he can avoid it, put the soft pedal on sex generally—in short, sugar-coat his pill as much as possible and perhaps the profession will swallow it." [92]

Some specialists became indignant at the involvement of laymen in the psychoanalytic movement, either as therapists or as patient enthusiasts. Janet remarked of patients: "They are happy because some one is occupied with them, that a new method of treatment is applied to them, a disputed treatment, strange and a trifle shocking in its apparent disdain of customary modesty." [93]

Burr was angered by the possibility that laymen, clergymen, and schoolteachers might take up the practice of psychoanalysis. Apparently referring to Oscar Pfister's *Psychoanalytic Method,* Burr argued that a discussion of intimate aspects of sexuality was likely to arouse indelicate feelings: "Will it be wise for an interesting and spiritual-looking young curate to discuss the sexual symbolism of dreams with girls rather susceptible to human passions? Would it not endanger his own welfare to do it with a woman who had reached the dangerous age?" [94]

Friedländer was irked by the interest among European laymen, an interest, he thought, that was deliberately fostered by Freud's disciples. It was wrong for lay publications to carry articles about psychoanalytic hypotheses that specialists in their journals had not yet reviewed or tested. Like psychotherapy, then, psychoanalysis threatened the professional role of the neurologist and psychiatrist. This important issue would provoke a characteristic response from the American psychoanalysts.

Psychoanalysis, Francis X. Dercum insisted in 1914, came from the same mystical wave that was producing "occultism and symbolism in art, music, literature and the drama—cubism, futurism, modernism, the problem play." Dercum had identified psychoanalysis with all the new movements of intellectual, artistic, and sexual rebellion after 1912. In fact, some of the same people were involved in both. [95]

The opposition of neurologists such as Dercum and Prince may have had undercurrents of class and cultural defensiveness. To men self-consciously upper class in origin, education, or current status, the psychoanalytic movement, and especially its most Freudian wing, may have seemed an invasion of southeast European Jews, other foreigners, and native Americans of humbler antecedents than their own. An analysis of the social background of the psychoanalysts in the next chapter will clarify the basis for this possibility. The American Neurological Association remained conservative in many ways. It was only in the 1890's and more rapidly after 1900 that public hospital alienists were admitted: Edward Cowles and Richard Dewey in 1891, Adolf Meyer in 1899; Smith Ely Jelliffe in 1900; William Alanson White in 1916. The psychoanalysts were attempting to establish themselves in the intensive, private treatment of nervous patients, hitherto a monopoly the neurologists had claimed on the basis of their superior specialized training. Yet the psychoanalysts from less exalted origins were almost as well trained as themselves, and they offered a new and radical technique of treatment.

Although the code of late nineteenth-century "civilized" morality and its accompanying reticence cut across class lines in America, upperclass groups in Boston and Philadelphia to which Prince and Dercum belonged were closely identified with it. In the meetings and publications of neurologists and other physicians, it was the psychoanalysts who first spoke out most frankly on behalf of Freud's sexual theories in 1908 and 1909, and it was Brill who denounced most vehemently the "Anglo-Saxon" prudery.

XII

The American Psychoanalytic Organizations 1910 - 1917

Of all the competing psychotherapeutic schools, only psychoanalysis created professional organizations formally responsible for training and doctrinal purity. Analysts and historians have debated whether these societies advanced or retarded psychoanalysis as a scientific discipline. Whatever one decides on that question, the organizations played a role in the formation of the psychoanalysts' professional identity.

After conversion some physicians, including James Jackson Putnam, whose case is the best documented, did not yet see Freud's system as a coherent doctrine to be accepted or rejected *en bloc*.[1] Then their exposition of Freud's theories to colleagues, especially its sexual elements, aroused opposition. This in turn increased the sense of identification with Freud's cause.

At this point, proposals to found professional organizations played a double role. They exacerbated the annoyance of opponents such as Morton Prince and apparently confirmed charges that psychoanalysis was not a technique and a theory but a cult. Once the organizations were established, the sense of each member's identity as a psychoanalyst became stronger. After the International Psychoanalytic Association had been

founded, Ernest Jones gave the impression that Freud's European followers were a thriving host. Yet, mere membership and mere organization
did not create faithful psychoanalysts. The nature of the organization
was essential. Had it not been for the New York Psychoanalytic Society,
which was cohesive for special reasons, it is unlikely that orthodox psychoanalysis would have gained a strong foothold in the United States.
Most important, perhaps, was the emotional relationship of each analyst
to Freud, the primal father of his often ungrateful horde. The most orthodox were those who felt a special bond of identity with Freud himself.

Freud took the initiative not only in organizing the movement in
Europe and America, but in defining a psychoanalyst as someone who
accepted his theories. This definition grew increasingly important as
Adler and Stekel broke away in 1911 and Jung in 1912. Freud's motives
perhaps never will be fully known, certainly not until the Freud archives
are finally opened. The conscientious Swiss psychiatrist Eugen Bleuler informed Freud that he was an artist who wanted to preserve his creation
intact and was passionately eager to secure its acceptance. Freud undoubtedly was sensitive to even the mildest criticism. He also distrusted
some of his followers and wrote to Bleuler that he feared their extravagance and needed a central headquarters to control their polemics.[2]

Freud's explanation to his psychoanalytic colleagues was somewhat
different. He argued that the International Psychoanalytic Association
was needed for one major reason: to claim a monopoly of the psychoanalytic technique so that members could repudiate unqualified practitioners of "wild" psychoanalysis who sometimes prescribed direct sexual expression as a treatment for the neuroses. Thus Freud hoped to prevent
the abuses he feared would arise as psychoanalysis became popular. Another reason was the need for training. The psychoanalytic technique
could be acquired solely from those who already were proficient. Formal
psychoanalytic organizations would train physicians whose work thus
would have some kind of a guarantee of competence. Finally, because of
the "boycott" of "official science," the adherents of psychoanalysis could
meet for "friendly communication" and "mutual support." [3]

At the second Psychoanalytic Congress held in Nuremberg March
30–31, 1910, the plan of an International Association with local
branches was adopted. Members, it was assumed, would be drawn from a
single city and its surroundings and thus be able to work conveniently
together. Sandor Ferenczi, a Hungarian psychoanalyst, and Freud drew

up detailed proposals which Ferenczi formally presented. After a quarrel between the Viennese and the Swiss, whose side Freud took, Carl Jung, then Bleuler's second in command at Zurich's Burghölzli Clinic, was elected president and Zurich chosen as international headquarters. For Freud the association represented an attempt to establish psychoanalysis as a recognized scientific and medical discipline. It would no longer be the possession of himself and a group of personal followers, as had been the case in Vienna.[4]

On April 1 Ferenczi wrote to James Jackson Putnam, professor of neurology at Harvard, omitting the political quarrels of the Congress, but mentioning branch societies in Vienna, Berlin, and Zurich. Could not one or more local groups be organized in America?

On April 14 Freud suggested to A. A. Brill that the time was ripe for him to form a branch society in New York and Ernest Jones, perhaps, to organize another in Boston. Then, on June 16, he urged Putnam to take the leadership of the American movement:

> In addition it has occurred to me—and I am writing this unofficially for I have no right to do so officially—that only you and only Boston could be the starting point for the formation of a psycho-analytic group to be joined by our friends in America, and which could then affiliate with the International Association that was planned in Nuremberg. I understand that all important intellectual movements in America have originated in Boston. I also know that no one else is as highly regarded as you; because of your unimpeachable reputation for integrity, no one else could protect so well the beleagured cause of psychoanalysis.[5]

Putnam hesitated. He wondered what effect the proposed psychoanalytic group would have on the new Psychopathological Association that just had been founded on May 2, with his friend, Morton Prince, the Boston neurologist, as president, and his former assistant, George Waterman, as vice president. Jones tried to assuage these doubts. On July 12, 1910 he wrote to Putnam:

> As to the branch society I agree that the American Psychopathological Association fills the bill at present. We are too few to make it worth while to organize a new society. On the other hand it would please Freud if we simply got up a *formal* branch of the Internat. Psychoanalytic Verein, and it might strengthen their hands a little. We could hold a short meeting once a year just before the American Psychopathological Asso-

ciation, and in a couple of years should be large enough to hold more definite meetings. It might also be of service in co-ordinating our work a little, and exchanging experiences and views. On the whole, therefore, I am in favor of forming a branch.[6]

In late July Putnam informed Freud that he and Jones would "bring about the Branch Association meeting . . . in connection with the next meeting of our new American Psychopathological Association," scheduled for the spring of 1911.[7]

Meanwhile Jones had sailed for Europe. He visited Freud, who was vacationing in Holland, and attended the International Congress of Medical Psychology in Brussels. On August 14 Jones wrote to Putnam from London that Freud had been puzzled at Putnam's initial silence:

> He was very clear, about our forming a local branch of the Verein, and produced the following argument which had struck me forcibly in America. We are so likely to have the work damaged by amateurs and charlatans that it becomes necessary to protect our interests by enrolling those with some proper knowledge of the subject in a rather official Verein which would therefore be *some kind* of guarantee in a general way that the members knew what they were talking or writing about. . . .[8]

Jones suggested to Putnam that together with Brill they send out an invitational circular to these psychologists and physicians: Stanley Hall, sixty-six, president of Clark University who had invited Freud to America; August Hoch, forty-three, who succeeded Adolf Meyer in 1910 as director of the Psychiatric Institute of the New York State Hospitals on Ward's Island; C. Macfie Campbell, thirty-four, a Scottish psychiatrist, trained at Edinburgh, who then was an associate at the Institute and taught at the Cornell University Medical College in New York; Edwin Bissell Holt, thirty-eight, assistant professor of psychology at Harvard; Bernard Hart, thirty-seven, a London medical psychologist; William Alanson White, forty, director of the Government Hospital for the Insane in Washington, D.C.; Isador Coriat, thirty-four, a Boston psychiatrist who had worked at the Worcester State Hospital under Meyer; and possibly Meyer himself, then forty-five, who had just been appointed professor of psychiatry at the John Hopkins University Medical School in Baltimore.[9]

Putnam replied September 14 that he agreed with Jones's proposals. He added, however, that Prince, from whom he had drifted a little apart, probably would not like the idea. Indeed, Prince told Putnam that if the Freudians could not bear discussion in a common society they surely must be cultists and fanatics.[10]

Once more Jones attempted to allay Putnam's misgivings. The new group, he wrote on December 11, 1910, would strengthen the older one "by arousing a greater interest in psychopathology. . . . All our members will join it, except perhaps the too pure psychologists. It will be practically, as you say, a wheel within a wheel, but it is impracticable to make it formally so owing to the need for union with Europe. Also there are so many questions that should be discussed, as Meyer puts it, en famille. Acceptances are coming in well, and it looks as if we shall have 25 members straight off. White, Meyer and Stanley Hall are very cordial in their support; especially Hall. The only refusal so far comes from Jelliffe," a New York neurologist who then was skeptical of psychogenic theories of the psychoses. Jones asked Putnam to be President and proposed either himself or Brill as secretary because of their European contacts. Brill's claims were stronger, he admitted, although Brill would not be in closer touch than himself with such men as Hoch, Campbell, and Meyer. Jones urged once more the danger from amateurs who, he insisted, already were beginning to spring up.[11]

Meanwhile, on February 12, 1911 Brill organized the New York Society. Membership would be limited to fifty physicians actively engaged in psychoanalytic work. Brill, thirty-seven, was elected president; Bronislaw Onuf, forty-five, a New York neurologist who had been abstracting Freud's articles since the 1890's was vice president; and Horace W. Frink, thirty-eight, an assistant at the Cornell University Medical College's out-patient clinic, was secretary and treasurer. Of the fifteen founding physicians present at Brill's apartment at 87 Central Park West, ten were or had been associated with Manhattan State Hospital.[12]

Jones had assumed that all psychoanalysts in the United States and Canada would join the American Association. According to Jones, Freud endorsed this plan in a letter to Jung of February 17, 1911. Thus Jones hoped that the New York Society would become affiliated. The American Association was founded May 9, 1911 in Baltimore, before the annual meetings of the American Psychopathological Association and the American Neurological Association. Only eight were present, either physicians who had come for the other meetings or who worked in Baltimore: Trigant Burrow, thirty-six, who had studied with Jung and was an assistant in clinical psychiatry at the Johns Hopkins University Hospital; Ralph Hamill, thirty-four, listed in the Social Register, who was an assistant to Hugh T. Patrick, the leading Chicago neurologist; John T. MacCurdy, twenty-five, then studying at the Johns Hopkins Medical School; Adolf Meyer, forty-four; James Jackson Putnam, sixty-five; G. Lane Taneyhill, thirty-one, an instructor in neurology at the Johns Hopkins University Medical School; G. Alexander Young, thirty-five, an Omaha neurologist born in England who also had studied in Zurich; and Ernest Jones, thirty-two. Membership was open to "those who have given evidence of a competent knowledge of psychoanalysis," but was not restricted to physicians or to practicing psychoanalysts.[13]

On June 27, 1911 the New York Society voted to remain independent of the American Association and elected Brill its official representative to the Third International Psychoanalytic Congress to be held at Weimar September 20–21. The Congress accepted the status quo and permitted membership in both groups. It also noted the unusual character of the American Association. Unlike the European branches, members of the Association were scattered over the United States and Canada and thus could not meet often. Members in fact were concentrated on the eastern seaboard, chiefly in New York, Baltimore, and Boston, with one member each in Omaha, Chicago, Cincinnati, Louisville, Kentucky, and Valley City, North Dakota. The Congress granted the New York Society and the American Association complete independence from each other and decreed that doubtful issues be decided by the International. In 1914 a Psychoanalytic Society was founded in Boston with Putnam as president and Isador Coriat as secretary.[14]

Several motives led to the founding of the American organizations. They would please Freud and strengthen the international movement. They would facilitate a stand against unqualified practitioners. They would also provide a means of exchanging information and experience. There was yet another reason, partly implicit, partly explicit. Judging from the debates of the American Psychopathological Association, the analysts needed a milieu where fundamentals need not always be questioned, and where their hypotheses, no matter what their nature, need not face continual ridicule and skepticism. As Adolf Meyer argued, the Freudians needed to discuss problems "en famille." Only in an exclusively psychoanalytic circle, where basic postulates were agreed upon, could the application of psychoanalytic principles to new material be exploited and appraised and the implications of the psychoanalytic model be worked out. Brill noted this function during an angry discussion in

1915 over charges by Prince and others that the psychoanalysts ignored the work of every other school but their own. Brill insisted that they gave credit to the achievements of others. But the "Freudian mechanisms" had opened up so "many new fields of investigation" that naturally they gave most of their time to that work.[15] This activity formed the basis for the professional culture of the psychoanalysts, and accounts in part for their unusual productivity.

Additional light can be thrown on their professional culture by looking at the ways in which the analysts were recruited, their social origins, their education, and the activities of the two organizations. Members of the American organizations had become interested in psychoanalysis in several ways: through teachers and associates in neurology and psychiatry here and in Europe; through reading; through listening to papers by the proselyting analysts, usually Putnam, Jones, or Brill; through an interest in psychology, sometimes acquired independently; often through academic work at Clark University under G. Stanley Hall.

The largest single group were students of August Hoch or Adolf Meyer at the Psychiatric Institute on Ward's Island, at the Cornell University Medical College in New York City, or of Meyer at the Johns Hopkins University Medical School. As teachers in positions of authority, Meyer and Hoch by fiat and example stimulated the interest of their students. Some physicians were recruited by colleagues who worked in the same out-patient clinics.

Four gained knowledge of psychoanalysis during study abroad. Ralph Hamill of Chicago was in Vienna around 1904, and Thaddeus Hoyt Ames around 1911. Brill and Charles Ricksher had worked at the Burghölzli in 1907. Trigant Burrow and G. Alexander Young were there in 1909 and 1910. Ralph Reed, a Cincinnati psychiatrist, had become interested in psychopathology while working in a state mental hospital. He corresponded with Boris Sidis and was familiar with the work of Janet, William James, and Morton Prince. Curran Pope, forty-five, who ran a sanitarium in Louisville, Kentucky, corresponded with Ernest Jones and heard him lecture to the Chicago Neurological Society in January 1911.[16]

Of the psychologists who joined the American Association, Rudolph Acher, thirty-seven, a professor of psychology at the State Normal School in Valley City, North Dakota, had been a Clark University fellow from 1908 to 1911; Harry Woodburn Chase, twenty-eight, professor of psychology at the University of North Carolina, also had been a Clark fellow and had translated Freud's "Five Lectures on Psychoanalysis." Louville E. Emerson, thirty-eight, received a doctorate in psychology at Harvard under William James and later worked at the new Psychopathic Hospital at Ann Arbor, Michigan as a psychologist.

The social origins of the psychoanalysts require inspection. Meaningful comparisons can be made between the presidents of the two psychoanalytic organizations and those of the American Medico-Psychological Association, representing the nation's psychiatrists, and those of the American Neurological Association from 1911 to 1918. Finally, a profile of the membership of the two psychoanalytic groups can be compared with a profile of contemporary scientists.

The presidents of the American Neurological Association still represented an elite social group. Most of them received their A.B.'s at Harvard, Princeton, or Yale; most had studied in Europe. Three were listed in the Social Register, as were several other neurologists (Charles L. Dana, Francis X. Dercum, John Kearsley Mitchell). At the time of assuming office their average age was fifty-two, and several taught at the major medical schools—Harvard, the University of Pennsylvania, the Johns Hopkins. Most had been born in the major cities or their suburbs: Baltimore, Boston, New York, Philadelphia.

The eight presidents of the psychiatric association still, as in the 1870's and 1880's, were primarily from small towns and provincial cities. Only four out of eight had earned a bachelor's degree or its equivalent; only two taught at major medical schools; only two had studied in Europe. Their average age was sixty.

The psychoanalysts resembled the neurologists in education and training, but in general their social origins were closer to the psychiatrists'. There were three presidents of the American Psychoanalytic Association between 1911 and 1918. Putnam and Hoch served two terms, and White, three. They were born respectively in Boston, Zurich, and Brooklyn, the son of a physician, a minister and hospital superintendent, and a small shopkeeper. Putnam taught at Harvard and Hoch at Cornell University Medical Schools, White at Georgetown and George Washington Universities. All had an A.B. or its equivalent and all had studied abroad.

The presidents of the New York society were closest to the psychiatrists. Only one was the son of a professional man, Bronislaw Onuf, whose father was a physician, although Frink had been raised by a physician uncle. Most were born in small towns or provincial cities: Brill, in Kanczuga, Austria Hungary; Frink, in Hillsdale, New York; Onuf, in Yeniseisk, Siberia; and Oberndorf in Selma, Alabama. Brill, Karpas, and Oberndorf were Jewish. None taught at a major medical school at the time they became officers of the Society, although all of them did so later. Four of the five, however, had studied medicine in Europe, and three had received the A.B. Their average age was thirty-eight. Thus, they were younger and better educated than the hospital psychiatrists; younger, less well trained, and of less exalted social origins than the neurologists.

The members of both psychoanalytic groups, as a whole, except for a higher proportion of foreign born, came from roughly the same social background as most American scientists, i.e. from professional or business families. One of the psychologists who had attended the Clark Conference in 1909, J. McKeen Cattell, was engaged in a statistical and eugenic study of the social origins of what he called "Homo Scientificus Americanus." By 1915 he had canvassed more than a thousand physical and social scientists—psychologists, anthropologists, chemists, physiologists, etc. About 43 per cent were sons of professional men— physicians, teachers, lawyers, clergymen. The fathers of nearly 30 per cent were engaged in "trade and manufacturing" enterprises ranging from a small shop in a university town to the control of a railroad system. About 21.2 per cent were farmers, mostly yeomen who owned their own farms. About 12.8 per cent were foreign born. None was a Negro, the son of a day laborer or a domestic servant. The sample for the psychoanalysts is too small to be statistically significant. Of the thirty-three different members of both psychoanalytic organizations in 1913, the occupations of the fathers of fifteen have been ascertained. Eleven were professional men; three were engaged in trade and manufacturing; only one was a farmer. Nine were foreign born, four of these in Eastern Europe; and seven were Jewish. The new immigration was most heavily represented in the New York Society.[17]

On the whole, those who joined the psychoanalytic movement were lower in the social scale than the business and professional men who led Theodore Roosevelt's Progressive Party in 1912 and very much lower than the solid phalanx of well-to-do families who directed the bulk of the nation's major business and banking enterprises. For many of the psychoanalysts, medicine represented an improvement over the status of their parents. Of the eight members of the New York Psychoanalytic Society in 1913 whose fathers' occupations are known, about five were moving up in the social scale.

As a mark of status, the bachelor's degree is significant. Until the 1920's only a small percentage of Americans were graduated from college. In 1907 only two American medical schools, Harvard and Johns Hopkins, required the bachelor's degree, and Cornell did so a year later. In 1914 at least six of the eleven members of the New York Society had the bachelor's degree, and fifteen of twenty-three members of the American Association.

In the early years psychoanalysts were largely self-selected and self-trained. Freud had made some of his cardinal discoveries, such as the Oedipus complex, largely through the interpretation of his own dreams and parapraxes. "Self-analysis" took on critical importance with his discovery of "counter-transference," the patient's influence on the analyst's unconscious feelings. In 1910 Freud warned the Nuremberg Congress that no analyst could go further than his own complexes and resistances permitted. The psychoanalyst must begin by analyzing himself and continually deepen his self-knowledge while he worked with patients. For Freud and for many of the first generation of psychoanalytic pioneers —among them Brill, Putnam, and Karl Abraham—self-analysis, reading, discussion, and correspondence constituted the major modes of psychoanalytic initiation. For men in isolated outposts such as Berlin or Boston, correspondence with Freud and other analysts and the long-distance epistolary interpretation of dreams and clinical material was a major aspect of apprenticeship and of the elaboration of the psychoanalytic model. This can be followed in detail in the letters exchanged between Freud and Karl Abraham or those Putnam exchanged with Freud and Ernest Jones.

A second method of training was originated by the Zurich school around 1907 when Burghölzli physicians began half seriously to analyze each others' dreams. Freud praised their growing insistence that a psychoanalyst first be analyzed by another analyst before treating patients. He should undergo, Freud wrote, a "purification" of his own "complexes." [18]

Putnam returned from the Weimar Conference in 1911 full of enthusiasm for the analysis of the analyst. However, according to Oberndorf, the Americans were reluctant to entrust one another with this delicate operation. A few, among them Oberndorf and Jelliffe, were analyzed by Paul Federn, who came briefly to New York in 1914 and returned when the war broke out. From 1919 to 1922 Adolf Stern, Oberndorf, and Frink went to Freud himself.[19]

Early training analyses were essentially an informal apprenticeship between analyst and analysand, and no distinction was maintained between a personal and a didactic analysis. Often these analyses were more casual and rapid than those of patients, which then were running from a few months to several years. Freud conducted Max Eitingon's analysis in 1910 during daily afternoon walks. With the founding of the Berlin Institute in 1920, and the International Training Committee in 1925, increasingly elaborate requirements and controls were established. These aimed at "super therapy"—a profound and deep analysis of the physician—and a close sense of identification with the movement. Michael Balint, a London analyst, has suggested that a need for self-discipline, authority, and the avoidance of schisms were the primary esoteric aims of the training analysis. In 1923 the New York Psychoanalytic Society, following precedent set by the Berlin and Vienna societies, required all candidates for active membership to have "undergone a satisfactory analysis at the hands of a competent analyst." The personal analysis of the analyst later was separated from the "training" analysis, which aimed specifically at controlled learning of treatment techniques.[20]

Largely because of the zeal of Brill, Frink, Oberndorf, and a few others, the New York Society became the most cohesive, active, and orthodox center of psychoanalysis in the United States. In social origins the members were more alike than those of the disparate American Association. It is not too far-fetched to suggest that the high proportion of Jewish members heightened emotional loyalty to Freud. Brill's transference and identification with him are clear. Possibly, Frink, too, orphaned early, developed an equally strong attachment, especially after his personal analysis with Freud. This emotional groundwork would have laid the basis for doctrinal orthodoxy. Deviance might have implied symbolic patricide to those who had read Freud's *Totem and Taboo.*

The New York Society's sessions could be lengthy and intense, and members met nearly every month, usually at one another's houses. Karl Menninger attended a discussion that lasted all night. Several papers of importance for the history of the movement first were presented there, among them, Brill's "Analysis of Anal Erotism and Character," Samuel Tannenbaum's "The Role of Sexual Abstinence in the Neuroses," Frink's

"The Analysis of a Severe Case of Compulsion Neurosis." Sometimes visitors such as Putnam or Federn attended, and for a time Isador Coriat, a Boston analyst, was a member.

Several members worked in the city's clinics and medical schools. Many were young and not yet established. In 1908 Brill wondered how he could make a living, because most of the psychiatric techniques in which he had been trained required a hospital setting. In his first book he made a point of naming the eminent neurologists who sent him patients—Morton Prince, Frederick Peterson, James Jackson Putnam. Oberndorf had a satisfactory private practice from the beginning because of referrals from his chief, Charles L. Dana. In 1911 Frink devoted four hours a week to the psychotherapeutic clinic of the Cornell Dispensary. Psychoanalysis was used far less often than hypnosis because of the time required and because the "average clinic patient" was of "inferior intelligence." Nevertheless, in three cases in which a complete psychoanalysis was made Frink reported cures.[21]

In 1913 Brill was appointed head of the psychiatric department of the Vanderbilt Clinic of Columbia University. In 1915 he became a lecturer in abnormal psychology and psychoanalysis at New York University. C. P. Oberndorf established an out-patient psychiatric department at Mt. Sinai Hospital. In 1916 Adolf Stern, an immigrant from Hungary, was applying psychoanalysis to disturbed children at the Mt. Sinai dispensary and two years later discussed Freud's theories before the medical section of the New York Academy of Medicine. Frink became an assistant professor of neurology in 1914 at Cornell University Medical College and an adjunct neurologist at Bellevue Hospital. Jelliffe gave up his teaching position at Fordham in 1912 to devote himself completely to editing and to private practice. Several medical journals published in New York carried articles on psychoanalysis, including the *New York Medical Journal, The Medical Record,* and *American Medicine.*[22]

Twice the New York Society rejected proposals to reply to the attacks of opponents. The first was turned down on May 23, 1911 after the unfavorable articles in the February–March issue of Morton Prince's *Journal of Abnormal Psychology* and the criticism by the Philadelphia neurologists Francis X. Dercum and James Hendrie Lloyd in the May 13 issue of the *Journal of the American Medical Association.*[23]

Membership in the New York Society fluctuated, and once or twice too few attended for a quorum. Membership rose to nineteen in 1913 with the addition of Frank Hallock, a psychiatrist who had been president of the Connecticut Medical Society; L. Pierce Clark, a specialist in epilepsy; and Charles E. Atwood, a state hospital psychiatrist. The latter two resigned a year later, and membership dropped by 1914 to eleven, according to an incomplete report in the *Zeitschrift*. After America's entry into the Great War, activity declined markedly. The New York Society's troubled early years ended in 1919, and it flourished throughout the 1920's. Some of those who joined after 1918, among them Frankwood Williams and Marion Kenworthy, gained national recognition in mental hygiene and child guidance.[24]

Because of its different membership requirements, the twenty-three members of the American Association in 1914 included only about nine practicing psychoanalysts. The rest were psychologists or hospital psychiatrists, such as Douglas Singer, director of the Illinois State Psychopathic Institute at Kankakee, or C. Macfie Campbell, professor of psychiatry at the Johns Hopkins University, who in 1920 succeeded Elmer Southard as director of the Boston Psychopathic Hospital and professor of psychiatry at Harvard University Medical School. The Association met annually the day before the American Psychopathological Association, which in turn met the day before the American Neurological Association.

Although by 1914 Brill and Frink had joined the American Association, a majority of its members were not orthodox Freudians. Stanley Hall remained eclectic. Adolf Meyer never identified himself as a psychoanalyst, and, according to Oberndorf, never used psychoanalysis as a therapeutic technique.

In 1914 the Association appointed a committee to draw up new by-laws, excluding any applicant "who had not identified himself publicly with psycho-analysis and whose work was considered to be sound. [sic]" In 1924 the American Association adopted the New York Society's requirement that members be physicians, but not its new stipulation that they have been satisfactorily psychoanalyzed.[25]

Out of both psychoanalytic groups a few physicians became recognized as spokesmen for the movement. To some extent at first this recognition depended on prior status. Putnam was one of the nation's foremost neurologists. William Alanson White headed one of the country's largest public mental hospitals. Smith Ely Jelliffe, as editor of the *Journal of Nervous and Mental Disease,* was well known to fellow neurologists. Their prestige was an important factor in gaining a hearing for psychoanalysis. These men were not "nobodies," a Washington, D.C. physician insisted during a discussion of psychoanalysis in 1916.[26]

Activity on behalf of the cause was another way to leadership. The leading American analysts were endowed with unusual verve, which their conversion seemed only to intensify. Their dedicated productivity annoyed their enemies, who were fond of pointing out that all the hullabaloo about psychoanalysis was raised by the prodigious output of a very few people. They treated patients and lectured and translated and wrote a stream of books and articles. For example, between 1908 and 1914 Brill wrote one major book of his own, translated four works of Freud and one of Jung, and turned out twenty-eight papers. Putnam published some thirty-three articles and a book between 1909 and his death in 1918. White ran St. Elizabeth's, a hospital of several thousand patients, dispatched forty letters each week, co-edited the *Psychoanalytic Review*, was consultant to the Army and Navy, and wrote thirty-four articles and five books, several of them major psychiatric texts, between 1914 and 1926. Because Ernest Jones so heartily detested Toronto's narrowness, "puritanism," and provinciality, he turned his energies to proselyting and writing. This period was one of the most productive of his life; many of the essays he wrote then entered the psychoanalytic canon. By 1912 he began to spend more and more time abroad, and finally he suggested that he divide his time between London and Toronto. This proposal was turned down by the University of Toronto, and he returned permanently to London in 1913.[27] By then he had written for the *American Journal of Educational Psychology,* The *American Journal of Psychology,* as well as the medical journals. His intelligence, clarity, and prodigious energy earned him a position as one of the most influential of the Freudians in both Europe and America. Jones helped to define what it meant to be a "psychoanalyst." He insisted that the term be limited to those who used Freud's method alone,[28] and he cherished the view that psychoanalysis was a body of scientific hypotheses which could be criticized fruitfully and consistently only from within. The *International Journal of Psychoanalysis* was founded in 1920 partly to promote these ends. Had Jones remained in America, he probably would have subjected American eclecticism to continual attack. By 1914 A. A. Brill had become the chief spokesman for Freudian psychoanalysis because of his translations, speeches, and articles, as well as his loyalty to Freud. The Philadelphia neurologist Charles W. Burr singled him out in his attack on psychoanalysis before the American Medico-Psychological Association in 1914.[29]

Although a full history of the European movement has not yet been written, tentative comparisons can be hazarded with its American counterpart. The European and the American movements differed in membership and social organization. Unlike the Americans, Freud was not what William James would have called a medical trades unionist. Psychoanalysis demanded not so much a medical education, Freud argued in 1913, as "psychological instruction and a free human outlook"; indeed, the majority of physicians were unfitted for it. At first he hoped that teachers, pastors, and social workers could help prevent the neuroses by timely intervention and treatment when symptoms first became apparent. Later, he insisted that practicing laymen be analysed because theoretical instruction—reading and lectures—failed to penetrate deeply enough. He once went so far as to suggest that an American might finance the psychoanalytic training of social workers, who would form a "Salvation Army" to "fight the neuroses of civilization." [30]

Some of Freud's closest early Viennese followers, such as Hans Sachs, Otto Rank, and Theodor Reik, began their careers as lawyers or literary intellectuals. With the exception of a few psychologists most of the Americans were physicians, chiefly neurologists and psychiatrists, many of whom had been active in the psychotherapy movement. The strong American tradition of lay mental healers remained a constant threat, and about 1917 the issue became increasingly acute, especially in New York. Several journalists, among them James Oppenheim and André Tridon, Emma Goldman's colleague at the Ferrer School, were treating patients. Tridon's activities gained wide publicity, and in 1920 he published a pirated edition of Freud's *General Introduction to Psychoanalysis*. Lay psychoanalytic study groups already had been formed. By 1924 the issue came to a head, and even Jelliffe, who for some years had employed lay assistants, now agreed that analytic practice must be restricted. This American insistence that analysts be physicians was protested by Freud and by most of the European movement.[31]

The Americans at the outset were more closely connected with community life than many of the Europeans, especially the Viennese. Many of the Americans were working in public and private institutions. Putnam taught at Harvard Medical School until he retired in 1912. Isador Coriat worked at Boston City Hospital, and later taught at Tuft's

College Medical School. G. Lane Taneyhill gave a course in psychoanalysis at the Johns Hopkins University Medical School. Trigant Burrow was associated with the Phipps Clinic. William Alanson White, because of his position as head of the Government Hospital for the Insane, had to get along with cabinet members and congressional committees and finally married a famous hostess, the widow of Senator John Thurston of Nebraska. Brill lectured to women's clubs. Karl Menninger taught at Washburn College in Topeka, Kansas.

Some of the Europeans, notably Ferenczi, Abraham, and the Swiss, also were involved in clinics and community projects. On the whole, however, their isolation from both the medical and academic communities seems to have been greater at first. Medical and university life in America was less centralized and hierarchical than in Europe, where positions were relatively fewer and were guarded jealously. Although poor youths—Charcot, Freud, and Rudolph Virchow among them—could rise to the top, the struggle seems to have been more difficult. William Alanson White believed that all competition was keener in Europe.[32]

Partly because of their institutional work, the Americans saw a far more varied group of patients than Freud and many of the Europeans, with the exception of the Swiss at the Burghölzli Clinic in Zurich or the German psychoanalysts at the Berlin Polyclinic founded in 1920. From the beginning the Americans were accustomed to treating tailors, chauffeurs, teachers, white-collar girls, factory workers, impecunious artists and writers, as well as upper-middle-class business and professional people. Those analysts who were young and not yet established had to take patients where they could find them, and treated poor but interesting patients at reduced fees, for example, writers like Floyd Dell or Max Eastman. Every Saturday Brill held what he called his "free clinic" for those who could not afford to pay. However, the bulk of private practice even of the younger men probably came from the middle and upper middle classes. Brill, finally, was treating children with governesses and a young lady who lived in "one of the finest apartment houses in New York." In 1915 patients were paying from $200 to $500 per month for analysis.[33]

The patterns of psychoanalytic therapy represented an innovation in neurological practice: the treatment of a single patient every day or several times a week for several months at the very least, and often for a year or more. Patients had been seen frequently, but seldom this regularly before for so long a period. The Americans usually followed Freud's practice of "leasing" an hour. Of necessity the patient paid for it, unless prevented by illness. Apparently at first patients found this lengthy procedure arduous, and in several recorded case histories left before treatment was completed to the satisfaction of the psychoanalyst. Some analysts have recalled that their patients believed it a disgrace to be treated by a psychoanalyst because this automatically indicated a sexual problem.[34]

A few Americans—notably Putnam, Brill, Frink, and the New York group—felt a deeply committed loyalty to Freud. For Putnam, Freud was one of those despised pioneers whom at Harvard he had been taught to revere. Even White and Jelliffe, although they had no sense of personal loyalty to Freud, did feel a close identification with the psychoanalytic cause and with "Freudian" psychology. Devotion to the "cause," however, just as in Europe, by no means obviated internal dissension.

Many of the Americans felt free to pick and choose what they liked from European authorities, as Americans always had done, without reference to European divisions and loyalties. Smith Ely Jelliffe believed that European "currents and cross-currents" were devious and distasteful. He practiced an eclecticism that he believed was entirely appropriate to America. The first issue of the *Psychoanalytic Review,* which he edited with White, printed Jung's Fordham lectures after Freud had refused to contribute. Dedicated to an "understanding of human conduct," the *Review* aimed "to be catholic in its tendencies, a faithful mirror of the psychoanalytic movement, and to represent no schisms or schools but a free forum for all." Accordingly it published contributions from every kind of psychoanalyst. It was a lively, iconoclastic journal, carrying original articles, excerpts, and full abstracts of the European literature, criticism from opponents, and arguments among the analysts. The New York Psychoanalytic Society adopted it as an official organ only as a personal favor to White and Jelliffe. The *Journal* had been founded after they had consulted a number of New York intellectuals. "Professors Dewey, Bish [sic], James Harvey Robinson, E. Boas" and others at Columbia had been "cordial and cooperative," Jelliffe wrote. G. Stanley Hall had been enthusiastic. William Alanson White wrote to Hall February 11, 1913, asking for an article on genetic psychology and psychoanalysis that would "point out the wide possibilities of the application of the new movement as you see it." Jelliffe had discussed the project of the new journal with Hall in New York, and, possibly, had turned down an offer of Hall's help. For White wrote, "I believe you understand from him that we feel that we must launch this thing ourselves, either sink or swim, without the responsibility of carrying any one else down with us if the former should be our fate." In fact, the journal stayed afloat, although its subscription list was small.[35]

In the *Review* Theodore Schroeder, a New York Lawyer, analyzed the erotic origins of religion, and Louise Brink and Isador Coriat discussed Christian love and sublimation. The *Review*'s intellectual level was not high. In 1933 the *Psychoanalytic Quarterly,* a far staider publication, was launched with the express purpose of promoting doctrinal orthodoxy.

The Americans were not as close-knit, formal, or exclusive a group as the Europeans. Probably few would have endorsed Wilhelm Stekel's proclamation in 1911: "La vérité est en marche. . . . We feel in these days like brothers of an order which demands from each single one sacrifice in the service of all. . . ." [36]

The published notes of the American organizations were scanty and informal compared with the meticulous records kept by Otto Rank of the Vienna Society's papers and debates. Relations between the European and American movements were cordial. The Europeans regarded the American movement as a lively and significant outpost. They devoted considerable space to American psychoanalytic literature and to meetings of the American analysts. From time to time sections devoted to "The Psychoanalytic Movement" noted the proceedings of non-psychoanalytic groups in which some analysts participated, such as the American Psychopathological Association or the New York Neurological Society. Jung and Ernest Jones both took pains to congratulate White and Jelliffe on the founding of the *Review.* Jones dilated on their lengthy experience in publishing neurological and psychiatric literature. Indeed, few manifestations of American psychoanalysis were beneath notice of the *Zeitschrift* and the *Zentralblatt,* including Morton Prince's "Psychoanalysis" of Theodore Roosevelt's motives in running for the presidency.[37]

During the Great War communication between the American and European analysts grew difficult. Notices in the *Zeitschrift,* especially after 1915, became sparser. Finally, in 1917 contact ceased, to be resumed in 1919. Whether the motive was wartime animosity, and hence less need for organized reinforcement, growing acceptance of psychoanalysis, or a persistent eclecticism, a proposal was endorsed by White, Mac-Curdy, and L. Pierce Clark in 1919 to dissolve the American Psychoanalytic Association. In 1920 the issue was raised again, and White argued, according to C. P. Oberndorf, that the time had come "to free American psychiatry from the domination of the Pope at Vienna." Brill and Adolf Meyer defeated the proposal. Meyer observed: "I think that the contributions of psychoanalysis to the understanding of psychiatry have been sufficiently great that there should be some organization in this country to guide its destinies." [38]

Freud viewed the American movement with pleasure yet growing distrust. He was pleased by its success and by the application of psychoanalysis to psychiatry. Yet, the increasing popularization annoyed him. He disliked American informality and eclecticism, which he regarded as symptomatic of a lack of intellectual integrity, scientific discipline, and feeling for authority. The decisive struggle over psychoanalysis would take place in Europe.[39] Americans came too easily by truths others had struggled desperately to discover. They failed to realize the importance of the great, unanswered questions and were too easily satisfied with superficial appearances.

Significantly, Freud did not ask a single American to join his secret inner committee which from 1912 on was to steer the international movement and preserve the purity of analytic doctrine. No American received, like the committee's members—Jones, Sachs, Ferenczi, Rank, Abraham, and Eitingon—a Greek intaglio from Freud. Very likely none of the Americans thought of themselves as those palladins of Charlemagne who fired Ernest Jones's Welsh imagination.[40]

American openness, mobility, and medical status exacted their price. The pluralism of American society fostered tolerance and accommodation; psychoanalysis had an easier hearing and penetrated more quickly than in Europe. Yet, American psychoanalysts seemed to feel a stronger pressure to synthesize psychoanalysis with received medical and moral opinions. They were less revolutionary than many of their European colleagues. From American social and cultural conditions a unique interpretation of psychoanalysis emerged.

XIII

The American Interpretation

of Psychoanalysis

1909 - 1920

Freud and the Americans

The Americans modified psychoanalysis to solve a conflict between the radical implications of Freud's views and the pulls of American culture. Their interpretation not only differed from Freud's but also clashed with important beliefs of the American progressive consensus.

They simplified psychoanalysis, taking little interest in Freud's system as a coherent theory. They muted sexuality and aggression, making both more amiable. They emphasized social conformity. They were more didactic, moralistic, and popular than Freud. They were also more optimistic and environmentalist. The factors already discussed were responsible for this American interpretation: the crisis of the somatic style and the nature of pre-Freudian psychotherapy, the unusual severity of American "civilized" morality, and the close links between popular and professional culture.

Nevertheless, the Americans did not manufacture their interpretation out of whole cloth. Rather, they emphasized congenial elements already present in Freud. These common elements have been disregarded, especially by those who stress Freud's tragic "European" stoicism. Yet, the therapeutic optimism, the environmentalism, the faith in the relatively simple resolution of conflict were unusually explicit in Freud's Clark lectures. This emphasis was exceptional for Freud even then, and while he moved away from it, the Americans tended to preserve it intact.

Nevertheless, on the normative elements that inevitably played a role in psychoanalysis, as in every medical psychology, Freud and the Americans remained in agreement. In Europe and America, psychoanalysis was a broad movement for the reform of attitudes and customs concerning sexuality, the family, child raising, and the treatment of nervous and mental disorder. Freud and the Americans both hoped to win that authority within the community from which they expected the ultimate extinction of the neuroses. Their rhetoric preserved the same polarity of values. On one side were reality, society, civilization, science, intellect, and something like unselfishness. At the opposite pole were the unconscious, the sexual, the primitive, the uncivilized, the asocial, the emotional, the selfish, and the childish.

Freud and the Americans held a number of similar intellectual as well as moral assumptions, the product of a common evolutionary outlook. History and cause were equivalent: to trace origins in the past was to discover the effective causes of the present. The development of the individual reflected in microcosm the development of the race: ontogeny recapitulated phylogeny. Freud and the Americans both believed in progress, to be won chiefly through the scientific and cultural achievements of great men. Both also believed that conflict was an inescapable aspect of the evolutionary process. At this point the common ground between Freud and the Americans fell sharply away.

To Freud, instincts never were fully tamed. They remained unruly and recalcitrant; regression and the return of the repressed always threatened. The intractability of instincts dictated continuing conflict. Just as stern necessity warred with the pleasure principle, so society seemed to require authority. As a Jew in an anti-Semitic Roman Catholic Empire, Freud's social models remained authoritarian: Le Bon's tract on mob psychology that preserved the European conservative animus against the French Revolution, the Roman Catholic Church, the Austrian Army, the patriarchal father and his sons. What John Stuart Mill Freud had read and translated as a young man did not markedly affect his social views. American assumptions were quite different. Society

resulted from a voluntary compact among reasonable men. Conflict issued in the general welfare. Even in evils, Herbert Spencer had taught
his American disciples, among whom several psychoanalysts counted
themselves, the student learned to recognize a "struggling beneficence."

No "struggling beneficence" presided over Freud's universe. He
shared only to a degree the moral emphasis Americans placed on distinctions between lower and higher, primitive and evolved. Freud's moral
sense had few religious overtones, and he put religious moralism first
among the traits he despised in Woodrow Wilson.

The American interpretation was not monolithic. American analysts
can be classified in two major ways, orthodox and eclectic, native and
foreign born. Usually, the orthodox and the foreign born were closer to
Freud. Among the former were A. A. Brill, Horace W. Frink, and James
Jackson Putnam. The eclectics included William Alanson White, Smith
Ely Jelliffe, and, to a lesser extent, Isador Coriat. Yet, the orthodox Putnam and Frink shared more implicit values with their native colleagues
than with Brill. Because Brill was educated in Europe to the age of fifteen, he escaped much of the moralism, optimism, and faith in progress
that marked the other Americans. These qualities characterized the progressive consensus before the Great War and remained with most of the
native analysts for the rest of their lives. Their outlook, formed in the
1890's and early 1900's, was about the same in 1910, in 1925, and in
retrospectives written at Freud's death in 1939. Putnam died in 1919
and Frink retired about 1924. The remaining analysts—Coriat, Brill,
White, and Jelliffe, with a few additions, such as Karl Menninger—
continued to dominate the American movement into the late 1930's.
Because leadership had remained constant, the first American interpretation of psychoanalysis continued to prevail with few changes through
the 1920's.

The texture of the American movement altered most markedly in
the 1930's. Then the optimism of the pre-war years and the 1920's gave
way to the bleakness and despair of the Depression. For perhaps the first
time the tragic implications of Freud's later theories were respectfully
and seriously taken into account. Refugee psychoanalysts from Europe
also helped to change the American movement.[1] Orthodox Freudians
such as Heinz Hartmann consolidated the new emphasis on ego psychology that Freud had begun. The ego's methods of defence and of control
over instinctual drives became the major focus of psychoanalytic interest.
Somewhat later a new schism was created by optimistic neo-Freudians,

such as Karen Horney and Erich Fromm. Ego psychology and neo-Freudianism had at least two elements in common—a shift of attention away from sexuality and from the unconscious. But this characterized American interpretations of psychoanalysis almost from the beginning, as Fred Matthews has suggested.

This chapter will contrast the views of Freud and the Americans on certain essential aspects of psychoanalysis—sexuality and the Oedipus complex, the nature of the child, techniques of treatment, attitudes toward the individual and society and toward theory and popularization. Finally it will assess the American interpretation in relation to the beliefs of the progressive American majority.

Freud's New Complexity and Popularization, 1914–1920

Psychoanalysis was simple enough in its early years for the average interested person to grasp and apply, William Alanson White nostalgically observed. This illusion disappeared in 1914 when Freud's views began to change again as drastically as they had between 1898 and 1905. The seventh chapter of *The Interpretation of Dreams* had been until then Freud's most abstruse writing. Now came a series of brilliant, dense essays dealing with the theory of instincts, metapsychology, and therapy.

The American reaction to these novelties was typical. Their complexity and theoretical nature offended some participants in the American movement. White began to welcome practical, clinical facts and an absence of theoretical speculation in reviewing new psychoanalytic writing. These virtues led him to commend Frink's *Morbid Fears and Compulsions*.[2] This American introduction to psychoanalysis, published in 1918, incorporated Freud's new departures. But Frink interpreted them largely according to the models and the vocabulary of Behaviorism. Brill, the most consistently orthodox Freudian, tended to ignore the new developments.

Yet some elements in Freud's new approach were welcomed, almost with relief: that old Darwinian saw, the "self-preservative" instinct, toned down the emphasis on sexuality. The "ego-ideal" was appreciated by Frink, who was especially sensitive to the tyrannical conscience, and by White as proving the constructiveness of psychoanalysis. The emphasis on the present also proved congenial. No American, however, echoed the new primacy of hate in the course of instinctual development.[3]

Freud included all these new departures yet sacrificed little of their complexity in *A General Introduction to Psychoanalysis.* He was, as G. Stanley Hall observed, an "effective and successful" popularizer. Yet his simplest expositions, like those of most of his European followers, assumed a more learned audience than did those of the Americans. *A General Introduction,* for example, had been intended for students at the University of Vienna and some of it solely for physicians and medical students.[4] Pastor Pfister's *The Technique of Psychoanalysis* was written for fellow clergymen and teachers, Edouard Hitschmann's *Freud's Theories of the Neuroses,* largely for physicians.

Freud himself experienced the pressures of American popularization. Encouraged by the handsome sales of the *General Introduction*— more than 11,700 copies in 1920 and 1921—he proposed a series of articles by which he hoped to recoup savings lost in the war and inflation. He proposed a title, "Don't Use Psychoanalysis in Polemics," and when the American editors countered with "The Woman's Mental Place in the Home," he withdrew the offer in disgust. Soon thereafter A. A. Brill collaborated with the journalist Bruce Barton in a typical *Saturday Evening Post* feature, "You Can't Fool Your Other Self." [5]

From the beginning the popularizations of the American psychoanalysts reflected the spirit of Chautauqua, the muckraking magazine, and the community lecture. Psychoanalysts often were invited to discuss their new speciality before medical and lay audiences, and the questions confronted in these exchanges shaped their presentations.

When James Jackson Putnam returned from the Psychoanalytic Congress at Weimar in 1911, he discussed psychoanalysis before the New England Hospital Medical Society. He was asked what "resistances" were and how physicians could overcome them in themselves. How long did psychoanalysis take? How did he keep his case records? Was psychoanalysis likely to make a patient morbid or self-centered? Could he treat ignorant as well as educated patients? What was the meaning of symbolism in dreams? Did he think that the sexual element was overemphasized? Was he in favor of instructing children about sexuality? These kinds of questions help to account for the nature of the American expositions and for their homely simplicity.[6]

From the beginning making psychoanalysis intelligible to non-experts and to uninformed colleagues presented troublesome problems. Although Ernest Jones thought Freud's writings were pellucid, Putnam, whose German was fluent, held a more common opinion. He believed that the physician and general reader found psychoanalytic principles so hard to grasp that there would be "room and welcome during many years to come" for books that expounded them clearly. To many laymen and physicians Freud bristled with new and unfamiliar difficulties. As late as the 1950's, when Freud had become something of an upper-middle-class household word, a doctoral dissertation in education classified him among the "dull, scientific, academic" writers, comprehensible to only "thirty-three percent of American adults."[7]

Between 1912 and 1921 six major American psychoanalysts had written introductory interpretations. These books were intended to disarm the prejudices of those to whom they were addressed. Several were intended for laymen and dealt little with the specifics of technique or theory: Putnam's *Human Motives,* published in 1915; White's *Mental Hygiene of Childhood,* 1916; Coriat's *What Is Psychoanalysis?,* 1917; and Brill's *Fundamental Conceptions of Psychoanalysis,* 1921.

The more popular the book or article, the closer the author to the dominant genteel standards, the more periphrastic was the exposition. Putnam, who worried over his "mischievous" New England background, could not bring himself to be frank about childhood sexuality in a book intended for the public. Coriat, who remained a consultant for the Reverend Elwood Worcester's religiously oriented psychotherapy movement, also was reserved. Yet, the strength of gentility is almost as clear in the popular expositions of Brill.[8]

None of these books except White's approached the frankness of those addressed primarily to physicians or those "ready to take the physician's point of view." These attempted to explain psychoanalytic theory and practice more systematically and included frank sexual detail. Among them were Brill's *Psychoanalysis: Its Theories and Practical Application* of 1912 and Smith Ely Jelliffe's *Technique of Psychoanalysis* published in 1918. The candor of these volumes was characteristic of the technical writing of most psychoanalysts. Not until several decades had passed, perhaps the end of the 1930's or 1940's, were popular expositions as frank as the medical books of the Progressive Era.

Sexuality: "Civilized" Morality, Sublimation, and the Oedipus Complex

Freud was more openly and severely critical of "civilized" morality than the Americans. Throughout his life he continued to insist that changes in custom were necessary—without specifying their nature. In 1909 he wrote to Putnam that such alterations of the mores would compensate for the therapeutic failures of psychoanalysis:

> I believe that your complaint that we are not able to compensate our neurotic patients for giving up their illness is quite justified. But it seems to me that this is not the fault of therapy but rather of social institutions. What would you have us do when a woman complains about her thwarted life, when, with youth gone, she notices that she has been deprived of the joy of loving for merely conventional reasons? She is quite right, and we stand helpless before her, for we cannot make her young again. But the recognition of our therapeutic limitations reinforces our determination to change other social factors so that men and women shall no longer be forced into hopeless situations.
>
> Out of our therapeutic impotence must come the prophylaxis of the neuroses. . . .[9]

Yet such changes did not imply license or "living it up." In *A General Introduction* Freud argued that strengthening a patient's asceticism or sensuality could only cause symptoms: neither relieved the internal conflict. Nevertheless, if after analysis patients deliberately decided on some "midway position between living a full life and absolute asceticism we feel our conscience clear whatever their choice. We tell ourselves that anyone who has succeeded in educating himself to truth about himself is permanently defended against the danger of immorality, even though his standard of morality may differ in some respects from that which is customary in society." [10]

With the exception of Samuel Tannenbaum, no prominent American analyst launched a broadside such as Freud's " 'Civilized' Sexual Morality and Modern Nervousness" of 1908. Significantly, Brill did not translate it.[11]

The American rebellion against "civilized" morality had taken at first a moralistic turn. The medical campaigns against venereal disease and against reticence advocated not license but purity. In literature as well as in the open discussion of such changes Americans were well behind Europeans, perhaps by a generation or two.

The American analysts reflected this native conservatism. They stood for the repeal of reticence, a franker recognition of sexual needs, especially in women, the systematic enlightenment of children, and more tolerant *attitudes* toward masturbation, homosexuality, etc. But the defiant overtones in Freud's essays, the equation for example of the "strong" man with the man who fulfilled his sexual desires, were absent.[12]

One of the clearest statements of American moralism came in a review of Edward Carpenter's *Love's Coming of Age* by Smith Ely Jelliffe. Condemning any relaxation of the marriage bonds, Jelliffe insisted that it was not a loosening of restraints that was needed but "the slow but inevitable process of reeducation and broadening and transformation of energies into activities racial in purpose and effect in the higher spheres, and these in turn purify and re-adjust the individual life." Jelliffe attacked Carpenter's attitude toward the "intermediate sex." "He seems to miss the fact that here too individual training and education must control the 'homogenic' tendency and direct it to the normal, well-adjusted sexual life where there need be no 'intermediate sex.' " [13]

The Americans would have sanctioned intercourse within marriage for pleasure as well as procreation. But they were divided about other changes in behavior. In 1920 the psychologist John B. Watson circulated a lengthy questionnaire to find out what physicians, including a number of urologists and psychoanalysts, would teach in sex education courses.

To the question, "Do you consider that absolute continence is always to be insisted upon or may it be taught that under certain conditions intercourse in the unmarried is harmless or beneficial?" the analysts gave different replies. The most orthodox were the least moralistic. White and Coriat stood firmly for the conventions, so, with some equivocation, did Jelliffe. Brill noted that he had given up an earlier practice of recommending intercourse to patients troubled by special sexual problems; after a successful analysis his patients could abstain for several years with no harmful results. Ernest Jones replied from London that continence, "if continued," was "rarely harmless." Frink argued that continence never should be unqualifiedly insisted upon and that indulgence sometimes might be "beneficial and very necessary." He added that teaching such "truths" would arouse "great and violent opposition." Both Brill and Frink asserted that premarital chastity for men and, Brill believed, even for women was not the best preparation for marriage. The urologists were more unreservedly for the conventions, except for one, who cynically remarked that sex was the only amusement that could not be legislated out of existence.[14]

The Americans, except for Brill, Frink, and Tannenbaum, gave sublimation a far more important role than Freud. Coriat defined it as "that unconscious conducting of the repressed emotions to a higher, less objectionable and more useful goal. . . ." [15] The term had been borrowed from chemistry and meant literally "the act of refining and purifying or freeing from baser qualities." White asserted that as treatment progressed primitive wishes and grossly sexual thoughts and dreams became more "spiritual." Frink and Brill, however, described sublimation largely as reaction formation and displacement, and illustrated it with vocational examples: the cruel, sadistic child became the useful butcher or surgeon.

Freud regarded sublimation as only one of the goals of psychoanalysis. He defined it as the diversion of "sexual impulses" from their "sexual aims . . . to others that are socially higher and no longer sexual." On May 14, 1911 he explained his position to Putnam, who had complained that he could find no theoretical justification in psychoanalysis for helping patients to achieve sublimation:

> . . . psychoanalytic theory really does cover this. It teaches that a drive cannot be sublimated as long as it is repressed and that this is equally true for every component of a drive. Therefore one must remove the repression by overcoming the resistances before achieving partial or complete sublimation. This is the goal of psychoanalytic therapy and the way in which it serves every form of higher development.
>
> If we are not satisfied with saying, "Be moral and philosophical," it is because that is too cheap and has been said too often without being of any help. Our art consists in making it possible for people to be moral and to deal with their wishes philosophically. . . . There are two reasons why we have said so little about sublimation. First, because it is irrelevant, and, second, because so many of the patients we really want to help are incapable of it. For the most part, these patients have inferior endowments and disproportionately strong drives. They would like to be better than they can be, yet this convulsive desire benefits neither themselves nor society. It is therefore more humane to establish this principle: "Be as moral as you can honestly be and do not strive for an ethical perfec-

tion for which you are not destined." Whoever is capable of sublimation will turn to it inevitably as soon as he is free of his neurosis. Those who are not capable of this at least will become more natural and more honest.[16]

Freud once admonished Ernest Jones not to deny to Putnam that "our sympathies side with individual freedom and that we find no improvement in the strictures of American chastity." [17] Almost from the beginning Jones had disdained this American emphasis. It was a commonplace, he wrote, that sublimated sexual energies were applied to higher cultural ends. This was but an aspect of the law of conservation of energy. But the ordinary meaning, that of consciously seeking a sort of consolation in art or religion for disappointed love, was absurd. As Freud argued, one might initiate sublimation by conscious choice, but the process itself worked spontaneously and unconsciously. Moreover, it was concerned chiefly with the childhood partial sexual drives rather than adult genital sexuality. Many people were seizing on sublimation as the only "respectable" side of psychoanalysis, and saddling "neurotics, perverts, criminals and social failures" with "all too heavy reconstruction programs," according to Monroe A. Meyer, a New York psychoanalyst who had worked with Freud. He thought that his countrymen delighted in sublimation because of their zeal for aspiring to "dizzy, puritanical heights." [18]

Jones, Pfister, and Hitschmann insisted, as did Freud, that there were limits to the capacity for sublimation and that a certain amount of sexual libido had to be directly satisfied. No reputable American psychoanalyst emphasized this except Tannenbaum. None of them cited so scandalously "European" a case as one described by Pastor Pfister, who believed in Luther's reasonable rehabilitation of the flesh. A drunken brutish husband, Pfister wrote, was reformed into a model of morality by a satisfying adultery with an affectionate widow.[19]

Freud insisted that patients spontaneously would discover their own modes of sublimation. In this and other respects, he exceeded the American faith in human capabilities. He did not regard psychoanalysis as a didactic exercise in the way Americans earlier regarded suggestion therapy. The changes psychoanalysis fostered occurred of themselves. Analysis alone sufficed. It worked like surgery or sculpture, cutting away neurotic accretions and leaving the patient what he would have been without them. As resistances were overcome, and symptoms analyzed into their component elements, "that great unity which we call his ego fuses into one all the instinctual trends which before had been split off and barred away from it," Freud wrote. "Automatically and inevitably" the patient's mind achieved its own synthesis and grew together of itself.[20]

Along with this faith in human capacities, Freud suggested a highly judicious and limited hedonism. He argued that the merely pleasurable aspects of sexuality were selfish and asocial. But he also argued that therapy included among its goals a surer use of reality in the service of reasonable, socially acceptable, and more lasting pleasure. He placed a value on individual happiness: it might be unattainable, but its pursuit could not be relinquished.[21]

In this period the Americans tended to ignore Freud's circumspect hedonism or to discuss it obliquely if at all. Rather, they stressed the selfish aspects of sexual pleasure. For example, Frink paraphrased extensively Freud's essay on the pleasure and reality principles. He omitted precisely the hedonistic element in Freud's argument: that in the process of development the ego learns to renounce immediate pleasure for the sake of an "assured pleasure at a later time," just as the sexual instincts learn to renounce auto-erotism for object love.[22]

In the tradition of Elizabeth Blackwell and the purity reformers, the American psychoanalysts fought what they insisted was a common prejudice that sexuality was shameful and nasty. It was "one of the universal properties and glories of all living things," Putnam wrote, denied only from "false and narrow pride." Its "ultimate ramifications" were "creative" and "unselfish" and "parental." To Jelliffe the word "sexual" meant any "human contact actual or symbolic" with another person by means of any "sensory area" which had *"productive creation* for its *purpose."*

Freud always was more prosaic. Sexuality was not shameful, but neither was it glorious. Like the American Catholic psychiatrist James J. Walsh, Freud took pains to note the "natural" disgust aroused by the location of the sexual organs. Brill, too, was prosaic. "Love and sex are one and the same thing; we cannot have the one without the other," he wrote. "The urge is there and whether the individual desires or no, it always manifests itself." However, Brill, following Freud, defined sexuality as the equivalent of the English word "love." He discovered that if he described it as "everything relating to and growing out of the love life," he met fewer objections.[23]

Other Americans—White, Putnam, Coriat, even the orthodox Frink—were in danger of stretching Freud's definition of sexuality in the direction of the Bergsonian Life Force. Following Jung, they tended to argue that a generalized energy preceded sexuality and was greater and more universal. Sexuality was not the source of "sublimated" energies; rather, evolution and progress were the sources of sublimation; sexuality was but one expression of primordial energy. Even Frink preferred Jung's term "hormé" to Freud's libido, which Frink argued had only one popular meaning—a directly sexual one. Frink coined the term "holophilic" to emphasize Freud's "broad" definition of sexual and proposed to use it whenever "sexual" might be construed in the "narrow," "popular" sense.[24]

The American psychoanalysts accepted Freud's theories of childhood sexuality and rejected their denial by Jung and Adler. The Americans also emphasized the innocence of childhood and placed extraordinary stress on the socializing function of the Oedipus complex. The child was not a "sensual" being in the sense of an adult indulgence in sinful pleasure. Rather, the child was amoral and had to learn to outgrow his auto-erotic and selfishly pleasurable engrossment.[25]

Freud introduced the Oedipus legend in *The Interpretation of Dreams* to illustrate his finding that certain typical dreams occurred in both patients and in normal human beings. These dreams often expressed intense hostility, including wishes for the death of persons who also were loved, notably parents and siblings. Boys desired to eliminate their fathers; girls, their mothers, as rivals for the parent of the opposite sex. Freud ranged through several kinds of examples in support of this hypothesis, among them the Biblical injunction to honor father and mother and Henrik Ibsen's plays. The latter dealt with a theme Freud often alluded to: the "sadly antiquated" notions of authority held by middle-class fathers who refused their sons "independence and the means necessary to secure it," and thus earned their hatred.[26] The myth of Oedipus represented a literal enactment of these universal dreams and fantasies of family love and hate. At first Freud insisted that one of the most significant results of overcoming the Oedipus complex was the choice of a heterosexual object; later, the dissolution of the complex was the precondition to the formation of the Superego.

Otto Rank provided a significant gloss on the Oedipal theme. He argued that the most important task of each individual was to achieve independence from his parents; social progress was "based upon this opposition between the two generations." Yet, the dissolution of Oedipal ties was exactly what all neurotics failed to accomplish. Rank also appended to Freud's Oedipal drama a "family romance." Because children felt neglected (Rank himself was orphaned), they substituted in fantasy noble parents—perhaps a king or an American millionaire—for their prosaic and actual father and mother. The Americans, especially White and Jelliffe, often quoted Rank's interpretation and united it with Jung's insistence that clinging to home and parents prevented adult self-realization.[27]

"Incest, then," White wrote, "from this broad standpoint is really the attraction to the home that keeps us infantile, it represents the anchor that must be weighed if we are ever to fulfill the best that is in us. . . . The symbols and mechanisms [of the Oedipus complex] are utilized to go onward and upward in the process of development . . . that takes the individual further and further away from the protection of the family group and more and more toward the goal of individual self-sufficiency. The process is but an exemplification of the unfolding of the creative energy which ever drives on in the path of development to complete self-realization and fulfilment." To Jelliffe, the Oedipus complex was an "instrument of precision" for testing the degree of socialization of the psyche.[28]

Brill observed a major impediment to loosening Oedipal ties that he found in his American patients—a mother's overly intense relationship to an only or favorite child. This in many instances was a far more crucial influence than the father's. Both the tyrannical and the too affectionate mother could be pathogenic. One American patient, a "fighting public figure," had had an unusually cruel mother. He often dreamed that a domineering, elderly wife kept him chained under a table by a ring in his nose. On the other hand, the overly indulgent mother spoiled an only or favorite child with too much love. Such children easily became fixated at the narcissistic stage of development and became paranoiac or homosexual. Only 93 out of 400 such children Brill studied had been married, although their average age was thirty-four. Thirty-six per cent of these children lived abnormal sexual lives and suffered from homosexuality, psychic impotence, sexual anaesthesia, exhibitionism, sadism, masochism; another 18 per cent suffered from dementia praecox.

For all the Americans, and especially for Brill, resolving Oedipal ties meant being able to compete as a socialized, yet self-reliant and autonomous individual in the social and economic "struggle for existence." The only child became a "very poor competitor," Brill wrote. He was devoid "of those qualities which characterize the real boy. He lacks independence, self-confidence and the practical skill which the average boy acquires through competition with other boys." [29] The only child associated only with grown-ups; he was petted and his every whim was gratified. In adult life the "slightest depreciation" could throw him into depression or rage and he became a "confirmed egotist." The elements in the Oedipus complex that Americans stressed were symptomatic of other differences in attitudes toward society and the individual.

Society and the Individual

The Americans came close to interpreting Freud's "reality principle" as social conformity. Freud believed, of course, that physicians must consider the environment as "normal" and try to enable patients to lead socially useful lives. Neuroses always implied a flight from reality, as Janet had suggested. But Freud's reality was, in an important sense, more biological than social. The child's biological state, for example, dictated the frustration of Oedipal drives.

The Americans tended to diminish this emphasis. Like many progressives, they sustained an uneasy tension between the claims of the individual and those of society. Individual and social good were compatible: self-realization could only benefit society, and adaptation to society could only sustain self-realization. Yet, the emphasis, for several reasons, fell increasingly on the claims of society over those of the individual.

The disparateness of America—the need, for example, of immigrants such as Brill to learn to "adapt"—made for a strong emphasis on "adjustment." The psychiatric experience of many analysts also was important. When White associated "laissez-faire" with leaving mental patients to languish without zealous treatment, the danger of exaggerating the claims of society could not have been overwhelming. Moreover, most psychoanalysts accepted Adolf Meyer's attempts to redefine nervous and mental disorder in terms of "adjustment" to reality. In mental hospitals failures to "adapt" were unusually glaring, and modest improvements in socialization were counted as major gains in mental health.

A recent fashionable interest in efficiency and scientific management intensified the psychoanalysts' emphasis on adjustment. White went so far as to suggest that the human body be treated as if it were a machine, with scientifically computed periods of work and rest, to assure maximum "output." H. Addington Bruce, the journalist, commended White's

Mental Hygiene of Childhood as a help in preventing the alarming increase of the "mentally and morally inefficient," as he defined delinquents and nervous and mental patients.[30]

Sometimes efficient adaptation was equated with "normalcy." Brill, perhaps the closest of all the Americans to Freud, announced to one of his patients, Mabel Dodge, that "normality" was a difficult achievement which she need not despise until she had attained it. To fulfill one's sexual needs in accordance with social custom indicated a "well-adjusted" libido. Is it conceivable that Freud would have delivered a similar sermon to his friend, Lou Andreas Salomé? [31]

All the Americans, especially the New Englanders, stressed the need to develop the "social instincts." This was particularly necessary in "narcissistic patients," Putnam believed. What he had in mind was not "adjustment" but willingness to sacrifice personal pleasure for a social end. His young psychologist colleague, L. E. Emerson, argued that because social relations fostered or upset the psychological balance, it was essential to decide which were "righteous." Despite Freud's insistence that ethics had no place in psychoanalysis, Emerson argued that the "law of reality" necessarily implied "ethics and ethical considerations." Abstracting "Instincts and Their Vicissitudes" for the *Psychoanalytic Review,* Emerson "ventured" a "word of criticism": Freud did not give sufficient emphasis to the fundamental social character of all psychic processes. . . ." He distinguished too sharply between ego and object, individual and society.[32]

Most of the Americans developed psychoanalysis into an ethical system. Its first commandment was to face "reality," to know one's own inner desires the better to control and sublimate them. This new code absorbed most of the equations of the old one. What had once been good was now "adapted," "conscious," "civilized," and "mature." What had been "bad" was "unconscious," "primitive," "childish," "emotional," "unadapted." Mere pleasure, sexual indulgence, passivity, laziness, and selfishness were "immature." Rationality, unselfishness, control of instinct, independence, were "evolved," "scientific," and "progressive."

A striking example of providing a psychoanalytic basis for an old American virtue was the new sanction given to hard work. In paraphrasing Freud's essay on the reality and pleasure principles, Frink added this note: "motor discharges" as the child matured became "purposeful actions, the forerunners of work." Brill went further. In *Fundamental Conceptions* he argued that the normally adjusted man never needs a vacation because he "works for work's sake." His vocation represents part of his "cosmic urge" and "hence he is unable to stop." [33] Here work for work's sake defined the "reality principle." By contrast with Weir Mitchell, who in the 1870's and 1880's criticized from an aristocratic viewpoint precisely the compulsive aspect of the American devotion to work, Brill praised the compulsion. Here, one more element present in Freud received from the Americans a stronger emphasis.

The problem of morality most deeply troubled Putnam and Trigant Burrow. Putnam's New England moralism and idealism and Burrow's Catholic training resulted in nearly identical emphases. Burrow feared that in the popular mind psychoanalysis was annihilating the "conventional incentives" for morals by exposing their primitive and biological origins. The Trinity was a sexual symbol; the Oedipus complex, the basis of religious emotion. "All great scientific theories," he wrote in 1912, "sooner or later filter through to the masses of mankind, modifying their opinions, altering their conduct, shaping their lives." Therefore, psychoanalysis should foster respect for the "permanent, the ulterior, and the social, as opposed to the immediate, the limited and the personal." With Putnam he insisted on the importance of unselfishness and collective, as opposed to individual, ends. It was almost as if for Burrow, Society replaced the Church. To Burrow, repression was a "biologically moral reaction" signifying the "moral value" of the unconscious personality." [34]

Except for Brill, the Americans tended to mute the conflict Freud posited between the individual and society. The reception of Freud's lugubrious "just-so" story, as he called *Totem and Taboo,* originally published in 1912, then translated by Brill and issued in the United States in 1918, illustrates the contrast. In Freud's text, humanity's "primal father" was murdered and eaten by his sons. Their guilt and need resulted finally in the renunciation of murder. White omitted the primal crime from his brief, respectful notice in the *Psychoanalytic Review.*[35] He quoted Frazer's *Golden Bough* for light on savage mores, not *Totem and Taboo.*

Like Freud in the Clark lectures, the Americans assumed that impulses could be readily controlled once they were made conscious. If one became aware of forbidden wishes, one would reject them unquestionably if they were truly immoral. Yet this solution assumed a sure moral sense. Perhaps, sensitive to the American emphasis on ethics, Ernest Jones, who had spent some time with Putnam and the Bostonians, proclaimed the easy control of impulse by the conscious will at the New Haven Conference on Psychotherapeutics in June 1909. What should a man do if he were attracted to the wife of a friend or relative? If he were shocked at his own sinful feelings, he might try to repress them, yet would become tormented by a fixed idea no longer under his control. What he should do would be to accept his wish as natural. But he should suppress it at once for obvious "social and ethical reasons." If he did so, he would think no more about it, except in the "most harmless way." This was a breathtakingly simple solution to a problem that had tortured St. Augustine, and no doubt numberless neurotics as well. Very likely following Jones's attractive example, White counseled that if a young girl wished her father to die because he opposed her marriage, she should dismiss the wish as "unethical." Horace W. Frink, in Freud's opinion the most brilliant and promising of the younger Americans, insisted that no clash need occur between wish and reality after a psychoanalysis. "Morality and wish fulfillment," he concluded, "are both easy and compatible when one's problems are simplified by the elimination of unreality and reduced solely to the question of adaptation to external facts." "Unreality" to Frink included internal inhibitions, and "unnatural ideals" inculcated by a too repressive early training.[36]

The Americans were political progressives, more radical than most of their medical colleagues. But they were also more conservative than segments of the European movement. A number of psychoanalysts, especially in Vienna and Berlin, were socialists, and the major attempts to reconcile Freud and Marx have been undertaken by Europeans.

Some of the first American psychoanalysts, notably White, Jelliffe, and Putnam, were sympathetic to the vague co-operative socialism of the pre-war period. White hoped for a redistribution of wealth according to the individual's contribution to society. Jelliffe displayed a distinctly middle-class animus. He warned psychoanalysts against rich patients, given to idle tittle-tattle about the analyst and to "definite" liaisons and perversions. He argued that psychoanalysis was democratic, "anti-aristocratic," and humanistic. The life force was moving toward the superman, not of Nietzsche, "but the true superman, the futuristic, socialistic ideal, more closely allied to the symbolic christian ideal than any as yet reached." Putnam, like many aristocratic Bostonians, had voted for the Democratic Party in the 1890's and 1900's. Just before the Weimar Congress he wrote that he was more inclined than ever before to throw in his lot with "reformers and socialists." Brill as a young man admired

William Jennings Bryan and thought him the greatest orator since Saint Paul. This kind of sympathy placed the psychoanalysts to the left of the leadership of the American Medical Association. Yet Brill also was a staunch supporter of competition and by the 1920's was suggesting that "chronic charity seekers" were neurotic.[37]

Although Freud stressed the need to win one's way in the world, and although he sometimes expressed contempt for the brutish masses, he did not attribute poverty to neuroses in the poor. Poverty was one of the harsh realities that made treatment by psychoanalysis especially difficult. Because bleak prospects awaited them, the poor had less incentive to recover than the rich. He later wrote that a socialist Utopia was impossible because men were innately aggressive and differently endowed with talent. Yet, because of his own early poverty, he sympathized with those who wished to reduce social inequality. A real change in the relation of human beings to possessions, he asserted, would be more effective than ethical preachments in reducing the tendency in men toward aggressions against one another.[38]

American Optimism

Some of the Americans were more hopeful than Freud about the range of ailments psychoanalysis could treat and the ease with which it could do so. Freud never tired of repeating the practical as well as the theoretical difficulties: heredity, early childhood experiences that never could be undone, poverty and other adverse social realities, the stubborn persistence of neurotic conditions, the incompleteness of analytic knowledge and techniques. All these precluded easy cure. Psychoanalysis, he wrote in 1916, was effective in hysterias, anxieties, and obsessions, but not in cases of paranoia, melancholia, or dementia praecox. Freud's opinion conservatively summarized more than a decade of experimentation by himself and his followers. They had been applying psychoanalytic principles to a wide variety of ills—alcoholism, drug addiction, paranoia, dementia praecox, manic depressive psychoses, even "psychogenic feeblemindedness." At the outset hopes were high. In 1909 at the New Haven Congress, Putnam asserted that dementia praecox patients might be helped by psychoanalysis. For a time these first hopes diminished, especially regarding the psychoses. Jung argued that schizophrenia did not yield to psychotherapy. Psychoanalytic psychotherapy could help some psychotic patients, including the aged, to adjust more effectively. Here and there a case might clear up. But the results with the narcissistic neuroses, i.e. dementia praecox and paranoia, where the libido was turned primarily inward and transference was impossible, were chiefly negative. Nevertheless, Freud insisted that for the first time in history, psychoanalysis had afforded insight into the psychoses and elucidated some of their psychological mechanisms.[39]

In brash and triumphant tones the Americans insisted that psychoanalysis was the most effective treatment for the psychoneuroses. They disagreed, however, about the results in the psychoses, Adolf Meyer early had established a category that softened the hopeless Kraepelinian prognosis. He suggested that many patients could be classified as "allied to dementia praecox." They were without hallucinations or delusions that interfered with their adjustment in daily life. In 1919 Brill reported a number of such cases to the American Medico-Psychological Association, but noted that most refused treatment. Publicly at least, Brill then agreed with Freud that the emotions of such patients were considerably blunted and that their transferences were either non-existent or "morbid and inadequate."[40]

Ten years later Brill reported to the same organization that since 1912 he had been treating such patients by modified psychoanalytic psychotherapy, if they were well enough to remain at home. In a number of cases he had achieved success: they were able to continue their lives without hospitalization. The important point of Brill's experiment was this: he had been motivated to disregard Freud's views on "narcissistic" conditions by hospital experience before his acquaintance with psychoanalysis. A catatonic schizophrenic patient at the Central Islip State Hospital, who had been tube-fed for about four years, unexpectedly recovered for a few days. At the Burghölzli Brill had successfully "analyzed" another patient who remained "perfectly adjusted." In the discussion of Brill's paper, Harry Stack Sullivan commented that if the leaders of American psychiatry had accepted Freud's doctrine on the narcissistic illnesses, American psychiatry would have become as dull as European.[41] Before the 1920's, however, several psychoanalysts, including Putnam, like the somaticists before them, looked to more results from the prevention of mental disorders than from their treatment. In 1917 Isador Coriat, the Boston analyst, was more sanguine than the rest. Psychoanalytic therapy was helpful in dementia praecox, paranoia, and manic-depressive insanity, after treatment of from one to six months, either daily or three times a week.[42]

American hopefulness was rooted, in part, in a more thoroughgoing environmentalism and a more ardent belief in the tractability and progressive tendencies of human nature. The Americans seemed to assert that the more one believed in the significance of environment, the more one could do to correct whatever condition the environment caused. By comparison with most neurologists and psychiatrists in the 1890's and early 1900's, Freud was a belligerent champion of environmental as opposed to hereditary explanations. But by comparison with important American psychiatrists and psychoanalysts after 1910, notably Adolf Meyer or Boris Sidis, Freud placed more emphasis on constitutional factors. In 1912 even Brill announced that one-third of his neurotic patients had parents with syphilitic background, while Freud had estimated the proportion of his own patients at one-half.

The child's mind, Brill insisted, was, like Locke's "tabula rasa," ready to be stamped by environment. The fact that "only children" with the most varied heredity developed quite similar difficulties testified to the overwhelming power of environment. Brill later urged that if mothers and teachers "understood" psychoanalytic principles—without insisting that they be psychoanalyzed—"we could reduce nervous and mental diseases as much as we have small-pox and typhoid." [43]

Freud's orthodox European followers—Jones, Ferenczi, Pfister, Abraham, Rank, and Sachs—were midway between Freud and the Americans in the degree of their optimism. Both European and American epigoni, as Pfister wrote of himself, at first looked at the world through rose-colored glasses. Fritz Wittels remembered that they hoped a psychoanalytic revolution would transmute the Victorian Era into a Golden Age. Freud, he wrote, remained the most sober and cautious of them all.[44]

Pfister in his hopefulness and his faith in education was perhaps the closest of all to the Americans. Freud considered both him and Putnam "optimists." Like Putnam, Pfister believed that psychoanalysis did not leave man a mere "sexual being" but also revealed his higher ethical nature. Pfister suggested that neurotics were "sick" from their inability to love and that in education the "sunshine of love and enlightenment" should replace pernicious punishments.[45]

Americans and Europeans both hoped the benefits of psychoanalysis could be made available to the masses. In 1919 Freud regretted that psychoanalysts like himself could afford to treat only the well-to-do. The neuroses were acute and widespread among the poor; the "pure gold" of psychoanalysis, diluted perhaps by the techniques of suggestion, should be available at public clinics. William Alanson White laid plans to experiment with psychoanalytic techniques in public mental hospitals.[46] Putnam campaigned for mental hygiene in Massachusetts, hoping that more widespread knowledge could prevent nervous and mental disorder. But most American analysts balked at the spread of psychoanalytic therapy by other than licensed physicians. The taunts about mind cures and the ubiquity of their practitioners led psychoanalysts to insist all the more on their medical identity.[47]

The Analytic Technique, 1912–1920

Until Freud wrote his series of papers on technique beginning in 1912, and in some instances long after that, the Americans presented accounts of the method that often differed from his in important details. No American analyst described analysis as the simple catharsis of the pre-1908 period. Nevertheless, they usually included an account of Breuer and Freud's first work and tended to describe Freud's later departures in less arresting detail. Of all the first expositions, Frink's was the most thorough about changes in technique.

The essentials of Freud's practice were clear by 1915: daily sessions, "leased hours," the couch, a relatively silent analyst. Apparently many Americans did not adopt the couch or daily sessions until the early 1920's when the entire movement, partly under the leadership of Ernest Jones, became more insistently "orthodox." At first Brill seated his patients in a comfortable chair. Putnam permitted them to walk about and do much as they pleased. A Baltimore psychoanalyst, G. Lane Taneyhill, according to C. P. Oberndorf, was the first to use the couch regularly. Taneyhill recommended it to Oberndorf in 1917, and he finally adopted it after his analysis with Freud in 1921 and 1922.[48]

In the tradition of pre-Freudian "educational treatment," the Americans, including Brill, apparently tended to question, explain, and interpret more than Freud, who described his own method of beginning treatment: "One lets the patient do nearly all the talking and explains nothing more than what is absolutely necessary to get him to go on with what he is saying." [49] Some Americans at the outset described the method and meaning of psychoanalysis to patients. Until 1912 Brill plied them with one hundred and fifty "psychoanalytic questions" to learn about their "milieu." Early in a case Jelliffe drew up a chart, schematically representing the "regressions and fixations" of the patient's libido.

The Americans, as befitted the training of many of them in Zurich, used Jung's association tests, while Freud did not. In 1909 he wrote to Pastor Pfister that he stubbornly adhered to free association except with psychotic or intractable cases because each new test word helped the patient to circumvent his resistances. Many Americans at first used the tests for diagnosis. Although Jelliffe did not recommend them, he regarded them as useful, especially with the "highly repressed" average Anglo-Saxon patient who could think of nothing when asked for free associations. The *Boston Evening Transcript* reporter at the Clark Conference had described the tests as a way of "asking delicate questions under laboratory conditions." [50] Gradually, analysts tended to use them primarily with psychotic patients.

By 1912 transference had become the major focus of treatment. A brief note on its origin and development will make the American interpretation clearer. Freud first defined transference in the *Studies in Hysteria* of 1895 as a "false connection" the patient made between a "past event and the physician." Along with feelings of being neglected or fears of becoming dependent on him, "transference" was one of the three ways in which patients "resisted" treatment. To overcome this "resistance," the patient had had to overcome the "distressing affect" aroused by having entertained a distasteful, improper wish. Freud handled transference as if it were any other symptom: he made the patient aware of it. Slowly the patient learned that these "false connections" were illusory compulsions which "melted away with the conclusion of the analysis." [51]

His first essay devoted exclusively to the subject, "On the Dynamics of Transference," was published in 1912. The "positive" and "negative" elements he had described in the patient's relationship to the physician in 1895 now were linked to his sexual theories and to his new therapeutic emphasis on not "acting out" past neurotic drives. Sexual impulses from the early years that could not be satisfied because of reality or inexpediency became unconscious or were elaborated only in fantasy. These unfulfilled elements easily could become transferred to the physician. The patient interpreted these drives, as they were discovered in treatment, as contemporaneous and real. "The doctor tries to compel him to fit these emotional impulses into the nexus of the treatment and of his life-history, to submit them to intellectual consideration and to understand them in the light of their psychical value. This struggle between the doctor and the patient, between intellect and instinctual life, between understanding and seeking to act, is played out almost exclusively in the phenomena of transference." [52] In the *Introductory Lectures* he wrote that transference was an "artificial neurosis" induced by the treatment, which aimed to dissolve it and thus create a permanent cure. "A person who has become normal and free from the operation of repressed instinctual impulses in his relation to the doctor will remain so in his own life after the doctor has once more withdrawn from it." [53]

Until Freud's essay in 1912 Americans paid transference scant attention. Then their interpretations attempted to mitigate what some of them apparently regarded as uncomfortable aspects of the conception: its implausibility and its emphasis on stereotyped repetition of the past.

Perhaps because they had more faith in the changeability of human character, many of the Americans ignored the element of repetition. In 1912 Brill defined transference as rapport, positive or negative. In 1920 it was the "displacement of affect from one idea to another or from one person to another." Both definitions omitted Freud's major point: the compulsive repetition of the past. However, two members of Brill's New York Society, C. P. Oberndorf and Adolf Stearn, wrote reasonably orthodox accounts of transference and resistance in 1917 and 1918.[54]

White prefaced his explanation with lines from Emerson exalting change: "This one fact the world hates that the soul becomes." He approached transference through Bergson's conception of memory, which was the opposite of Freud's. Bergson argued that the "cerebral mechanism" drove "back into the unconscious almost the whole of the past" and admitted "beyond the threshold only that which can cast light on the present situation." Yet this was exactly what neurotic repetition failed to do. In Freud's view transference obscured the present and cast light only on the past. White argued in conclusion that through transference the patient's libido was "transferred" from "himself" to "reality," a large oversimplification of Freud's argument.[55]

Jelliffe derived the major point of his interpretation from G. Stanley

Hall and Alfred Adler. The physician's social role symbolized "security" from disease and from "the fear of death—physical, financial or social." Anxious patients craved security because of failure to realize the wish of self-maximization and transferred the protective aspects of the Oedipus complex to the analyst.[56]

To make transference plausible, White and Frink borrowed the language of behaviorism, which became influential as a hard-boiled scientific psychology after 1913. Its founder, John Broadus Watson, had become interested in psychoanalysis by Adolf Meyer at the Johns Hopkins University in 1910. Watson was "convinced of the truth of Freud's work," but the psychoanalysts' terminology must be discarded. In teaching "the Freudian movement to my classes," he wrote, "I drop out the crude vitalistic and psychological terminology and stick to what I believe are the biological factors involved in his theories. (Freud himself admits the possibility of this.)"

"The central truth I think Freud has given us is that *youthful, outgrown and partially discarded habit and instinctive systems of reaction can and possibly always do influence the functioning of our adult system of reactions, and influence to a certain extent even the possibility of our forming the new habit systems which we must reasonably be expected to form."* [57] Habit systems need never have been conscious, and borrowing from Adolf Meyer, Watson insisted that "habit disturbances" caused the psychoneuroses.

What especially irked him in psychoanalysis was the Freudians' conception of affective processes that hung in mid-air, disembodied from any response, and that could be transferred to different stimuli. "Transference" could be explained entirely in terms of "conditioning." The unconscious, which had been made into a "metaphysical" entity, in fact signified merely a group of habits.[58]

Perhaps unconsciously echoing the sternness of his fundamentalist background, Watson absorbed and exaggerated Freud's and Brill's emphasis on not spoiling children. "If Freud has taught us anything," he wrote, "it is to give heed to our relations with our children. Overindulgence in caresses is far worse for their future happiness and poise than is overindulgence in material things." [59] Too much fondling encouraged too close a parent-child relationship. Watson also suggested that all men in high government posts should be selected after they had been psychoanalyzed.

Adopting a behaviorist stance, Frink argued that only studies of animal conditioning made transference phenomena plausible, and, much as Morton Prince had done years before in describing association neuroses, Frink summarized Pavlov's experiments.[60] A dog's gastric juices were conditioned to flow at the sound of a bell because over a period of time the bell had signaled the arrival of food. Thus the sound became a stimulus that set off the whole pattern, and a person in the patient's past could set off reactions that belonged to that past. By making the patient aware of the origins of these habitual reactions, which were re-enacted in relation to the physician, the patient's "adaptive flexibility" was enhanced.

Jung and Adler, 1911–1920

Some Americans selected congenial contributions from the first schismatics, Alfred Adler and Carl Jung, both of whom played down sexuality and exalted moral and social goals. A Viennese general physician and an ardent socialist, Adler became associated with Freud in 1902, succeeded him as chairman of the Vienna Psychoanalytic Society in 1910, and resigned a year later. Growing differences had led to the break, and in 1914 Adler founded the *International Journal of Individual Psychology.* From Adler's early formulations White, Jelliffe, and to a degree Coriat adopted the inferiority feeling, the aggressive drive, and the theory of compensation, all characteristically modified.

Adler believed that every neurosis could be traced to a somatic condition—a weak bodily organ or function, usually involving the "sexual apparatus." This defect lowered the child's self-esteem and made him feel unsure and inferior. Usually very early the aggressive drive to compensate came into play. The child blamed his parents and fate for his inferiority, which he consistently exaggerated. To achieve security he constructed a "life plan," a "guiding fiction" to which he rigidly adhered. His fictive goal was a sense of mastery; it was as if he directed his life by saying to himself, "I wish to be a complete man." All the Oedipal fantasies, as well as stubbornness and sexuality, served this "masculine protest" against the sense of impotence and inferiority.[61]

The Americans sweetened Adler in two respects. First, engaged in rehabilitating the public image of the neurotic, they ignored Adler's insistence that neurotics were envious, distrustful, and malicious. To most analysts, especially to Frink and Burrow, the neurotic was the victim of too idealistic standards imposed by misguided parents.

Second, the Americans welcomed the will to power as a happy psychic impulse to self-maximization, much in the tradition of popular success manuals. G. Stanley Hall provided White and Jelliffe with their major Adlerian theme. "The existence of sub or abnormal organs or functions," Hall wrote, "always brings Janet's sense of incompleteness or insufficiency, and this arouses a countervailing impulsion to *be* complete and efficient which those to whom nature gives lives of balanced harmony do not feel. The ideal goal is always to be a whole man or woman in mind and body. . . ." [62] White cited the case of a eunichoid patient of Lewellys Barker, who worked as a cowboy and a boiler-maker and fought in the Boer War to compensate for his bodily inferiority. Isador Coriat suggested that the man who had overcome youthful weakness —the reference to Theodore Roosevelt was obvious—advocated the "strenuous life." Almost as an afterthought White appended Adler's argument that the individual's attempt to compensate caused neurosis because it was unrealistic.

A number of Americans, including Glueck, White, Jelliffe, Hall, and Coriat, were interested in Adler's contributions. But only a few became formally committed to Individual Psychology, among them, Ira S. Wile of New York's Mt. Sinai Hospital and a leader in mental hygiene. André Tridon, a lay analyst in New York who publicized a strident and aggressively shocking version of psychoanalysis in the early 1920's, claimed membership in "The International Association for Individual Psychology of Vienna, Austria." [63]

When Adler came to the United States in 1926 on the first of several visits, he found American psychoanalysts hostile to his social philosophy. They had been corrupted, he explained, by the "spoiled child" psychology of Freud. Toward the end of the 1920's, White found Adler's books increasingly meager and thin. But in 1916 he had seized on Adler's theories as a means of reconciling functionalists and organicists, and thus solving the crisis of the somatic style.

By arguing that an inferior somatic organ or function gave rise to compensatory psychological behavior, Adler had established an indisputable link between these two sets of phenomena. The link destroyed "that old bugaboo" of "psycho-physical parallelism," White wrote. He cited Walter B. Cannon's *Bodily Changes in Pain, Hunger, Fear and Rage*, published in 1915, for evidence that fear or anger caused specific physiological reactions. "In this situation," White explained, "we see a perfectly clear and understandable relation between certain physiological reactions on the one hand and certain psychological reactions on the other. In the last analysis why should not every physiological reaction have its psychological co-ordinate?" [64]

If Adler reinforced the American tendency to physiologize, Jung abetted the asexualization of the libido. Jung's differences with Freud on this and other issues were exacerbated by a trip to America in 1912, which included a visit to a patient in Chicago and a series of lectures in September at Fordham University. Jelliffe had issued a similar invitation to lecture there to Ernest Jones, who had declined because he thought a Jesuit University an "unsuitable" platform for a discussion of psychoanalysis. [65]

At Fordham, Jung outlined his "modest and moderate criticism" that could hardly be regarded as schismatic, he wrote, unless psychoanalysis were a "faith." He rejected infantile sexuality, the Oedipus complex, and the primarily sexual nature of libido. [66]

Against conventional critics, Jung insisted that Freud had observed correctly the sexual content of dreams and fantasies. Yet, these had no etiological significance; patients used them to evade the true cause of neurosis—the failure to fulfill "life tasks." Jung did not define these very clearly, but they were not identical with conventional duties. They involved the harmonization and development of "individual trends."

Jung criticized Freud's concept of libido as too narrow to account for the withdrawal not merely of sexual energy but of the psychotic patient's entire interest in the external world. Only a view of libido broad enough to explain this withdrawal could be useful. Why not extend Freud's definition to include Bergson's generalized *élan vital?* Indeed, the child's libido remained pre-sexual and primarily "nutritive" until puberty; the mother was not a sexual object but a provider and protector. At puberty and adolescence came a major life task—escape from the mother who could act as both destroyer and nourisher. Marriage and adult sexual life also depended on this necessary escape.

Brill and others who heard the Fordham lectures reported the drift of Jung's departures to Freud. When Jung returned from America in

1912, he wrote to Freud that he had found he could make psychoanalysis more acceptable by leaving out the sexual themes. All he had to do, Freud replied, was to leave out more still, and psychoanalysis would become even more acceptable.

White and Jelliffe adopted Jung's Bergsonized libido, and White diplomatically "resolved" the differences between Freud and Jung by arguing that the libido could be either nutritive (i.e. self-preservative) or sexual (i.e. race preservative). They also agreed with Jung that difficulties in the present caused regression into fantasy and neurosis. But they did not reject infantile sexuality.

The Americans also mistakenly welcomed an apparent Jungian reinforcement of the value Americans traditionally placed on action in the real world—Jung's division of human personalities into two fundamental types, introvert and extrovert, which Jung derived partly from William James's tender and tough-minded. Introverts directed libido toward the inner world of their own feelings, the extroverted toward the objective world of action. Jung remained respectful of the introvert. His American expositors, notably White, while insisting on their neutrality, often described the introvert by examples drawn from the psychoses. It was but a short step to the semi-popular use of introvert as a term of reproach associated with un-manly shyness, inefficiency, day-dreaming, and schizophrenia.[67]

The American eclectics adopted only one theory from the complex system Jung slowly developed after his final break with Freud in 1914—the unconscious as corrective or compensatory of the conscious. This view was close to James's view of the unconscious as a positive creative force. The details of Jung's development of the archetypes of the collective unconscious have been described in Henri Ellenberger's brilliant *The Discovery of the Unconscious.* The archetypes represented the "deepest thought-feelings of mankind," and the possibility of their representation was transmitted by the brain. Their existence explained the identical forms of legends the world over and the reproduction by the mentally ill of images and associations from the distant historical past. Major archetypes were the fructifying and destroying mother, the animus and anima or masculine and feminine polarities of the psyche, the shadow or destructive impulses, the eternal child, the wise old man.

One should take toward the archetypes, Jung advised, an attitude at once detached, observant, and respectful. To free one's perceptions from their domination was a prerequisite to the fullest realization of the Self. Sometimes unconscious figures exercised a compensatory, cautioning function. If a too passive and cerebral intellectual dreamed of a shadow figure, perhaps he should take account of his feelings and his aggressive impulses. The psyche was self-correcting. The eclectics did not incorporate Jung's "primordial images" of the collective psyche into their version of psychoanalysis.

White welcomed the principle that the unconscious and especially dreams were "corrective." As a symbol, the dream brought the elements of neurotic conflict to consciousness so that an adjustment on a "higher plane" could be reached. A young man dreamed that the body of his dead grandfather moved. According to White, this expressed a conflict between his longings for idleness, regression, and death and "the opposite" tendency, the desire to be on the road to progress, to be active and constructive. Later on this young man was very much better and happier as a result of going into business and being quite successful. His grandfather had been a successful man, so he "reached a solution of his conflict by success in business thus identifying himself with his grandfather. . . ." [68]

There were distinct limits to the enthusiasm of most American psychoanalysts for Jung. He added "weight and dignity" to the cause, White argued. Yet only a few physicians became committed Jungians, among them S. P. Goodhardt, a former collaborator of Boris Sidis on studies of multiple personality, and Beatrice Hinkle, who tended to minimize Jung's growing differences from Freud. Jung himself continued to assert that Freud's or Adler's limited approach covered the relevant facts in some cases. Jung's greatest appeal was to artists and literary intellectuals such as Floyd Dell or Louis Untermeyer, who translated the poetry for Hinkle's edition of *The Psychology of the Unconscious.*

Away from Progressive Moralism

The psychoanalytic outlook of the Americans, diluted, eclectic, and responsive to fashion and criticism as it often was, separated them from important attitudes of what Henry May has called the true blue majority of progressives. The overt Protestant moralism that filled the public speeches of Theodore Roosevelt and Woodrow Wilson was absent from the rhetoric of the psychoanalysts. God had receded in time and space.

No longer the father-god, he had become faith in a "power not ourselves that makes for righteousness," the power of an Ideal.[69] Roosevelt and Wilson explicitly had identified the moral order with their own fathers whom they worshiped. By contrast several of the analysts—Putnam, Burrow, White—were closer to their mothers and far closer to the tender-minded than the tough-minded variety of moral judgment. Implicitly they denied the faith of Wilson and Roosevelt that the ordinary man controlled his character by conscious will and must judge and be judged according to absolute, unchanging standards.

Most analysts were moving in the currents of dissent in the company of John Dewey, whose ethics several admired, the sociologists and anthropologists who emphasized the relativity of human mores, and that segment of progressives who were antithetical to "Puritan" judgments.

Something of the tradition of righteous denunciation remained with the analysts, but the objects of denunciation had changed—from sin to "Puritanism" and "over-moral repression." Frink expressed precisely the new analytic righteousness, partly radical, yet also deeply conservative. Patients were not to learn "new standards" but to unlearn old, mistaken, and distorted ones:

> Ignorant and too rigorous early repressive training, bad family influences, or unnatural ideals establish in the child such habits of "feeling and believing about actions" that when he becomes an adult he can not without pain and horror see himself as he actually is. Tendencies which are innate and inevitable become sources for the development of tormenting effects of guilt; energies that are normal and deserving of direct expression, and energies that, though not normal, could become so, are fruitlessly and wastefully confined and repressed. Instead of securing outlets in the form of activities that are compatible with the requirements of social existence, and which could be given them once they were faced and understood, many of these fundamentally normal and natural impulses remain as skeletons in the closets of the individual's psychic household, which, whenever he gets a glimpse of them, excite him to spasms of morbid fear. Analysis . . . enables him to see the whole of himself as it actually is; to face his defects, whatever they may be, without horror or self reproach, but simply as matters of biological fact; and to develop the energies at his disposal along the lines that will most fully adapt him to his place in life.[70]

Most of the analysts were moral relativists and pragmatists. Even Putnam believed in immanent ideals, but did not identify these with existing customs. Other analysts insisted that mores changed and developed according to time and place. The psychoanalysts' evolutionary assumptions necessarily implied that morality and culture had emerged from savagery just as the adult developed from the child. The voice of conscience was the voice of the parents.

But conscience and culture did not always represent, as most progressive Americans assumed, the culmination of an upward course. Both could be too severely repressive, especially in the realm of sexuality. Some analysts criticized this cultural ambitiousness by exploiting social anthropology. In Japan, Frink pointed out, it was immoral and disgraceful for a girl to come to her husband without sexual initiation and training.[71] The doubt about the wisdom of premarital chastity expressed by Frink and Brill, although not proclaimed very loudly or publicly, was an important departure from American "civilized" morality. So was the insistence that purity of thought was an impossible, indeed, a dishonest demand.

No sharp, scientific distinction could be drawn between normal and abnormal, well and neurotic, criminal and conventional. The analysts condemned harsh uncompromising judgments for exacerbating nervous and mental disorder. These judgments tended to alienate people from themselves and their society. The motives of those who judged people "mean, cowardly, selfish, criminal, perverse," Putnam wrote, needed scrutiny.[72] Harsh judgment could hide one's own worst impulses from oneself. Therefore analysts fostered tolerance and understanding in both patients and the community.

Despite the commitment to sublimation, in practice analysts agreed that they could not solve their patients' moral problems. After hearing a paper in which Burrow argued that the "neurotic patient is striving towards some higher morality," Putnam suggested that the "tendency of psychoanalytic treatment ought to be in the direction of a better social synthesis on the patient's part." Freud believed that such a result need not necessarily come from analysis, although instances in which it did not were rare. Jelliffe observed that although the patient's struggle for a "higher integration of the personality" might fail, the effort might prevent deeper regression. He asked whether Burrow was thinking in terms of an "absolutistic" conception of morality, with which Jelliffe often expressed little sympathy, or a "pragmatic" one. William Alanson White argued that there was no criterion for determining whether a patient were capable of sublimation or not.[73]

The outward behavior that conformed to custom might arise from pathological motives. For the psychoanalysts, motives assumed more importance than behavior itself, which became a "symptom." White was interested, he wrote, in the "root of appearances" not in "surface indications." [74]

Reaction formation—the theory that one might compensate for particularly strong and unacceptable instinctual drives by overdeveloping their opposites—was used to question or discredit traditional virtues and moral attitudes. For example, the prudish were sexually obsessed. The compassionate surgeon, the good butcher, the anti-vivisectionist may have been childhood sadists. The compulsively neat housewife or the parsimonious capitalist might be anally fixated.[75]

Although Freud had speculated that in America anal eroticism must have taken special forms, few American analysts studied it. Brill, however, in a much-cited essay, analyzed an unusually neat, gentlemanly businessman. He was wealthy, ruthless, and a remarkable success because of his orderliness and reliability. All these characteristics, Brill argued, could be traced to an unusually strong early preoccupation with feces.

One other traditional virtue, especially conspicuous in Wilson and Roosevelt, was debunked by the psychoanalysts—filial piety. The American analysts emphasized the psychoanalytic "discovery" of ambivalence and hostility between parents and children. The Americans often illustrated the existence of hostility, especially between fathers and sons, by the standard English novelists Meredith or George Eliot, for instance, or by more "modern" authors such as Bernard Shaw, J. M. Synge, H. G. Wells.

In describing the unconscious the Americans were pulled in two contrary directions—toward optimistic confidence and Freudian distrust. Sometimes they expressed paradoxes of which they may not have been aware and which they did not try to resolve. They made the unconscious more agreeable than did Freud. But they also broke with the beneficent and inspiring subconscious of the medical psychologists, notably Prince, who assumed that if the subconscious created neurotic symptoms, this was because it was acting in an atypical, pathological fashion. Above all, Prince and Sidis insisted that the subconscious was highly suggestible and therefore educable.

This ready pliability of the unconscious the Freudians denied. The unconscious was intractable. Through the lifetime of an individual it remained infantile and primitive; it had no sense of time, logic, inhibition, or conflict. It was a Titan, mysterious and savage, the source of primal impulses and the mainspring of conduct. Frink described the unconscious as all "wish energy," continually "urging and pressing for outlet like steam within a boiler." It was as if the doctrine of "original sin" which the religion of "healthy-mindedness" had largely done away with had returned in a welter of "complexes," "repressions," and "sublimations," John MacCurdy, a Baltimore psychoanalyst, wrote in 1915. The unconscious contained, Putnam argued, those "half-gods" who kept men back from sublimation and progress—pleasure-loving tendencies that pulled toward sloth and selfishness.[76]

Some Americans, notably Putnam, also tended to make the unconscious constructive. He asserted that the analyst's task was to bring out the patient's ideal tendencies, which were unconscious but real aspects of the human mind. "Moral intuition," the sense of "ideal obligations," the upward drive of the life force were all as "immanent" as the infantile drives to pleasure and domination. White worked out no very systematic description of the unconscious, but he tended to a frankly Bergsonian view. The unconscious represented the historical past of the race and the individual. It acted as a kind of regressive agent and mysterious "stabilizer." Yet, it also was a source of energy and progress.[77]

To control the unconscious by repression or denial was unreliable and unstable, as dreams, slips, and neuroses plainly indicated. Coriat remarked that we are all "emotional volcanoes" and that repressed desires continued to "play a potent, active role in the unconscious." [78] Thus the importance of the unconscious became far greater than it had been for either Prince or Sidis, far more of a danger to conscious, everyday life.

The determining force of the unconscious seemed to leave little room for "will power" or "free will" as most Americans understood them. Apart from Putnam's writings, and some other scattered references, "will" almost vanished from the psychoanalytic vocabulary. Analysts asserted that the subjective sense of freedom to choose, which James considered irrefutable evidence of the reality of freedom, was illusory. James had defined will partly as an act of focusing attention, partly as a motivating force identified with instinct and emotion. The analysts tended to identify will solely with instinct. In the sense of concentration by conscious—sometimes wrenching—effort, will, they asserted, availed nothing against the neuroses.[79]

How then were the instincts, represented by the unconscious, to be controlled? They could not be eradicated, and they required expression.

Yet, they must not be expressed in crude and primitive ways but in sublimated and socially useful ones. The surest way to control instinct was by sublimation and by conscious understanding and insight, attained through the analytic process.

Obliquely, but unquestionably, psychoanalysts restored what they seemed to deny—free will. This new discipline promised to control the underworld of human thought. This was not to be accomplished directly. Rather, the restoration of norms was achieved obliquely by their temporary relaxation, by lifting repressions and acknowledging the existence of instinctual forces. Just as the rest cure preceded the reimposition of social duties, so analytic freedom to acknowledge instinctual desires preceded their sublimation and redirection. Patients should assume a passive attitude toward their neuroses and should work at breaking down resistances to the free acknowledgment of their unconscious wishes. Insight, not "will power" or wrestling with sin, made free choice possible.[80]

Yet, the lifting of repressions also assumed the existence of deeply ingrained ethical controls, that "ego ideal" which psychoanalysts were beginning to investigate systematically. Ernest Jones explained that, if the patient were sufficiently high-minded to repress his wishes, surely his ethical standards would make him control them even more successfully when they were made conscious.[81]

The psychoanalysts favored some aspects of progressive reform and deplored others. As already suggested, greater frankness and tolerance about sexuality were to play a part in the prophylaxis of nervous and mental disorder. All the analysts advocated the judicious sexual enlightenment of children. Brill advocated birth control. White advised against vengeful punishment of criminals and delinquents. He was the most interested of all the analysts in the relation of social conditions to neurotic illness. Thus he emphasized modifying the environment. He approved recent progressive legislation governing factories, tenements, food, drugs, and child labor. He was pleased by the efforts of mental hygiene societies, and movements for "moral prophylaxis" against venereal disease and alcohol. But like all the analysts he was hostile to governmental attempts to legislate morality or promote eugenics.[82] Laws should reflect the sentiments of the community, not create them. Although control, not chance, should direct social policy, the control must come from social consciousness, not coercion. In 1915 John MacCurdy, a young physician trained by Adolf Meyer, wrote:

> Think of the pass we have come to in America, from our prudery, with our white slave traffic and our inconsistent divorce laws! When a knowledge of psychoanalytic principles has become general, our laws will correspond to the moral sentiment of the community . . . rather than the extravagant, hysterical repressions of the law-makers who satisfy their consciences by making illegal what they most want.[83]

White insisted tht harsh and punitive moral legislation represented the search for a scapegoat: those with a sense of sin wished to feel pure by regulating the conduct of others. Brill suggested that censors were at heart prurient. With an element of Old Testament fervor he denounced the "Puritanism" that condemned dancing and other innocent pleasures which he regarded as healthy sublimations of the sexual instincts. Thus the moralistic aspects of progressive reform were rejected by the analysts.

The analysts also "analyzed" reformers in a manner that suggested Lombroso's linking of the genius and the neurotic. Horace Frink traced the motives of an ardent feminist to reaction formation against a masochistic wish to be dominated by a man. "The very normal people who have no trouble in adjusting themselves to their environment," he wrote, "are as a rule too sleek in their own contentment to fight hard for any radical changes, or even to take much interest in seeking such changes made. To lead and carry through successfully some new movement or reform, a person requires the constant stimulus of a chronic discontent (at least it often seems so) and this in a certain number of instances is surely of neurotic origin and signifies an imperfect adaptation of that individual to his environment. Genius and neuroses are perhaps never very far apart, and in many instances are expressions of the same tendency." [84] Other analysts traced anarchistic and revolutionary sentiments to murderous hatred of the father. Brill tempered his reductionism, as had Frink, by arguing that neurotics who selflessly supported causes were the "salt of the earth," the peculiar yet worthwhile instruments of change.[85] At least publicly, the analysts failed to apply this reductionism to their own motives for professional reforms.

Because the analysts tended to narrow their focus to the family, they considered correct child raising to be the surest guarantee of progress. All of them assigned a greater importance to parents than did Freud. Early he had suggested that parental influences could be more direct and crucial than heredity. But he also emphasized the child's own wishes and desires. The Americans made the child passive, asserting that what parents did to their children was decisive. Thus parents were allotted a new, more difficult and demanding role. It was they who transformed the child into a mature, socially responsible human being.[86]

The analysts' view of the child's nature was full of contradictions. Following Freud and Adler, the child was a selfish, domineering savage, pursuing pleasure and power. Yet, the child, White wrote, tended to grow toward higher levels of socialization, toward parenthood and the complete unfolding of the personality. Putnam insisted that the child was a "natural idealist," endowed with "glorious self-forgetting spontaneity." Only Brill failed to sweeten the child's "insatiable" and stubbornly asocial nature.[87]

The analysts by no means prescribed unrestricted love and permissiveness. The best way to raise children was neither to smother them with too much love nor to stunt them with too little. The analysts insisted with Freud that each child spontaneously developed curbs on instinct such as shame and disgust and that education should gently reinforce these as they appeared. Freud had noted that too much unearned love was bad for children. The Americans were afraid of spoiling boys, of making them unmanly and incapable. Children showered with too much love, like only children, failed to develop independence; they could not stand life's inevitable disappointments and were prone to sexual abnormalities. Parents should avoid caresses or other attentions that might arouse premature sexual impulses. Mothers were warned against developing pathological Oedipus complexes in their sons.[88]

The psychoanalysts were creating a role that was unprecedently inclusive. Some psychiatrists, especially in the mid-nineteenth century, had used their medical theories to criticize social customs. Some clergymen had dealt with the intimate moral dilemmas of their charges. Some neurologists, psychiatrists, and family physicians had come to have profound knowledge of their patients' emotional and sexual lives. Janet observed in 1913 that Freud's psychology touched those aspects of sexuality that had psychological consequences. It was the inclusion of these functions in the role of the psychoanalyst—social criticism, moral cauistry, medical knowledge—that was unique.

Their role necessitated, Coriat wrote, that the analyst have as "clean a mind as a surgeon has clean hands." He also had to be able to control his own sexual impulses and his anger. Analysts had to have the toughness to withstand their patients' "negative transference." Jelliffe asserted that the smallest details—the cut of a suit, the books on the waiting-room table, the decor of the office, the smile of the maid—could be used by patients as excuses to vent their hatred.[89]

The analysts saw themselves as eminently scientific healers and explorers. They probed the unconscious the better to map and control it. Psychoanalysis was a daring treatment that touched the most explosive human forces. But the analysts believed themselves courageous enough to face down these hazards. Because only psychoanalysts really understood human nature, they best could suggest ways for meeting human needs. As one sympathetic critic noted, they seemed to be "planning and directing individual and national self improvement enterprises." Jelliffe, not unexpectedly, held out the most exalted view of the therapist's role: the analyst leagued himself with the unconscious, with "cosmic progress." [90] Other analysts were more modest, but still their role was grand. Freud fulfilled it best. As the movement's hero, he was presented as painstaking, profoundly erudite, truly scientific, the "Darwin" of a new dispensation of the mind.

XIV

American Psychoanalysts
1909 - 1920

Within a roughly common American interpretation, the core of crusading physicians shaped psychoanalysis to their vigorous, highly disparate personalities. They also adapted it to three major American creeds: nineteenth-century idealism, Darwinism with overtones of Bergson, and the relativistic naturalism of the twentieth century. What united them was the sense of participation in a common cause, springing from a variety of personal and ideational motives.

Here they can be divided into New Englanders by birth or adoption and New Yorkers, representatives of the new multi-racial capital of American culture. To Putnam, a member of the old New England upper class, psychoanalysis was a radical new technique for uncovering painful truths that kept men back from the service of the common good. The New Englanders—Isador Coriat, William Alanson White, Putnam —kept the progressive moral emphasis. To the New Yorkers, psychoanalysis took on a different cast. To Horace W. Frink psychoanalysis provided a tool for adjusting to society through a nice balancing of instinctive clamors and conventional demands. Brill, the successful Jewish immigrant, preached competitive struggle and, finally, resignation to forces more powerful than the individual self, and he almost entirely dropped progress and ethics from discussion. This study omits important psychoanalysts who were less broadly representative or played a less active role among lay and professional groups, as well as many lesser figures.

James Jackson Putnam, 1846–1918

Full of the "rugged" courage he admired in Freud, James Jackson Putnam probably did the most to ensure psychoanalysis an initial hearing in the United States. No one else, Freud realized, could have dispelled so successfully the deeply rooted American "resistance" to psychoanalysis as a disreputable, immoral cult. Freud considered Putnam honest, clever, and gallant and in 1914 wrote that Putnam was the chief pillar of the movement in America.[1]

Perhaps the most gifted, cultivated, and critical of Freud's American followers, surely the finest writer, Putnam fought all his life to resolve the tensions between "physics and metaphysics," the claims of science and his own New England tradition of social service and moral responsibility. He struggled to reconcile psychoanalysis with philosophical idealism. More important still, his vigorous quarrels with Freud rested on some of the same points of disagreement as those of Jung, Adler, and the later neo-Freudians. Yet Putnam remained loyal because he believed that Freud did the most rigorous justice to the truth.

Putnam was descended, Ernest Jones took pains to record, from some of the "most notable families of New England." Ancestors on both sides had come over in the seventeenth century. His paternal grandfather, Samuel Putnam, had been a justice of the Massachusetts Supreme Court; his father was a distinguished Boston physician, and his father's father-in-law one of the greatest of America's nineteenth-century doctors.[2]

Scholarly and persistent, rather than a fighting controversialist, Putnam was indignant at flippant criticism. He consistently fought for principles, and he enjoyed "everything animate and inanimate"—nature, hiking, sailing off Cotuit, his friends, and his garden. He always had a book in his pocket, and his conversation seldom lingered on minor topics. He had an almost child-like enthusiasm for new discoveries in medicine and all his life worked for a series of causes. A younger colleague, Edward W. Taylor, who explored psychotherapy with him and also taught neurology at Harvard, wrote: "Any movement, in fact, which made a human appeal and which offered the slightest opportunity of helping his unfortunate fellowmen enlisted forthwith his warmest sympathies." [3] After his Harvard medical degree he studied with Theodore Meynert and later with Hughlings Jackson, Europeans who also had deeply influenced Adolf Meyer and Freud. He returned home, full of enthusiasm for his first cause, the young discipline of neurology, which he helped to establish in the 1870's. According to tradition he set up a laboratory in his home. He worked on the localization of brain function, the physiology of the cortex, pathological anatomy, syphilis, paresthesia, and electrical therapy. In 1879 he defended the right of women to a medical education, a departure just undertaken in Zurich and then under fierce criticism. He waged a "vigorous campaign" against industrial diseases, among them, lead poisoning. His incrimination of wallpaper as a common agent in arsenic poisoning, Taylor wrote, caused "something approaching a panic, at least in the vicinity of Boston."

Putnam became interested in the functional nervous disorders, and, some years after his first major paper on the subject in 1895, devoted himself almost exclusively to them.

From his first meeting with Freud at Worcester in 1909 he was troubled by the possible clash between his deepest convictions and the implications of psychoanalysis. Putnam, then sixty-three, had largely accepted his friend Josiah Royce's restatement of idealism, and had rejected the scientific determinism of the Darwinists and the materialists. His concern was an old and racking one that took on a special urgency in the late nineteenth century: how ask men to endure pain and suffering in an impersonal, amoral, relativistic universe? How could science guide the moral life? In *Psychotherapy, a Course of Reading* (1909) he declared his position:

> If the doctrine of free will is, in every sense a mockery and it is really true that "science ends where liberty begins"; if spontaneity, the conscious choice of "purposes," and conscious effort to attain them are but dreams; if "adaptation" to an environment which in the last analysis is a physical environment is the principle by which we finally are governed; if we have not the right to consider it—I will not say proved, but possible that the universe is, in some sense the expression of a purpose, of the will of a moral personality, then the lot of those on whom the stress of life falls heavily is indeed harder than, in my judgment, it should be held.[4]

Putnam believed that Freud's determinism was mistaken, just as he believed the pluralism of his adored William James mistaken. As a Hegelian monist, Putnam believed that everything men know must reflect the nature of mind. Even the infant could not perceive without using mental categories of "time, space and causality" that antedated experience. Macrocosm and microcosm reflected each other; the structure and working of the universe were revealed in that of the human mind.[5]

After all the empirical observations were in, there was still something left over and unaccounted for. This residium, Putnam explained, in his popular exposition of psychoanalysis, *Human Motives,* was a "self-active energy" emanating from God, the Eternal Renewer of Evolution. Putnam found Bergson useful in shoring up his idealist beliefs. The *élan vital* permeated every "detail and part" of the universe and afforded a conception of God that scientifically restored spontaneity, will, and intelligence. Putnam cited the very latest speculations of C. J. Keyser, professor of mathematics at Columbia University, that reason and logic proved the existence of an Overworld of Omniscience, Beauty, Eternality, Omnipotence, Universal Harmony.[6]

The reconciliation of psychoanalysis to this *Weltanschauung* was not as difficult as it might seem. Putnam's belief in a life force seeking higher forms of expression made it possible for him to be critical of actual social conventions and definitions of normalcy. With the fervor of a New England reformer he argued that conventions could be cruel and narrow, and that society was almost fanatic in its demands for conformity. Conformity, then, was in itself a kind of slavery. Traditional customs, the "filtered wisdom of the ages," were sacred, but so was the "right to revolt against them." [7] The necessity of adaptation to social conventions could reinforce the unhealthy and morbid aspects of conscience and self reproach. Far more than any other American analyst except possibly Frink, Putnam was alert to the cruelty and irrationality of what Freud later called the Über-Ich, or Super-Ego.

Custom also blinded men to the importance of sexuality. "The essential function of all life is to reproduce and to perpetuate itself," Putnam wrote.[8] Sexual emotions provided the richest source of energy and excitement, of the most significant, indeed, sacred tensions. Putnam's clinical papers dealt fully and explicitly with infantile sexuality, stages of sexual development, fixations and their survivals in adult life, and he helped translate Von Hug-Hellmuth's pioneering psychoanalytic study of the sexual life of the child.

But Putnam's attitude toward sexuality was not one of romantic liberation. Sexuality and the child could be both progressive and regressive, socially sublimated or selfishly fixated. He flatly disagreed with Emerson's charming conviction that our "best selves" rush to meet us. Struggle and conflict, Putnam believed, were prerequisites to the triumph of the good.[9]

The child possessed, as his most real aspect, creative energy, immanent ideals of service, poetry, and curiosity. For all these reasons he should not be crushed into conformity. Too harsh repression, especially of sexual curiosity, could lead directly to delinquency, he wrote in an appreciative review of William Healy's classic, *The Individual Delinquent*.[10]

But sexuality, as mere pleasure seeking, dangerously could block progress. "Too easy contentment is the condition mainly to be dreaded," he warned. Any fixation on selfish and self-centered sensuality could destroy growth toward social responsibility. Here psychoanalysis offered a unique remedy. It unearthed these hidden fixations, removed inhibitions, and thus made possible the fullest development of character, the "normal birthright of every individual."

Gradually, by paying attention to motives and tendencies rather than outward conduct, parents could lead the child away from selfish pleasure toward unselfish love. If necessary, they could hold intimate psychoanalytic talks with their children, during which the child could express all his feelings for adults. Putnam often likened psychoanalysis to Boston's favorite poem, Dante's *Divine Comedy:* first a descent into the repressed emotional life under the guidance of the "rational physician." After learning from Reason and Self Study, memories of the past were washed away in the waters of Lethe. Finally, the patient passed from the "dominion of his infantile life" and advanced toward a realization of his true self. This realization included the patient's own ideals and ethical aspirations, which Putnam considered as much a part of him as his sexuality. It was not so much, as Freud supposed, that Putnam wanted to inculcate idealism, but rather that he believed the patient should develop values of his own. Every analysis, he insisted, involved making important decisions about truth and falsity, right and wrong.

Human Motives was published in 1915 as the first volume in H. Addington Bruce's "Mind and Health" series. Putnam's allusions to sexuality in it were guarded. White welcomed it as testimony to Putnam's enthusiasm for Emerson and as a corrective to the "silly criticisms" current against psychoanalysis. Richard Cabot, Putnam's colleague at the Massachusetts General Hospital, dismissed the book in the social workers' *Survey* as an expurgated conglomeration of "Emerson, Bergson and Freud in 175 pages." The *Dial* found it "consoling when not convincing"; the *Nation's* critic thought it confused.[11]

Human Motives contributed to a nightmare-like dream Freud had on July 8, 1915. Before falling asleep he had been reading Putnam, and he dreamed that his oldest son Martin had died on the Russian front. Freud had been thinking in the past few days about the occult, and about his superstitious belief in his own impending death. Somehow, Putnam's idealistic religion had become associated with occult powers. So Freud interpreted the dream as a challenge to the supernatural to do its very worst. Two days later Freud wrote Ferenczi that the book was "good and loyal," but "filled with the sense of religion I am irresistibly impelled to reject."

The day before the dream Freud had written Putnam a famous letter revealing his own faith in the self-evident nature of morality, a deeply engrained moralism Philip Rieff has associated with Freud's Jewishness. Freud frankly accepted one of Putnam's strictures in *Human Motives* as applying to himself and quoted it in full: "To accustom ourselves to the study of immaturity and childhood before proceeding to the study of maturity and manhood is often to habituate ourselves to an undesirable limitation of our vision with reference to the scope of the enterprise on which we enter." [12]

Freud failed to understand Putnam's idealist philosophy. He was fond of Putnam, but somewhat ungraciously attributed these preoccupations to a tendency to obsessional neurosis.[13] Futilely, Putnam had tired to interest other psychoanalysts in these issues. At the Weimar Psychoanalytic Congress in 1911 he had pleaded for "The Importance of Philosophy for the Further Development of Psychoanalysis." The address reminded Freud of a "decorative centerpiece; everyone admires it but no one touches it." He wrote later that Putnam's attempt to place psychoanalysis "in the service of a particular philosophical outlook on the world . . . and urge . . . this upon the patient in order to ennoble him" is "after all . . . only tyranny, even though disguised by the most honorable motives." Putnam argued the issue eloquently and at length with Ferenczi but failed to convince him. Even Oskar Pfister thought Putnam was trying haughtily to give directions to empirical science.[14]

Putnam's attitude toward Jung and Adler is important. Putnam rejected Jung on clinical grounds and Adler on grounds both clinical and moral. What chiefly emerges from his detailed critique of them both is his acceptance of Freud's sexual theories.

A year after the publication of *Human Motives,* Putnam frankly recognized in himself a tendency to tone down and to seek a convenient way out of the unpalatable aspects of psychoanalysis. On some occasions it was difficult to describe bluntly the whole of Freud's doctrines, he wrote. Nevertheless, Putnam insisted, as he had in 1911, that infantile sexuality, repression, fixation reflected Freud's "sleuth-hound" genius and his observations, which were as "reliable as the wax impression of a coin." The emphasis on sexuality came not from Freud but from patients; it was not, as Adler asserted, a mere "form of speech" but a literal truth. Putnam insisted that sexuality was far more essential than Adler's will to power, which he considered a cold and negative doctrine. Putnam admitted that perhaps "complexes" of his own kept him from doing Adler justice and that he had therefore carefully reviewed his work several times. These "complexes" could have included what he regarded as his own timidity, his lack of interest in raucous competition, and his sense of loyalty, reinforced perhaps by the teachings of his friend Josiah Royce. To Putnam, Adler was a schismatic who had forged weapons to strike down doctrines of Freud without putting much of value in their place.

The "will to power" was simply not as irrational or as repressed a force as sexuality: "surely patients are not torn and thrilled by their desire for supremacy (regarded as free—if one can so regard it—from the sex-feelings that attend it) at all as they are torn and thrilled by their (unrecognized and unacknowledged) sex-passions." [15] Adler's goal receded toward Nietzsche's repulsive procession of "super, and super-super-men." Adler left no room for love. But he conceded that Adler's theories of organ inferiority, compensation, and self-assertion were valuable, as Freud, too, found them to be.

Jung he dismissed more sympathetically—and more summarily. He could "not sympathize in the least" with Jung's rejection of some of Freud's finest and most important contributions, infantile sexuality for example. He appreciated Jung's stress on the "present issue," but not at the price of ignoring the essential and painstaking psychoanalytic investigation of the patient's past. Jung's doctrine of an unfilled life-task invited the physician to impose his personality upon the patient. All these tendencies to make psychoanalysis sweeter and more superficial could readily degenerate into the old "mentor and advisor types of psychotherapy," whose inadequacy had led him to take up psychoanalysis.[16]

Yet, despite his outright rejection of Jung and Adler, he sympathized with some of the grounds for their rebellion. Increasingly after 1915 Putnam became disturbed by the reductionism and the materialist and behaviorist trends increasingly apparent within orthodox psychoanalysis. He could accept the utility of Freud's genetic method, which he described as "tracing out the workings of the apparently blind, will-less primary instinct in even the highest, that is, the most complex manifestations of conscious human life." [17] But genetic description, like science, was narrow. One had not said all there was to say about religion, for example, by tracing these genetic origins. Putnam could perfectly easily accept a bell tower as a phallic symbol. But that did not exhaust the significance of the tower. It also had, properly, a real religious meaning on another level. He reacted similarly to the psychoanalytic biographies of great men, fictional and real, such as Leonardo and Hamlet. These treated what other biographers had neglected. But the whole approach was too monographic. Exactly this one-sided genius in Freud was responsible for the creation of much that was valuable in psychoanalysis. Yet, Putnam argued, when Freud left the role of clinician and became a sociologist, or a philosopher, as he seemed to be doing, he became subject to all the strictures of disciplined, logical, and philosophical criticism. He could regard Freud, then, as a clinical genius, but a poor philosopher.

Putnam could not agree that civilization represented only the "sublimation of sexual instincts." Culture, rather, was a product of the life force, of which sexuality was one manifestation. Psychoanalysis, he warned, was weaning too many observers away from the "study of the conscious life and the ultimate intuitions." Without examining these, one could not even gain a just impression of the unconscious. Putnam argued that if Freud himself had taken inventory of his own "best tendencies"—his "literary aspirations, political liberality, zeal for widening the bounds of knowledge, sense of obligation to the calls of fellowship and duty,"—he might have come to realize that sublimation was not merely the "by-product of libido" combined with social pressure.[18]

Nevertheless, in an assessment of the strengths and weaknesses in psychoanalysis which he was writing when he died, he insisted that Freud's great work had come as a "refreshing breeze." Freud had substituted knowledge for "unreflective emotional reaction," had helped eliminate "passion, misunderstanding and misery, even if only in a somewhat greater measure than before. . . ." Freud had helped to realize that social ideal Putnam cherished, a vision of a world of: "far less prejudice and cruelty . . . far less envy, jealousy and suspicion; far less terror, disappointment, depression of spirits and suicide; far less disorders of the nervous system, far less inability to realize our best destinies." [19]

In 1917 he was delighted by the progress of psychoanalysis in America: "Who would have dreamed, a decade or more ago, that today, college professors would be teaching Freud's doctrines to students of both sexes, scientific men turning to them for light on the nature of the instincts, and educators for hints on the training of the young." [20]

In 1917 Putnam developed, like his friend William James, a fatal heart disease, but suffered no diminution of intellectual vigor. Then, November 4, 1918, suddenly, after a happy outing the day before in Concord, Emerson's home, he died. He had just extended his hand to greet his physician, a gesture, Taylor wrote, "symbolic of the graciousness of his life." Ernest Jones lamented that in Putnam's death the "infant science of psychoanalysis" had suffered one of its greatest blows. A. Lawrence Lowell mourned him as a "man of genius, eminent in his field, a philosopher, and a saint." [21]

Isador Coriat, 1875–1943

After Putnam's death, Isador Coriat remained until the late 1920's one of the very few psychoanalysts in Boston and perhaps the only fairly orthodox Freudian. He was a positivist and a determinist, rejecting Putnam's nineteenth-century loyalties to Emerson and immanent ideals. Yet he was full of the New England emphasis on morals and community service. He was not interested in religious dogmas, yet endorsed religion's sublimating virtues. Coriat emphasized environment over heredity as an assimilationist Jew committed to erasing theories of the characteristics of race.

Photographs show a small, anxious-looking, fastidiously dressed man, with the ascetic, aquiline profile of an El Greco cardinal. He was born in Philadelphia, attended high school in Boston, and was graduated from Tufts College Medical School. For two years he worked at the Worcester State Hospital under Adolf Meyer. As part of Morton Prince's circle, he was heavily influenced by Janet.

Jung's association tests first interested him in psychoanalysis, but his final commitment came gradually and cautiously. Freud's Clark lectures, and especially the doctrine of infantile sexuality, seemed at first too revolutionary. Gradually, however, through discussions with Brill, debates at the American Psychopathological Association, and his own clinical work, he became convinced. Psychoanalysis began to seem more useful and explanatory than the alternatives offered by Prince and the French.

Coriat came to take special pride in his role as a trail-blazer and rather looked down on later epigoni. Like the other psychoanalysts, he turned out an enormous amount of clinical work as well as popularizations. After 1924 he continued the informal seminars on psychoanalysis Putnam had initiated in his own home, with Jungians, Freudians, and Rankians. They got along more amicably than the entirely Freudian group which set up the Boston Institute in 1935. Coriat was president of the American Psychoanalytic Association in 1924 and again in 1937, and a vice president of the International Association in 1936 and 1937. Toward the end of his life he recalled the lonely early days in Boston: "In view of the changed attitude towards psychoanalysis, it is very difficult at present to reanimate the antagonism, resistance, and worst of all, the ambivalence which psychoanalysis met here among the medical profession; but I went on convinced more and more of the truth of Freud's work and could demonstrate its verification in the analysis of every case." [22] In fact, he exaggerated his own early orthodoxy. From 1917 to 1920 he considered Adler "one of the greatest thinkers of the Freudian school," and used freely Jung's distinction between introvert and extrovert.[23]

As consultant to the Reverend Elwood Worcester's Emmanuel Movement, Coriat had a profound respect for the useful sublimations afforded by "emotional religion." A patient of Coriat's described in the *Psychoanalytic Review* the religious overtones of her analysis. The trust in the doctor necessary for overcoming resistances was akin, was it not, "to the constraining power of the divine love?" [24]

Coriat combatted, in a period of intense racism in some quarters, hereditary theories and the stereotypes of the Jew. The parsimony, the obstinacy, the vengefulness of Shakespeare's Shylock, for example, could be satisfactorily explained by anal eroticism. Psychoanalysis offered causal theories based on early experience in place of the hereditary influences Coriat once believed predisposed to neurosis.[25]

Coriat was one of the first Americans to apply psychoanalysis to literature and to champion the social role of the artist. He denied the claim of the degeneracy-mongers that art and genius were the product of insanity and hereditary taint. Rather the artist tapped the universals of human experience buried in the subconscious and transmuted these into symbolizations. He expressed humanity's deeply repressed desires and conscious protests. For this reason, works of art could be interpreted like the neuroses; one could seek the same determinisms in either. Coriat's criticism, which was fairly extensive, was scarcely literary or critical. His analysis of *Macbeth,* published in 1912, stripped the participants in the tragedy of all responsibility. Lady Macbeth was neither degenerate nor criminal; she was a pitiable coward, whose outward ambition to be queen compensated for a repressed wish to have a child. While proclaiming the rigid, clinical determination of human action, Coriat resorted to the witches' enchantment of Macbeth to explain the denouement of the tragedy. Like the younger intellectuals, he welcomed the Russian writers, but less for their passion than for their superb descriptions of nervous and mental disorder.

Despite considerable crudity of approach, Coriat displayed a real sympathy with the new literary movements. Joyce, Proust, and Eliot were among those he believed to have come under psychoanalytic influence. In 1917 he welcomed Conrad Aiken's "Jig of Forslin" as a "genuine psychoanalytic poem" of the highest merit, which beautifully described a cycle of unrealizable wish dreams. Like the analyst, the artist, he believed, was a gifted seer who fearlessly explored the unconscious.[26]

William Alanson White, 1870–1937

While Putnam represented by training and Coriat by adoption the traditions of New England moralism, William Alanson White represented the popular Bergsonian faith in evolution of the Progressive Era. White was one of the most influential of the psychoanalysts because of what he made of his position as administrator of St. Elizabeth's, the federal hospital for the insane in Washington, D.C. He survived changes in Presidents and cabinets, political pressure and congressional investigations, and held the job until his death in 1937. He was tactful, charming, tough, and a facile popularizer who wrote compulsively. It must have taken courage for a man who lived, as Ray Lyman Wilbur remarked, a "gold fish bowl existence," to champion Freud.[27]

White worked hard to bring psychoanalysis to the attention of the "intelligent classes," and lived in a whirl of lecturing, reviewing, and battling for causes—mental hygiene, progressive education, and, above all, the reform of criminology. He testified for seven hours without interruption in the Loeb-Leopold case and was able, an admirer wrote, to put into general circulation ideas that had belonged only to the "enlightened minority."

White's round, bald head, cheerful tolerance, inscrutability, and aloofness suggested the image of a tonsured cleric to a lady journalist.[28] He probably regarded himself as a priest-physician of the new order. The study of man, White argued, had escaped from its traditional moorings in philosophy, theology and morals, and had become a biological science. The psychiatrist therefore should lead man's "first attempt" to come to grips with his own instinctual nature. Psychoanalysis was a kind of higher realism which presented human beings in the raw, stripped them of illusions and "artificial haloes," yet took account of their aspirations. Greater tolerance of instinct was the first prescription of this scientific faith. White expressed compassion for a young girl tortured by guilt over contraception and for a man who had murdered his wife.[29]

Of all the psychoanalysts, White and his friend Jelliffe best projected the practical idealism of the Progressive Era. An ideal only had dynamic worth to psychoanalysts, White wrote, if it could be put into practice.[30] White derived his conceptions of evolution and energy from Spencer and Bergson, whom he managed to reconcile without a qualm. He never was seriously shaken from Spencer's hopeful evolutionary catechism, which at the age of thirteen he had accepted as the key to all knowledge. He had been prepared for Spencer by a youthful diet of Henry Ward Beecher's sermons and by discussions at the local Unitarian Church among a group of intellectuals who later formed the nucleus of Felix Adler's Ethical Culture Society. Buckle's *History of Civilization* and Darwin's *Origin of Species* made him ready to extend the law of cause and effect to human actions. When he wrote his autobiography in the 1930's, he realized that Spencer's nineteenth-century universe of simple determinism and inexorable evolutionary processes had been destroyed. Yet he never seemed to renounce his own faith in regular laws of progress. Determinism paradoxically enlarged the possibilities of free choice, the "field of conscious control." If one knew the motives of conduct, and he defined motive as the history of instinctual expression, conduct could be controlled.

From Bergson, White learned that the nineteenth-century god of energy was always dynamic and that evolution could be far more exciting than it had seemed in the Spencerian dispensation. Bergson had united body and brain, man and the cosmos, in a single functioning process. It was even possible to see psychoanalysis as one of those triumphant, unexpected leaps of the Life Force.

White's vague generalities and his syncretism may have been exacerbated by his role as hospital administrator. Always of what he himself considered a "philosophical turn of mind," he thought that administration cut him off from his former close contact with patients.[31] Case histories occupied a diminishing place in his writing; generalities flourished. It is easy to imagine reflections of the administrator's need for smooth reconciliation in his attempts to fit comfortably together the most opposed points of view. He believed that if people only understood each other's motives, their differences of opinion would melt away. He once cited Pavlov, Watson, Fraser, Sherrington, James, Darwin, Kant, and Hegel in the same defense of psychoanalysis. Sharp-witted and skeptical, Elmer Southard suggested that White might be suffering from a "phagocytosis" or overingestion of theories.[32]

From his New England parents, White learned the virtues of hard work and persistence; he distrusted the new emphasis on leisure in America in the 1920's and 1930's: it seemed as if people were trying to get more pay for doing less work. His parents moved from Boston to Brooklyn, where he was born and brought up, except for summers in Massachusetts. He was sensitive to the fact that his mother seemed to come from a higher social class than his father, who talked of nothing but business and politics. Wealth, particularly unearned wealth, was a positive evil, White early believed. Ten years younger than his brother, he felt as if he had been raised as an only child and diagnosed himself as an introvert who had been forced to lead an extrovert's life. At fifteen he won a scholarship to Cornell, then earned a medical degree from Long Island Medical College. He supported himself through school by long, grueling hours of outside work. He started out as an ambulance surgeon, and he was so poor that when he took a job at the New York State mental hospital at Binghamton, he spent his first pay for a tailcoat to cover the patch in his pants. He saw firsthand the sufferings of the insane poor, who, drained of hope, waited for death. One of his first acts at St. Elizabeth's was to abolish all forcible restraint. Full of initiative, he was learning French and German by correspondence course at Binghamton, where he met Smith Ely Jelliffe in 1896. They became lifelong friends and collaborators, translated books on psychotherapy and psychoanalysis together, and wrote neurological and psychiatric texts.

White was more interested in the environmental and social causes of nervous and mental illness than some of the other psychoanalysts. One of his early papers analyzed the geographical incidence of insanity in America, and later he investigated social strains and tensions. He fought the eugenics movement's demands for mass sterilization of the defective, the criminal, and the insane. Slyly he argued that the Puritan divine, Jonathan Edwards, would not have been born had his beautiful promiscuous grandmother been sterilized. She logically would have been, had measures eugenicists were then proposing been in effect in her lifetime. Moreover, feeble-mindedness and defect, like insanity or criminality, were not absolute, formal categories but relative states. Scientific knowledge simply was not advanced enough to make eugenics a reliable guide.

White was at his indignant and humanitarian best when advocating reforms in criminology. Americans must wipe out once and for all, he urged, that "hang-over from our medieval theologies"—the notion that antisocial conduct was a sin to be punished.[33] Criminals were hated and punished because they performed acts everyone wanted to carry out. The criminal was not sinful, but sick. White advocated the indefinite sentence, to be set by judges, lawyers, and juries with the help of psychiatrists. He considered murder likely to occur only under rare and very special circumstances; so murderers often could be safely discharged into the community. He abhorred the barbarism of capital punishment and believed that psychiatrists should not testify for the prosecution. White was able to justify destructive tendencies in men by arguing that these were once useful virtues in the evolutionary struggle. He urged that the old-fashioned moralistic terms—delinquent, dependent, and defective—be dropped in favor of "socially inadequate." In place of hateful judgment, society should substitute that up-to-date, successful, scientific, and progressive emotion—Love. Greater freedom would foster a sounder morality and not an orgy of indulgence, as pessimists claimed. But he took pains to define "true freedom" not as "license" but as control of instincts and conformity to social customs.

Psychoanalysis, White believed, was of more practical benefit than all the years of laboratory psychology that had preceded it. "Born of the heartaches and sufferings" of the sick, psychoanalysis was useful and humanistic.[34] In 1914 he defended it before the nation's psychiatrists against concerted attacks by conservatives such as Carlos Macdonald, then head of the New York State Commission in Lunacy, and Charles W. Burr, the Philadelphia organicist.[35]

Like many of the other psychoanalysts, White combined optimism with shock tactics. His *Mental Hygiene of Childhood,* published as one of H. Addington Bruce's "Mind and Health" series, contained a frank discussion of childhood sexuality and stressed the ambivalence and hatred of children toward parents. Above all, White's message was environmentalist, as Bruce's preface suggested. Psychoanalysis supplied evidence that traits were impressed on the child by his early environment, and doctrines of heredity served merely to "block efforts at improvement."

White first welcomed, then regretted the popularization of psychoanalysis he himself had fostered. In 1918 he was delighted to find psychoanalysis permeating the American social fabric through the stage, novels, schools, and universities. Later he decided that the early enthusiasm of the public was superficial. The average "convert" had been attracted for emotional, not intellectual, reasons and had been so "bucked up" by psychoanalysis that he had felt impelled to apply it.[36] Even after these strictures White continued to popularize psychoanalysis in books, articles, press interviews, and lectures. Without men like him, Karl Menninger once remarked, psychoanalysis wouldn't have had a chance in America.[37] Among those whose interest in psychoanalysis White stimulated were Bernard Glueck, Edward Kempf, and Harry Stack Sullivan.

White's popularizations generally were well received. Although the *Dial'*s reviewer found *The Mental Hygiene of Childhood* "not simple enough" for the average parent, John MacCurdy thought it contained the soundest psychoanalytic advice then available for parents. It became a standard item recommended by the National Committee for Mental Hygiene. His *Mechanisms of Character Formation* sold modestly, about 4690 copies between 1916 and 1935. But its influence probably was greater than these figures indicate. Partly as a result of his own efforts, White lived to observe that the "psychiatrist is received with open arms in almost every direction and asked to help solve the problems of almost everybody." [38]

Smith Ely Jelliffe, 1866–1945

A Niagara talker, Smith Ely Jelliffe impressed his patient, Max Eastman, as hearty, friendly, and bursting at the seams with miscellaneous knowledge. He knew something about paleontology, botany, painting, mysticism, the drama, neurology, and psychiatry. His house was filled with thousands of books, homemade wines and liqueurs, collections of fungi, mosses, and pressed botanical specimens.[39]

He was neither a major educator nor a public figure, but his influence on medical colleagues was considerable. He was a clever extemporaneous speaker, with a prodigious memory and a Rabelaisian wit. His energy was as torrential as his conversation, and he published some four hundred books and articles.

He was born in a brownstone house on West 38th Street, New York, and brought up in Brooklyn. His grandfather was a hat maker, his father, a high school principal. As a child and a young man, Jelliffe avoided "feuds and partisanships." His parents were not church-goers, but he went through what he called a "typical adolescent conversion" and was for several years an active Baptist. He later dismissed this religiosity as a "fractionated concomitant of my adolescent love object finding." But he enjoyed the atmosphere of a Catholic hospital and of Fordham University, where he taught for a while.[40]

His real religion was evolution; he was a Lamarckian and later a passionate Bergsonian. Still quoting Bergson in 1939, he asserted that men were set apart from static forms of life by a unique "mammalian liberation." [41] It was this special freedom that allowed him rosy hopes for human progress through psychiatric enlightenment. Jelliffe enjoyed emphasizing G. Stanley Hall's point that the individual recapitulated all the stages of racial development. Drawing on his early love of geology, he compared the human mind to the Grand Canyon, whose walls preserved a record of the "vast accumulations" of past experience. From birth to five years of age, the child went through a development comparable to that from the "anthropoidal ape to man of the agricultural period. . . ." Properly directed, primitive energies, in themselves neither constructive or destructive, could be harnessed for social progress.

His curiosity about the relation of thought and emotion to illness was first prompted by a friend who violently sneezed when told that ragweed abounded at Lake George, where Jelliffe spent his summers. In his early hospital work he saw a "mythomaniac" who had joyfully undergone twenty-eight different surgical operations. In 1890 Jelliffe spent a *wanderjahr* in Europe, attended a few of Charcot's clinics, but devoted more time to art galleries. In 1894, at twenty-eight, he married his adolescent sweetheart, after waiting to complete his medical education. He earned only $75 from his first year's practice and began to supplement his income by medical writing. This covered an enormous variety of subjects from "Notes on some Microscopical Organisms Found at World's Fair Water Supply" to "Insect Powders." In 1896 he went to Binghamton State Hospital, partly to support his growing family. There he met White, who soon started work with Boris Sidis on hypnosis. Jelliffe recalled that he himself was "mulish and would not understand" this radical departure. In 1902 he became managing editor of the *Journal of Nervous and Mental Diseases* and began to translate Paul Dubois's *Psychic Treatment of Nervous Disorders.* After testifying for the defense at the Harry Thaw trial, he returned to Europe in 1907 to study with Kraepelin. He met Dubois, as well as Jung, and other members of the Zurich school and first heard of psychoanalysis. In 1908 he listened to Dejerine, Babinski, and Janet in Paris. He heard Dejerine argue that 80 per cent of the ailments treated by Paris physicians were really neuroses. By 1908 Jelliffe was trying out a number of psychotherapeutic methods. In 1910 he was working on Joseph Collins's service at the New York Neurological Institute with A. A. Brill: "After our clinics three times a week Brill and I walked homewards together through the park and as formerly with Dr. White we argued and argued and he persisted and thus I became a convinced Freudian." [42] His principal psychoanalytic collaborator, White, spent ten summers with the Jelliffes at Lake George. Jelliffe recalled: "We were continuously at each other, our dreams, our daily acts and aberrations, not for hours but sometimes all day." Jelliffe developed an extensive psychoanalytic practice in New York, and for a long time worked with lay assistants. Like A. A. Brill, he knew some of the younger intellectuals, among them Mabel Dodge Luhan and Leo Stein.

Jelliffe may have been more conservative than White about morality and society. Jelliffe angered his socialist patient, Max Eastman, by interpreting his neurosis as partly the result of "hostility to the father working itself out in prejudiced radicalism." [43] Jelliffe offered sublimation to patients as the solution to conflicts involving immoral sexual drives. He told his colleagues in the Psychopathological Association: "Energy must flow through psychical avenues and spiritual children must be born. The task, therefore, is to find the type of spiritual child which will be of value in the world in which the patient lives. Here philanthropy and science offer large opportunities." [44] He disliked intensely the promiscuity of the "new morality," and he considered the "new woman" mannish and decadent.[45]

He often attended the theater and sometimes psychoanalyzed plays. These critiques were chiefly sermons in psychoanalytic uplift. He believed, as did most polite critics, that literature should instruct and improve. In a study of James Barrie's *The Willow Tree* he argued that the artist revealed ways of release from the "fetters" of heredity or early circumstances that kept one back from "the complete exercise of one's powers." [46]

Jelliffe coined the phrase "psychosomatic." His exploration of the emotional factors in illness were both uncritically speculative and full of "flashes of insight," in the opinion of Karl Menninger, whom Jelliffe first interested in psychoanalysis. In the flu epidemic of 1918 Jelliffe warned, rather absurdly, that "the manner in which each individual is going to react to the grippe virus is going to be determined by his [personality structure] and the way in which he has handled or is handling his conflicts." [47] He was both laughed at and imitated. Adolf Meyer considered Jelliffe consistently open-minded and stimulating. Freud thought enough of his psychosomatic pioneering to write that he was preparing the way for the medicine of the future.

Horace W. Frink, 1883–1936

Almost a full generation away from Jelliffe in both years and outlook, Horace W. Frink was unconcerned with grand evolutionary visions. He assimilated psychoanalysis to the next popular scientific model, a hybrid behaviorism derived from Pavlov, Irwin Bissell Holt, and John B. Watson. Within this environmentalist context, he made a special point of the relativity of sexual morals and may have been more permissive about sexual conduct than the other analysts under discussion

He came from a rural American colonial family and was raised by a physician uncle in Hillsdale, New York. He was graduated from Cornell Medical College in 1905 and worked at Bellevue and at the Cornell outpatient clinic under Charles L. Dana. He experimented with hypnosis, and after 1909 began to try psychoanalysis with clinic patients. In 1914 he became a professor of neurology at Cornell University Medical College and was a lucid expositor of psychoanalysis. One of the founding members of the New York Psychoanalytic Society, he was its first secretary in 1911, and president in 1913 and 1923.[48]

He would have played a greater role in the American movement but for personal tragedy. In 1920 and again in 1923 he went to Freud for analysis. He developed a psychosis, recovered to Freud's satisfaction, and returned to New York. There he aroused resentment by informing the older analysts how out of date they were. He considered himself Freud's American nuncio, responsible for orthodoxy, and criticized scathingly some of the work of Brill and Coriat. In 1923 his psychosis recurred. He was treated at Adolf Meyer's Phipps Clinic but never fully recovered.

Frink's *Morbid Fears and Compulsions,* published in 1918, was one of the soundest and most entertaining of all the early American expositions. His style was fluent and colloquial, and the book included illustrative anecdotes, newspaper cartoons, and long, informative case histories. Putnam provided an introduction in which he insisted that men believed in innate social ideals of "life and conduct and the good." Frink's text explicitly rejected this claim.

Both Putnam and Frink were acutely sensitive to the pathological and overscrupulous conscience but regarded this inner guide in two different ways. Citing John Dewey's and James Hayden Tufts's popular *Ethics,* Frink defined conscience as a system of acquired habits of approval and disapproval.[49] Conscience developed first around the age of four, when the capacities for shame, disgust, and sympathy first checked the amoral infantile impulses. The drives and impulses that then were repressed formed the first nucleus of the unconscious. Next, narcissism and the approval or disapproval of parents combined to form the "ego ideal," which also was largely unconscious. Since the apparatus of conscience was formed early, it could function in adult life in a childish and anachronistic way. Psychoanalysis scrutinized its operations to determine whether it furthered or hindered adaptation to instinctive needs and social requirements. The actual modification of conscience was not a rational, but an emotional process, and therefore was achieved through transference. Mere intellectual conviction was not enough. Frink cited the case of an ardent feminist who adopted intellectually the cause of free love. Yet when she put this conviction into practice, she developed spasms of moral repugnance and finally a neurosis.

Frink believed that sexuality held great potentialities for happiness and good. Since the sexual instinct was the natural instinct most warped and deformed by society, the analyst's job was to help patients accept sexuality more rationally. Therapy should promote a freer flow of instinctual energy and foster more satisfactory sublimation of partial and infantile tendencies.[50]

Frink invoked behaviorism to explain why and how psychoanalysis cured, a process Freud had said little about. Until psychologists began to repudiate all instinct theory in the early 1920's, the terms of each system could be translated very roughly into those of the other. The basic conception was that of habit, in William James's sense of the repetition of behavior. Frink defined a "complex," just as Morton Prince had done, as a pattern of associations.

Frink argued that it was possible to define Freud's unconscious as the "instinct" of the behaviorists. The behaviorists' "habit" became the equivalent of the Freudian "foreconscious," the locus of acquired resistances and controls over instinct. These controls created memory gaps in neurotic patients. The neurotic was like a prisoner; he was burdened by unfilled desires and an inappropriate sense of guilt. Yet he could be unaware of the causes of the desires or the guilt, or even of their existence.

Analysis focused awareness on habitual patterns. With the analyst's help the patient filled in memory gaps. He thus reconstructed the exact historical origin of each habit and the circumstances under which his neurotic reactions first were formed. With this information, he no longer need be habit's prisoner. He now could deliberate, value, and "choose anew." Thus, analysis fostered mature decisions adapted to reality. Just as the demands of conscience might be accepted, rejected, or modified, so wishes and desires once expressed in habitual symptoms could be either sublimated or fulfilled within the demands of society.

Frink asserted the American confidence in the results of correct child raising, but he made the process far more complex than did most of the other analysts. Parents, he argued, tended to identify themselves with their own parents and, unconsciously to imitate them in handling their children. Thus, Frink wrote, they "not only reproduce . . . practically all the mistakes of their own bringing up, but fail to take advantage of what opportunities for advance are offered to the newer generation." [51]

Morbid Fears and Compulsions was prepared with the help of Wilfred Lay, a Quaker schoolteacher and journalist. The *Boston Evening Transcript* welcomed the book for its anticipated "large educational and curative effects." James Harvey Robinson in the *New Republic* pronounced it "clear and attractive." [52]

Frink in a limited way foreshadowed later analytic studies of the irrational nature of political opinions. He was astonished to discover how often a patient's political and social ideas, although eminently reasonable on the surface, were in fact conditioned by "complexes." Frink insisted that one patient favored women's suffrage because he was a Jew. His interest in the equality of the sexes manifested his more personal interest in the equality of races. Another patient advocated Woodrow Wilson's re-election in 1916 ostensibly because "he kept us out of war." In fact, however, the patient believed himself to be a coward and defended Wilson because he believed Wilson resembled himself in that respect. This critique of the basis of social action was important and, in some ways, ominous, and broke with Putnam's nineteenth-century democratic faith in the inner lights of the average man. Frink never drew the consequences of his analysis or attempted to untangle its contradictions. All social reforms, he insisted, were based on neurotic discontent, but this was not in itself bad. Reformers might be motivated by a desire to escape a reality to which they could not adjust. The only way Frink justified this contradiction was to revive the old argument of Cesare Lombroso and Max Nordau that genius and neurosis were closely linked. Did neurotics see reality more clearly than normal patients because of their discontents? And if so, how did that affect the obliteration of sharp distinctions between normal and neurotic?

In effect Frink substituted Dewey's worship of rational intelligence for White and Jelliffe's faith in evolution and progress. It was no doubt a deep identity of ends that made Freud consider John Dewey one of the few great men of his time.[53] Psychoanalysis was more readily adapted to Dewey's relativism than to Putnam's painful salvaging of nineteenth-century idealism.

A. A. Brill, 1874–1948

Brill's sense of harsh reality, his insistence on renunciation as the price of culture, and his European Jewish roots drew him closer than any other American to Freud. Yet, the hard work and moral control demanded by Brill's Jewish tradition allowed him to identify himself with the pioneering Pilgrim of the New World. Probably without realizing it, he successfully merged the dedication of Putnam's New England tradition with Freud's tragic vision.

Brill published almost nothing about the first fourteen years of his life in Austria-Hungary; nor did he describe in much detail the years just after he landed in New York in 1889. In a psychoanalyst these reticences are ominous. Perhaps both periods were painful to recall. The first may have reminded him of the Old World family he had fled. The second may have been too full of the wrenching conflicts of the immigrant.

At home he had felt "literally stifled," he wrote Smith Ely Jelliffe. A non-commissioned commissary officer in the Austrian Army, his father behaved toward him like a Top Sergeant. Orphaned and uneducated, he was eager for his son to become a doctor. His mother wanted him to be a "learned man," a rabbi, "the ambition of every Jewish woman at that time for her son." The family moved to many parts of the Empire; and young Brill learned several languages and received a good Hebrew and secular education. Like Freud, he once saw his father humiliated. When he was four or five, his father took him to the regimental surgeon to have a finger treated. Brill was scared to death. This infuriated the doctor; he took out his anger on Brill's father, who stood by stiffly at attention. "I never forgot this experience," Brill wrote. "The only impression left was that for once I saw my father being bullied and accepting it and by a physician. I must have resolved then and there that someday I should be one." [54]

His father had served with Maximilian in Mexico, so Brill read W. H. Prescott's history of the Spanish conquest, and began to think of America as a "promising, clement wilderness with plenty of room for an adolescent lad. . . ." He decided to see it for himself:

> Whether I was driven from home by the Sturm und Drang period of life, or by the fact that my father always treated me in his Top-Sergeant manner is hard to say, but before I was thirteen, I became obsessed with the idea of leaving home and coming to America, spurred by Robinson Crusoe, Fenimore Cooper and Prescott to meet some adventures. My home environment literally stifled me. I finally annoyed my parents to such a degree that my father consented to letting me come here. He hoped that the experience would cure me and that I would be very glad to return and stay home. [55]

Instead of the clement garden, Brill found the Lower East Side of New York. He did not go home but bent all his energies toward becoming an American. He once eloquently described the conflicts of the immigrant without making any personal allusions. Although he was dealing with the orthodox, nevertheless he may have been writing of his own experience. On one side in these lacerating inner wars were the cherished authority of parents and the austere Jewish "inhibitions" of which the Jewish boy became ashamed but which constituted his chief controls over impulse. On the other side were "wild emotion," the yearning for "freedom and new life" as an American.[56] In many Jews, already sensitive as a race, this conflict could lead to nervous and mental breakdown, or, especially on the Lower East Side where the American examples were already bad, to crime. Brill was relieved of direct parental pressures, for he had come here alone. Yet with only one set of protagonists objectively present to fight, the inner struggle may have been even more tormenting.

Paula Fass has suggested that Brill's repudiation of home and father left him with a deep need for a total new identity, for which he could sacrifice himself in expiation. After many years of search and after an emotional crisis he found in psychoanalysis a cause, and in Freud a father.[57]

Often, Brill wrote, the Jew gave up his faith and searched vainly for another religion. Many became Christian Science practitioners in a "flight from reality." At sixteen Brill rejected "Jewish theology" and his mother's ambition that he become a rabbi. He flirted with Methodism and Catholicism for a time and might have become a priest or a minister.

Jews needed to escape, Brill believed, from the suffocation of Jewish family devotion. Once he was asked by a colleague to treat a very nervous young son in a wealthy, cultivated Jewish family, so devoted that when any member fell even slightly ill they all became upset. His colleague was delighted by this charming solidarity. "Instead of sharing his enthusiasm," Brill wrote, "I felt depressed." Excessive attachment to family was psychologically bad. Progress consisted in *not* "following in one's parents' footsteps." [58]

If his relation with his father had been troubled and if he rejected close family ties, he preserved the Jewish reverence for teachers and sages. He was seeking a guide, not a Top Sergeant. He found a spiritual father in Spinoza, who also had rejected Judaism. He preserved something of filial affection for Frederick Peterson, professor of neurology at Columbia. Freud became, of course, his lifelong father and teacher. Brill guessed that the similarities he unconsciously perceived between Spinoza and Freud explained the immediate attraction psychoanalysis held for him. Both taught that human actions should be neither ridiculed nor scorned, but understood. With European punctilio Brill dedicated his first book to "My Esteemed Teacher, Professor Doctor Sigmund Freud, LLD, whose ideas are herein reproduced." [59]

Philip Rieff has suggested that Freud's sense of his own Jewishness may have sustained his necessary self-image as a solitary, defiant fighter. Brill believed that Freud's greatness partly reflected his Jewish heritage. Brill felt that in a lesser way he, too, fought for truth among the hospitable but sometimes scornful Americans. Brill once wrote of Freud what he could have written of himself: "Most of his years were punctuated with hardships and annoyances; but being a realist, he learned to adapt himself to circumstances. He never complained in adversity, he accepted his lot with fortitude." Just once Brill complained. "Well, you are young," Freud replied. "You should not complain, but act." [60] Pain endured, Brill told his close friend Theodore Dreiser, made sympathy and kindly action possible.[61]

Possibly the puritanical bent of Jewish morality provoked Brill's contradictory attitude toward sexuality. He was at once rebellious and conforming, shocking and delicate. The orthodox Jewish tradition was so rigid, he wrote, that a "normal love outlet" was hardly possible. "Woman and everything that goes with her are under the severest religious taboos." [62] Marriage for love was not tolerated, and love in the Western sense scarcely known.

Perhaps Brill's dislike of this rigidity was expressed in his frank insistence on sexual details and his delight in risqué jokes. He may have identified Jewish puritanism with the American "Puritanism" the cultural rebels were attacking.[63] Brill shrewdly detected sexual wishes his patients had repressed because of religious or social prohibitions. He wrote very frankly about sexuality for physicians and insisted that the subject be taken with high seriousness. In lectures to education students at New York University, however, he was a model of discretion. In 1910 Brill was shocked to hear Freud grant permission to his fifteen-year old daughter, Anna, to read the *Leonardo da Vinci.*

Brill never treated morality as an easy burden. Society, he told the student teachers, made sexual gratification, especially for young people, very difficult. He argued that from the earliest years the child had to renounce instinctual pleasures, and this left a residue of irremediable dissatisfaction. By 1921 and probably earlier, he argued that total happiness and freedom were impossible because of the nature of man. The degree of his emphasis on hardship and renunciation was shared by few of the other American analysts and differed conspicuously from Frink's assumption of the easy compatibility of instinct and morality.

Brill was a classical American success. Probably he would have explained this by what he considered the Jew's "flexible adaptibility to difficult" [64] realities. His experience probably led him to insist with a Darwinian flourish that the normal man wanted most to triumph over obstacles in the struggle for existence. He himself had been winnowed and selected. He believed therefore in rugged individualism, both political and social. Because he identified with men of action and moral courage, he supported Theodore Roosevelt. Because he preferred practicality to "idealism," he respected Woodrow Wilson's intelligence but considered him naïve. Reformers were neurotic, but like artists they were also the "salt of the earth." They were moved to seek pensions for the aged, prison reform, and slum improvement because they themselves were sensitive and dissatisfied. Socialists, he argued, exaggerated the unselfishness of human nature.

If Brill supported the traditional American middle-class belief in the individual, his emphasis on practicality, his relativism, and his tolerant cosmopolitanism attuned him perfectly to the major directions of cultural change. To him as to many other Americans, "reform" after 1914 became identified less with anti-trust legislation than with blue laws, literary censorship, and prohibition drives. He warned that the "self-righteous" person madly pursued virtue because unconsciously he was the reverse of good. A teetotaler, Brill condemned prohibition for denying people the useful comfort of alcohol.[65]

Brill's ideal was a cosmopolitan, tolerant, multi-racial culture. He had sufficiently come to terms with his Jewishness by 1917 to caution that rejecting racial identity was as futile as trying to change religions.[66] He believed, he told Dreiser, not in "absolute" but in "relative" good and evil. These had to be defined by each individual for himself; what was good for one person might be wrong for another. He urged each patient to follow his own moral code. This plurality of standards was appropriate to the city, the source of stimulation and of tolerance. The cosmopolite was more "broadminded" than his provincial brother. For this reason "scientific and more liberal-minded" city people became "honest and outspoken in their attitude to sex," while the old "hypocritical reserve" lingered on only in the "remote mountain fastnesses," he wrote in 1930.[67] The city also gave more outlets to both neurotic and normal people. Friends, gossip, and seeing pretty girls at the theater could make monogamy bearable. A thirst for excitement could be satisfied by a roller coaster. But the end of variety was not pleasure as such. By providing more "outlets" the city facilitated progress and made the individual's control over his feelings and emotions easier. The degree of this control measured the whole progress of civilization, Brill believed. His idea of "progress" differed from Putnam's or Jelliffe's. Brill did not conceive of it as caused by an inevitable biological process or by an unexpected leap of the Life Force. Progress was an entirely human achievement, the result of effort and renunciation. Progress was not inevitable, and Brill's version of it preserved no metaphysical overtones of a vanishing religious faith.

The city also facilitated progress by offering the "tonic challenge" of a fierce struggle for existence. The city dweller was a hero, Brill believed. To have abandoned the small town for the metropolis was to be a daring and combative pioneer made of the same stuff as the Pilgrim Fathers. Brill's very Jewishness made this identification possible. For the Jew, he believed, was distinguished by a devotion to learning and culture, to renunciation and control, to adjustment to reality because of thousands of years of "ruthless hardship" and persecution. Like Coriat, Brill insisted that traits considered specifically Jewish were in truth universal human attributes. For all these reasons Brill could consider the city dweller and perhaps himself a true Pilgrim, if not a Puritan. Thus Brill could fuse in a single identity the best of his own and that of his adopted tradition.

Brill had found tolerance, painful struggle, and progress for himself in New York. His first home there had been next to the jangling elevated railway, and he loved the city's noisy bustle. His passion to attain culture and status as a scientist showed wistfulness and bravery, for the odds against him were enormous. For fourteen years he painfully worked his way through school, often sleeping less than five hours a night. Time and again he dropped out because of poverty. He looked very much what he was, the eastern European Jewish immigrant. Yet his eagerness, his good nature, his devotion made people like him. And he must have had a profound feeling for other people's suffering. At the Columbia College of Physicians and Surgeons he had been fascinated by Moses Allen Starr's clinics in neurasthenia. Neurasthenic patients, Brill recalled, were sympathetic because they talked feelingly of their symptoms and "apparently wanted to be helped." [68]

By 1917 Brill had become the American leader of the psychoanalytic movement. "Psychoanalysis was unknown in this country until I introduced it in 1908," he wrote, opening his preface to Freud's *Basic Writings*.[69] This was not the case, but was an understandable exaggeration. For Brill contributed more than any other single man to the success of psychoanalysis in America.

His translations of Freud were a remarkable achievement in themselves. But they were seriously criticized, and this caused the sensitive Brill the few years of discomfiture he experienced in his relationship with Freud. Brill apparently feared that Freud, too, disliked the translations. After 1914, according to Ernest Jones, Brill maintained a sulky silence. In 1920 he raised $1000 for the Vienna Verlag, the psychoanalytic publishing house, then in serious financial straits. But he also sent a letter to Freud, who in turn, explained the situation to Jones: "I have received a letter from Brill, a long, tender, crazy letter not mentioning a word about the money but explaining away the mystery of his behavior. It was all jealousy, hurt sensibility and the like. I will do my best to soothe him." [70] Freud had not despised the translations, and he reassured Brill, who from then on, Jones wrote, became his old "loyal and friendly self," a Rock of Gibraltar. This strength was useful. By 1919 Frederick Peterson was denouncing psychoanalysis as vehemently as he had welcomed it. Bernard Sachs, who employed a footman to hold his top hat during bedside consultations, remained an ardent, abusive enemy. "Dr. Brill," he once remarked, "I want you to keep your hands off children. Your doctrines already have done outrageous damage." [71]

Brill almost regretted the finally established respectability of psychoanalysis. What would become, he wondered, of the crusader's aggressions? "Hitherto one of the greatest functions of the Freudian was to combat the resistances of the outer world." [72]

Yet, despite the hostility psychoanalysis aroused, Brill had a genuine capacity for keeping the respect and admiration even of people who opposed him. Many years later Brill planned to include a tribute to Peterson at a meeting of the American Psychiatric Association. Peterson wired Bernard Sachs, "It is possible I may be accused of having introduced psychoanalysis into this country. If I did, I apologize." Sachs brandished the telegram at the meeting. Peterson wrote Brill a few days later: ". . . despite our differences of opinion, I have always admired you for your honest faith in [psychoanalysis], for your genuine ability and fine attainments, and cherish a real affection for you as a loyal friend to me, and a good aide in the old days." Brill replied: "I always have considered you one of the greatest influences in my life. You were not only my teacher and guide when I was just a hospital interne, but a generous friend in need later." [73]

A superb clinician, Brill apparently had considerable success using psychoanalysis with patients other physicians had failed to cure. He was an equally able propagandist and frequently lectured to medical and lay audiences. He often gave interviews to journalists, and the *New York Times* respectfully covered his opinions. In 1916 he addressed the Child Study Association of America on "Masturbation, Its Causes and Sequellae." [74] He argued that, although it was nearly universal and would do no physical harm, it could stifle the development of an aggressive, extroverted personality. He argued for a cautious, gentle correction of the habit in children instead of the "medieval" policy of "threatening and punishment."

At times Brill expressed attitudes close to Dreiser's naturalism. Sexuality, for example, was a great force, and an outlet for the "love-life" in the broad psychoanalytic sense was as necessary as "pure air and food." Without such an outlet people eventually suffered. However, Brill was far more hopeful than Dreiser. In an affectionate sketch for the *American Mercury,* Dreiser presented his friend as a believer in the mercy of Nature. The essay reads as if Dreiser were daring Brill to prove during one of their long arguments that life was not senselessly painful. Brill considered pain an "incentive" and "urge to life." He cited the case history of an unusually ugly girl who had a breakdown and in her insanity believed herself to be beautiful. Following Spinoza, Brill argued that the universe was a machine with a built-in impulse to create compensatory balances; it was not purposely cruel. It would not respond to wishes and prayers as hopeful Christian Scientists and New Thoughters believed. Yet, the girl's very delusion illustrated the niceties of compensation. Later Brill approached Dreiser's sense of tragedy, or perhaps reverted to the fatalism he had always associated with the long history of Jewish suffering.

Possibly the horrors of the 1930's and 1940's and Freud's exile and death also may have contributed to this change of outlook. About a year before he died, he addressed the Vidonian Club, an informal group of New York psychiatrists to which he had belonged for many years. He suggested a connection between Spinoza's pantheism and Bryant's "Thanatopsis," which he quoted. He remarked: "And so, whether we have passed through life as Vidonians or as criminals, we have all done our best with the means at our disposal. For whatever we did depended, not as we think, on ourselves, but on accidental factors which we ourselves did not control." [75]

XV

Popular Psychoanalysis
1909 - 1918

Ingredients of Success

Within six years of the Clark Conference psychoanalysis had eclipsed all other psychotherapies in the nation's magazines. By 1911 the Emmanuel Movement had disappeared from their pages, and by 1914 hypnotism and suggestion were almost, but not quite, out of fashion. Psychoanalysis received three-fifths as much attention as birth control, more attention than divorce, and nearly four times more than mental hygiene between 1915 and 1918.[1]

The figures are impressive: between 1910 and 1914 eleven articles about Freud and psychoanalysis were published, most of them favorable; between 1915 and 1918, about thirty-one, of which twenty-five were favorable. Perhaps one-fourth reviewed psychoanalytic books, more than another fourth were full-dress popularizations; the rest were short serious expositions or criticism, sometimes summarized from medical journals. In 1915 psychoanalysis reached the mass circulation women's magazines, and the first American psychoanalytic novel, *Mrs. Marden's Ordeal,* was published two years later.

Freud's progress as a celebrity was rapid. In 1912 the 425,000 readers of *McClure's Magazine* were informed that more and more "progressive and successful physicians" were adopting his radical method. Freud's indefatigable ingenuity, his "persistent courage" in the face of opposition, painstaking care, and conquest of new "territories on the frontiers of the mind" quickly became the clichés of his public image. In 1913 Freud's dream analysis was taken up by the nation's most popular fictional detective, "Craig Kennedy," in Hearst's *Cosmopolitan.* In 1918 the Chicago *Daily News* described Freud simply as the "world-renowned father of psychoanalysis." [2]

Several factors and the timing of each account for this remarkable success. As the Clark Conference demonstrated, the connections between professional and popular culture were unusually intimate in the United States. The most important single reason for popularization was the growing interest of physicians. The medical interest preceded the popular and exceeded it in intensity. After the Clark Conference there was a sudden rise in articles in medical journals; between 1912 and 1914 a hundred were published, mostly favorable; between 1915 and 1918, more than seventy, outnumbering those in lay magazines by more than two to one.[3] Journalists introduced psychoanalysis with the insistence that it was not a fad but a science; its practitioners were eminent and respectable physicians, "renowned" in neurological and psychiatric circles "throughout the world." Psychologists such as G. Stanley Hall and E. W. Scripture, educators, and the religious leaders of the Emmanuel Movement added to the growing professional approval.

A second factor was the lively public interest in mental healing that became national in scope between 1906 and 1910. Its characteristics were carried over into the popularization of psychoanalysis. The analysts themselves, organized and militant, eager to spread the gospel, co-operated enthusiastically with journalists. The editors of *Everybody's Magazine* vouched that Max Eastman's articles had been "examined and approved by one of the most experienced practical 'psycho-analysts' in America." Every aspect of the American interpretation was exaggerated —the environmentalism, the moralism, the high claims, the eclecticism. In some stories Freud, Jung, Adler, G. Stanley Hall, A. A. Brill, White, Jelliffe, and Ernest Jones—the major American dramatis personae, except for Putnam—were seen as members of a single ongoing movement.

In the spread and popularization of psychoanalysis the network of analysts and patients played a key role, much of it unknown, and because of medical ethics, probably unknowable. In 1912 a new group of enthusiasts and patients joined the movement. These were the young intellectuals in Chicago and New York, who were engaged in what Henry May has called an "innocent rebellion" against the true blue standards of majority progressivism. Some psychoanalysts, notably A. A. Brill, Smith Ely Jelliffe, and the Jungian Beatrice Hinkle, drew patients from this group. Several of them became major publicists for psychoanalysis.

The young intellectuals will be dealt with only cursorily in this volume for one major reason. Their neurotic problems differed in significant ways from the classical neuroses that made up the bulk of the analysts' practice. The young intellectuals were initiating new life styles that broke with major values of the past. Although they interpreted Freud in various ways, they united in exploiting him in their attack on "New England Puritanism."

This attack coincided, as did the growth of popular psychoanalysis, with the repeal of reticence. Gradually, especially after 1914, articles and books about psychoanalysis became franker, reflecting a change in public tastes.The periphrasis of the leaders of the Emmanuel Movement was replaced, finally, in a middlebrow magazine by the frankness of Havelock Ellis.

In their popular articles the young intellectuals, some of whom had been successfully treated, wrote with the fervor of the analysts themselves. Alfred Booth Kuttner, who discussed pscyhoanalysis in the *New York Times* and the *New Republic,* had been analyzed by Brill. Floyd Dell, who wrote about psychoanalysis in *Vanity Fair,* had been treated by Samuel Tannenbaum. Mabel Dodge Luhan, who serialized her therapy for the Hearst papers, had seen Brill and Smith Ely Jelliffe. Max Eastman became an expert in psychoanalysis in the same way he had become an expert in suggestion therapy. He consulted Jelliffe and Brill. He "read Freud and every book on Freud then available in English, rehearsing the doctrine point by point" with Jelliffe, its "agile-minded apostle, and becoming once more a kind of amateur specialist in mental healing." [4] Journalistic clichés and techniques added exaggeration to enthusiasm.

Powerful new elements in psychoanalysis readily lent themselves to popularization and created a new genre of cure literature. The analysis of dreams, hints of infantile and adult sexuality, ferreting out the slips of everyday life, the 1890's model of trauma and catharsis, furnished the major ingredients. Freud's first hysterical cases were irresistibly dramatic, and Elizabeth R. appeared again and again in the popular press. Beneath the clichés and simplification, important themes emerged: the "menace" of suppressed desires, the necessity of breaking from family attachments to succeed in adult careers and sexual roles, franker attitudes toward sexuality, and, finally, a new view of the therapist's role and a search for new ways in which to recast traditional social norms.

Coverage in magazines and newspapers is only one indication of the spreading vogue. By 1915, according to an opponent, courses on sexual hygiene and the "newer psychology," i.e. psychoanalysis, were being given in extension classes, summer schools, and "girl colleges" in the New York area. These topics, largely ignored by the medical schools, had a deep appeal to young men and women. One graduate, exposed to such a course, remarked, "If it's so, every girl ought to know about these things." [5] The lay interest weakened and strengthened the movement. It created a sympathetic public and willing patients—some sought psychoanalysis after reading about it. Yet, the interest also aroused the hostility of physicians, angered by the publicity and by lay meddling in matters previously considered a medical monopoly. This chapter will review the popular images of the psychoanalytic method: sexuality, the unconscious, dreams, and the new role of the therapist. Finally, it will survey the controversies among laymen that Freud aroused and an important new semi-popular ethic of happiness a few sociologists and psychologists created out of their interpretations of Freud.

Miracles

Popularization, like the American interpretations, selectively exaggerated elements present in psychoanalysis and added a miraculous dimension drawn from the folklore of happiness and success that had pervaded pre-Freudian therapies and mind cures. Freud's cathartic method was transformed by the popularizers into a scientific miracle "almost beyond belief." The confident claims of Freud himself and the more glowing claims of the American analysts were pushed still further: "You see," Max Eastman wrote, "it is a kind of 'magic' that is rapidly winning the attention of the most scientific minds in the world of medicine."

He opened his two part popularization in *Everybody's Magazine* in 1915 by an apology:

In describing a method of treating disease which I believe may be of value to hundreds of thousands of people, it is difficult to avoid the language of the patent mdicine advertisement.

Are you worried? Are you worried when there is nothing to worry about? Have you lost confidence in yourself? Are you afraid? Are you depressed, nervous, irritable, unable to be decent-tempered around the house? . . . Do you suffer from headaches, nausea, "neuralgia," paralysis, or any other mysterious disorder. . . ?

If so, then Freud had found a way to sink a "shaft into the subconscious region and tap it of its mischievous elements" without hypnosis, hypnoidization, or suggestion.[6]

By 1915 self-realization, character improvement, and success had become goals of popular psychoanalysis as important as the cure of neuroses. Floyd Dell, a young writer and journalist living in Greenwich Village, described the faults that psychoanalysis was designed to cure: "inability to achieve results in one's work commensurate with the efforts put forth; infelicity in personal relationships, and a sense of not being able to get at grips with the realities of life." [7]

Psychoanalysis, wrote an analysand half satirically, was a new "beauty parlor" that smoothed away the "lines of mental and moral stress" and imparted "the roses and lilies of a rejuvenated spiritual complexion." Freud and his followers promised to make the patient a new person and to correct super-sensitivity, overconscientiousness, restlessness, incompetence, moodiness, suspicions, irritabilities, lack of confidence, indecision, flabby-mindedness, forgetfulness, lack of initiative, headache, nausea, worry. Psychoanalysis brought "new powers of self-knowledge, happiness and achievement," because Freud had discovered that the greatest possibilities of self-improvement "lie in the control of the unconscious, automatic part of one's mental machinery. It is the things you do without knowing or asking why that make the difference between success and failure." [8]

As a rule the method was as automatic as the claims were inflated. It might take a very long time, hour after hour, day after day, three days a week, month after month of "quiet interviews." But once the cause was unearthed, permanent cure followed.

The imagery was surgical, as befitted a psychotherapy emerging from a somatic tradition. Analysts, including Freud, often likened their treatment to a major operation because of the skill, seriousness, and thoroughness required. In popular journalism the cause of the neuroses, a forgotten memory, was an "appendix of the soul," an abscess, a "mental cancer" that could be dissected out. "Lance an abscess and relief is instantaneous. Tell your painful memory and you will begin to forget it." [9] Of resistance, transference, and working through there was not a trace.

When the image was not surgical, it was religious. Psychoanalytic catharsis was a confession; the physician was a "priest in a confessional of science." [10] In both the religious and surgical images cure was almost instantaneous.

The process of exaggeration can be traced directly from Freud to *Everybody's Magazine.* In 1910, in an address on "The Future Prospects of Psychoanalytic Therapy," Freud remarked, "There is hardly anything like this in medicine, though in fairy tales you hear of evil spirits whose power is broken as soon as you can tell them their name—the name which they have kept secret." Eastman simplified the analogy: "We have but to name these nervous diseases with their true name, it seems, and they dissolve like the charms in a fairy story." [11] He added vaguely that this was not always true, because in some cases a "total re-education of the mind" was required, a process he did not describe. The case of a hysterically blind husband who recovered as soon as he realized that he hated his wife was used over and over.[12]

Psychoanalysis completed that dethronement of the will already begun by theorists of heredity and by suggestion therapists. For Prince, Sidis, and the Emmanuel Movement, strengthening "will power" had been a major therapeutic goal, although the patient temporarily surrendered his will to healing suggestions from the physician. Most advocates of hypnosis defended both the morality and utility of this temporary surrender. In popular psychoanalysis, however, a radical change occurred.

The automatic nature of trauma and catharsis obviated the use of "will power" entirely. Trying to fight one's neuroses by will power was a waste of energy, a point the American interpreters of Freud already were making. Journalists carried this point a step further. The patient recovered not by a "wearied exertion of the will, but as simply and naturally as the switching on of the light drives away the dark." [13]

Just as the *Ladies' Home Journal* had drawn hope for the insane from Morton Prince's hypnotic treatment of Sally Beauchamp, so the *New York Times* announced "new hope" from the "remarkable studies of Sigmund Freud of Zurich [sic]." In view of the bleak reviews of Beers's *Mind That Found Itself,* the optimistic Freudian outlook was a significant departure. Psychoanalysis had proven that all the "morbid" phases of insanity could be detected as "through a reversed telescope in states of mind which we accept as normal." A full page review of Brill's *Psychoanalysis: Its Theories and Practical Application* offered these glowing prospects: "Patients hovering on the borderline of insanity are made to see the absurdity of their fears, delusions and obsessions, and by the force of logic are restored to mental health. Even in cases of real insanity the persistent application of the Freud methods results in a great deal of benefit to large numbers of patients." [14] At Ward's Island under Dr. Adolf Meyer and in New York's out-patient clinics, patients were being treated effectively.

"Suppressed Desires"

Psychoanalysis altered the popular image of the cause of neurosis in three ways. It discarded environmental stress, "the strain of modern living," except as this represented repressive customs. It strengthened the existing emphasis on trauma and located this almost entirely in early childhood. It added a new cause—the "menace of suppressed desires," sexual and aggressive.

These "secret springs" of neurosis set up conflicts with the "mainstream of conscious life." It is clear that aggressive drives, as James Jackson Putnam argued, were more socially acceptable than sexual ones. The popularizers' handling of Freud's sexual theories reflected even more strongly than that of the analysts the conventions of "civilized" morality and the gradual, ambiguous repeal of reticence in which psychoanalysis played a role. Journalists warned their readers that to describe Freud's sexual theories in magazines that circulated in the home was impossible. Max Eastman handled the problem delicately. The most troublesome desires were related to the "sex life because sex desires are so strong and . . . because they are urgently repressed by culture and conscience and convention." [15]

The slow repeal of reticence can be traced in the descriptions of sexuality from 1912 to 1917. In 1912 the Reverend Samuel McComb explained in the staid *Century* that Freud "does not limit the word 'sex' to the instinct which is necessary to the preservation of the race. He uses it with a broader consideration which need not be further discussed here." [16]

By 1915 Floyd Dell was insisting that psychoanalysis conferred an air of "ponderous German scientific propriety" on subjects that a few years before would have been deemed "horrific" and on a frankness "not native to our culture." [17] Such psychoanalytic talk, Dell wrote, already had spread from Greenwich Village to the suburbs. Yet, despite his brave words, he left the subject quite vague, except for references to "very dreadful animal instincts, such as gluttony, lust, hate, jealousy and vanity." That same year Eastman used examples that defined sexuality largely in terms of romantic love or a wish for the protection of home and mother. Often Freud's Oedipus complex was softened by the assertion that every "proper youth" first fell in love with his mother and later with women who resembled her. Mamma's boys, wrote Eastman, were likely to grow up to be nervous men, or they remained bachelors, unconsciously seeking their mothers' protection. Women became old maids because they cherished an infantile ideal of father. A young secretary who was obsessed by fears of death "had become desirous of the love of her employer, but . . . refused to *face* it, deeming it a shameful and impossible thing. But that desire, which she would not let come outward into any form of expression, turned and delved inward into her soul, and it revived and inflamed all the old and more or less forgotten desires of childhood, which her education had repressed, and they all conspired together in secret against her sanity and health." [18]

The popular novelist Peter Clark Macfarlane was franker and more orthodox in two stories in *Good Housekeeping* that, judging from their content, might have been inspired by A. A. Brill or another member of the New York group.

Psychoanalysis, founded by an "Austrian Jew, Sigmund Freud," steps "into the holy of holies of the domestic relation and indicates profound and significant truths that have to do with married happiness. . . ." Neurotic fears such as agoraphobia, the fear of open spaces, clustered about an impulse rejected by the moral nature, an act desired but repressed. The desire originated in a "set of facts and experiences . . . so fundamental as to be described only as biological." [19]

The sex life, Macfarlane wrote, following Brill, could be scientifically equated with the "love life" and began before the age of six. Floundering in ambiguities that no American analyst had resolved, Macfarlane described the instinct as "race preservation" and "self-preservation," including "almost the whole range of life activities, beginning with the desire for food and shelter, mating, protection of the young, maintenance of the family, home-building, wage-earning, and the whole struggle to survive."

The sex instinct was an "appetite for life," determined partly by constitution, partly by environment, and sought "every kind of sensory gratification"—warmth, food, being coddled, petted, and loved. "If it gets its yearning it is as contented as a nursing infant. If it does not, beware! It will never be stopped except with satisfactions."

Macfarlane attempted to deal with polymorphous perverse sexuality. The child's infantile libido was "blind." It did not respect "persons, forms, proprieties. Anything that gratifies is good. . . ." The child's body "is a small bundle of tow that may be inflamed in various ways. It is in the broadness of this range of possible satisfactions that the dangers lie." He alluded to zones of erotic interest, to thumb-sucking, to a son sleeping with his mother, to a fixation on the same sex. However, the details of fixation or of sublimation, he insisted, could not be described fully because they involved an "intimate consideration of subjects perfectly legitimate and wholesome in themselves but which, because of their relationships and associations in the public thought are not suitable for discussion in a journal of this kind." [20]

Havelock Ellis in 1917 finally described Freud's infantile sexuality more explicitly in the middle-brow *Bookman.* The infant's sex life consisted in "simple tactile pleasures, in thumb sucking, in friction of the various body openings, or of other sensitive spots. It develops into a special interest in the activity of the excretory functions. Extending to other persons, it tends to attach itself in the boy's case to his mother, in the girl's case to her father, and it also tends to ignore the adult distinction of sex: 'You will not be wrong,' Freud says, 'in attributing to every child a fragment of homosexual aptitude.' These special attractions may easily become special aversions. Fundamentally, however, they are wishes." [21]

Repressed desires could be not only sexual, but hostile and aggressive. Thus a patient who dreamed of choking a little dog in fact wished to kill a backbiting neighbor. In 1913 a review in *Current Opinion* of White's translation of Karl Abraham's *Dreams and Myths* was titled, "Why We Long for the Death of Our Relatives." A young attorney found that increasingly he "tore round energetically but without accomplishing anything. This had spread to his social life, even to his attempts at conversation; and in short, he felt himself going to pieces." He had been placed in charge of managing estates by an old lawyer who had just died. All the young man's dreams contained the idea of collapse, of failing.

"You are tempted to loot those estates," blurted out the psychoanalyst when his offhand diagnosis was complete. "Consciously, of course, you have rejected it because you are a man of integrity, but unconsciously you want to do it, and it is that struggle which is breaking down your nervous system." Macfarlane argued that the lawyer should have thanked him "for his warning of the terrible danger of yielding to the temptation in which he stood," but instead, he left in a rage.[22]

Journalists moralized the psychoanalytic message even more than some analysts. In dealing with repressed hostile and sexual impulses the first aim was to overcome repressions fostered by polite convention and high moral ambitions. This was not in order to gratify wishes but to face them frankly. Sometimes the mere acknowledgement of such wishes caused symptoms to "begin to disappear."

Acknowledgment might create a sense of conversion or a "conviction of sin." "Again the accepted cure for blindness is to go and sin no more," an anonymous analysand wrote in 1918. "The conversions that the Rev. Mr. Sunday and his less notable peers are wont to accomplish in an hour, these painstaking scientists patiently bring about in from some scores to some thousands of hours of equally strenuous labor." [23] Thus the revelations of the unconscious were expected to intensify therapeutic shame. To this end the popularizers strengthened the analysts' distinction between adult and infantile, primitive and refined. To Eastman the "nervous person" was a grown-up infant, someone who had never broken away in the depths of his heart from the "family situation." Neurosis prevented men and women from being "grown-up and efficient human beings." Those most likely to remain neurotics had been either coddled or "too hideously nagged" in childhood. Another less frequent resolution was realistic renunciation. Eastman described a case of Brill's in which a girl decided "to take her disappointment straight, and not to long unconsciously for a proposal from a man who did not love her."

Most important, the analyst directed the patient toward sublimation and the fulfillment of social norms. Thus the man who became hysterically blind stayed married by verbally expressing his anger at his wife. Macfarlane explained sublimation in just these terms:

With adult life the man or the woman finds release for his or her excess of libido in legitimate social interests, in the assumption of responsibility, in the battle for bread, in the care of the home, in providing for the comfort of the children, if there be any; or, compensating for lack of any or all of these, by interest in some public concern, in social service, in church work, in politics, in athletic sports, in travel, even in fads and hobbies—any interest, in fact, that serves to work off the vital energies and relieve the nervous system from a heavy pressure of steam which means, sooner or later, explosion, shock, and the disruption of the normal forces of life.[24]

The animal instincts could be lifted up, purified, and refined.

The element of economic discipline was especially clear in advice to parents to bring up males who would successfully compete. Instincts, sexual and aggressive, were not to be suppressed but directed. Parents should be particularly careful not to stifle their sons' pugnacious instincts, and every city should provide playgrounds for their outlet. For in the hard world of the adult, in "real life," there was plenty of room for the fighting spirit. Above all, the child should not be coddled, and he should be carefully protected from sexual traumas. Macfarlane summed up this double message: "Parents, guard your children; especially from yourselves; from over-loving; from over-coddling; from under-consideration of the wax-like plasticity and marble-like retentiveness of the child-soul, from shocks, from scenes that have a meaning to the unconscious life at a time when the unconscious life is so much the more powerful element in character determination." [25]

Journalists were careful to counter interpretations that might be socially dangerous. Thus, the evils of repression did not imply that sexual instincts must be directly satisfied. Macfarlane wrote: "To avoid misconstruction I hasten to say that nothing written or quoted here is to be taken as meaning that in order to live a normal life each individual must seek and obtain the gratification of every demand of the libido—neither that every individual must marry, nor that every marriage must be a perfect union in order to produce a normal nervous life." [26]

Yet, psychoanalysis had revolutionary implications. To a man in love with a woman beneath his "social status" the psychoanalyst remarked: "When it comes to squaring your conduct with the standards of a social circle and the standards of your own nervous system you are too sensible a man to halt long, I think." Floyd Dell observed in *Vanity Fair:* "There is, very unfortunately, an extent to which the instincts fundamental in mankind seem to be incompatible with civilization as it now exists. After we have sublimated and sublimated, the world still remains in some respects a fitter place for the beings we unsuccessfully pretend to be than for the beings we actually are." [27] This problem not even the psychoanalysts had been able to solve as yet.

The Unconscious

Psychoanalysis altered the image of the unconscious or the subconscious, as it was interchangeably called, from an uncanny subliminal self to a Darwinian Titan. At first Freud's unconscious was the mysterious entity of the Emmanual Movement, a place of "strange things, dreams and forgotten memories, mediumships and telepathies and doublings of personality, uncouth and primitive impulses, inspirations of genius." [28]

Soon the primitive "troglodytic" impulses became more conspicuous. To Harry Woodburn Chase, the psychologist who translated the Clark lectures, Freud's unconscious harbored the murderous and sexual impluses characteristic of the early savage stage of racial development. This unconscious was childish: "All the unethical acts and unsocial ways of thought of the child, repugnant to us today still exist in the lowest dark chamber of the soul, not strong enough to break out into action, but alive." [29]

An analysand observed that one became as easily resigned to psychoanalytic theories of sexuality as to the evolutionary hypothesis: "At first it comes as somewhat of a shock to the beginner to hear 'all emotion summarily classified as 'sexual,' normal filial or parental affection designated as 'incestuous,' friendship as 'homosexual,' self-respect as 'narcissistic,' and the life force or will to power as 'the libido.' " But it was "no more unpalatable that all emotion is derived from sex than that all human beings are descended from an apelike ancestor." [30] The unconscious, beginning with Chase's vulgarization of Freud's topography of the mind, became a cellar or a cavern beneath the "tidy parlors and drawing rooms" of the soul.

Yet the American tendency to make the unconscious beneficent kept breaking through. The unconscious was all-powerful and pulled the strings that "managed the puppets of the conscious self," yet, with therapeutic help, it could be tamed.[31] If no longer beneficent *per se*, it remained uncanny and powerful. If domesticated, it became a new source of power for success and happiness. Once wasteful repressions were overcome, new energies flowed freely. Thus, in effect, Freud's unconscious was assimilated to the earlier tradition of a helpful subliminal self, at the same time that it lost its innate ethical goodness.

Dreams and Determinism

Dream analysis, Freud's royal road to the unconscious, inevitably became one of his most widely known achievements. In 1912 two popular articles explained this "flood of light." To the Reverend Samuel McComb, Freud was a "modern Daniel" who had made all previous theories of dreams obsolete. "For the first time dreams are seen to be in continuous relations to our waking experiences,—nay, to be the inevitable outcome of psychological antecedents bound to each in accordance with laws that are as rigorous as the regularities of the physical universe." Even the "most fantastic imagery, the most incongruous absurdity can be reduced to an intelligible and orderly scheme." [32]

In most popularizations the essentials of Freud's "orderly scheme" were lost, notably the systematic collection of associations to each element of a dream; the insistence that each dream represented the fulfillment of an *infantile* wish; the difference between manifest and latent content. Sometimes dreams were treated like allegories. Sometimes, in the tradition of Janet, they represented fearful traumas.

Freud's dream analysis was seen chiefly as a way of baring the soul's "innermost secrets," usually hostile or sexual desires that the conscience forbade. One of the most open allusions to sexuality described not infantile wishes but adulterous desires.

Craig Kennedy, a dashing fictional detective, used Freud's "dream analysis" to solve a murder for the readers of *Cosmopolitan* in 1913. "The Dream Doctor" by Arthur B. Reeve included suggestive allusions similar to those that had characterized that novel of multiple personality, *The White Cat*. A distinguished citizen was murdered on the street; his beautiful wife suffered from harrowing nightmares.

Kennedy applied to the case two aphorisms he attributed to Freud: "love is nothing but a cold scientific fact"; "there is no neurosis in a normal sex life." Because the widow dreamed that she had been attacked by a bull, which changed into a serpent, Kennedy accused her of harboring a passion for his chief suspect, a notorious bachelor who had just returned from Vienna. Her indignant denial merely increased his certainty. "The Freudists," he explained, "insist that 'people often become indignant when you strike the main complex.' " Kennedy discovered, however, that in fact the bachelor had learned details of the widow's nightmares from her physician rather than from her. The young man, knowing that he might be able to seduce her because of the unconscious desire her dreams revealed, poisoned her husband with cobra venom. Kennedy used Freud's dream analysis to solve several more murders, and by 1918 Freud had "reduced the workings of the soul to an exact science." [33]

In 1914 *Technical World* reported that Freud's dream analysis was being applied by a "black-bearded, deep-eyed . . . dissector of madness, a pocket edition of Svengali," the psychiatrist Adolf Meyer. At the Phipps Clinic in Baltimore physicians woke patients up to find out what they had dreamed. "After the doctor has heard every detail, he goes away, carrying in his mind the recital of the vision as carefully as he would hold some specimen for the microscope. When something definite has been found, a plan of occupational treatment is determined upon in which these choked-off desires are given outlet. Even now it can be said that good results have been attained." [34]

Not only dreams but word associations and the slips and errors of daily life revealed the unconscious. Psychoanalysis provided a means of "reading the inmost secrets of heart and soul," in spite of and "sometimes without the least suspicion of the person who is the subject of investigation." In 1914 the *New York Times* devoted three quarters of a page to Alfred Kuttner's long summary of Brill's translation of *The Psychopathology of Everyday Life*.[35]

Psychoanalysis provided a higher degree of participation by the patient in his own treatment than any other method. With hypnosis and suggestion the operator performed the central role. In psychoanalysis, it was pointed out again and again, the patient's recovery of memory and his insight into desire effected the cure; he must minister to himself. One of the first articles on dream theories in 1912 urged readers to test the method themselves. Everyone could keep a dream diary and by bringing up associations to each element of a remembered dream penetrate its secret meaning. This was made plausible by giving the "censor" and resistance a less important and powerful role than did Freud. If one knew the censor's disguises, it was implied, one could understand one's dreams fairly easily.[36]

Freud's determinism was one of his most popular contributions. His science had revealed that the mind was governed by laws as binding as those that ruled the body or "the sweep of the planets through space." Stephen S. Colvin, a professor of psychology at the University of Illinois, described "real Mind Reading" based on Jung's association tests. The "laws" of mental association could no more be "broken at will than . . . [the laws of] gravity." [37] One of the first popular allusions to the tests and to Freud's theories of hysteria had come from Hugo Münsterberg in 1908 and 1909. He suggested that the tests were useful in discovering crime and in checking the truth of confessions. They were far more effective and less brutal than the third degree, and he described his use of them in what probably was the case of Harry Orchard, who confessed to murder and dynamiting in connection with the bitter labor disputes at the Cripple Creek Mine in Colorado.[38] Thus, psychoanalysis was filling a void in conventional academic psychology and bringing every aspect of daily life under the "control of science."

The determination of symptoms, dreams, accidents, and parapraxes made it possible to turn psychoanalysis into a parlor game, as automatic writing had become a decade or so before. Psychoanalysis was the source of a comforting set of explanatory rules that banished the mystery from mind and revealed what people usually wished to keep hidden. A. A. Brill supplied a classic story of a parapraxis:

> A wealthy but not very generous host invited his friends for an evening dance. . . . To the great disappointment of the guests there was no supper. Instead they were regaled with thin sandwiches and lemonade. As it was close to election day the conversation centered on the different candidates; and as the discussion grew warmer one of the guests, an ardent admirer of the Progressive party candidate, remarked to the host: "You may say what you please about Teddy, but there is one thing— he can always be relied upon; he always gives you a square meal"— wishing to say "square deal." [39]

The Priestly Analyst

Mysterious overtones kept cropping up, especially in the images of the lovable yet uncanny psychoanalyst. He became endowed with some of the disquieting powers of the hypnotist. He was described as a surgeon of the soul or a detective who could read one's blackest secrets and reveal corners of the soul unknown even to oneself. In this sense he was faintly sinister. Yet, this anxious image was immediately coupled with another, which had gentle, churchly overtones. The analyst was also a paragon, "earnest, sincere, conscientious, highminded, of enormous persistence, with rare gifts of what may be termed spiritual insight, of high personal character. . . ." [40] He was, above all, calm and patient. He was not out to make money because he could treat too few patients by this new, slow, painstaking method.

He became a "priest of the new order." In the *Psychoanalytic Review* Florence Kiper Frank, poet, social worker, and wife of the lawyer Jerome Frank, began a poem about the psychoanalyst:

> With piercing questioning, relentlessly,
> He haunts the startled fugitives of the soul. . . . [41]

Lay images of the analyst went to the heart of the paradoxes of American psychoanalysis. The analyst revealed discreditable secret desires. Yet he did so in a style that broke with traditional Protestant moralism, the thundering judgments of Roosevelt, for example. The analyst might reveal to a patient his secret wishes, but he never played the condemning preacher or the absolute judge. He combined the qualities of physician and priest. The most complete embodiment of this wish for a non-judging figure who would assuage guilt occurred in the poem of a fifteen-year-old girl printed in 1916 in Margaret Anderson's avant garde *The Little Review*:

> I wish there were Someone
> Who would hear confession:
> Not a priest—I do not want to be told of my sins;
> Not a mother—I do not want to give sorrow;
> Not a friend—she would not know enough;
> Not a lover—he would be too partial;
> Not God—he is far away;
> But someone that should be friend, lover, mother, priest,
>   God all in one,
> And a stranger besides—who would not condemn nor
>   interfere,
> Who when everything is said from beginning to end
> Would show the reason of it all
> And tell you to go ahead
> And work it out your own way. [42]

If the analyst was a saintly detective, surgeon, and priest, his patient was often a victim of virtue. He was too good, too moral, and hence "nervous and high strung." Usually patients (perhaps journalists were describing their self-image) were intelligent and industrious and suffered from their own idealism. The blind laborer was too kind to think of divorcing his hated wife; the widow of the murder victim could not believe she harbored a passion for the dissolute bachelor. Yet, like the figure of the analyst, at once uncanny and benign, the patient was both childish and selfish as well as idealistic. "Bluntly stated," Bruce wrote, "nervous attacks frequently are sheer manifestations of selfishness, although the victim might not be aware of it." [43]

Psychoanalysis furthered that rehabilitation of the neurotic that marked the progressive enthusiasm for mind cures. Neuroses, although still more frequent among women, caused serious suffering in men as well. The fact that traumas and repressed desires were unconscious and universal mitigated the shock of their content. Journalists insisted, time and again, that the line between normal and abnormal had been proven by Freud to be a fiction.

Moreover, psychoanalysis was clearly intended to be a treatment for those with "stable" characters, high ideals, and a good education. Except for the hysterically blind mechanic, most patients were white-collar or professional people. Admittedly, psychoanalysis was expensive. It was becoming so popular with professional invalids, wrote Eastman, that they were paying "from two to five hundred dollars a month to have their souls analyzed—and frequently with the happy result that they take up some other line of business." [44] Psychoanalysis was worth all the time and expense because, wrote an analysand: "The strangest thing about this extraordinary process is that it really does cure the mind diseased." [45]

In most popularizations the same major theme recurred. Neuroses prevented the successful fulfillment of a social role, be it the characteristics of the ideal male or female, or a man's career or a woman's marriage. Precisely this kind of failure psychoanalysis could cure.

The first American psychoanalytic novel, *Mrs. Marden's Ordeal*, combined all these elements: traumatic childhood experiences, the detective story, the miraculous cure, a virtuous upper-class neurotic, a wise and compassionate psychoanalyst, the fulfillment of social roles. The author, a Washington journalist, James Hay, Jr., later contributed a glowing profile of William Alanson White to *World's Work*.

Mrs. Marden, an exquisite, proper, sensitive young society matron, was struck dumb after witnessing the murder of a pretty young friend she thought her husband loved. A famous psychoanalyst, who happened to be a family confidant, undertook to cure her by analyzing her dreams—"those hieroglyphic records" of past troubles when one has been "hurt and sent stumbling into the battle of life."

The analyst had "none of the manner of a preacher, no platitudes, no outbursts of blame or censure—merely a friendly, discriminating, yet convincing way of discussing things with unerring accuracy and unqualified frankness." He had infinite patience, a deep knowledge of art, literature, history, mythology, crime, science, and current events. A social lion, who was famous in Europe and America, he spoke in a deep bass voice that resounded like a cathedral organ. His patients were fine men no longer able to work and tormented women whom the world passed by as "foolish victims of nerves." The distraught Mrs. Marden promised to tell him about "sex things" and dreams.

Soon she was on the verge of collapse. When speech and memory returned, she learned that it was she who had strangled the girl in a fit of jealousy, then blotted out all memory of the hideous act. Yet Mrs. Marden was not to blame, because her mother had taught her to hate: "You are no more responsible . . . than was your dead mother," the analyst insisted. "We are all fond of saying we are the masters of our fates. We prate about free will. And yet, how little we know of what we say. . . . All of us come into the world endowed with freedom of will, crowned with the glory of doing as we choose. But this free will, this divine independence, may be taken from us, today, yesterday, or tomorrow. . . ."

This brilliant testimony by the psychoanalyst left judge, jury, and reporters "spellbound": "I do not find it hard to believe that Benedict Arnold was a traitor because in his childhood he heard, through some half-open door, his father saying strange, ugly things for money—that Judas bartered away his soul because, as a little boy, he saw his own mother sell her scarlet lips for gold." [46] Mrs. Marden was acquitted. She took back her estranged husband, who now wanted her to have what he had always disliked—children.

This preposterous novel, spiced by hints of adultery, placed the psychoanalyst in the service of motherhood. The *Nation* thought it a clever application of Freud's theories, and the *Springfield Republican* felt it demonstrated an interesting view of the influence of childhood on adult character.

The popular images of the psychoanalytic process, of patient and therapist, exaggerated the distortions of the American analysts. By confining psychoanalysis to the catharsis of traumatic experiences or the automatic disappearance of inconvenient and forbidden desires once they were made conscious, these images eviscerated the content of Freud's view of human instinct. Instead of a child or an adult whose unconscious fantasies were inordinate and irrational, the human being became, far more clearly than in the version of the native analysts, the victim of environmental circumstances, of his parents, of unfortunate social conventions.

The Analyst as Social Critic

Like the neurologists before them, the analysts took upon themselves the role of social commentator. Instead of criticizing the stresses of commerce, they criticized America's "Puritanical" customs or its failure to live up to proclaimed ideals. Both themes were mingled in the Progressive Era, but the second gradually became weaker. It was exemplified, however, in the first American "psycho-analysis" of a living public figure, Morton Prince's dissection of Theodore Roosevelt in "the manner of the new psychology."

Although Freud's name was not mentioned, Prince applied his version of Freud's methods of explaining dreams and slips in a page and a quarter in the *New York Times* on March 24, 1912.[47] The headline summarized Prince's message: "Famous neurologist says colonel will go down in history as one of the most illustrious examples of the distortion of conscious mental processes through the force of subconscious wishes."

Prince analyzed Roosevelt's conduct as a conflict between his desire for the presidency and the honor and loyalty one expected from a "high-bred gentleman" of his "family and class traditions." If Roosevelt resented the analysis, Prince wrote, this would merely indicate its probable truth. Roosevelt had left office, despite his promises never to accept a third term, with a "lurking liking to be President again" that he "sternly repressed." His conscience was victorious at first. Gradually, however, the wish began to overcome his personal loyalty, and he was able to believe that Taft had been a mistake and that the grievances against him were "grievances of the people." Roosevelt's vehement denial that he was a candidate demonstrated the principle that "an inappropriate and intense dislike" reveals a "suppressed unconscious desire," just as a man might detest pink carnations because his rival for a woman's affections habitually wears one.

Gradually, Roosevelt's wish overcame his scruples, and he was able to interpret the ban on a third term as merely a prohibition against three consecutive terms. At first he took no part in his own candidacy and urged his friends not to do so. Finally, however, the wish was victorious, and he decided to run.

Current Literature more than corrected the silence about Freud by calling Prince one of Freud's most brilliant American "disciples." [48] Ernest Jones described Prince's "analysis" in the *Zentralblatt* as one of the amusing episodes of the presidential campaign. He considered the analysis superficial, insisting that Prince had told him privately that Roosevelt's conflict probably had been conscious. The editors added this solemn note: "We should like to emphasize that we do not at all agree with the tendency to exploit psychoanalysis for the invasion of privacy." [49]

In 1916 a famous New York divorce case associated Carl Jung with free love. A physician sued his wife for divorce and for custody of their child because she was living with another doctor. This physician claimed that he and the doctor's wife had studied under Jung in Zurich and wanted to be left alone to pursue "higher and nobler knowledge." The editors of the *Times* condemned the adulterers and insisted that psychoanalysis was not an excuse for loose morals. Brill praised the editors for their perspicacity and added that the sole aim of psychoanalysis was to "adjust the individual to his environment." [50]

In 1914 in the *Times* Brill defended the new dances—the tango and the turkey trot—in a report about an obscure new neurological malady, "tango foot," the result of too much dancing. Brill argued that the dance mania sweeping the country represented the breaking through of repressed sexual emotions and feelings. "Puritan prudery and Anglo-Saxon hypocrisy have for centuries acted the part of the ostrich and refused to acknowledge the existence of the sexual impulse."

The dances "are as soothing to the populace as rocking is to an infant. Indeed, I have lately seen about a dozen nervous cases which were very much helped. . . . I know two timid and shut in persons who were completely changed by the new dances. They no longer fear to meet persons of the opposite sex and are thinking seriously of matrimony. . . .

"The new dances offer good exercise and enjoyment to thousands of people, and serve besides, as an excellent sublimation." [51]

Controversy

After 1912 but especially after 1915, Freud's sexual theories became increasingly involved in the disputes over fundamental values in American culture that were appearing in the magazines of opinion. The use of psychoanalysis to discredit cultural heroes made it especially repulsive to conservatives. The discovery of pathology in men of genius had been a stock in trade of the degeneracy theorists, and the psychoanalysts brought new methods to this old pursuit. In 1916 the *American Journal of Psychology* printed a series of studies of historical figures, many of them translated from the journal for applied psychoanalysis, *Imago*. Andrea del Sarto, the painter, was a masochist; Dante was bisexual; Napoleon suffered from an Oedipus complex. In a letter to the *Nation* in 1916 the Columbia University psychologist Robert S. Woodworth argued that psychoanalytic writers were insinuating that Socrates was motivated by an unconscious homosexual tendency, that Hamlet expressed Shakespeare's "assumed incestuous inclination toward his own mother. And so it goes, every conspicuous product of human invention in literature, art or industry is made on 'analysis' to give up the same secret." [52]

In a lengthy review of Jack London's *John Barleycorn*, Wilfred Lay in the *Bookman* detected strong homosexual elements in the way London "flaunted" his masculinity. Lay defined homosexuality psychoanalytically as taking "more than the average satisfaction from the company and in the activities" of men. The life of the sea, the hobo, and the monastery were equally homosexual. London, he argued, was a sado-masochist who liked to shock people and who understood dogs better than women. Wandering up and down the face of the earth was "extraversion gone mad," he insisted. Lay argued that in going to sea and taking to drink London was actually seeking mother and oblivion. Lay's attack was directed above all at the stereotype of the rugged male and the excessive cultivation of the daring, strenuous life. [53]

By 1916, the year of Wilson's second successful campaign for the presidency, controversy grew fiercer. Already a play, *The Boomerang,* had appeared on Broadway, inspired by "psychoanalysis," according to *Current Opinion.* The theories of Professor Freud already "had done yeoman's duty" in the laboratory and the "pot-boiler's workshop." [54]

The more popular psychoanalysis became, the more hostility it aroused. J. Victor Haberman's attack in the *Journal of Abnormal Psychology* was summarized in *Current Opinion* in January 1916; psychoanalysis was unscientific, its sexual theories were pure fantasy, its overblown claims had been repudiated in Europe.

The *Nation,* edited until 1914 by Paul Elmer More, a leader of the neo-Humanists, had opened a period of qualified opposition to Freud by a review of Brill's translation of *The Interpretation of Dreams* in 1913.[55] The humanists detested irrationality, Bergson, pragmatism, and utilitarian ethics—all the elements that characterized the native American interpretation of psychoanalysis. They argued that Freud was obscene and unscientific and that his theories destroyed the foundations of moral choice.

There were two major reasons for the reviewer's attitude. The first was Freud's "forced and repulsive" findings concerning the Oedipus complex. Second, Freud's theory of the unconscious seemed to deny the capacity of reason to control the animal instincts. Such control could be assured only by insisting that consciousness essentially was a unity and that every aspect of awareness influenced every other aspect, but Freud's unconscious seemed too sharply cut off from consciousness to be influenced by it.

The reviewer granted the probable importance of the "erotic element" in infantile and adult life. But Freud overgeneralized. Readers should be warned against accepting his "clever interpretations" without question merely because of his "abounding self-confidence and enthusiasm."

Nevertheless, Freud's therapeutic discoveries had earned him the respect and attention of physicians, and, because of his findings, psychologists no longer could ignore the subconscious. "The value of his work in relation to hysteria and kindred problems will be remembered long after his theory of dreams has been forgotten," the reviewer concluded.

In 1916 Warner Fite, professor of philosophy at Princeton, and a sympathetic colleague of More's, took up the same argument. Fite objected to what he interpreted as Freud's denigration of a rational, discriminating self and his insistence that the unconscious alone revealed the deepest tendencies of human nature.

"Who is the real I?" Fite asked. "Not the personality expressed in sober and reflective judgment; not the conscious self, but the demon that rules when I am off my guard." Psychoanalysis converted the "leavings" of this conscious self into a second, fully organized personality, a "demoniacal Alter Ego, crafty and mysterious, lustful and malevolent, a demon of vileness and indecency." If this demon were as powerful as the Freudians insisted, how could psychoanalytic therapy be effective? How could learning about one's desires make it easier to control them? Sublimation was merely an absurd, make-believe gratification.

"If psychoanalysis is efficient, the underworld becomes rather a tame affair and the demon hardly worth personifying," he concluded. The sexual content of the underworld especially annoyed Fite. He denounced Jung's *Psychology of the Unconscious* as "500 pages of incoherence and obscenity in the form of a psychoanalytic interpretation of the experiences of a sentimental young American woman. . . . Whether Brill talks about love or Freud about sex or Jung about the life impulse, their illustrations . . . are of the same order; not merely sexual, but abnormal and obscene." [36]

Yet Fite admitted some good things in psychoanalysis, none of them Freud's contributions, he insisted. Conflict, the wish, repression, and the censor came from William James, Dewey, Stout, Schopenhauer, Wundt, Kant, Leibnitz, and Aristotle. Reflecting the growing acceptance of psychoanalysis by physicians, Fite admitted that it might revolutionize the treatment of mental disorder, not because it was scientific but because it was "humane." Freud was an artist, not a scientist, and "scattered among the fruits of an ingeniously obscene imagination, there are evidences here and there of fine feeling."

Fite provoked indignant replies. On August 13 Samuel E. Eliot, Jr., the grandson of Charles William Eliot, president of Harvard, defended psychoanalysis as a patient who had experienced its benefits. From Asticou, Maine, listed in the Social Register as the Eliots' summer home, he wrote that psychoanalysis cured by "freeing the sufferer's own perception and will. . . . Each patient is made aware of lusts he is repressing. . . . There follows what a theologian would call a conviction of sin, lightened by the consciousness of greater knowledge than most of mankind possesses, and easily transformed into a healthy will to use the passion that has been wasted on repressed lusts . . . for some larger and more helpful end. This is called sublimation." [57]

The representative of a totally opposite point of view, Samuel Tannenbaum, a member of the New York Psychoanalytic Society, wrote to the *Nation* that Fite was biased and ignorant. Freud's conceptions were no more obscene than Darwin's and Wallace's were heretical. Fite answered Tannenbaum with citations from the German edition of *Die Traumdeutung* and took him to task for an unperceptive interpretation of the review.[58]

Two months after Fite's essay Christine Ladd Franklin, a retired psychologist and philosopher, protested that psychoanalysis was "intimately bound up with German kultur" and that unless means were found to prevent its spread, "the prognosis for civilization is unfavorable." [59] Harvard was Mrs. Franklin's prime target, because in "psychology seminaries and departments in medical schools" psychoanalysis had become a required subject of discussion. Edwin Bissell Holt and James Jackson Putnam probably were the teachers she had in mind.

In retrospect the major object of Fite's attack, the problem of internal controls, had been one of-the least systematically developed elements in psychoanalysis up to about 1914. In one sense, the American interpretations of Freud, apart from the imposition of American assumptions, can be seen as attempts to correct some of these imbalances. Not until Freud began to construct his ego psychology and to study the nature of internal controls did the major focus of psychoanalysis begin to change.

Yet even before the elaboration of ego psychology Americans could interpret Freud as a traditional moralist by exploiting his distinction between pleasure and reality. "Freud accepts the popular, for that matter, the philosophic, religious conceptions of human nature as a conflict between a higher and a lower part," Horace Kallen, the philosopher, wrote in 1913 in a review of *The Interpretation of Dreams.* "It is a sociological commonplace that civilization is in no small degree the 'sublimation' of sexual energy." [60]

Stressing elements already present in psychoanalysis, the popularizers produced an American caricature of it that exaggerated still further the distortions of the native analysts. The analytic process became a surgical miracle, a scientific confession obviating the need for effort, except to overcome repressions. This was a therapy not only for the neuroses but for unhappiness and failure, those traditional goals of American mind cures. The claims to success, already strong in Freud and his followers, especially in the application of psychoanalysis to insanity, were carried to extremes. Americans stressed the ease of sublimation and toned down the unruly nature of the sexual instincts. One common version of the analytic process went back to Protestantism, with its "conviction of sin" as a prelude to salvation. For the most part Freud's sexual theories were delicately veiled, although a growing frankness was apparent by 1918, the result of a general American repeal of reticence in which psychoanalysis played a part.

The two figures of the analytic transaction also were transformed. The patient, for the most part, became a victim of his own high ideals or of suppressed desires he failed to acknowledge. The most significant new figure was the analyst, the priest of a new secular order. He was assimilated to the uncanny tradition of the hypnotist, and his detective powers were enhanced by Freud's analysis of dreams and the slips of daily life. The analyst broke with the condemning moral judgments of the period. He listened and forgave, yet subtly restored the capacity of his patient to fulfill his normative obligations. This combination of seeming defiance yet deeply ingrained conservatism is apparent in the ethical implications that were explicitly drawn from Freud.

The Ethics of Happiness

Three American academic popularizers seized on the moralistic elements in Freud to restate traditional Protestant values in the guise of a new "scientific" ethic. These elements were Freud's emphasis on work, the renunciation of pleasure, and the importance of rational controls. The three Americans were raised in deeply religious homes and one was the son of a clergyman. Although they loudly rejected "prudery" and "ascetisicm," they preserved many of the values they had absorbed as children.

They were also interested in "practical" applications—simple guides for making moral choices, mental hygiene, personal counseling, education for marriage, child raising. During the 1920's these developments became part of an important, new scientific utopianism.

The sociologist Ernest Rutherford Groves inaugurated the first college course on marriage in the late 1920's at the University of North

Carolina, after receiving permission from president Harry Woodburn Chase, who had translated Freud's Clark Lectures. Groves's course was frank, but impeccably uplifting, the logical culmination of a development that began in 1916 with his first psychoanalytic book, published for the Young Men's Christian Association, *Moral Sanitation.*[61]

Like the aim of marriage, the aim of life was wholesome happiness, to be attained by a "moral sanitation" that Freud finally had made possible. The first step was for parents to develop "efficient homes" in which repressive attitudes toward sexuality were renounced. Repression created "moral sores" in the form of persistent cravings. Groves left the relation between sexuality and cravings distressingly vague; but by making parental repression the cause of the cravings rather than the instincts themselves, he added one more example of the American tendency to blame the environment.

In adults such cravings had pathological results. They caused unnatural Puritanism, the dwelling on sin and repentence, the hostility to wholesome pleasure that characterized small country churches. Cravings also gave rise to daydreams and thus threatened useful work, the sacred barrier against the "free, wild capricious life" of the primitive savage.

By making possible an understanding of the history of these cravings, psychoanalysis provided a way to extirpate them with finality. In efficient homes, suffused with psychoanalytic insight, "wholesome happiness" became "in a supremely high and noble sense" the result of "right acting." Psychoanalysis ensured the virtues Groves had been taught to revere.

He grew up in a poor family in Saxonville, a Massachusetts mill town. The "conventional New England dogma" of the Congregational Church dominated his early religious life. In adolescence an itinerant revivalist stirred him to a declaration of faith. He was graduated from Dartmouth and Yale Divinity School, then rejected the ministry for sociology. Soon he found himself consulted by students, and he began to puzzle over the hidden cravings that suddenly cropped out in ostensibly good people. The Protestant churches, he was convinced, failed to provide sufficient personal guidance.

During a summer course in psychiatry at the University of Pennsylvania, he first heard of Freud's new psychology. He avidly read Adler, Freud, and Jung, whom he liked best. Precisely the guidance the churches failed to give, he found in the new role of the psychiatrist, who had become a "moral specialist" versed in human temptation and weakness.

Moral Sanitation, inspired by Putnam's *Human Motives,* was not widely reviewed and may not have been widely read. William Alanson White congratulated Groves for shedding psychoanalytic light on the "springs of human conduct" and thus advancing "efficient living." The New York psychiatrist Ira S. Wile detested Groves's Christian moralism, its suggestion of easy cure through the revelation of "inner cravings." He attacked Groves's insistence that all moral problems arose from repressions in early childhood to the exclusion of other social factors.[62]

Next, Groves informed fellow sociologists of the psychoanalytic message. His first article, "Freud and Sociology," was returned by Albion Small, the editor of the *American Journal of Sociology,* because no sociologist knew anything about Freud. The *Psychoanalytic Review* published it in 1916, and Groves became a close friend of one of the editors, William Alanson White. However, at Small's request, Groves submitted a more general article, "Sociology and Psychoanalytic Psychology," which appeared in the journal in July 1917.[63]

Groves argued that psychoanalysis had constructed casual laws of mental experience that obviated the reliance of academic psychology on neurology. As a result, psychology had been forced "out of its comfortable little niche" to assume "in the open field of science the more heroic task of handling all mental phenomena."

This psychological level marked Freud's emphasis on sexuality, a subject too often ignored by psychologists and sociologists. Freud did not build his system on the "mere reproductive instinct" of the animal world, Grove argued. Rather, for Freud the "significance of sex life is mental in origin." On a technical level Groves's observation was correct: psychoanalysis was largely a psychology without neurology. Yet, it was false to suggest that Freud somehow ennobled sexuality by making it something other than the "mere reproductive" instinct.

Ignoring Freud's criticism of "civilized" morality, Groves accepted existing social custom without reservations. Society defined morality and sanctioned it by "social opinion." These together constituted "social control." Groves was the first and perhaps the only participant in the American movement explicitly to associate psychoanalysis with this major preoccupation of the sociologist E. A. Ross.

Social control and the capacity for sublimation, he insisted, must have developed together in the course of human progress. Sublimation was the most interesting aspect of psychoanalysis for the sociologist. Its usefulness had found "abundant proof" in the adverse effects of "precocious sex experience." Had not Professor Ross demonstrated that Peruvian boys became mentally arrested at sixteen or seventeen because of "precocious indulgences?"

Groves held out a grand vision of the future to be won by an élite which would sacrifice the pleasure to the reality principle: "Freud's theory of progress allows no misunderstanding. It maintains that human advancement comes from the work of those who, laying aside childish cravings, master the real world by investigations that recognize no desires that would distort the fact for the advantage of a wish, conscious or unconscious. . . . Only . . . those who with real courage throw aside human prejudices in order to know the real world must alone assume the heroic task of furthering human development." [64]

This vision of a scientific elect, purged of childish cravings, came close to Freud's own. Groves caught with unusual sensitivity exactly the notes of austerity in Freud's underlying Jewish moralism. By exploiting Freud's dichotomy between the pleasure and reality principles, it became possible to arouse a responsive echo in Protestant American hearts. In effect Groves reverted to the Puritan asceticism he purported to have renounced.

Groves obviously regarded himself as a member of the elect, and he devoted the rest of his life to elaborating moral sanitation. Believing that sexuality should be faced with candor, he organized a course in "family preparation" at Boston University and devoted himself to mental hygiene, social psychiatry, and parent education.

By making the environment responsible for failures in adjustment, Groves stripped Freud's views of a major component: the recalcitrance of human instincts. Wilfred Lay carried the argument further by altering the nature of Freud's unconscious, as well as the relation between the unconscious and conscious life.

Lay had been a high school teacher in Flushing, New York since 1898. A Democrat and a Quaker, he had received a doctorate in psychology from Columbia University and helped Horace W. Frink with *Morbid Fears and Compulsions.* Lay believed that neuroses were caused by a "moral struggle" between "archaic" infantile desires and the censor, which represented duties imposed by environment and the "spirit of social evolution which alone makes for progress." Lay made sexual life the center of this moral struggle and argued that sublimation was its only acceptable outcome.

Even more flatly than Groves, Lay insisted that parents almost literally created the characters of their children during the first five years of life. The content of children's minds was determined by their early experience. This proposition required a change in the nature of Freud's unconscious.

Lay transformed the unconscious into a semblance of William James's powerful, constructive subliminal self; the unconscious could be easily tamed by conscious purposes. Borrowing an image from *The Interpretation of Dreams,* Lay described the unconscious as a "Titan," representative of the "primal impulses of animal life." At first it might seem hopelessly savage. But in fact it behaved like an awkward, docile child, "willing to follow directions and gain the reward held out to it." The unconscious was like a "hose with a powerful stream of water flowing through it which needs only to be aimed in the direction where it will do the most good." Its power ought to be "as completely at our command as is the power in an automobile." With this primordial force at our disposal we could be raised to the 'nth power.[65]

Lay also domesticated Freud's sexual theories, less than the mass magazines or than Putnam or Coriat in their popular books, but far more than Brill or Frink. Lay combined a Bergsonized libido with animadversions against prudery. Lay, too, hoped to disarm anticipated repugnance to infantile sexuality by arguing that Freud understood sexuality in "a very broad sense." "In ascribing sexual feelings to childhood, the Freudians do not intend that the innocence and purity of the child shall be doubted," he insisted. Similarly, if a young virgin dreamed of burglars, she was expressing an unconscious sexual craving, but this was "absolutely no derogation of her character." [66]

Like the telephone, wireless telegraphy, and aviation, psychoanalysis had resulted from "scientific progress." Freud was the first "aeronaut" of the human mind. The physician of the not far distant future would give this advice to anyone in poor health but without organic disease: "What you need is to be analyzed. Go to Dr. Blank and I think he will be able to help you."

Lay's tract was welcomed as a "more acceptable" interpretation of Freud than most others because it eliminated much "seeming offensiveness." The *Boston Evening Transcript* praised its "clarity, charm and high moral force, its power to instruct and improve." William Alanson

White hailed it as an accurate and entirely worthwhile popularization.[67]

If Lay emphasized the tractability of the unconscious, Edwin Bissell
Holt altered Freud's view of reality. He accomplished this by changing
the nature of instinctual drives and by choosing examples of the successful solution of innocuous conflicts.

Holt's *The Freudian Wish and Its Place in Ethics,* published in 1915
and 1916, was one of the most popular early books on psychoanalysis. It
was addressed not to the mass public but to the cultivated reader, and
was so well known among New York intellectuals that the young publisher B. W. Huebsch was inspired to stake André Tridon to the first of
his deliberately provocative popularizations of psychoanalysis that began
to appear in 1918.[68]

The reason for Holt's success is not far to seek. He attacked civil and
religious authority and social convention—the major sources of "ethics from above." Instead, he proposed a morality of hard facts, an "ethic
from below." He saw Freud as a classical moralist and a scientific behaviorist. Holt's work sums up several major themes of this study—the
relation of neurology to psychology, the American search for a "scientific" ethic, the ambiguous appeal of psychoanalysis, at once iconoclastic,
yet deeply conservative.

Holt's father was a Congregational minister at the Bangor Theological Seminary and died when Holt was a boy. Holt received his A.B. and
a doctorate in psychology from Harvard, where he encountered the deepest influence of his intellectual life, William James. He became an intimate of the James family and the household of Morton Prince, the Boston neurologist. Holt was also close to the young psychiatrist Elmer
Southard, the physiologist Walter B. Cannon, and the philosopher Ralph
Barton Perry, James's intellectual biographer. Holt was assistant professor of psychology at Harvard from 1905 to 1918 and supervised much
of the work in the psychology laboratory. Inspired by James's essays on
radical empiricism, Holt and Perry founded the "neo-realist" movement
around 1912. They argued that all perceived objects of consciousness, including such secondary qualities as intensity, really existed "out there,"
and were not artifacts of the perceiving mind, as the idealists supposed.

Holt attended the Clark Conference, and, one of a pride of American
psychologists, was photographed along with Freud and the visiting psychoanalysts. In 1910 Ernest Jones suggested Holt as a member of the
proposed American Psychoanalytic Association.

The basis of Holt's interpretation of Freud was established in 1914 in *The Concept of Consciousness,* which reflected major changes in American psychology since the 1870's. Like Putnam and Meyer, Holt rejected psycho-physical parallelism and defined consciousness in behaviorist terms as the "specific responsiveness" of the organism to its environment. Holt also reflected the revolution in definitions of the will that had occurred since the passing of Scottish Common Sense philosophy. Instead of a disembodied function whose action proved the direct and instantaneous impingement of mind on matter, the will was identical with desires, yearnings, and purposes. These were as objectively verifiable as the operations of the nervous system. One could apprehend one's own purposes by reflection and perceive other people's by what they did and said.

In *The Freudian Wish* Holt argued that in the wish Freud had provided psychology with "the first key which psychology has ever had which fitted and . . . the only one that psychology will ever need." This achievement was making "established professors" look "hopelessly incompetent."

Holt dismissed the "ill-founded" and "widespread" rumor that to read Freud's own works would be "fairly to immerse oneself in the licentious and the illicit." "It is true," he conceded, "that the unco prudish would experience a *mauvais quart d'heure* if they ever permitted themselves to read Freud on the source and significance of prudishness, but it is also true that the pruriently curious would be baffled to the point of tears if they were to search in Freud for a stimulus to their own peculiar type of imagination."

Assuming that he was drawing out the central implications of Freud's position, Holt equated the good with psychological integration and the moral sanction with "facts." He defined Freud's wish as an "impulse, tendency, desire, attitude, a course of action which some mechanism of the body is *set* to carry out, whether it actually does so or not." [69]

Holt ignored almost entirely the nature of the internal "need" that activated the wish. He did not link it with instinctual drives but referred vaguely to "nutritive and sexual" and possibly other "elemental appetites." He also omitted the regressive, hallucinatory mode of satisfying wishes.

Yet, he purported to take full account of the unlovely aspects of human nature Freud had revealed. "Freud has amply demonstrated," he wrote, "that 'possession by devils' is not merely a literary figure of the New Testament." Unhallowed motives smouldered in the heart.

Yet, because "higher wishes" were internalized in the conscience or censor, the "dark submerged wishes"—even if they were realized only in dreams—aroused the "upper personality" to "horror and remorse." Such conflicts could be resolved by conscious discrimination. One need not suppress wishes, but control them by a calm factual analysis: "Right is that conduct attained through discrimination of the facts, which fulfills all of a man's wishes at once, suppressing none. The moral sanction is the fact." [70]

Holt's argument depended on examples. He resurrected Meynert's famous diagram of the baby and the candle to illustrate the neurological learning process. One mother anxiously prohibited her child from even coming near enough to feel the candle's warmth. This illustrated "suppression." The child associated the prohibition not with the flame but with his mother, and so he developed an irresistible desire to play with fire when she was absent. An enlightened mother allowed her child to put his hand toward the flame; he naturally withdrew it when he felt its heat and thus learned his lesson, not by suppression but by learning the fact that fire was hot. Nevertheless, a boy must learn from others, and not from his own experience, that tobacco is harmful.

More instructive was the case of a young woman raised to condemn the theater as a place of abominations. She came to the city and learned that her new friends attended plays. She had these alternatives. She could resist temptation, cut herself off from friends, and be doomed to a nervously diseased spinsterhood. She could suppress the precepts of home, indulge in every pleasure, and die a drunken prostitute. Or she could give in to the impulse yet despise herself. But there was another way: she could attend good plays. Here she suppressed no wish; she followed out her "best possibilities" and became neither a "careless butterfly nor a grim ascetic." [71]

Each of Holt's examples offered one alternative that was both rational and desirable. But what if the wish were murder or adultery? Such examples were conspicuously absent from Holt's repertoire of factual discriminations. Even more revealing of his unexpressed assumptions were examples of what happened to real wickedness. In these cases something very much like Supernatural Judgment intervened:

> We often think of the wicked as supremely young, just taken in some
> act of dashing highwaymanship, and while envying them their youthful

vigor we forget that they are true cases of arrested development. In short, the picture we make to ourselves is of the brief heyday of wickedness, and we fail to see that this very wickedness reveals a now arrested integration, and that the next phase will be a fearful display of suppressions, anxieties, and mental incapacity (and all this apart from any artificial penalties which the wickedness may incur.) [72]

By omitting the uncongenial essentials of Freud's view of instincts as well as the repetitive and almost hallucinatory nature of wishes, and by limiting "reality" to congenial examples of wise choice or automatic punishment, Holt made the solution to every ethical problem seem simple and easy. Reality, then, to accommodate Freud's revelation of morbid wishes, was moulded to do away with their consequences.

The Freudian Wish recorded a vogue of parlor Freudianism already flourishing. Holt described a party at which a man with a penchant for stretching the truth innocently began to describe his dreams. Another guest interrupted: "I beg your pardon, Sir, but since your are not a Freudian, you are unwittingly making the most intimate revelations. I do not wish to be an eavesdropper, even in such a way." [73]

Putnam, who believed in innate ethical impulses, judged Holt's ethics inadequate to cope with the "warmth and richness" of "life's conflicts." The *Nation's* anonymous reviewer, possibly Warner Fite, dismissed Holt's morality of discrimination as merely "what all the self-realizationists, both in ethics and education, have been saying for the generation past. We need not go to Freud to be assured of this." Moreover, Holt's behaviorism denied any value to one's own subjective view of one's purposes. John B. Watson argued that in preserving such terms as "subconscious" or "thought" Holt was at heart a subjectivist and not sufficiently behavioristic. Watson's worst strictures were reserved for Holt's "unduly extravagant" praise of Freud. Freud's wish, Watson argued, was a mystical entity, "never defined," whether in "terms of psychology, biology or the well known principles of habit formation." [74]

The most significant criticism, however, came from a young intellectual better informed about Freud's meaning than Holt. Alfred Booth Kuttner, a friend of Walter Lippmann, had helped Brill with the translation of *The Interpretation of Dreams*. In the *New Republic* he accepted Holt's behaviorism but rejected his ethical formula: "The wishes of the unconscious are boundless and often mutually exclusive," Kuttner wrote. Not even the most searching discrimination of the facts would enable a man to fulfil all his wishes at once. "Freud has repeatedly shown that the whole progress [sic] depends upon the suppression of certain fundamental wishes which if indulged would always keep us at the level of savagery. . . . Involuntarily we speak of the war as a bad dream until we suddenly realize that the dream, as Freud interpreted it, is the prototype of an ideal unmorality, because in it, though only in fancy, we do fulfill all our wishes at once, suppressing none." [75]

A New Moral Guide

That Freud's insights had been confirmed by the Great War also had occurred to a few other Americans. In 1917 the psychologist Joseph Jastrow argued that the "baser side of living assumed in Freudian psychology" was reflected in the "horrors of murder, rapine and pillage . . . perpetrated daily by persons who a few years ago were at home a body of tender-minded citizens. . . ." The psychiatrist Elmer Southard dismissed Freud as an example of darkest German pessimism. It was Freud's gloomy view of human nature that chiefly interested the Chicago *Daily News,* which insisted that such qualities were found in Germans alone. Yet, the image of Freud as the proponent of a disturbing and pessimistic view of human nature was held by relatively few Americans. [76]

To a larger group Freud was primarily the proponent of new and daring theories of sexuality. By far the largest number of Americans— the millions of readers of *Everybody's, McClure's, Ladies' Home Journal, American Magazine, Good Housekeeping*—knew Freud as the creator of a new scientific miracle of healing that had vague, yet insistent sexual elements.

Despite the broad dissemination of some knowledge of psychoanalysis, books by Freud and by the American analysts were read by very few people. No psychoanalytic book sold more than one-tenth of the 900,-000 copies that qualified a book as a best seller in the period from 1910 to 1919. Freud's most popular book in translation was *The Interpretation of Dreams,* of which only 5250 copies were sold in Great Britain and America between 1910 and 1919, and another 11,000 copies between 1920 and 1932. *Totem and Taboo* sold less than 1000 copies between 1914 and 1931. Freud's first work that became relatively popular was *A General Introduction to Psychoanalysis* published in America in 1920. However, *The Basic Writings* sold about 174,805 copies in the decade after it first was published in 1938, the date from which Freud's real popularity can be said to have started.

The American analysts had slightly larger but still modest sales: about 3810 copies of White's *Mechanisms of Character Formation* between 1916 and 1926; about 8750 copies of Brill's *Psychoanalysis: Its Theories and Practical Application* between 1912 and 1931. Wilfred Lay's *Man's Unconscious Conflict* sold 10,000 copies between 1917 and 1927.[77]

The popularizations, especially those in mass magazines, demonstrate the eager acceptance by laymen, some of them patients, of the analysts' new role and self-image. Because of the arresting and novel subject matter of psychoanalysis—sexuality, dreams, childhood, etc.—Freud came to symbolize all the new developments in medical psychology. It was Freud who became most closely identified with the shift from a somatic to a psychological style. Havelock Ellis observed in 1918 that instead of the close description of physical symptoms that characterized older methods, Freud had made hysteria a human mystery to be patiently investigated. Its "key often lay far back and forgotten in the patient's history, and when skilfully used, with knowledge and insight, the patient's medical history acquired not only psychological significance, but something of the interest of a novel. . . . Freud's art is the poetry of psychic processes which lie in the deepest and most mysterious recesses of the soul." [78]

The popularizations suggest a search for a new moral guide and a new style of ethic, perhaps, indeed, for a new substitute for religion. The popularization of psychoanalysis coincided with a long term decline in interest in what has been called "traditional Christianity," a decline well under way among intellectuals that was spreading to larger groups of Americans. One aspect of this is clear in the popularizations—a rejection of absolute moral judgments and of asceticism, both associated with Protestantism. All the major popularizers were in rebellion against this moral style, which they condemned as morbid denunciation.

Yet, there were religious overtones to their version of psychoanalysis, and it is no accident that many of them were from deeply religious homes. Psychoanalysis readily lent itself to this interpretation. It was perhaps the only new medical psychology in which moral conflict was an essential part of the theoretical structure. Psychoanalysis dealt with "suppressed desires" not by blotting them out by suggestion but by "facing" them. It insisted on the universalization of "cravings," much as

Protestantism once insisted on the universal nature of sin. This conviction was won through a therapeutic, cathartic confession, a highly un-Protestant device. But the method was less important than the attitude of the confessor.

Over and over again the popularizers emphasized the benevolence and compassion of the therapist toward the patient's confession. Insight and forgiveness take on special significance in view of the massive changes in American sexual behavior that were just beginning and that will be examined in the conclusion to this study. The analyst provided an ideal new guide for a period of changing mores, a guide who would assuage rather than exacerbate guilt.

The popularizers also were careful to insist that psychoanalysis was not an automatic remedy for any kind of wrongdoing. Max Eastman caught precisely the ambivalence of the new spirit: sin was unseemly, but repressed desires, remediable by psychoanalysis, were most likely to give rise to it: "I do not want to give the impression that a man can sit up to the table like a pig, or swill liquor till the small hours of the night, and then go and be cured of his deserts by a psychoanalyst. This treatment, like all others, has its own field of application; it has its limits, and they are as yet undetermined. I would be willing to say, however, that a great many people who do sit up to the table like pigs, do this just because they are driven on by repressed desires in the unconscious." [79]

The psychoanalysts were pleased by the extent to which Freud's theories were becoming known in the United States. William Alanson White observed that *Mrs. Marden's Ordeal* was "another concrete evidence of the extent to which the psychoanalytic idea is filtering through the social fabric. We hear it mentioned on the stage, we see it referred to in the short stories in the magazines, and here is a novel which incorporates it. It really looks as if the psychoanalysts were coming into their own [and] gaining steadily that recognition . . . which is due." Introducing Frink's *Morbid Fears and Compulsions,* James Jackson Putnam argued that Freud was so familiar that "you feel a sense of lack, if on looking through a volume or a magazine where human motives are discussed one does not find some reference to the doctrines here at stake." [80]

Even sympathetic observers found this popularity disquieting. The anthropologist Edward Sapir deplored the "not altogether healthy over-popularity of the subject, while assessing one of the reasons for it: 'Psychoanalysis takes hold of chunks of mental life as they present themselves in experience for the purpose of clarifying them and examining them under the microscope. . . . Whitmanesque poets sing poems to Jung's libido . . . while half-baked doctors fearlessly disentangle homosexual complexes at the end of a first half-hour's consultation with hysterical patients.' " [81]

Freud was distressed by his growing popularity in America, a discomfiture that waxed in the 1920's. The spread of psychoanalysis in America, he came to believe, indicated knowledge that was not profound, critical, or accurate. "America is a mistake; a gigantic mistake, it is true, but none the less a mistake," he once wrote to Ernest Jones. [82]

XVI

A New Model in Neurology and Psychiatry 1909 - 1918

In the 1920's Morton Prince, the gentlemanly founder of the Boston school of psychotherapy, would write: "Freudian psychology had flooded the field like a full rising tide, and the rest of us were left submerged like clams buried in the sands at low water." [1]

What accounts for the success that Prince so ruefully described? The major reason, it has been suggested, is that psychoanalysis became important in America during the crystallization of crises in two subtly interconnected fields—the social realm of sexual morals and the professional realm of the treatment of nervous and mental disorder. This chapter will examine Freud's influence on the crisis in neurology and psychiatry, and the next the social conditions that created a unique resonance for Freud's message.

Psychoanalysis achieved two results in neurology and psychiatry. First, it provided a new etiological model and a new style of treatment. Second, it strongly reinforced already existing trends in the reconstruction of these specialties. Confined to a small avant garde in 1909, these trends by 1918 had influenced a rapidly growing and powerful minority in both professions.

The reasons for this professional success are clear. Psychoanalysis resolved central issues in the crisis of the somatic style and filled important gaps in the theory and practice of American psychotherapists. In 1909 no single method dominated American practice. Although important beginings had been made, especially in Europe, no coherent theories linked sexuality, childhood, and family relationships, although these seemed an inextricable part of the raw data of nervous and mental disorder. Although there had been increasingly psychological interpretations of the psychoses, the theories of Adolf Meyer, for example, were tentative and vague before his contact with psychoanalysis. Neurologists, and to a considerable extent psychiatrists, were troubled over the nature of their social role and the techniques they legitimately should employ.

The Impact Within the Psychoanalytic Movement: A New Model of Nervous and Mental Disorder

The way in which psychoanalysis resolved the crisis of the somatic style can be observed most clearly among those associated with the movement, whose militant convictions were the major factor in the diffusion of the psychoanalytic model. The psychoanalysts treated as one converging novelty the theories of Kraepelin and the psychopathology of Janet, Prince, and Sidis. Whether deliberately or not, they tended to identify all these tendencies with psychoanalysis and with Freud. It was he who had moved from mere "description" to "cause" in psychiatry, who had replaced despair with hope, and who had founded a truly scientific method. Partly through this insistent propaganda, Freud has come to symbolize all these trends, just as he came to symbolize, rather more accidentally, pervasive changes in attitudes toward sexuality.

The year 1914 marked a turning point in the confidence of the psychoanalysts. Already they had published major texts and had founded a journal, as well as two organizations with distinguished members. They insisted they possessed a scientific technique of investigation and treatment that could be learned only by careful apprenticeship. For them psychoanalysis functioned uniquely as a therapy, a psychology, and a critique of society. At the annual meeting of the American Medico-Psychological Association, two Philadelphia neurologists, Francis X. Dercum and Charles W. Burr, arraigned the "Freudian sect." Psychoanalysis was the unscientific moonshine of madmen. The true hope for psychiatry lay in biochemical studies of "auto-intoxications, metabolism . . . the internal secretions." [2]

William Alanson White and August Hoch replied with surprising mildness. A careful investigation of the psychoses, Hoch insisted, corroborated "the most fundamental facts of Freudian psychology, namely the existence of unconscious infantile motives . . . identical with those . . . of the neuroses." White asked his colleagues for scientific open-mindedness and reminded them of Copernicus and Galileo. Psychoanalysis was a method, and "as a result of the method, certain things have been uncovered. . . . I ask the people who criticize the movement to come forward and tell us what these things mean. We offer our explanation; we are willing to withdraw if we are wrong." [3]

Two problems must be disposed of before giving an account of the elaboration and diffusion of the psychoanalytic model. First in both neurology and psychiatry major elements still were unscientific and could not be resolved by existing techniques—a situation almost as true today. For this reason the psychoanalytic style was a mixture of somatic and psychological constructs like the classical somaticism that preceded it.

Second, the model must be defined by what the Americans said it was. The relation of their version to Freud's own complex and developing views already has been explored. Recently psychologists and some psychoanalysts have attempted to elucidate a series of orderly propositions that they believe constitute the central, most important elements of Freud's theory. To project this recent conception back into the early years of the psychoanalytic movement is to create a violent anachronism.

Freud was deeply interested in developing a satisfying and consistent set of propositions that did justice to his clinical findings. But, as he wrote to Putnam in 1909 and as he repeated again and again, he considered his own theoretical formulations to be less than sacred. They were incomplete, and they were to be discarded or modified when they proved inadequate in the light of new material.[4]

Psychoanalytic orthodoxy hardened after the splits with Jung and Adler, the founding of the *International Journal of Psychoanalysis* in 1920, and the beginning of formal training institutes. By 1930 the institutionalization of the movement had become relatively fixed.

Up to 1917 psychoanalysis was more flexible, more experimental, and more open. Its adherents were self-chosen, not winnowed candidates who successfully had completed analysis and passed boards of review.

Because of this openness and experimental character, it is important to survey the conceptions physicians actually *thought* were "psychoanalytic." Because psychoanalysis relieved and explained conditions previously believed to be hereditary and somatic, it swung the emphasis to environmental and psychological factors, thus facilitating a new interaction of mind and body. What were the major tenets of the model the first American analysts constructed?

First, all nervous and mental disorders, except those with definite somatic causes, were to be interpreted according to the model of the psychoneuroses, i.e. hysteria and the obsessions. Psychic determinism made possible an understanding of the psychoses on the basis of the neuroses. Because these disorders could be cured by psychoanalysis, so, too, psychoanalysis could assist recovery in the psychoses. This proposition was new, and its closest precedent had been Adolf Meyer's emphasis on habit training.

Second, nervous and mental disorders were caused by conflicts between wishes, the result of instinctual drives and internal repressing forces, the products of environment and education. Third, these causes could be traced to the patient's earliest childhood and involved powerful emotional relationships, usually sexually tinged, with parents and siblings. Fourth, sexuality was the most important of all the instinctual drives and developed in distinct stages from the earliest years. Fifth, drives, conflicts, and repressing forces operated unconsciously. Finally, nervous and mental disorders must be attacked psychologically, by psychoanalysis. This new technique involved free association, overcoming resistance to the verbalization of instinctual wishes, and the transference of early affective relationships to the physician. All these propositions created a new role for the physician, whose major treatment was now psychoanalysis.

The influence of this model on physicians who either joined or were sympathetic to the movement can be traced in three influential medical texts written between 1907 and 1915: William Alanson White's *Outlines of Psychiatry,* one of the most popular short texts in American psychiatric history; *The Modern Treatment of Nervous and Mental Diseases,* edited by White and Smith Ely Jelliffe and published in 1913; and White and Jelliffe's *Diseases of the Nervous System, a Textbook of Neurology and Psychiatry,* published in 1915 and reprinted in 1917. Each self-consciously included the "latest" developments.

The Modern Treatment of Nervous and Mental Diseases, a two-volume, 1652-page compendium, assembled distinguished contributors, such as Havelock Ellis and Henry Goddard, one of the nation's foremost experts on the feeble-minded and the new Binet intelligence tests. Even more important were contributors just rising to eminence in 1913 who would be among the major influences in American neurology and psychiatry for the next generation: Adolf Meyer, professor of psychiatry at the Johns Hopkins University and director of the Phipps Clinic, the first American psychiatric teaching hospital on the European pattern; William Healy, head of the Juvenile Psychopathic Institute in Chicago, who became the nation's foremost, indeed its first, psychiatric expert in juvenile delinquency; Bernard Glueck, a young assistant physician at St. Elizabeth's Hospital, who would shortly publish the first intensive American psychiatric studies of crime and prisons; Elmer Southard, a brilliant young professor of neurology at Harvard, who directed the new Boston Psychopathic Hospital. Thomas W. Salmon, then working in the immigration section of the United States Public Health Service, would establish the psychiatric services in the American Expeditionary Forces in the Great War. C. Macfie Campbell, a Scottish psychiatrist, taught at the Johns Hopkins Medical School and later at Harvard.

To illustrate the impact of the psychoanalytic model, the neuroses will be taken up first, and next the psychoses, as they were understood in 1907 and then as psychoanalysis progressively modified these views. In 1907, full of enthusiasm for Boris Sidis, Pierre Janet, and Emil Kraepelin, White, superintendent of St. Elizabeth's Hospital, wrote *Outlines of Psychiatry,* probably for the classes he was teaching at Georgetown and George Washington Universities. He divided the neuroses into three major categories—hysteria, neurasthenia, and psychasthenia. The symptoms of hysteria included anaesthesia or hypersensitivity of the body, paralyses, dream states, deliria, amnesias, all of mental origin. Adopting Janet's contributions, White argued that hysterics suffered from a narrowing of the field of consciousness. They had split off, or dissociated from ordinary awareness, painful emotional shocks. These forgotten subconscious ideas were dynamic and could grow until they became organized subsystems and, sometimes, second personalities. Hysteria was expecially a disease of the young associated with growing awareness of sexuality. Treatment consisted in restoring the continuity of the broken associations by psychotherapy, usually administered in Sidis's hypnoid or semi-hypnotic state. The split-off contents were linked once more with conscious awareness. Neurasthenia White described almost as broadly as had George Beard forty years earlier. Either inherited or acquired, neurasthenia was characterized by such mental and physical symptoms as pervasive fatigue, emotional irritability, headache, backache, depression, hypochondriasis, an inability to concentrate. Rational psychotherapy and the Weir Mitchell rest cure were the most effective treatments.

White adopted Janet's new syndrome of "psychasthenia"—intense nervous disorders on the border of insanity, characterized by a uniform lowering of "psychological tension" and a lessened ability to perceive reality. These patients suffered from obsessions, compulsions, morbid desires, chronic doubts. The causes of psychasthenia were hereditary and the prognosis was poor. In this complex nervous disease, treatment required "the most detailed regulation and re-education of the mental life." [5]

In the 1911 text White's outlook had somewhat altered, chiefly in his exposition of hysteria. Now he introduced Freud's theory alongside Janet's and treatment now included Freud's "cathartic method." Psychasthenia was much as before, except that White followed Freud in separating out compulsion and anxiety neuroses from psychasthenia. The prognosis remained poor.[6]

Freud's revolution in the classification, prognosis, and treatment of the psychoneuroses was fully reflected in the 1913 White and Jelliffe text. Ernest Jones contributed a brilliant, exhaustive survey of all the new methods of psychotherapy. With customary lucidity and belligerence he described the neuroses as among the most widespread illnesses, in some respects more distressing than "most forms of bodily disease."

Jones limited the symptoms of neurasthenia to inordinate fatigue, difficulty in concentration, a "sense of pressure on the head," an "irritable spine," various paresthesias. Masturbation and emotional and moral conflicts about it were the cause. Psychotherapy was the best treatment. Again, following Freud, Jones classified anxiety neurosis as a separate syndrome. It resulted directly from unrelieved sexual tension and also was best treated by psychotherapy.

Finally, Jones tore apart Janet's psychasthenia as a chaos of miscellaneous symptoms, and thus destroyed the largest single group of psychoneuroses attributed largely to heredity. The most severe cases in Janet's category, Jones argued, actually were victims of dementia praecox or manic depressive insanity. The rest suffered from anxiety, hysteria, or obsessional neuroses, whose origin was chiefly environmental. For these conditions psychoanalysis offered the most efficient treatment. It might take from four months for an "average case" to a year or even more in exceptional cases. But the results were likely to be permanent.[7]

Accepted views on the etiology and treatment of the insanities changed far less radically. Nevertheless, new methods of exploration were opened, and psychoanalysis reinforced a program of prophylaxis and intensive treatment. Again, this impact can best be illustrated by White's texts of 1907 and 1911 and the compendium of 1913.

By 1907 White already had learned from Meyer and Sidis to study intensively the individual patient's character, "mental make-up," and life experience, a study White called "psycho-analytical." Like Meyer, he argued that insanity was not a disease entity and that it resulted in a "lessened capacity for adaptation to the environment." Otherwise White remained fairly orthodox.[8]

He followed Kraepelin's classifications for the insanities and kept the traditional underlying causes—alcohol, syphilis, and, in 90 per cent of all cases, heredity. Nevertheless, White insisted that something more than these predisposing causes had to operate. "Exciting" causes such as illness, injury, mental and emotional strain or shock also were necessary for insanity to occur. Paranoia and dementia praecox resulted from hereditary predisposition and were incurable. A tendency to manic depressive psychosis was directly transmitted, and ultimate recovery was unlikely. The outlook was relatively gloomy and constricted.

By 1911 psychoanalytic studies already had had a considerable effect. For the first time White introduced Freud's theory of psychic determinism. Every emotional or mental symptom in insanity, regardless of its underlying organic causes, also had its own appropriate psychic cause. Without making clear the usefulness of psychotherapy or psychoanalysis for the insanities, White now suggested that they be used for illnesses in which mental causes operated. For the first time, he denounced suggestion therapy as superficial. He defined psychotherapy as a process in which the physician took the patient into his confidence and pointed out the "exact mechanism" that caused his illness. This usually was an unsatisfactory attempt to satisfy an impossible longing or desire.

Psychoanalysis was far more effective than all previous psychotherapies, and the patient's role was central. During an initial interview the physician exhaustively probed the patient's entire life and the history of his illness. Probably certain problems would stand out for further investigation. Then the patient was left alone with the physician in a quiet room; a monotonous noise, such as the "buzz of a faradic coil," would induce relaxation. The patient was asked to associate freely to those points the initial interview had disclosed. Thus, each symptom, as well as previously undiscovered material, would be dealt with, and the "submerged" complexes would emerge. Since Freud had proven that the mechanisms of dreams and symptoms were alike, dreams too must be studied by the same associative technique. If the patient were blocked, association tests with a stop watch might be tried. The process was long and difficult but effective. Above all, the physician should take an uncritical attitude and relieve the patient's tendency to blame himself.

White altered only slightly his earlier description of etiology. Significantly, the statement that heredity was a factor in 90 per cent of all cases of insanity was dropped. To paranoia he added a new category of mild, recoverable cases that were directly caused by mental conflicts. He also included Hoch's recent studies of the "shut-in" schizophrenic personality and Jung's theory of complex formation. The prognoses, however, remained about the same.[9]

The 1913 text registered a mild revolution in the etiology and treatment of insanity. White argued that psychoanalytic cures of hysteria destroyed the argument that environment was unimportant. Just as tuberculosis could be prevented and helped, so, now, perhaps, could some of the insanities. Dementia praecox, which he had described as an almost hopeless illness six years before, might be "favorably modified" or "prevented by early recognizing the danger signals" and minimizing environmental stress.[10] C. Macfie Campbell argued that Kraepelin's theory of an organic cause for dementia praecox was absurdly vague and that Kraepelin had ignored all the relevant studies "bearing on the actual meaning and development of the delusions, hallucinations and bizarre activity of patients." Campbell suggested a therapy best described as an attempt to give the patient gradual insight in an atmosphere of "warm enthusiasm." Above all, the physician must try to help the patient accept his sexual impulses to which many symptoms were closely related. The therapist explained in a broad way Freud's theory of childhood sexuality and tried to persuade the patient to take a robust common-sense attitude. The bogies of these early years were to be brought into the "sunlight of clear, honest thought. . . ."[11] It was extraordinary, Campbell noted, how much relief patients obtained from this kind of treatment. It helped in a "large number of cases" to modify or remove symptoms. The article on manic depressive insanity was conventional and took no notice of Karl Abraham's recent psychoanalytic forays into this new frontier.

Adolf Meyer, with typical caution, ventured a slightly more sanguine prognosis for paranoia. He examined at length all the psychoanalytic studies and suggested that in exceptional cases Freud's method could be tried. Most important, physicians were getting some insight into the way the disease developed and its symptoms were formed. No "law of prognosis" could be laid down, but the disease had a constitutional basis and was likely to be relatively incurable. Nevertheless, it was no longer quite the hopeless illness it had been six years before in White's first text.[12]

By far the most optimistic contributor to the White and Jelliffe symposium was James V. May, a state hospital superintendent, born in Lawrence, Kansas, who had served with the American Army in the Philippines. "It is only recently that the value of early supervision in mental diseases has received proper attention," he wrote. "The researches of Kraepelin, Freud, and Jung, and the investigations of Meyer and Hoch, show that under proper guidance and control many cases of functional nervous disease and insanity are of such a nature as to be manageable and preventable. Of the manageable and preventable conditions, dementia praecox, manic-depressive insanity, involution melancholia, and perhaps paranoic conditions are the most hopeful. . . ." "Mental conflicts must be brought to the surface and adjusted without brooding and introspection; sexual problems and relations must be disposed of fearlessly. . . ." "Early advice," he argued, "might have prevented an undetermined number of cases." [13]

Thus, within a few years, psychoanalysis had profoundly changed the attitude of a few neurologists and psychiatrists. It offered them a comprehensive, genetic psychological theory in which the more accessible psychoneuroses were, as Ernest Jones wrote, an "indispensable path to the psychoses." [14] Psychoanalysis was itself a therapy for the former and provided a basis for investigation and treatment of the latter. Finally, it was associated with a social critique, chiefly of sexual morality. All these elements, implicit in Freud's lectures to the Clark Conference in 1909, had been developed by 1917.

The social criticism had long-range and immediate implications. White placed the new psychology, by which he meant chiefly psychoanalysis, among the contemporary movements for improving the environment. Laws were regulating child labor and sanitation in factories and tenements. Educational campaigns were bringing mental hygiene and psychiatry to the public. The new psychology was redefining the nature of man, altering the treatment of nervous and mental disease, and transforming education and the schools.

White fitted eugenics into this program of progessive reforms. Condemning sterilization, urging a compromise between the neo-Lamarckians and neo-Darwinians, he suggested that public sentiment for the marriage of the fit, "positive" eugenics, he called it, should be fostered by educational campaigns. Finally, White suggested that psychoanalysis, as well as the text on neurology and psychiatry, were intended not only for the physician but for the educator, legislator, judge and lawyer, the student of criminology, of immigration, of dangerous trades, the social worker, the military man, the hospital superintendent, the intelligent layman.[15] More grandly, Ernest Jones argued that psychoanalytic knowledge of the neuroses gave the physician a new, uniquely authoritative social role. He must become, of necessity, the paramount critic of civilization. Marriage and education, genius and crime, religion and philosophy, art and mythology, anthropology, history, philology, and human character all were coming within his purview. The reviews were generally favorable.[16]

The psychoanalytic model was elaborated more radically in White and Jelliffe's *Diseases of the Nervous System*. The term "diseases" indicates at once the mixed character of the new style. First, Jelliffe and White broke with the limitations of classical somaticism, what they called "sensori-motor neurology," with its description of "gross lesions of the brain, spinal cord, cranial and peripheral nerves." [17] Neurology had been transformed by an interactionism based on a growing knowledge of the sympathetic and the autonomic nervous systems, pioneered by Walter B. Cannon, George Crile, and others, and by a knowledge of "mechanisms operating at psychic or mental levels," i.e. the psychoanalytic model. White and Jelliffe attempted to reformulate the mind-body issue on evolutionary principles reminiscent of Herbert Spencer and Hughlings Jackson.

The nervous system was organized according to evolutionary principles from the older and simpler levels to the newer, more complex and more integrated. The simplest and oldest system was the "vegetative" or autonomic and sympathetic nervous system, just then being investigated; next, was the sensori-motor system of the brain and spinal cord, the traditional province of neurology; finally came the psychic or symbolic level, the province of psychoanalysis. Each level increased integration in the service of broadly defined goals for both the individual and the race. Laws of opposites—ambivalences—governed all three systems. The first system regulated growth, nutrition, and development; the second, the functions of the whole body. The psychic level integrated the individual as well as his relations to the social environment. This level carried energies expressed in symbols, and these energies could be transformed and translated into conduct. A disturbance in each system could affect the others: constipation, for example, could have an origin at either the physiological or psychic level.

The highest system must be investigated by psychoanalysis. It was as "important for the understanding of the construction of the psyche as dissection was for the understanding of the structure of the body, or chemical analysis for the understanding of the constitution of the molecule." The psyche had its embryology and its comparative anatomy—in short—its "history." This history equated with cause could be used to understand nervous and mental diseases in the human being viewed as a "biological unit."

Psychoanalytic investigation and treatment were recommended not only for the psychoneuroses but for mental diseases as well, from dementia praecox to manic depressive insanity. Moreover, psychological etiology, the "psychogenesis" that psychoanalysis disclosed, was more important than physical or hereditary factors. Rest, massage, electricity, exercise that once were the major modes of treatment—still regularly prescribed by Francis X. Dercum and other somaticists—occupied a very small place.

Dementia praecox became a psychogenic disorder. "So far as our observation goes, the etiological factors," they insisted, "lie almost, if not quite entirely in the mental sphere, and one must therefore conceive of the physical changes as superadded." [18] This was a more exclusively psychological interpretation than Adolf Meyer's.

At the "higher psychological levels" psychoanalysis was the only mode of treatment. Although it could not cure psychotic patients, as in the psychoneuroses, nevertheless "many of the symptoms may be largely relieved, if not dispersed altogether." Moreover psychoanalysis provided information that made it possible for the physician to deal intelligently with the patient. For psychoneurotics and psychotics alike, interviews in a quiet room alone with the physician were prescribed. By 1917 White had two full-time psychotherapists on the staff at St. Elizabeth's Hospital.

Thus the authors moved far toward a psychoanalytic psychiatry. Yet, as they realized, the psychoanalytic model also left unsolved issues in the psychoses. Just as classical somaticism had included subordinate elements of psychological and social factors, so now, heredity and somatic factors were retained in the new framework. The psychoanalytic model, for all its advantages, had not solved the scientific problem and could not, therefore, remain pure.

These considerations are explicit at critical points in the text. First, if the model adopted from the neuroses also held for the psychoses, what decreed that one patient would develop hysteria and another, dementia praecox? This was precisely the question Jung had raised in 1907. Second, was it possible to predict that a given kind of psychological conflict or make-up, such as the "shut-in" personality, would result in a distinct kind of mental disorder? Could an invariable sequence be established? The answer again was that it could not, and for one central reason. The exact nature of the relationship of psychological and physical had not been solved and even could not be stated precisely. After insisting that hard and fast lines could not and should not be drawn between mind and body in dementia praecox, they cautioned:

> In our present state of knowledge, however, we are quite unable to make any specific correlation between the physical findings and the mental symptoms, while on the other hand, it is quite possible to express the symptomatology of the disease, to describe it, to, so to speak, reconstruct the psychosis purely in psychological terms. For the present, therefore, the disease must be described psychologically, and the explanation of the mental symptoms must be sought psychogenetically, without, however, forgetting that there are certain somatic changes which are pretty generally attached to the symptomatology of the disease process and which must ultimately be made to fit into the general rubric before a complete understanding of the situation is had.[19]

This text, it is safe to say, was the most radical yet to appear in American psychiatry and neurology. This time the reception was more mixed. Elmer Southard judged the text suitable for third-year neurology students: it showed what some "leading neurologists" thought about psychoanalysis. Meyer Solmon, an anti-Freudian psychiatrist in Chicago, considered it "right up to the minute in every respect," although he disapproved of its frankly psychoanalytic approach. The *Boston Medical and Surgical Journal* decided it was judicious and balanced, and that the "description of the psychic origin of the neuroses is very striking and illuminating. . . ." [20] It was revised and reprinted two years later, and twice in the 1920's.

Thus White and Jelliffe had broken with major postulates of the somatic style. By applying and elaborating the psychoanalytic model they had moved further than they ever had before toward a functional psychiatry. Even heredity, which they kept as an etiological factor, was far less important than it had been for White in 1907, eight years earlier. The model dictated a new role for the physician whose chief mode of treatment had become psychoanalysis.

Some physicians learned of these new developments in medical schools in general courses in neurology, psychiatry, or psychotherapy and occasionally in special courses devoted exclusively to psychoanalysis. As John Burnham has suggested, catalogues often are uninformative, and more students probably were taught something about psychoanalysis than the record reveals. In addition to formal courses at the Universities of Colorado, Boston, Fordham, Michigan, Putnam taught one at Harvard Medical School, G. Lane Taneyhill another at the Johns Hopkins; there were others at the University of California Medical School and at the University of Nebraska. [21]

Probably the analysts included as much as they could. The active core of the movement taught at the major medical schools—Jelliffe at the New York Post Graduate Medical School and Fordham; Brill and Louis Casamajor at Columbia; Trigant Burrow at the Johns Hopkins; Frink at Cornell; White at George Washington. Uncommitted members of the American Psychoanalytic Association such as Adolf Meyer and C. Macfie Campbell also taught at the Johns Hopkins; Ralph Hamill at Northwestern; Alfred Reginald Allen at the University of Pennsylvania. Sympathetic non-members such as Sidney I. Schwab taught at Washington University in St. Louis; Edward Wyllys Taylor at Harvard; and C. C. Wholey at Pittsburgh.

The attitude of Adolf Meyer, the major influence in American psychiatric teaching, requires comment. Psychoanalysis strengthened his own functional approach by providing him with specific facts and hypotheses. In turn, the psychoanalysts used his conceptions in their new psychiatry. Jelliffe and White relied heavily on his theory of habit disorganization in recommendations for the treatment of dementia praecox. At. St. Elizabeth's Hospital, White's staff conducted research on the "habit training" of schizophrenic patients, partly on the basis of work by Edwin G. Boring, then a staff clinical psychologist.

By 1915 Meyer was much less sympathetic than he had been in 1910. Perhaps because of the increase of lay interest, perhaps because of contact with the behaviorists at the Johns Hopkins—Knight Dunlap and John Broadus Watson. In a discussion of psychology for medical students before the American Medical Association, Meyer relegated psychoanalysis to the domain of the "specialized and more or less unfamiliar" along with hypnotism and the unconscious. He hoped to discourage "stagy psychology" and the "ever-lurking interest in the occult and semi-occult." Instead he wanted "solid confidence in reliable methods" of "objective study." He urged students to "trace the plain life history of a person." What were his assets, his "affective and expressive activity," his "conations, affections, cognition, discrimination, reconstructive and constructive imagination?" [22]

The Psychoanalytic Model
Among Physicians Outside the Movement

Between 1909 and 1918 the number of interested physicians markedly increased across the nation from New York to Seattle. Most of them felt free to try out psychoanalysis as they wished, without reference to its theoretical structure or technique. Each element from infantile sexuality to free association could be interpreted and combined idiosyncratically. Some rejected the sexual etiology of the neuroses; others accepted it but rejected transference; still others rejected all the techniques, but accepted the mechanisms of conflict, repression, childhood trauma.

A few symptomatic pronouncements marked the stages of this disorderly assimilation. In 1915 L. E. Emerson, the Boston psychologist and psychoanalyst, remarked that psychoanalysis was just beginning to be "respectable." That year Julius Grinker, the Chicago neurologist who had been interested in psychotherapy before 1909, observed that even the enemies of this "virile" system were "unconsciously practicing psychoanalysis" while "protesting violently against Freud's over-accentuation of sex." Soon Freud was able to make one of his rarely favorable observations about America. "Leading psychiatrists" there, in contrast to those in Europe, were expounding psychoanalysis and studying patients in its light. In 1919 Horace Frink claimed that the initial battles for psychoanalysis had been won. It was no longer "worth one's life" to read a paper about it before a medical group. Yet, the really dangerous period was beginning, he warned—psychoanalysis had become too popular, too many physicians were practicing it without training or qualifications.[23]

One index of assimilation was the progress of psychoanalytic conceptions from unorthodox new ideas to familiar parts of medical explanation. In 1909 an article on psychotherapy described Freud's method as possibly useful for cases of functional disorder. In 1917 a physician not entirely sympathetic to psychoanalysis simply described hysteria as a means for the "fulfillment of unconscious wishes," many of them "unethical." [24]

The unusual productivity and the strategic position of some of the members of the psychoanalytic organizations were responsible for a large number of favorable articles, reviews, and abstracts in the two official psychiatric and neurological journals. Up to 1917 favorable material substantially outweighed unfavorable, in proportion of at least three or four to one. In the *American Journal of Insanity* from 1909 to 1917 there were at least twenty-four favorable references or discussions about psychoanalysis, and of these, twelve were articles by psychoanalysts. Of five unfavorable references, two were major articles. Books by Freud and other analysts were often, although not always, reviewed sympathetically, sometimes by psychoanalysts or their friends.

The official journal of the American Neurological Association was even more favorably disposed, probably because Jelliffe had been managing editor since 1902. The *Journal of Nervous and Mental Disease* ran fewer long articles and abstracts until 1913, when it printed a translation of Otto Rank's *Myth of the Birth of the Hero.* Again, reviews of psychoanalytic work were usually favorable. Jelliffe tended to praise or condemn other neurologists or psychotherapists in proportion to the amount of psychoanalysis they had absorbed. Thus, Boris Sidis's *Foundations of Normal and Abnormal Psychology* contained "nothing fundamentally new." Because Sidis denied that people repressed painful complexes, the book had "only limited bearing on the practical problems confronting the psychiatrist." [25] In 1916 the *Journal* began to include long abstracts from the *Psychoanalytic Review,* which Jelliffe also edited. It is no wonder that opponents protested that they were being called "back numbers" because they failed to accept Freud. Frederick Tilney, who succeeded Jelliffe as president of the New York Neurological Society, in 1917 asserted that no review of neurological endeavor was complete without a mention of psychoanalysis.[26]

Jelliffe's pro-psychoanalytic editing apparently lost the journal its status as official organ of the American Neurological Association. Charles L. Dana, C. K. Mills, Moses Allen Starr, Hugh T. Patrick, and Frederick Peterson—all by then more or less anti-Freudian— resigned from the advisory board; but Adolf Meyer, Lewellys F. Barker, Harvey Cushing, William Osler, Edward Wyllys Taylor, and Elmer E. Southard, all of whom were more favorably disposed, remained. In 1919 the *Archives of Neurology and Psychiatry* was founded as the new official journal. Articles about psychoanalysis appeared now and again in a wide variety of general medical journals, but chiefly in the *Boston Medical and Surgical Journal,* the *New York Medical Journal,* the *Medical Record* (New York).

Neurological and psychiatric texts were as conservative in relation to psychoanalysis by 1917 as they had been in relation to the new psychotherapy and psychopathology by 1909. Yet, even in texts, the psychoanalytic impact was significant and can be measured in two ways: first, by the discussions of psychoanalysis and recommendations for psychoanalytic treatment; second, by the adoption of a more functional and psychological approach, which resulted from the competition of the psychoanalysts.

Of thirteen texts reviewed in the major journals between 1909 and 1917, four were by psychoanalysts, notably those of Jelliffe and White. Psychoanalysis was discussed favorably in four more. It was criticized in five more, three by Philadelphia and two by New York neurologists.

The influence on those unfavorably disposed or highly ambivalent is even more significant. Often their major premises remained unchanged, yet, each moved to a more sympathetic consideration of psychopathology and psychotherapy. The case of Francis Xavier Dercum is noteworthy because he was one of the most outspoken opponents of psychoanalysis. In 1903 in a text on therapeutics Dercum had relied on orthodox somatic methods, with some use of suggestion, and had condemned hypnosis as merely the artificial creation of a hysterical state. Janet had no place in this conservative scheme. By 1914, however, Janet had become the creator of "brilliant," satisfying generalizations about psychasthenia. These embodied a "kernel of truth," a "modicum of actual scientific fact," instead of the "mystic methods" of the psychoanalysts. The source of this new-found enthusiasm can perhaps be explained by an entirely new chapter on "The Psychological Interpretation of Symptoms." It included a scathing denunciation of the new school in Vienna, as well as a brief description of Jung's association tests without comment, and a theory of "complexes" attributed in part to Janet, but in fact Dercum's own. A complex was a group of "associated ideas," either normal or abnormal. If complexes were painful, they might be "suppressed altogether." Dercum again denounced psychoanalysis and its loathsome sexual theories.[27]

It is reasonable to conclude that without the prodding of psychoanalysts Dercum would not have taken so readily to Janet, whom he had largely ignored earlier. The rest of his outlook in both the *Manual* and the text on therapeutics was primarily somatic, and a reviewer in the *American Journal of Insanity* criticized his reliance on hypnotic drugs.[28]

The attitude of Charles L. Dana, professor of neurology at the Cornell University Medical school, who taught several of the younger psychoanalysts, also changed in contradictory ways. His theories of etiology, including degenerate heredity, altered scarcely at all. He continued to insist that "psychic factors" were often "only an extra and minor element" in the neuroses.

In the 1915 edition he judged Freud's method more "poetical" than safe and "scientific"; the "subconscious psychic factor" had only "an academic and metaphysical importance." He relied on isolation, re-education, and the physician's skill in being simple, impressive, and sincere.[29]

In 1917 he argued that, while psychoanalysis was a "most valuable contribution to our technical methods of studying and treating mental disorders," it was also a cult, devoid of scientific standards. It was being practiced by laymen partly as a parlor game and by "inexperienced," perhaps not altogether honest physicians.[30]

In 1920, however, Dana's text included a chapter on medical or dynamic psychology. It would be inexcusable, he suggested, for a man surrounded by so many younger workers to omit this latest aspect of neurology. He briefly discussed instincts, complexes, conflicts, and repressions drawn from sources ranging from Freud to William McDougall. Although he cautioned that psychoanalysis was not always safe or adequate, he no longer described it as "poetic." Thus, within ten years he had moved from relatively orthodox somaticism to a serious consideration of psychological factors, forced on him, the context suggests, by the zeal of the young psychoanalysts he knew.[31]

Moses Allen Starr, who had denounced Freud publicly in 1912, also combined disapproval with assimilation. He was more sympathetic to psychopathology and psychotherapy than Dana. His text in 1909, however, included few references to Janet beyond the admonition to decipher the patient's "fixed idea" and enter into his "train of thought." In the edition of 1913 Starr included a larger section on Janet and a discussion of psychoanalysis, which he regarded as a mode of suggestion, like the Catholic confessional. Despite its errors, chiefly the sexual emphasis, it included elements of truth.[32]

Psychoanalysis also reinforced the psychogenic attitude toward the insanities of Church and Peterson. They were perhaps the most favorably disposed to psychotherapy and psychopathology of the major American textbook writers. Their 1908 text had included a section on Janet's psychasthenia and they insisted that treatment of the neuroses be chiefly "psychic." By 1911 Peterson argued that the old "moral causes" of an earlier psychiatry were being re-examined and rechristened "psychogenic." He included Jung's association tests and Freud's psychoanalysis as diagnostic methods. While denouncing the overemphasis on sexuality, he argued that Freud had opened up new fields for exploration and especially praised the interpretation of dreams. Thus psychoanalysis reinforced a direction in which Church and Peterson had been moving.[33]

Hospitals and Clinics

A few neurological and psychiatric hospitals and clinics also were taking up psychoanalysis. James Jackson Putnam was responsible for the appointment of the psychologist Louville Eugene Emerson as psychoanalyst at the Massachusetts General Hospital in 1911; [34] Lydiard Horton received a similar appointment from Elmer Southard at the Boston Psychopathic Hospital in 1914. Trigant Burrow, C. Macfie Campbell, and Adolf Meyer worked at the Johns Hopkins Hospital and the Phipps Clinic in Baltimore; Albert Barrett headed the Psychopathic Hospital at the University of Michigan and Douglas Singer the Illinois State Psychopathic Institute. Physicians at public mental hospitals in Kansas, Illinois, Washington, D.C., and Massachusetts contributed psychoanalytic studies. At St. Elizabeth's, for example, Mary O'Malley investigated

Negro patients' frequent dreams that they were white, and interpreted these as an "obvious wish fulfilment." Several other case studies from St. Elizabeth's dealt with the application of psychoanalytic concepts to dementia praecox.[35]

The New York State hospital system was the most heavily influenced. Adolf Meyer's course in dynamic psychology at the Psychiatric Institute was continued by his successor Adolph Hoch. At the Buffalo State Hospital Helen J. C. Kuhlmann studied the "once startling and novel" Oedipus complex in a number of patients.[36]

The role symbolism and repression could play in explaining psychotic symptoms and behavior was demonstrated in 1917 by a paper by Heyman L. Levin, a senior assistant at the Utica hospital. His patient had been routinely questioned during nine years of hospitalization for longer periods than her psychoanalytic treatment required, yet no one had elucidated any of the motives for her behavior. She was a forty-year-old spinster who made periodic attempts to escape in order to search for relatives who she insisted were alive, but who in fact had died. She had been brought up on an isolated farm and had nursed her father, whose favorite she had been, through a fatal illness. At first she seemed unaffected by his death. Then one morning "she awoke with the idea that her father must be alive and in the vicinity." Her delusion no longer seemed bizarre because psychoanalysis had demonstrated that it was motivated by the "quite understandable wish and fond hope that she would soon return to the life she lived with her father before his death." [37]

Levin's work was based partly on Jelliffe and White's 1915 text. He applied similar techniques to ten or twelve other chronic cases. Another patient had led a miserable life with her husband and had been unable to face the fact that he had abused her. She had "attempted constantly to repress or crowd out of her conscious mind all such thoughts, with the result that they reappeared in the shape of manic symptoms, attempting to kill husband, etc." [sic], "believing herself never married to him." After the psychoanalytic interviews she began to "take a more frank attitude toward the wishes she had for years repressed." She had been allowed to return home, had been separated legally from her husband, and had remained well in the two years since the separation.

Levin's study focused strikingly on those elements that usually had been omitted from both psychiatric and psychotherapeutic investigation —the nature of relationships with parents and spouses, childhood, and a motivation interpreted as a "wish." Here the psychoanalytic technique, based on an approach to mental disorder taken directly from the model of the neuroses, had made possible a new understanding of symptoms. This was an important result on the level of hospital practice.

The prolonged discussion of Levin's paper indicates the kind of opposition and support psychoanalysis provoked. August Hoch argued that its major contributions to psychiatry were "the wish as a dynamic factor in mental life, the discovery of motives of which we are not aware," and the indirect, distorted, transposed, or symbolic expression of ideas or tendencies. The Oedipus wish, for example, had been unknown before psychoanalysis, yet could be demonstrated in "case after case." [38]

A. J. Rosanoff, who had been applying Mendelian ratios to heredity in mental disease, had been repelled by the extreme claims and tactics of the psychoanalysts. But he had come to recognize psychoanalysis as essential to the "full development" of psychiatry. Psychoanalysis gave a more complete understanding than other methods and went to the heart of things "beneath surface appearances."

Charles R. Wagner, who just had denounced psychoanalysis as a "salted mine" of superstition in his presidential address to the nation's psychiatrists, argued that hydrotherapy was just as effective a treatment. If psychoanalysis were so scientific, why had it had such a hard time in the country of its origin? "The infant was born eleven years ago in Worcester, Massachusetts, and has quite outgrown its parent in Europe. . . ." [39]

A social worker, identified only as Mrs. Frank, spoke up to praise psychoanalysis on the basis of personal experience. It had benefited her, and she described its almost magical properties: "I . . . feel that everything that occurs in darkness and in repression becomes better when full light is put upon it, and that is the way psychoanalysis acts. . . . Perhaps Vienna and Paris have not accepted psychoanalysis as readily as we have in the United States, because they have more repression. . . ."

Charles W. Pilgrim, the chairman of the New York State Hospital Commission, who had allowed the debate to continue without adjourning for lunch, concluded: "I speak as one to whom the subject appeals very strongly, for my experience leads me to believe there is much good in it, and as Dr. Hoch says, Why discard it? Why not follow the Biblical injunction, 'to prove all things and hold fast to that which is good!' " [40]

Probably psychoanalysis made little change in the lives of most mental patients. Yet, here and there, a few patients treated by some version of it did recover. Between 1909 and 1917 psychoanalysis occupied a growing place in the interpretation of the symptoms of mental disorder and sometimes in their treatment. Yet, in 1917 of a total of about one hundred and thirty-five articles on psychiatry and insanity, perhaps six clearly indicated the assimilation of aspects of the psychoanalytic model. Psychoanalysis touched only a small minority of psychiatrists; most still were assimilating Kraepelin and utilizing traditional somatic approaches in diagnosis and treatment.

Psychotherapy

Gradually during the years 1909 to 1917 psychoanalysis came to occupy the most prominent place in the literature of psychotherapy and the treatment of the neuroses. The number of articles on hypnosis and suggestion listed in the *Index Medicus* dropped from one hundred and five in 1909 to six in 1917. Those on psychoanalysis rose from one in 1909, when it became a separate topic in the *Index,* to twenty-two in 1915 and twelve in 1917. If these figures are corrected to eliminate errors in classification, in 1917 about thirty-five articles on psychoanalysis were listed and twenty-five on psychotherapy.

Sidney I. Schwab, a St. Louis neurologist who had studied with Janet and Freud, contributed important testimony about the degree of penetration of psychoanalytic conceptions into neurological practice by 1917:

> Whether the [clinic or hospital] neurologist or his staff are accredited Freudians or not, there is to be observed a curious automatic insistence on the part of the clinical inquirers in regard to the presence of psychical traumata in the past history of the patient. This inquiry does not stop with the ordinary conventional questions, it drives back into the realm of the patient's earliest memories. It is surprising, indeed, to observe how many facts are in this way obtained and how much richer the personal story of the patient becomes.
>
> The conclusion seems irresistible, therefore, that the body of Freudian doctrine has penetrated largely into the general neurological mind irrespective of personal differences of opinions as to the correctness of many of its separate units.[41]

Schwab was a sympathetic critic of Freud. He did not join either psychoanalytic organization, yet continued to read the literature and discuss it with psychoanalysts. For some years he had been testing the methods of Freud, Janet, Babinski, and Adler and concluded that Freud's remained the most promising. Janet's theories were useful, but his therapy was pure hypnosis and suggestion.

Repression, wish fulfillment, infantile traumata, dream interpretation were conceptions of such practical value that they should be applied whether or not the physician were an "accredited Freudian." So valuable a body of doctrine "must contain sufficient truth to admit of partial belief and partial mastery." The techniques could be acquired through reading, contact with "well-known Freudian students and actual experience and study."

Psychoanalysis worked, Schwab insisted, in a contradictory passage, because it was a "determinate thing." It gave patients a "more luminous view" of their neuroses, regardless of whether a particular hypothesis were true or not: "The pertinent fact remains in the patient's mind that there is an explanation and not the mystery that before enveloped him and his disease. The cure or improvement of the neuroses is intellectual, as is the method used, and the great strength of the Freudian school in my opinion is due chiefly to this fact." [42]

In 1917 Edward Wyllys Taylor also argued that psychoanalysis was too valuable to be restricted by the dictates of Freud's theories. This was not a plea to *supersede* the orthodox technique, he wrote, but to broaden the application of a useful method. Even enemies of psychoanalysis were being affected "by its permeating influence in spite of their protestations." [43]

The New "Psychoanalytic" Style

Although the major centers of psychoanalysis remained on the eastern seaboard, chiefly New York, Boston, Baltimore and Washington, not only neurologists but general physicians across the country became increasingly interested. Psychoanalysis was more widely discussed in medical meetings, and a growing number of papers were devoted to it. By 1917 Freud's theories had achieved perhaps a wider currency among physicians in America than those in any other country.

At the most general level of this haphazard, yet surprisingly widespread assimilation, psychoanalysis meant a detailed investigation of patients' lives, almost identical with Adolf Meyer's study of the patient's life history. This inquiry was new in the context of the general questions physicians were likely to ask. The diagnostic blanks of neurologists, for example, had space for a rich variety of questions about smell, sight, eye movements, hearing, reflexes, heredity in ancestors near and remote, childhood diseases, menses, etc., but nothing about the patient's emotional life. On a more specifically psychoanalytic level the new inquiry included a search for trauma in the earliest years, perhaps elicited by direct questions.

Second, psychoanalysis supplied a conceptual model of the neuroses that also could be applied to the psychoses. This included that elastic term "the complex," meaning an organized constellation of ideas and affects. This had ancestors in American adaptations of French psychopathology, but came chiefly from Jung and the first essays of the American Freudians. Both neuroses and psychoses were caused by repression— still often defined as a half deliberate "crowding out" from awareness of complexes. These operated from the subconscious or the unconscious, terms used interchangeably to mean whatever was outside awareness. Painful conflict between instinctual wishes and conscience or custom was the major cause of repression. Unconscious complexes or ideational-affective components could be elicited by interpretation, although few agreed to a universal meaning for given symbols. Dream analysis, no matter how defined, had become a part of this interpretative process. Third, the complexes often represented wishes, usually but not invariably of a sexual nature. The wish provided an important teleological mode of interpreting behavior, especially the more bizarre behavior of psychotics.

The investigation of family relationships, often although not always seen in terms of the Oedipus complex, assumed a new and important place in diagnosis. Sexuality included adult sexual behavior, and sometimes but not as often Freud's theories of infantile sexuality. The investigation of childhood experience assumed major importance, especially early relationships to parents and regression to childish attitudes. Here the Oedipus complex could be interpreted as an intense attachment, not necessarily sexual, to the parent of the opposite sex.

Finally, psychoanalysis included techniques of investigation and treatment—word association tests, free associations, catharsis, the working-off of emotions that had been repressed or not consciously faced. Transference played a lesser role, and resistance almost none.

The elements of the psychoanalytic model that were absorbed by those outside the psychoanalytic organizations were close to the version

Freud had described at Clark University eight years before: trauma, catharsis, repression, conflict, wish fulfilment, and to a lesser extent infantile sexuality and the repressive nature of prudish customs. Usually these elements were not defined, except in a rough way, as Freud defined them, nor was the "psychoanalysis" these physicians practiced Freud's psychoanalysis. Yet, without Freud and Jung these conceptions would not have existed. They were new, and a number of them were gaining acceptance among dozens of neurologists, psychiatrists, and general physicians.

For many physicians psychoanalysis had resolved the crisis of the somatic style by forcing the development and acceptance of a functional neurology and psychiatry. Where the emphasis before had been primarily somatic, psychological and emotional factors had come to play an equal role.

The model of nervous and mental disorder was no longer general paralysis but the neuroses interpreted according to psychoanalytic conceptions. Neuroses were viewed as primarily the results of environment and experience and thus could be "cured." Many physicians and some psychoanalysts, including Freud, continued to give an etiological role to heredity. Yet diagnostic and therapeutic efforts now were directed primarily toward the interpretation of life experiences and family relationships.

Moreover, the major therapeutic techniques were no longer somatic but psychological, i.e. psychotherapeutic. The nature of psychotherapy also had changed. It was no longer hypnosis and suggestion but usually a hybrid form of psychoanalysis.

Telling evidence that psychoanalysis, by its militant radicalism, forced the issue of a functional approach emerged in the early 1920's. At a symposium before the American Neurological Association in 1921, psychoanalysis again was vehemently attacked by Charles K. Mills, a conservative Philadelphia somaticist, who ridiculed mysticism and the Oedipus complex. Yet Mills also admitted that psychoanalytic "propaganda" had directed the attention of physicians to a more minute study of sex and of "energic phenomena."

As if with a single reflex, Meyer and Prince defended the psychoanalysts for acting as effective crusaders for a functional and dynamic psychology. Prince recalled that before the campaigns of Freud's "enthusiastic followers" the American Neurological Association had paid scant attention to dynamic or subconscious psychological processes. As long as the American Medical Association *Journal* condemned "honest work on dynamic psychology," Meyer noted, just so long would laymen exploit psychoanalysis. Meyer defended the Oedipus complex against Mills's sarcasm. The conception touched "principles and fears" of a vital "life situation" as no other description had done. George Kirby insisted that mental disease could not be understood by studying "brain processes" and that psychoanalysis had thrown a "flood of light" on nervous and mental disorders. C. Macfie Campbell asked if any psychiatrist now studied the deteriorating psychoses without using Jung's *Psychology of Dementia Praecox.*[44] In paradoxical ways that cannot be discussed here the Great War also had facilitated the general change in paradigm.

Psychoanalysis had helped decisively to alter the structure of neurology and psychiatry: first among an avant garde of neurologists and physicians, next among a few general physicians, and by the 1920's and early 1930's among writers of textbooks who were not psychoanalysts.

The second piece of evidence is an unpublished lecture by Meyer to students at the Johns Hopkins University Medical School in 1924. Meyer astutely yet critically assessed four major reasons for the success of psychoanalysis.

Above all, Freud had created more communicable, "telling terms and formulae" organized in a system than any one else in the previous thirty years. These formulas dealt directly with the "pulsating heart of passion and conflict and were close to the "very core of life and fate. . . ." Precisely such meanings, which Edward Bradford Titchener had ruled out of scientific psychology, Meyer regarded as essential to any psychobiology physicians could use.

Psychoanalysis succeeded for another closely related reason. Many physicians still insisted that they could alter a patient's symptoms biochemically without dealing at all with his emotional problems. Yet the latter had to be faced, and this psychoanalysis did directly.

Meyer also argued that in the popular view psychoanalysis gave a "scientific vindication" of wish fulfilment. He apparently half believed that psychoanalysis taught the "absolute necessity of gratification," yet he also quoted Freud's essay in Marcuse's *Encyclopedia* of 1922 that included a sharp condemnation of sexual license. Clearly Meyer was referring to the popular psychoanalysis of the intellectuals of the 1920's, a version quite different from the pre-war popularizations.

Finally, the attitude of the psychoanalysts had been an important factor. They were "radical" and "cocksure," qualities Meyer detested. He cautioned against accepting their therapeutic claims. Some of his friends had been psychoanalyzed with good results, others had not. At that very moment he was treating a favorite disciple of Freud's, whom Freud twice had analyzed but who then had undergone a disastrous breakdown.[45]

In arraigning the analyst's optimism, Meyer might well have had in mind a report made seven years earlier. In April 1917, the month America declared war on the Central Powers, Isador Coriat published a statistical study of the results of psychoanalysis in ninety-three cases. The results were extraordinary. The cases ranged from mild to severe—hysterias, compulsion neuroses, sexual neuroses, paranoias, early manic depressive insanity and early dementia praecox. All had been treated by other methods, and in many of the cases psychoanalysis was used as a "last resort" after drugs, hypnosis, rest, diets, electricity, suggestion, and explanation had been tried in vain. Coriat saw his patients three times a week for one to six months. He used no explanations, his results came from transference, breaking down resistances and "setting free . . . infantile limitations which were the prime factors in developing the neurosis and giving it its automatic character." [46]

His criteria for recovery were demanding: in anxiety, the disappearance of fears during the day and anxiety dreams at night; an end to compulsions; the complete disappearance of homo-erotic feelings during the day and wish fulfillment dreams, literal or symbolized, at night; in dementia praecox, the attainment of "complete touch with reality." Out of ninety-three patients, forty-six recovered, twenty-seven were much improved, eleven were improved, and nine were not. The best results were obtained in homosexuality and hysteria, and he had gratifying success in early cases of dementia praecox. A child of eleven was cured of hysterical blindness; a severe insomnia cleared up entirely in the space of two months of psychoanalysis; a case of alcoholism of twenty-years standing recovered.

Psychoanalysis, Coriat triumphantly concluded, was the "most efficient therapeutic method yet known to medicine." At the Ann State Hospital in Illinois, David H. Keller, who was not a psychoanalyst but who tried his own version of Freud's method, reported a case of hysterical insanity cured by twelve hours of analysis and the revelation of a sexual etiology. In 1912 at Massachusetts General Hospital, the multiple contractures of a nineteen-year-old girl were almost eliminated after a month of in-patient psychoanalysis by Louville Eugene Emerson.[47]

On what basis can these successes be explained? Every new method, psychiatrists believe, works at the outset.[48] Although all the reasons for these initial results remain unknown, the enthusiasm and faith of the therapist is one factor. More important may be the appropriateness of a given method for a given social milieu and historical period. Freud's sexual theory of the neuroses had an unusual relevance for Americans in the years from 1900 to 1917.

It is significant that some general physicians welcomed the psychoanalytic emphasis on sexuality and united this with a new emphasis on psychotherapy. In 1912 Henry S. Munro, a self-styled "unsophisticated country practitioner" from Omaha, denounced prudery, reticence, and the "theological" doctrine that sexuality was shameful. Like Sir James Paget, Queen Victoria's physician, Munro testified to all the disabilities associated with "civilized" morality—ignorance of sexuality among the married, premature ejaculation, venereal disease, women's disgust for sexuality, the evil consequences of masturbation.[49]

Cecil E. Reynolds, a Los Angeles physician who had been a hospital resident with Ernest Jones, noted in 1915 the widespread shame and secrecy in "sexual matters," the inevitable distortions learned from peers, the frequency of sexual symptoms, particularly of repressed homosexual drives that had not been sufficiently sublimated. "I could quote by the dozen cases of hysteria, of psychoses and psychoneuroses, giving rise to all kinds of bizarre symptoms which I have proved circumstantially as well an analytically were of purely repressed sexual origin, or sexual replacements." Carl Renz, an Alsatian immigrant who had practiced for twenty years in San Francisco, reported the case of a man who had benefited from the candid discussion of his "vita sexualis," which had been consciously or unconsciously repressed so long. By 1915 Renz argued that Freud's views, even if distasteful, were true and therefore should be fought for.[50]

Frankness about sexuality was associated more closely with psychoanalysis, because of its entire theoretical structure, than with any other mode of psychotherapy. On the whole, those who defended traditional somatic views also defended reticence, while those like Adolf Meyer, who espoused a functional dynamic psychiatry, also subscribed to a new candor about sexuality.

Psychoanalysis played an important role in forming this new candor, just as it had formed, because of its militant followers, the adoption of a more dynamic point of view. Thus, change on the professional and the moral level became inextricably linked, and Freud came to be the legendary symbol of both.

The penetration of psychoanalysis coincided with the crisis of American "civilized" morality and the repeal of reticence. The penetration also coincided with the maturity of men and women who had been reared during the years of its most stringent development. These two facts created the hidden, underlying social reality that accounted in part for the unusual psychoanalytic success.

XVII

Conclusion: "Civilized" Morality
and the Classical Neuroses,
1880-1920,
the Social Basis of Psychoanalysis

Writing the history of the psychoanalytic movement in the wake of schism and defection, Freud looked back in 1914 with benign ambivalence to his warm reception at Clark University five years before. Then, on American soil, psychoanalysis, he believed, had changed status from fantasy to reality. Freud himself posed the fundamental paradox of his American reception.

"In prudish America," he wrote, "it was possible in academic circles at least, to discuss freely and scientifically everything that in ordinary life is regarded as objectionable." Of the psychologist G. Stanley Hall, his host, Freud mused: "Who could have known that over there in America, only an hour away from Boston, there was a respectable old gentleman waiting impatiently for the next number of the *Jahrbuch,* reading and understanding it all, and who would then, as he expressed it himself, 'ring the bells for us.' " [1]

Hall was typical of the oldest generation whose social experience provided the basis for the American welcome to psychoanalysis. Goaded by early moral traumas and by confrontation with the easier ways of Europe, he had been one of the first to initiate the repeal of reticence and the exploration of sexual problems. The "old ideals of absolute purity in thought, word and deed for all our boys and girls today," he wrote in 1911, were impossible.[2] However, his own ingrained reticence was such that in 1920 in his preface to Freud's *A General Introduction to Psychoanalysis* he alluded only in the most general terms to the sexual theories of the "most original and creative mind in psychology in our generation." "Freudian themes," he wrote, had "given the world a new conception of both infancy and adolescence, and shed much new light upon characterology; given us a new and clearer view of sleep, dreams, reveries, and revealed hitherto unknown mental mechanisms common to normal and pathological states and processes, showing that the law of causation extends to the most incoherent acts and even verbigerations in insanity; gone far to clear up the *terra incognito* of hysteria; taught us to recognize morbid symptoms, often neurotic and psychotic in their germ; revealed the operations of the primitive mind so overlaid and repressed that we had almost lost sight of them; fashioned and used the key of symbolism to unlock many mysticisms of the past; and in addition to all this, affected thousands of cures, established a new prophylaxis, and suggested new tests for character, disposition, and ability, in all combining the practical and theoretic to a degree salutary as it is rare." [3]

The therapeutic success and the explanatory power that Hall praised had won the loyalty of the first American psychoanalysts. With monotonous regularity they insisted that psychoanalysis had shed a "flood of light" on their patients' symptoms. How can this claim be interpreted historically? The somatic disciplines of the first neurologists and the suggestion therapies of the first psychopathologists had reinforced reticence and control by the conscious will.

The psychoanalytic method revolutionized therapy in one overwhelmingly important respect. The analysis of dreams, the encouragement of free associations, and the recovery of repressed experiences broke through the norm of reticence. The psychoanalysts elicited the fullest expression of their patients' feelings and listened more sensitively and systematically than most other therapists.

The psychoanalysts became convinced that their technique allowed them to uncover and to cure some of the most disturbing and widespread anxieties of their time. Their new sense of professional identity required success and to a remarkable degree Freud's conceptions of trauma and catharsis and of overcoming resistances to the expression of normatively forbidden feelings were appropriate for the problems of their first patients.

The image of the volcano was a frequent self-description in the last decade of the nineteenth century and the first decade of the twentieth; it occurred in Woodrow Wilson, in patients, in novels. The image presumes forces ready for explosion. What was their nature? James Jackson Putnam argued in defense of Freud against Alfred Adler that competition, rivalry, the "will to power," what today would be called "aggression," was far less repressed than sexuality.[4]

In 1908 Freud argued that the repression of sexuality was the specific way in which modern civilization structured nervous illness. The regulation of sexuality had developed in three evolutionary stages: first, a stage of unrestricted freedom; a second, of the repression of all partial sexual drives, auto-erotic, homosexual, etc., that did not subserve the purpose of reproduction; the third stage, modern "civilized" morality, added other rules. It prohibited sexual activity outside monogamous marriage and required premarital chastity. These prohibitions were exacted from women, but not invariably from men, and thus arose the modern "double" standard.

In expanding Freud's thesis, it has been suggested that "civilized" morality was enforced with special vigor by Protestants, Catholics, and Jews who formed the rising middle classes in Europe and the middle and upper classes in America. These groups tenaciously valued a stable family, the accumulation of capital, business success, and professional status.

New conditions in the nineteenth century made sexuality more threatening and the moral code more severe. Desire for professional training and a rising standard of living fostered fewer children and later marriage. These, in turn, in a period when contraceptive techniques were crude, required control of the passions. The alternatives to marriage—fornication, adultery, prostitution—violated religious canons. To secure premarital chastity and monogamy, the core of the code, prudish reticence was exacted about all sexuality; women were kept ignorant of their sexual role and described as not feeling sexual passion. The stereotypes of the manly man and the innocent woman accompanied this shift in the sexual code.

"Civilized" morality structured the conscience, the internal control over conduct. Because of the added requirements of reticence and purity of thought, the code increased the impulses that had to be kept outside awareness. Repression acted as a sensitive inner mechanism that accomplished this end, and it was matched on a social level by disapproval and ostracism. This moral system formed attitudes not only toward adult sexual acts but toward all manifestations of sexuality in childhood. Parents tended to treat the latter with the severity with which they viewed the former.

Psychoanalysis was at once a protest against and a specific therapy for the disorders induced by "civilized" morality. Freud and his first patients, the American psychoanalysts and their first patients, for the most part, had been raised according to its tenets. More stringent than the system Freud knew, American morality bore the stamp of Anglo-Saxon culture, evangelical Protestantism, the absence of aristocratic or popular traditions of hedonism. The American code formally required mental chastity of both men and women, placed a special emphasis on female purity, and installed the mother as the guardian of morals. The sexual secretions, it was taught, must be conserved, lest character and intellect be weakened or destroyed.

In minute and telling detail the operations of this moral system were reflected in the psychoanalysts' first patients. Most of them were born between 1870 and 1900, the high point of "civilized" morality in the United States. A comparison with Freud's case of the hysterical Dora illuminates the differences between European and American patients. Freud emphasized the discovery of Dora's instinctual drives as these were expressed in her symptoms. On the whole, he spent little time on the normative sources of her sense of shame and disgust, apart from their "biological" aspects—their association, for example, with excretions and with eneuresis. Dora was not disturbed by the immorality of her father's affair with Frau K, but rather by the jealousy it aroused in herself. What she most deeply repressed were her own heterosexual desires and, more deeply still, her homosexual impulses.[5]

The American case histories often seem superficial by comparison. Frequently they described the way in which parental instruction about moral norms created the repressive shame and disgust associated with sexuality. The Americans emphasized less fully the partial sexual drives, as if they were uncovering in sexuality itself an important area of anxiety just below the surface of their patients' awareness. Their case histories are historical documents that demonstrate how American "civilized" morality structured the forces of repression, operating as conscience.

As already suggested, the American psychoanalysts' patients at first represented a wider cross section of society than Freud's. American case histories indicate that, far more than in Europe, "civilized" morality cut across lines of class, occupation, and education. Often American patients had been raised in deeply religious families by strict, puritanical mothers, some of whom had instructed their children in the sexual hygiene of "civilized" morality.

The American patients seemed untroubled about what standards to follow. They knew how they ought to feel and act. But they had failed in their own eyes to fulfill the conditions of a clearly outlined morality, internalized in a strong, scrutinizing conscience. They were assailed by intense guilt, shame, and fear of punishment. They felt that any forbidden impulse or act, past or present, would become publicly known and be followed by swift, certain condemnation. They seemed to take over in their feelings about themselves the sharp moral judgments of the period.

Obviously millions of people raised in the sternest traditions of American morality either broke the code without much remorse or conformed to it without becoming ill. Many neurotics, however, seemed to possess the peculiar combination of intensely strong drives and equally intense consciences. They were involuntary victims of, yet rebels against "civilized" morality. Typical of these cases were a New York secretary, a young businessman, a middle-class married woman, a sailor from a farming community, and a New England spinster.

The sudden onset of acute anxiety attacks kept the secretary, who was twenty and unmarried, confined to her home. She suffered from panicky fainting spells during which she felt as if she were dying. She had been referred to Horace Frink for hypnotic treatment, but her fear of it was so intense, he decided to try psychoanalysis. Her mother, strict, puritanical, but kindly, said the girl had been raised with the greatest care and shielded from all knowledge of evil. She herself insisted her thoughts always had been pure. Frink was sparing of interpretations and encouraged her free associations.

Her first dream, about being bitten by a long-nosed dog who drew blood, recalled sexual thoughts of which she was ashamed. She had been upset by a recent newspaper account of a white slavery investigation and a girl friend's insistence that sexual intercourse was bloody and brutish and that men were "wild animals" full of "bad diseases." This impression reinforced some early childhood fantasies, which, combined with her parents' attitude, made her believe sexuality was dirty and sinful, depraved and base. The only feelings she should have for a man must be spiritual, she insisted. If she had experienced physical sexual sensations, she would have felt so guilty she would have had to stop seeing any man who aroused them, she said. She also felt guilty whenever she had allowed herself to be kissed.

After a long and painstaking analysis of her resistances, it became apparent that she had associated fainting spells with sexual assaults. It turned out that she was strongly attracted sexually to a young man with whom she worked, for whom the dog in her first dream had stood. The anxiety attacks signified her terror at these feelings and her fear that, like a hypnotized person, she would let herself go and give in to them. Her attacks ceased after she was able, with considerable encouragement from the analyst, to accept her sexual desires as natural and something she could satisfy legitimately within marriage.[6]

Her case reflected with remarkable accuracy the "civilized" doctrine that women should not have lustful feelings and that sexuality was animalistic and evil.[7] Here the attempt to discipline sexuality by silence and repression made it at once sinister and fascinating.

Another of Frink's patients illustrates the male equivalent, the dissociation of sexuality from affection. A young businessman complained of his lack of confidence in his own decisions, depression, and discontent with his recent marriage. His wife was beautiful, intelligent, refined, and not frigid. However, he felt that the sexual relations he had had with "bad" girls before his marriage were more satisfying. He was obsessed by fantasies of flamboyant, "sporty" women.

Quite early in treatment he confided that at ten another boy had described intercourse to him. The incredulous patient had replied, "That may be true of *some* people, but I know my mother would never have done such a dirty thing." Frink then explained that probably the patient's attitude toward his wife reflected his early training. The young man agreed, provided a few details, then suddenly developed a severe, negative transference. He lost all confidence in the analyst and became avidly curious about his sex life. Yet he felt like a "nasty minded little boy" and feared the analyst would ruin him or harm him physically because of this curiosity. As he painfully made these confessions, he would raise his hand to his face and turn his head away. It dawned on him that he unconsciously was warding off a blow.

Then he was able to recall some early experiences. At five a little girl had volunteered to show him what "pappa and mamma" did. They were discovered and severely punished. Later his mother found him hid-ing under her bed, suspected the nature of his curiosity, and threatened to tell his father.

Not until he remembered this material could he understand his feelings about his wife. It became clear to him that he could not bear to look at her nude or display any sexual passion for her because he feared she would scold and punish him. He had identified her with his mother. He also realized that he had come to believe that only "bad" women, like the little girl, were bold enough to enjoy sexual intercourse. With this realization his fantasies ceased, and he was able to resume satisfying relations with his wife. Again the correspondence with the teachings of "civilized" morality is direct. Only "bad" women, usually prostitutes, could feel sexual passion. Mothers and refined women were pure.[8]

The severity of childhood sexual discipline against which an entire generation of American psychoanalysts and educators rebelled is illustrated by two psychotic patients. Both were diagnosed as suffering from dementia praecox at St. Elizabeth's Hospital, Washington, D.C. The first, a middle-class, intelligent woman, had been made to feel "deplorably shameful" at the age of four for asking about a bull's scrotum during a country walk with her embarrassed parents. Later she had been scolded for inquiring about her sister's menstruation. Her father criticized her for indecency when she crossed her legs and objected to seeing her in kimonos. Her mother forbade all discussions of sexuality and refused to receive her "intimate confidences." Later, during a difficult marriage, she tried to commit suicide. She had been told as a child that masturbation would kill people, and she believed that because she had practiced it that she was degenerate, unworthy to raise her child. Edward Kempf, her therapist, found it "astonishing" that after two years of therapy she recovered and that two years after her discharge she was still well.[9]

The second patient, treated by a lay therapist, had been born in 1896 on a farm. His mother had given him a "thorough moral training," warning him that she would know if he did anything bad. When he was eight, she told him masturbation would kill him. He began to masturbate, believed everyone could see it in his face, and, finally, in despair, quit school and ran away. In the city he had a homosexual experience. Later, he enlisted in the Navy, hoping it would make a man of him or that he might be killed. Then he contracted gonorrhea from a prostitute and believed he was unworthy to marry his fiancée or associate with decent people. He felt "utterly, irrevocably disgraced, a social outcast, the scum of the earth, an insect." He heard voices calling him a sinner. His therapist believed that the confession of his sexual problems presaged the start of a successful recovery.[10]

The religious element in "civilized" morality can be illustrated by other examples. One of Brill's most successful analyses was of a Jewish immigrant who suffered from paralyzing depressions on the Day of Atonement. Brill doggedly discovered a sexual transgression she had suppressed from memory. Frink treated a Catholic girl who developed an addiction to taking medicine as a defense against impure wishes and thoughts.[11]

James Jackson Putnam chose a case to illustrate what he considered the "morbid" aspects of the New England conscience, rooted in the "old-fashioned" evangelical emphasis on sin. The case describes exactly the kind of conscience and style of moral judgment that began to die in the Progressive Era.

His patient believed that an angry God would punish her for her sexual desires. Her conscience, cruelly condemning of other people, was cruelest to herself. It reflected, moreover, an intricate relationship between her religious training, her ties to her parents, and her sexuality. Her passionate love for her father, an equally intense jealous hatred of her mother, made her think guiltily of God as a harsh avenger, the thundering Jehovah of the Old Testament, about whom she had been taught in Sunday School. The severity of her conscience was exacerbated by her need to control conflicts between her sexual impulses and religious canons of sexual purity.

Putnam's patient was an unhappy, forty-nine-year old spinster, a semi-invalid much of her life who suffered from bad eyes and migraine headaches. She was raised in a community distant from Boston, in an atmosphere of revivals and hell-fire tracts. Outwardly she seemed noncommittal, remote and shy, almost cold; inwardly she was overwrought and excruciatingly sensitive. She judged herself to be selfish, morbid, and egoistic, and had developed an uncanny ability to do or say things that defeated her own wishes. The people she wanted to attract she alienated; her voice would grow harsh and loud, her hands would tremble, her manner dismiss them. Yet she longed for friendship and affection.

She had been born prematurely, the result of a secret marriage. Her mother, a schoolteacher, loathed her; her father adored her. At four she had begun to take part in daily Bible readings and was exposed to a "narrow, bigoted" publication of the American Tract Society, outlining the proper behavior for a young lady. She was forbidden any amusement other than hymn-singing. "Work for the Night is Coming" was a favorite of her Sunday School teacher. Wasting time was sinful. "We are accountable to God for every moment," she was told. This warning was "like a lash driving us to duty." She came to believe that her body hindered the life of the soul and could blacken the soul with sin.

She became harsh toward both herself and her schoolmates. Forbidden a musical education, she reacted, Putnam wrote, in "Calvinistic fashion" and became a "wet-blanket" to enjoyment. Yet in her heart she yearned for the very pleasures she condemned. Her father's love inspired her to excel in every game and every school subject. "I tried to be as hard with other people as I was on myself," she recalled.

Her affection for her father grew so strong that at the age of eight she began to wonder why she should not become her father's wife, since Joseph had had seven wives. About that time some neighborhood boys tried to molest her and several other little girls. She was frightened yet fascinated to an overwhelming and disastrous degree. Her religious upbringing had taught her that her true self was free of earthly passions. Soon she began to dream that she was being pursued by ogres.

Then her mother began to reproach her for her passionate attachment to her father. In the midst of these cumulative crises, she was taken to a great revival, which lasted for two months. The preacher dilated on the bliss of heaven and the tortures of hell. Suddenly she dreamt that her mother had died. In an agony of remorse, she became even more devoted to duty and was as cruel to herself as if she had been striking her body with a lash. She began to imagine God as an all-seeing and avenging eye, like the image on her father's Masonic diploma. When her father died, she felt as if God had forsaken her.

Her psychoanalysis uncovered the bitterest resentment against what she called her "Puritan training," which had come to symbolize to her a "hateful hindrance to full self-expression." "Fear pursued me every moment of the day, fear lest I swerve from the straight path of duty," she recalled. Therapy seemed a rebirth. She went to work as one of Putnam's aides, perhaps as one of the first medical social workers at the Massachusetts General Hospital.

The rebellion she discovered within herself foreshadowed that of an entire generation of young American intellectuals, who also were rejecting religious backgrounds—evangelical, Catholic or Jewish. "My . . . disappointment," she wrote, "was the result of obedience to parents, faithfulness to duty, utter elimination of self. It did not pay to be so self-sacrificing. If I had my life to live over again I would be reckless in self-gratification." [12]

Some therapists, like Edward Kempf, were astonished by the results of catharsis and reassurance. Because of the sense of isolation caused by guilt, they were careful to assure patients that everyone had roughly similar sexual impulses. Rid of "morbid remorse," patients became filled with "courage and self-respect."

The fact that most of the psychoanalysts had been raised in accordance with the canons of "civilized" morality made them sensitive to the problems it created in their patients. Their own attitude toward this prevailing moral system was in some respects ambiguous; they protested reticence, yet on the whole conformed. In general, they were far less frank about themselves than about their patients. They were not as candid as G. Stanley Hall or William James. White's goal of seeing human beings "in the raw" did not include self-exposure. Nevertheless, there is some evidence that the first generation of analysts, like some of the psychologists, found "civilized" morality burdensome. Jelliffe, who fell deeply in love at sixteen, was forced by lack of money and the necessities of medical training to postpone marriage until he was twenty-eight. Similar considerations operated for White, Putnam, and Brill. Indeed, Putnam's protest against the extremists who advocated marriage for procreation only may have mirrored personal drives and needs.

Once convinced, the psychoanalysts unearthed the sexual demon with a leveling zest that also characterized muckraking, Beard's discovery of the self-interest of the Founding Fathers, or Veblen's corrosive exposé of pecuniary emulation. This truculent emphasis on sexuality aroused furious opposition and fostered the psychoanalysts' sense of themselves as fighters against irrational, oppressive conventions.

The social origins and personalities of the psychoanalysts became fused with the sexual issue to generate among them a special élan. Several of the most militant psychoanalysts had begun as outsiders by reason of class or personality or religion. Only Putnam and Ralph Hamill had come from the established upper class. For all his impeccable social credentials, Putnam felt himself to be a rather anachronistic idealist by comparison with his friends James, Prince, or the younger analysts. Ernest Jones, a Welshman, had a fiery personality that got him into trouble. For many of the psychoanalysts medicine represented a rise from middle- or lower-middle-class status to the professions. Brill was an immigrant Jew, the son of a non-commissioned commissary officer in the Austrian Army; White's father was a storekeeper, Tannenbuam's a tailor.

The older established authorities, especially the several opposing neurologists listed in the *Social Register,* stood for the orthodox methods and genteel social conventions.[13] These the analysts had to attack, if they were to be heard. They were convinced they had at their command the latest, most daring technique, one that uncovered and cured the sexual traumas of their patients. With this cutting sense of superiority, they could look down on organicists who still were dispensing pills, rest, and diets. This ardor and solidarity were heightened by their own shock-tactics and the contempt of their opponents.

The American psychoanalysts, like most progressive reformers, were more revolutionary in rhetoric than in program. They believed that the therapeutic destruction of inhibitions and the acknowledgment of sexual impulses would result in a more genuine morality. As Frink put it, psychoanalysis attempted to "step in and do for the individual what the ordinary moral forces had tried to do for him and failed." These had failed, he and Putnam would have agreed, because they were needlessly repressive. The analysts advocated not free love and trial marriage but more satisfying monogamy, the control and sublimation of sexual impulses, not their indiscriminate expression. Yet this message had ambiguous overtones.

Except for Brill and the other Europeans, the analysts had been raised in the American version of "civilized" morality and were imbued with the American substitute for religion—faith in progress and human goodness. The European revolt against both faith and morality had begun earlier and gone further. For these reasons, as well as the importance they placed on their status as physicians, the American analysts were more conservative than Freud.

Like other progressives, they wished to strengthen, not to destroy, fundamental institutions. Thus, by repealing nineteenth-century reticence, they hoped to preserve the goals of "civilized" morality. Among these traditionally had been a regulation of the sexual life that would ensure hard work, family stability, and rising status. The loudest appeal of Woodrow Wilson's campaign for the New Freedom in 1912 was to keep open the paths of upward mobility. Some psychoanalysts believed that the traditional discipline of effort was being undermined in new ways.

To Jelliffe the rich were setting an example of idleness, luxury, and sometimes of open sexual immorality. American mothers were spoiling their sons. Doctrines of heredity and eugenics were fostering fatalism. The other psychotherapists inadvertently strengthened dependence. Against all these threats, psychoanalysis offered a new therapeutic discipline.

By limiting the influence of heredity they widened the possibilities of successful therapy. Putnam, for one, had become convinced of the merits of psychoanalysis partly for this reason. He could demonstrate that what Janet had interpreted to be signs of inherited nervous degeneracy—for example, excessive shyness or precocious sexuality in early childhood—were in fact the result of specific experiences or the expression of universal human drives.

Freud tended to see the therapeutic dissolution of Oedipal ties as a prerequisite to sexual maturity, the Americans, except for Putnam, as a preliminary to successful achievement in social life. The American analysts especially stressed independence from the mother. The role of the father traditionally had been weaker in the American family and had been eroded further by the demands of urban business and professional life.

By 1900 there had been a noticeable shift in family authority. Mothers had come to play a more decisive part than ever before, especially in education and morals. Brill's life demonstrates the differences between America and Europe in this respect. He hated what he considered the suffocating closeness of the European Jewish family. Whatever the influence of the mother in such households, for the most part European analysts stressed independence from the father. (Philip Rieff has argued that the mother's influence was so pervasive yet so sacred that Freud failed to see it clearly.) Brill came to America chiefly to escape what he called his father's "top sergeant" manner. Here, however, he attacked the protective mother and the privileged position of the spoiled only child. His case histories and those of other analysts described homes where the father was often absent, perhaps a traveling salesman, or a professional or businessman so absorbed in his work that he paid little attention to his family. No other system of psychotherapy placed this systematic emphasis on family structure and the autonomy of the child.

Increasingly between 1909 and 1917 some psychoanalysts identified themselves with social and especially with cultural change and reform. Jelliffe, for instance, insisted on the anti-aristocratic stance of psychoanalysis. He defended modern painting as an important mode of expressing hitherto forbidden wishes.[14] Brill attacked American "Puritanism." William Alanson White was actively working for the reform of criminal procedures, and Putnam for the extension of psychiatric social work. All these attitudes and actions tended to give psychoanalysis a progressive coloration. By 1917 one of the most conservative of all the Philadelphia organicists, Charles W. Burr, who believed that a combination of heredity and environment were degenerating the race, thundered against uplifters and the growing meddling of laymen in medical matters:

> It is one of the results of the wild fury of altruism which has overwhelmed the country, and has produced much unwise legislation, brought about by the public being misled by incompetent newspaper physicians, whilst those most competent to speak hold themselves silent and aloof. . . .
>
> Suppose a reactionary like myself, who believes in jails and hanging and such brutal things, should be on a commission with an "uplifter," who thinks that everyone may be reformed, how could we come to any agreement either as to diagnosis or treatment.

Eugenics, sterilization, life-long segregation were the only answers to crime and alcoholism, Burr insisted. William Alanson White replied that he belonged to the "pseudo-psychiatrists" Burr had attacked:

> I should be very sorry to agree with a program that was so statically concrete and so filled with repressive measures . . . for I believe we make a great mistake when we lay at the door of heredity so many things. . . . We know perfectly well that some actual conditions of misconduct have been traced to social and environmental circumstance that could be remedied. . . .[15]

Yet, by comparison with the attitude of Sidney I. Schwab, who looked to industrial conditions as a cause of nervous disorder, the psychoanalysts were conservative. Their attention to the family and to the individual broke with important elements of progressive reform.

There were also elements of cultural revolution in psychoanalysis. Upward mobility and achievement have remained important goals in American life. But sexual behavior and attitudes about it have changed profoundly, and so, apparently, have neurotic symptoms. Did psychoanalysis foster these massive shifts in the mores?

Freud's initial major impact coincided with the first stage of America's moral revolution, the repeal of reticence that occurred roughly from 1911 to 1917. This initial breakdown in "civilized" morality had been prepared by at least two decades of study and discussion among physicians, sociologists and psychologists. Independently of psychoanalysis, the destruction of what the anthropologist Elsie Clews Parsons called the "taboo of direct reference" was chiefly the work of purity crusaders, earnest doctors, and women bent on stamping out prostitution and the double standard.[16] Their zeal inadvertently fostered an explosive public consideration of sexuality. Discussion was soon followed by real changes in behavior. This historic shift, documented by Alfred Kinsey's interviews, began around 1916 among women born around 1900: a far higher proportion of women in this generation experienced orgasm, and from two to three times as many had premarital intercourse as those born before 1900. Premarital experience for men did not increase, but occurred less with prostitutes and correspondingly more with other women. This change began among city girls and spread to the country; it grew progressively more marked until by 1930 it had become a firmly established pattern. Basically, then, it was a women's revolution.[17] What explains it?

Kinsey attributed the new pattern partly to a change in attitudes, which he associated with the doctrines of Freud and Havelock Ellis. Ironically, White and Jelliffe, among the most moralistic of the American analysts, had linked these two authorities by having Ellis contribute to their 1913 text, *The Modern Treatment of Nervous and Mental Diseases.* Often citing Freud, Ellis attacked all the nineteenth-century additions to "civilized" morality—reticence, the older sexual hygiene, late marriage, the asexualization of women, the doctrine of sexual intercourse for procreation only. Ellis blamed the mistaken high-mindedness of civilized education for frigidity in women and for male ineptness in intercourse. He urged that women be granted the same erotic privileges as men and suggested that the art of love be cultivated. Completely reversing the opinion of many British and American medical authorities, he insisted that sexual abstinence, although not physically harmful, yet could be difficult and irksome. The doctrine that continence was perfectly compatible with health, he urged, was an argument of extremists who would limit intercourse to two or three acts in a lifetime.

Ellis's position was ambiguous. On the one hand he described human behavior with anthropological detachment, noting differences in frequencies and practices in much the style Kinsey later developed. He clearly implied that the "traditional influences of morality and convention were not biological," that is, not biologically sound or desirable. Yet, as a physician, he could not suggest their therapeutic violation. Finally, with that strain of romanticism that runs through his writing, he insisted that sexual intercourse was capable of becoming a "fine ecstasy." In an ambiguous passage on the pitfalls of marriage and celibacy, Ernest Jones argued "the impossibility of enforcing without harm a uniform standard and mode of living." Thus, closely associated with psychoanalysis, a medical justification for changes in sexual behavior could be construed before the revolution occurred.[18] Freud, ironically, became the symbol of this entire change.

What had this specific text or psychoanalysis in general to do with the changes in conduct? The question cannot be answered directly. The 1913 text was favorably reviewed across the nation by psychiatric, neurological, and medical journals. But the psychologists seem to have ignored it. Probably it received little attention from the intelligent lay audience that its editors had hoped to attract. Like most other psychoanalytic books, its sales probably were modest, its readers comparatively few.

Nevertheless, a revolution in American attitudes as well as conduct did occur. According to surveys of American magazines, approval of extramarital relations, of divorce and birth control, grew quickly after 1918 and reached a peak between 1925 and 1929, especially among intellectuals. By 1918 some 23 per cent of all intellectual magazines favored "sex freedom," by 1928 some 56 per cent. The corresponding percentages for the mass magazines were 13 per cent in 1918 and 40 per cent in 1928.[19] These shifts closely parallel those in behavior. Twenty-three per cent of all girls born between 1900 and 1909 were having premarital intercourse by the age of twenty, for those born between 1920 and 1929 the percentage rose to thirty-one.

But there is an even more intimate correlation. Kinsey suggested that the most vital social factor controlling the sexual behavior of women was not occupation or education but religion. In just these years there was a growing disapproval of "religious sanctions" for sexual conduct. Fewer people than ever before endorsed the churches' opposition to divorce and birth control or invoked Biblical authority in sexual matters. Thus, religious controls were weakening at the time conduct was changing. There are some qualifications that need to be made, however. Periodicals reflected editorial tastes in New York; most magazines then were read by the better educated public, meaning largely the middle and upper classes. Paradoxically, the pulp magazines, which had a wider mass audience, paid more attention to sex than the ladies' magazines but were more conventional about it.[20] The movies, of all the mass media, presented sexual irregularities in the most approving light. Kinsey has suggested that the change in sexual behavior was identical for people of every social group and every degree of education.

What, then, was the role of psychoanalysis in this change in attitudes? The number of articles about psychoanalysis also reached a peak between 1925 and 1928. Yet the bulk of these articles disapproved of psychoanalysis. Those that favored it treated it primarily as a new kind of psychotherapy and took conventional attitudes toward sexual behavior. So, too, did the books most psychoanalysts wrote. Moreover, by 1930 the influence of Freud and psychoanalysis probably still was confined to comparatively small groups of Americans.[21]

The change in attitudes, however, does coincide with profound alterations in exactly those factors that caused the nineteenth-century additions to "civilized" morality. The decline of religious controls over sexuality already has been noted. By 1900 American observers had become aware of decisive changes in the economic system. The American sociologist Simon Patten argued in 1907 that America was moving from an economy of deficit and saving to one of surplus and abundance.[22] A new kind of character had to emerge, no longer dedicated to austerity and sacrifice but to leisure and rational enjoyment. Repression would give way to release. The new economy was giving a new place to women outside the home and family. Some of the first changes in sexual behavior, the increase of intercourse among the unmarried, had been initiated by girls and women who worked in factories and offices alongside men. Even the marriage age was beginning to decline, perhaps partly as a result of the fact that more women could marry and still work.

These changes were conspicuous in the rapidly growing cities, which presented the immigrant from Europe or the American countryside with widely varied patterns of behavior. The sharp moral controls of the small town—close-knit neighbors, churches, "society"—were replaced by relative anonymity and isolation.

By 1917 the prominent psychoanalysts, like other medical specialists, were living in the better sections of the growing cities, on Central Park West in New York, where Freud had visited Brill, or Marlborough

Street in Boston, where Coriat had his office. Their first patients, on whom they had based their unique American interpretations of psychoanalysis, had reached maturity before the moral revolution. Their case histories represented some of the frankest and most bitter of all contemporary criticism of "civilized" morality.

These patients exhibited the classical phobias, obsessions, and hysterias which were beyond the control of the conscious will. They erupted as inexplicable intrusions into otherwise integrated personalities. Such clear-cut neuroses expressed conflicts over instinctual drives which could be neither acknowledged nor discussed according to the canons of "civilized" morality.

From the beginning there were "cool and callous" patients, untroubled by the conscience of "civilized" morality. Some had no strong desire for sublimation or, indeed, any strong purpose in life, Putnam informed Freud.[23] There were still others from what could be called Freud's second level of sexual morality. Case histories reveal a considerable number of patients displaying homosexual behavior, latent or overt, sadism, masochism. Others displayed the excessive sexual drives that George Beard and William Hammond had observed. Mabel Dodge Luhan, for example, described her compulsive search for a sense of security and power through love and sexuality.[24]

Possibly physicians sent to psychoanalysts patients who seemed openly or covertly to suffer from symptoms involving sexuality or such patients themselves sought psychoanalytic treatment. How important this element of possible selection may have been for the psychoanalysts' sense of successful conviction is impossible now to determine.

These were not the classical neuroses that the model of trauma and catharsis had healed. Indeed, many physicians who were not psychoanalysts already noted that catharsis and the frank discussion of sexual behavior did *not* result in the cure of symptoms, as the cathartic model had led them to expect. Yet all these cases involved sexual impulses.

Psychoanalysts today are divided on the relation between human drives and human culture and the degree to which the latter modify the former. There are those who argue that the family constellation and the sexual instincts exist in every human being, relatively alike, regardless of time or place. Other psychoanalysts suggest that such drives are strongly modified by culture.

Regardless of the merits of either argument, it seems quite certain that historical culture in fact determines the ways in which drives are expressed or, rather, the ways in which society channels this expression. Philippe Ariès has demonstrated that in the seventeenth century every later taboo concerning childhood sexual talk and behavior was non-existent at the court of France. Fondling, ribaldry, utter frankness were acceptable. Perhaps the rise of psychoanalysis resulted from the rise of the small, bourgeois family. It may be noted that one of the "revelations" of psychoanalysis that most strongly struck not only the analysts themselves, but psychologists such as Harry Woodburn Chase, was the existence of intense hostility within the family group.[25] Children's hatred of their parents and parents' hatred of their children were shocking violations of normatively prescribed behavior. Thus psychoanalysis revealed in its first American decade not only sexuality but hostility within the family.

Gradually the classical neuroses gave way to less sharply defined disorders of character or identity. Sidney I. Schwab, the neurologist, observed in 1917 that hysteria was becoming a rarity.[26] Insight and catharsis which astonished the first psychoanalysts by their power no longer were so effective. Psychoanalytic treatment began to take three, four, or five years. Where Freud had assumed that an integrated character of high integrity was a prerequisite to psychoanalysis, now patients were presenting characters that seemed patently torn or malformed. Their controls over impulse seemed shaky; they put forth less will and effort; their consciences were less clearly structured. The domain of the unconscious itself had narrowed, and impulses once kept outside awareness were commonly acknowledged and discussed.

When these new problems first appeared is uncertain. The new symptoms can be observed among some of the intellectuals, writers, and artists who were psychoanalyzed before the Great War. In open rebellion against every canon of "civilized" morality, they complained to their psychoanalysts chiefly of unhappiness, unstable sexual relationships, longings for meaning and permanence. Because their problems are in many important ways so different from the striking symptomatology of the first decade of psychoanalysis, they will be discussed at length in the second volume of this study.

Psychoanalysts have been grappling with these new problems on a significant scale at least since the middle 1930's. Indeed, psychoanalysts have turned more and more from the study of sexuality and the unconscious to explorations of the ego and its synthesizing and executive functions. Perhaps, as choice has widened, the mechanisms for making it have become subject to greater strain. Perhaps Durkheim's anomie, or normlessness, or, more exactly, a confusion of norms, is as likely as too rigid and stringent control to create its own pathology.

The evidence from psychoanalysis and from case histories suggests that neurosis may be a sensitive indicator of strains within a system of social values. If neurosis does reflect this kind of social stress, then case histories become important historical documents.

Psychoanalysis initially was successful in America because it met profound needs generated by "civilized" morality. Its introduction coincided with the first stage of the moral revolution, the repeal of reticence. The Great War hastened the changes in sexual behavior and intensified the psychoanalytic impact.

Putnam's reaction to that war is illuminating. He slept badly. He sought to understand what had gone wrong, but failed. Then he blamed himself and felt somehow obscurely responsible for the "evil deeds of his fellow men." [27] Satisfying explanations for the disaster occurred readily to Freud, to his own deep regret. He wrote his good friend, Lou Andreas Salome:

> I cannot be an optimist and I believe I differ from the pessimists only in so far as wicked, stupid, senseless things don't upset me because I have accepted them from the beginning as part of what the world is composed of. My friend Putnam maintained in a recent book which is based on psychoanalysis that perfection has not only a psychic but also a material reality. That man can't be helped, he must become a pessimist! [28]

The War opened new and darker directions in Freud's *Weltanschauung*. It dampened his earlier hopes for the reformation of society. However, Putnam and most of the other American analysts failed to become pessimists, and Freud's differences with them deepened. Freud's essay on narcissism, written in 1914, explored in part the irrationalities of conscience, the kind of conscience that had kept Putnam sleepless. Freud suggested that the common ideal of a family or a nation represented an extension of the personal conscience, or "ego ideal." A year later he was to conclude that the common ideal and conscience of Western civilization had been shattered, perhaps forever. Yet, ironically, the War hastened the acceptance of a dynamic, functional psychiatry among wider groups of physicians, and of a psychoanalysis partly shorn of its sexual emphasis.

Abbreviations

Collections:

AMP   Adolf Meyer papers, the William Henry Welch Library of Medicine, The Johns Hopkins University, Baltimore

CUP   Clark University Papers, Worcester, Mass.

EBTP   Edward Bradford Titchener papers, John M. Olin Research Library, Cornell University, Ithaca, New York

PP   James Jackson Putnam papers, The Francis A. Countway Library of Medicine, Boston

WJP   The William James papers, Houghton Library, Harvard University, Cambridge, Massachusetts

Titles:

Ad Psa   James Jackson Putnam, *Addresses on Psycho-Analysis* (London, Vienna, New York: International Psycho-Analytical Press, 1921)

AJI   American Journal of Insanity

AJP   American Journal of Psychology

Bost J   Boston Medical and Surgical Journal

C   Zentralblatt für Psychoanalyse und Psychotherapie

CPF   Sigmund Freud, *Collected Papers* (London: The Hogarth Press and the Institute of Psycho-Analysis, 1957)

CPM   The Collected Papers of Adolf Meyer, edited by Eunice Winters, 4 vols. (Baltimore: The Johns Hopkins University Press, 1952)

J   International Journal of Psychoanalysis

J Ab P   Journal of Abnormal Psychology
JAMA   Journal of the American Medical Association
JNMD   Journal of Nervous and Mental Disease
JPP   Nathan G. Hale, Jr., ed., *James Jackson Putnam and Psychoanalysis: Letters Between Putnam and Sigmund Freud, William James, Ernest Jones, Sandor Ferenczi, and Morton Prince, 1877–1917* (Cambridge, Mass.: Harvard University Press, 1971)
Q   The Psychoanalytic Quarterly
R   The Psychoanalytic Review
SE   The standard edition of the *Complete Psychological Works of Sigmund Freud*, translated from the German under the general editorship of James Strachey in collaboration with Anna Freud, assisted by Alix Strachey and Alan Tyson (London: The Hogarth Press and the Institute of Psycho-Analysis, 1955–1966)
Z   Internationale Zeitschrift für ärztliche Psychoanalyse

Notes

Chapter I. The Clark University Conference

1. Freud, "An Autobiographical Study," SE, XX, 52.
2. "Three Men," *The New Yorker,* 32:10 (April 28, 1956), 34–35.
3. Edward Bradford Titchener, "The Past Decade in Experimental Psychology," *Lectures and Addresses Delivered Before the Departments of Psychology and Pedagogy in Celebration of the Twentieth Anniversary of the Opening of Clark University, September, 1909* (Worcester, Mass., 1910), pp. 160–162.
4. Emma Goldman, "Adventures in the Desert of American Liberty," *Mother Earth,* 4 (September 1909), 210–215; Rev. Eliot White, "In Worcester," *ibid.,* pp. 216–217.
5. Hall to Freud, October 7, 1909, Clark University Papers.
6. AJP, *21* (April 1910), 181–218.
7. Worcester *Telegram,* September 8, 1909, Clark University, Scrapbook of the Clark Conference.
8. Freud, "The Origin and Development of Psychoanalysis," *Clark Lectures and Addresses,* p. 2.
9. *Ibid.,* p. 5.
10. *Ibid.,* p. 14.
11. *Ibid.,* p. 23.
12. Worcester *Telegram,* September 12, 1909, Clark University, Scrapbook of the Clark Conference.

13. Freud, "The Origin and Development of Psychoanalysis," *Clark Lectures and Addresses,* p. 27.

14. *Ibid.,* pp. 27–28.

15. *Ibid.,* p. 38.

16. Freud, " 'Civilized' Sexual Morality and Modern Nervousness" (1908), CPF, II, 76–99.

17. Quoted in Ernest Jones, *The Life and Work of Sigmund Freud,* 3 vols. (New York: Basic Books, 1953–1957), II, 417–418.

18. Lewis Feuer, *The Scientific Intellectual* (New York: Basic Books, 1963), pp. 311–314; Jones, *Freud,* I, 128–141, 2.

19. Henri Ellenberger, *The Discovery of the Unconscious* (New York: Basic Books, 1970), pp. 291–303; Stephen Kern, "Sigmund Freud and the Emergence of Child Psychology, 1880–1910," Doctoral Dissertation, Department of History, Columbia University, 1970.

20. Herman Nunberg and Ernst Federn, eds., M. Nunberg, trans., *Minutes of the Vienna Psychoanalytic Society,* I, 1906–1908, (New York: International Universities Press, 1962), p. 190.

21. Jones, *Freud,* II, 417–418.

22. Freud, "Thoughts for the Times on War and Death" (1915), CPF, IV, 296.

23. James Jackson Putnam, "Elements of Strength and Elements of Weakness in Psychoanalytic Doctrines" (1910), Ad Psa, pp. 449–450; see also William Ernest Hocking, "The Holt-Freudian Ethics," *The Philosophical Review, 25* (May 1916), 479–506.

24. "Recent Freudian Literature," AJP, *22* (July 1911), 425–426.

25. Jones, *Freud,* III, 126.

26. Freud to Maria Montessori, December 20, 1917, *The Letters of Sigmund Freud* (New York: Basic Books, 1960), pp. 319–320.

27. Freud, "Civilization and Its Discontents" (1930), SE, XXI, 145.

28. G. Stanley Hall to Sigmund Freud, October 7, 1909, Clark University Papers.

29. Charles G. Hill, Presidential Address to the American Medico-Psychological Association, AJI, *64* (July 1907), 6.

30. Meyer to Prince, November 4, 1909, AMP.

31. Adolf Meyer, "The Dynamic Interpretation of Dementia Praecox," *Clark Lectures and Addresses,* pp. 152–153.

32. Putnam to Freud, November 17, 1909; Freud to Putnam, December 5, 1909, PP.

33. Hugo Münsterberg, "School Reform," *Atlantic Monthly, 85* (May 1900), 661.

34. Adolf Meyer to E. B. Titchener, September 18, 1909, September 23, 1909, October 30, 1909, April 6, 1918, EBTP.

35. Jones, *Freud,* II, 57; Carl Jung to Virginia Payne, July 23, 1949, cited in Virginia Payne, "Psychology and Psychiatry at the Clark Conference of 1909," Dissertation for the degree, Doctor of Medicine, University of Wisconsin, 1950.

36. Dorothy Ross, *G. Stanley Hall: The Psychologist as Prophet* (Chicago: University of Chicago Press, 1971), pp. 542–543.

37. William James to Mary Calkins, September 19, 1909, in Ralph Barton Perry, *The Thought and Character of William James* (Boston: Little, Brown and Co., 1934), II, 123.

38. William James to Theodore Flournoy, September 28, 1909, in Henry James, ed., *The Letters of William James* (Boston: Atlantic Monthly Press, 1920), II, 327–328.

39. Freud, "An Autobiographical Study," SE, XX, 52.

40. Lewis Terman to G. Stanley Hall, April 27, 1911, Clark University Papers.

41. "The Twentieth Anniversary of Clark University," *Nation, 89* (September 23, 1909), 284–285.

42. *Boston Evening Transcript,* September 8, 1909, p. 2. The interview, published September 11, 1909 Part 3, p. 3, was by Adelbert Albrecht, probably the *Transcript*'s correspondent at Clark, a translator and in 1913 associate editor of the *Journal of the American Institute of Criminal Law and Criminology.*

43. Harold Bolce, "Blasting at the Rock of Ages," *Cosmopolitan, 46* (May 1909), 665–676; "The Alleged Decay of Responsibility in America," *Current Literature, 45* (October 1908), 424–426; "The Blurring of the Vision," *Outlook, 90* (October 3, 1908), 244–245; "Is Progress an Illusion?", *Current Literature, 44* (January 1908), 70–72.

44. G. Stanley Hall, *Life and Confessions of a Psychologist* (New York: D. Appleton and Co., 1923), pp. 223, 406–409; James Mark Baldwin, *Between Two Wars* (Boston: The Stratford Co., 1926), pp. 2, 10–12, 14–17.

45. Jones, *Freud,* II, 57.

46. André Tridon, *Psychoanalysis and Behavior* (New York: Alfred A. Knopf, 1920), pp. 239–249, 251; *Psychoanalysis and Love* (Garden City: Garden City Publishing Co., 1922), p. 267; Emma Goldman, *Living My Life* (New York: Alfred A. Knopf, 1931), I, 455.

47. G. Stanley Hall, *Adolescence* (New York: D. Appleton, 1904), II, 547.

48. Quoted in Ray Stannard Baker, *New Ideals in Healing* (New York: Frederick Stokes, 1909), p. viii; "The Marvelous in Medical Science," *Review of Reviews, 38* (September 1908), 363–364; Moses Allen Starr "Recent Discoveries in Medicine," *Harper's Monthly, 117* (July 1908), 259–262.

49. Frederick C. Howe, *Confessions of a Reformer* (New York: Charles Scribner's Sons, 1925), p. 7.

Chapter II. American "Civilized" Morality

1. Anthony Comstock, "The Work of the New York Society for the Prevention of Vice and Its Bearing on the Morals of the Young," *Proceedings of the Child Conference for Research and Welfare, 1, 1909* (New York: G. E. Stechert, 1910), pp. 91–109. The quotations are from Anthony

Comstock, Address, in *The National Purity Congress, Its Papers, Addresses and Portraits* (New York: American Purity Alliance, 1896), pp. 420–421.

2. Freud, " 'Civilized' Sexual Morality and Modern Nervous Disorder" (1908), SE, IX, 182, 186–188, 192–193, 196–198; *A General Introduction to Psycho-Analysis* (Garden City: Garden City Publishing Co., 1920), pp. 308–309.

3. Walter E. Houghton, *The Victorian Frame of Mind* (New Haven: Yale University Press, 1957), pp. 340–393, esp. p. 359, and Peter T. Cominos, "Late Victorian Sexual Respectability and the Social System," *International Review of Social History, 8* (1963), 18–48, 216–250.

4. Charles Francis Adams, "Some Phases of Sexual Morality and Church Discipline in Colonial New England," Massachusetts Historical Society, *Proceedings,* second series, 6 (June 1891), 478, 514.

5. Alexis de Tocqueville, *Democracy in America* (New York: Vintage Books, 1945), II, 166, 210–225, 247–249; J. P. Mayer, ed., *Alexis de Tocqueville, Journey to America* (New Haven: Yale University Press, 1959), pp. 222–224.

6. Amariah Brigham, *Remarks upon the Influence of Mental Cultivation and Mental Excitement upon Health* (Boston: Marsh, Capen and Lyon, 1833), p. 14.

7. Tocqueville, *Democracy,* II, 204–205.

8. *Ibid.,* p. 210.

9. Theodore Roosevelt, *The Winning of the West* (New York: The Review of Reviews Co., 1914), I, 138.

10. David V. Glass and D. E. C. Eversley, eds., *Population in History, Essays in Historical Demography* (Chicago: Aldine Publishing Co., 1965), pp. 41, 663, 675–678; Philip J. Greven, "Family Structure in Seventeenth Century Andover, Massachusetts," *William and Mary Quarterly, 23* (April 1966), 241–244; Thomas P. Monahan, *The Pattern of Age at Marriage in the United States* (Philadelphia: Stephenson Brothers, 1951), esp. pp. 103, 329, 345; William B. Bailey, "A Statistical Study of Yale Graduates, 1701–1792," *Yale Review, 16* (February 1908), 400–426; G. Stanley Hall and Theodate L. Smith, "Marriage and Fecundity of College Men and Women," *Pedagogical Seminary, 10* (September 1903), 275–314, esp. pp. 279–280 and 300–305; John Cowan, *The Science of a New Life* (New York: J. S. Ogilvie and Co., 1869), pp. 134–135.

11. William B. Bailey, *Modern Social Conditions* (New York: The Century Co., 1906), pp. 152–162; Monahan, *The Pattern of Age,* pp. 208–211; J. McKeen Cattell, "Families of American Men of Science," *Scientific Monthly, 4* (March 1917), 251–252; U.S. Department of Commerce, Bureau of the Census, *Historical Statistics of the United States* (Washington, D.C., 1960), p. 15.

12. Hall and Smith, "Marriage and Fecundity"; Bailey, "A Statistical Study."

13. Junius Henri Brown, "To Marry or Not to Marry," *Forum,* 6 (December 1888), 438.

14. G. Stanley Hall, *Adolescence* (New York: D. Appleton and Co., 1904), II, 606; Bailey, *Modern Social Conditions,* p. 156.

15. Hall and Smith, "Marriage and Fecundity," 277.

16. Norman Himes, *The Medical History of Contraception* (New York: Gamut Press, 1963, originally published by William Wood & Co., 1936), pp. 334–394; Heywood Broun and Margaret Leech, *Anthony Comstock* (New York: Literary Guild of America, 1927), p. 132; George Napheys, *The Physical Life of Woman* (Walthamstow, E.: Frederick Mayhew, 1879), pp. 97–98; Alice B. Stockham, *Tokology, a Book for Every Woman* (New York: R. F. Fenno and Co., 1893), pp. 152–157, 326; R. T. Trall, *Sexual Physiology: A Scientific and Popular Exposition of the Fundamental Problems in Sociology,* 13th ed., (New York: Wood and Holbrook, 1872), p. 206; Cowan, *The Science of a New Life,* pp. 279–305; J. B. Engelmann, "Decreasing Fecundity," *Philadelphia Medical Journal, 1* (January 18, 1902), 121–127, esp. p. 126.

17. Jennie G. Drennan, "Sexual Intemperance," *New York Medical Journal, 73* (January 5, 1901), 20.

18. Sarah E. Wiltse, "A Preliminary Sketch of the History of Child Study in America," *Pedagogical Seminary, 3* (1895), 212; J. Belangee, "Sexual Purity and the Double Standard," *Arena, 11* (February 1895), 372–373; J. McKeen Cattell, "The School and the Family," *Popular Science Monthly, 74* (January 1909), 92–93.

19. C. P. Selden, "The Rule of the Mother," *North American Review, 161* (1895), 638; G. Stanley Hall, *Educational Problems* (New York: D. Appleton & Co., 1911) I, 245, 258–259; Walter Rauschenbusch, *Christianity and the Social Crisis* (New York: Macmillan, 1907), p. 275; Mrs. Julia M. Bradley (pseud.), *Modern Manners and Social Forms* (Chicago: James B. Smiley, 1890), pp. 292, 315–317; John A. Ruth, *Decorum, a Practical Treatise on Etiquette and Dress of the Best American Society* (Detroit: F. B. Dickerson, 1879), pp. 62–63.

20. A. J. Ingersoll, *In Health,* 4th ed., revised (Boston: Lee and Shepard, 1892), pp. 108–111.

21. Henry Ward Beecher, "The Strange Woman," in *Lectures to Young Men* (New York: J. B. Ford and Co., 1873), pp. 137, 158; Phillips Brooks, "The Choice Young Man," in *The Light of the World and Other Sermons* (New York: E. P. Dutton Co., 1910), pp. 93–94.

22. Charles Franklin Thwing and Carrie F. Butler Thwing, *The Family, an Historical and Social Study* (Boston: Lee and Shepard, 1887), pp. 101, 145–151, 170–171; G. Stanley Hall, *Life and Confessions,* p. 132; William G. McLoughlin, Jr., *Billy Sunday Was His Real Name* (Chicago: University of Chicago Press, 1955), p. 133; Anton Boisen, *The Exploration of the Inner World* (New York: Willett, Clark, Co., 1936), pp. 272–274.

23. *Nation, 10* (June 9, 1870), 367.

24. John Foster Scott, *The Sexual Instinct* (New York: E. B. Treat, 1899), p. 34; Howard Atwood Kelly, *A Scientific Man and the Bible* (New York: Harper and Bros., 1925), pp. 41–61.

25. Sylvester Graham, *A Lecture to Young Men on Chastity,* 10th stereotype ed. (Boston: Charles H. Peirce, 1848), pp. 74, 83–84, 181–182; Stephen Nissenbaum, "Careful Love: Sylvester Graham and the Emergence of Victorian Sexual Theory in America, 1830–1840," Doctoral Dissertation, University of Wisconsin, 1968.

26. Scott, *The Sexual Instinct,* p. 35; William Acton, *Functions and Disorders of the Reproductive Organs,* 8th ed. (Philadelphia: P. Blakiston, 1894), pp. 15, 134, 182.

27. Hall, *Educational Problems,* I, 478.

28. Quoted in Scott, *The Sexual Instinct,* p. 138.

29. Leo Jacobi, "Sexual Intercourse: A Physiological Interpretation," *American Journal of Urology,* 6 (February 1910), 75.

30. T. S. Clouston, *The Hygiene of Mind* (New York: E. P. Dutton Co., 1907), p. 246.

31. Elizabeth Blackwell, *Counsel to Parents on the Moral Education of Their Children* (New York: Brentano Bros., 1883), pp. 21–25, 100–105; William Lee Howard, *Sex Problems in Worry and Work* (New York: Edward J. Clode Co., 1915), p. 27; Scott, *The Sexual Instinct,* pp. 69–75; Acton, *Functions and Disorders,* p. 197; Cowan, *The Science of a New Life,* pp. 30–35; William Hammond, *Sexual Impotence in the Male and Female* (Detroit: George S. Davis, 1887), pp. 129–130; T. S. Clouston, *The Neuroses of Development* (Edinburgh: Oliver and Boyd, 1891), p. 19.

32. Acton, *Functions and Disorders,* p. 126; *Medical Record* (New York), 2 (March 1, 1867), 17; J. H. Kellogg, *Plain Facts for Both Sexes* (Battle Creek, Michigan: Good Health Publishing Co., 1910), pp. 309, 314–315, 340.

33. Cowan, *The Science of a New Life,* pp. 25, 97–98, 107; James C. Jackson, *The Sexual Organism: Its Healthful Management* (Boston: B. Leverett Emerson, 1865), pp. 256–258; Stockham, *Tokology,* pp. 150–162; Kellogg, *Plain Facts for Both Sexes,* pp. 259–260, 31.

34. Joseph Howe, *Excessive Venery, Masturbation and Continence* (New York: Bermingham and Co., 1883), pp. 72–144; C. F. Lallemand, *A Practical Treatise on the Causes, Symptoms and Treatment of Spermatorrhoea,* ed. Henry J. McDougall, 3rd American ed. (Philadelphia: Blanchard and Lea, 1858), pp. 33, 292–297; Louis E. Schmidt, *Genito-Urinary and Venereal Diseases* (Philadelphia: Lea Brothers, 1902), p. 216; Kellogg, *Plain Facts for Both Sexes,* pp. 292–320. For the museums, see A. A. Brill, "Masturbation, Its Causes and Sequellae," *American Journal of Urology and Sexology,* 12 (1916), 217; Charles L. Dana, *Textbook of Nervous Diseases* (New York: William Wood, 1892 ed.), pp. 458–460, (1915 ed.), pp. 530–532; Archibald Church and Frederick Peterson, *Nervous and Mental Diseases* (Philadelphia: W. B. Saunders, 1914), p. 781; Henry Morris, *Injuries and Diseases of the Genital and Urinary Organs* (New York: William Wood, 1897), pp. 38–39.

35. Sir James Paget, *Selected Essays and Addresses* (New York: Longmans, Green and Co., 1902), pp. 33–42, 48, 51–52; E. L. Keyes and W. H. Van Buren, *A Practical Treatise on Surgical Diseases of the Genito-Uri-*

nary Organs Including Syphilis (New York: D. Appleton, 1878), p. 455; E. L. Keyes, *The Surgical Diseases of the Genito-Urinary Organs* (New York: D. Appleton, 1896), p. 442; George M. Beard, *The New Cyclopedia of Family Medicine* (New York: E. B. Treat, 1879), pp. 882–886 and *Sexual Neurasthenia,* edited with notes by George A. Rockwell, 5th ed. (New York: E. B. Treat, 1900), pp. 117–123; William T. Belfield, *Diseases of the Urinary and Male Sexual Organs* (New York: William Wood, 1884), pp. 323–325.

36. See Francis H. A. Marshall, *The Physiology of Reproduction* (New York: Longmans, Green, 1910), pp. 282–309, 349–356. For sex education essays that still carried something of the older warnings see Howard, *Sex Problems,* pp. 32, 110–111, 183–185; Maurice A. Bigelow, *Sex Education* (New York: Macmillan, 1916), pp. 143–144; Ira S. Wile, *Sex Education* (New York: Duffield and Co., 1912), pp. 82–83; Winfield S. Hall, *From Youth into Manhood,* 13th ed. (New York: Association Press, 1919), pp. 54–55; William Trufant Foster, ed., *The Social Emergency* (Boston: Houghton, Mifflin), pp. 98–99, 143–145; Paul S. Achilles, *The Effectiveness of Certain Social Hygiene Literature* (New York: American Social Hygiene Association, 1923), pp. 12–13, 15.

37. Howard A. Kelly, *Medical Gynecology* (New York: D. Appleton, 1908), pp. 296–297. Kelly was familiar with Freud and Jung; see p. 559.

38. Cowan, *The Science of a New Life,* pp. 116–117; Acton, *Functions and Disorders,* p. 189; Kellogg, *Plain Facts for Both Sexes,* pp. 506–515; Hammond, *Sexual Impotence,* p. 129; see also Beard, *Cyclopedia of Family Medicine,* p. 888, *Sexual Neurasthenia,* pp. 40, 118.

39. Acton, *Functions and Disorders,* p. 196; Beard, *Cyclopedia of Family Medicine,* p. 887; *Sexual Neurasthenia,* pp. 106, 131–132; Van Buren and Keyes, *A Practical Treatise on Surgical Diseases of the Genito-Urinary Organs,* pp. 455–456.

40. Anna M. Galbraith, *The Four Epochs of Woman's Life, a Study in Hygiene* (Philadelphia: W. B. Saunders, 1911), pp. 92–93.

41. George H. Napheys, *The Physical Life of Woman, Advice to the Maiden, Wife and Mother,* 7th ed. (London: Frederick Mayhew, 1879), pp. 74–76; compare Kelly, *Medical Gynecology,* with Edward John Tilt, *A Handbook of Uterine Therapeutics and Diseases of Women,* 4th ed. (New York: William Wood, 1881), pp. 88–90; see also Henry J. Garrigues, *A Textbook of the Diseases of Women* (Philadelphia: W. B. Saunders, 1894), p. 120; William H. Howell, ed., *An American Text-Book of Physiology* (Philadelphia: W. B. Saunders, 1897), p. 902; Howard, *Sex Problems,* p. 201; Kellogg, *Plain Facts for Both Sexes,* pp. 418–421; E. L. Keyes, *The Surgical Diseases of the Genito-Urinary Organs,* p. 435; Eugene Fuller, *Disorders of the Male Sexual Organs* (Philadelphia: Lea Brothers, 1895), p. 135.

42. Ferdinand C. Valentine, "Education in Sexual Subjects," *New York Medical Journal, 83* (February 10, 1906), 276.

43. H. Newell Martin, *The Human Body* (New York: Henry Holt, 1899), pp. 663–664.

44. H. D. Trall, *Sexual Physiology: A Scientific and Popular Exposition,* 13th ed. (New York: Wood and Holbrook Publishers, 1871), p. 69; J. Wesley

490 ‡ Notes for Pages 40–43

Bovee, ed., *The Practice of Gynecology in Original Contributions by American Authors* (Philadelphia: Lea Brothers and Co., 1906), p. 282; W. Symington Brown, *A Clinical Handbook on the Diseases of Woman* (New York: William Wood, 1882), p. 235; J. Matthews Duncan, *Clinical Lectures on the Diseases of Women*, 3rd ed. (London: J. & A. Churchill, 1886), p. 155; Kelly, *Medical Gynecology*, p. 349; Garrigues, *A Textbook of the Diseases of Women*, p. 120; William B. Atkinson, *The Therapeutics of Gynecology and Obstetrics* (Philadelphia: D. G. Brinton, 1881), p. 214; Grailey Hewitt, *The Pathology, Diagnosis and Treatment of Diseases of Women*, 2nd American ed. (Philadelphia: Lindsay and Blakiston, 1872), pp. 739–740.

45. Tilt, *Uterine Therapeutics,* p. 89; Napheys, *The Physical Life of Woman,* p. 74.

46. William Goodell, *Lessons in Gynecology* (Philadelphia: D. C. Brinton, 1887), p. 567.

47. Stockham, *Tokology,* pp. 152–153; Smith Baker, "Conjugal Aversion," JNMD, *19* (September 1892), 673; Caroline Wormeley Latimer, *Girl and Woman, a Book for Mothers and Daughters* (New York: D. Appleton, 1910), pp. 138–141.

48. A. C. McClanahan, "A Plea for Physiology of the Sexual System," *New York Medical Journal, 64* (July 4, 1896), 16.

49. Smith Baker, "Conjugal Aversion," JNMD, *19* (September 1892), 674.

50. Scott, *The Sexual Instinct,* p. 44; Bradley, *Modern Manners* pp. 371–372.

51. E. L. Keyes, "If Education upon Sexual Matters Is To Be Offered to Youth, What Should Be Its Nature and Scope . . . at What Age Should It Commence," *New York Medical Journal, 83* (1906), 275.

52. Howard, *Sex Problems,* p. 36.

53. Scott, *The Sexual Instinct,* pp. 25, 177; Bigelow, *Sex Education,* pp. 160–162.

54. Hall, *Educational Problems,* I, 469; Scott, *The Sexual Instinct,* pp. 95–99; A few physicians considered that prolonged continence might cause nervous and other disorders; see Howe, *Excessive Venery,* p. 184; Tilt, *Uterine Therapeutics,* p. 291; Napheys, *The Physical Life of Woman,* p. 77. This view became less frequent after 1900.

55. Howard, *Sex Problems,* p. 10.

56. Hall, *Life and Confessions,* p. 84.

57. Silas Weir Mitchell, *Dr. North and His Friends* (New York: The Century Co., 1903), p. 385.

58. Mitchell, *Dr. North and His Friends,* p. 103; Blackwell, *Counsel to Parents,* p. 33. Wile, *Sex Education,* pp. 5, 106.

59. James Bryce, *The American Commonwealth,* ed. Louis Hacker (New York: G. P. Putnam's Sons, 1959), II, 518; Annie Randall White, *Twentieth Century Etiquette* (1900), pp. 118, 371–372; 62; Bradley, *Modern Manners,* pp. 315–317.

60. William Dean Howells, *A Modern Instance* (Boston: James R. Osgood & Co., 1882), p. 472.

61. Henry Ward Beecher, "The Strange Woman," in *Lectures to Young Men,* p. 131.

62. Ernest Earnest, *Silas Weir Mitchell, Novelist and Physician* (Philadelphia: University of Pennsylvania Press, 1950), p. 174.

63. George S. Viereck, *Roosevelt, a Study in Ambivalence* (New York: The Jackson Press, 1919), p. 96.

64. Mabel Dodge Luhan, *Intimate Memories,* III, *Movers and Shakers* (New York: Harcourt, Brace and Co., 1936), pp. 508–509.

65. Maurice Parmelee, *Personality and Conduct* (New York: Moffat, Yard and Co., 1918), p. 130; Margaret Anderson, "Mrs. Ellis's Failure," *The Little Review, 2* (March 1915), 18; Jones, *Freud,* II, 109.

66. Stockham, *Tokology,* p. 342; Blackwell, *Counsel to Parents,* p. 33; Acton, *Functions and Disorders,* pp. 23–26; Wile, *Sex Education,* pp. 5, 106; Henry Maudsley, "The Physical Basis of Will" in *A Selection of Lectures Delivered Before the Sunday Lecture Society* (London: Sunday Lecture Society, 1886), pp. 18–24; Henry Maudsley, *Body and Will* (New York: D. Appleton, 1887), pp. 109, 264–266.

67. Hall, *Educational, Problems,* I, 229.

68. Alan Wheelis, *The Quest for Identity* (New York: W. W. Norton, 1958), p. 19.

69. Roosevelt to Hugo Münsterberg, June 3, 1901, *Letters of Theodore Roosevelt,* Elting E. Morison, ed. (Cambridge: Harvard University Press, 1951–1954), III, 86; Roosevelt to Albert Shaw, April 3, 1907, *Letters of Theodore Roosevelt,* V, 638; Roosevelt to G. Stanley Hall, November 27, 1899, *Letters of Theodore Roosevelt,* II, 1100; Theodore Roosevelt, "Social Evolution," *American Ideals and other Essays, Social and Political* (New York: G. P. Putnam's Sons, 1907), p. 326.

70. Theodore Roosevelt, "Character and Success," *Outlook, 64* (March 1900), 727.

71. Quoted in William Henry Harbaugh, *Power and Responsibility, the Life and Times of Theodore Roosevelt* (New York: Farrar, Strauss and Cudahy, 1961), p. 15.

72. Mitchell, *Dr. North and His Friends,* p. 294.

Chapter III. The Somatic Style

1. Massachusetts General Hospital, *Archives,* Ward G. Nerve, Vol. I, March 29, 1906; May 5, 1906.

2. Edward Spitzka, "Reform in the Scientific Study of Psychiatry," JNMD, 5 (April 1878), 218.

3. William James, *The Principles of Psychology* (New York: Dover Publications, 1950), I, 12–80. Moses Allen Starr, "Where and How We Remember," *Popular Science Monthly, 25* (September 1884), 609–620; Joseph Jastrow, "Localization of Function in the Cortex of the Brain," *Science, 8* (October 29, 1886), 398–399; H. A. Buttolph, "On the Physiology of the Brain and Its Relation in Health and Disease to the Faculties of the Mind," AJI, *42* (January 1886), 277–316.

4. William Hammond, "The Relations Between the Mind and the Nervous System," *Popular Science Monthly, 26* (November 1884), 7; Edward Cowles, "Neurasthenia and Its Mental Symptoms," Massachusetts Medical Society, *Medical Communications, 15* (1891), 285–388; E. C. Seguin, "An Outline of the Physiology of the Nervous System," *Medical Record* (New York), 9 (November 16, 1874), 617–622.

5. Alfred Meyer, "Emergent Patterns of the Pathology of Mental Disease," *Journal of Mental Science, 106* (July 1960), 793; W. Griesinger, *Mental Pathology and Therapeutics* (London: The New Sydenham Society, 1867), pp. 412–414, 393–394; E. C. Spitzka, *Insanity: Its Classification, Diagnosis and Treatment, a Manual for Students and Practitioners of Medicine* (New York: Bermingham & Co., 1883), pp. 100–101.

6. Comment of Francis X. Dercum on Joseph Collins and Joseph Fraenkel, "Reflections on the Nosology of the So-Called Functional Diseases," JNMD, 26 (January 1899), 29; James Hendrie Lloyd, "Hysteria: A Study in Psychology," JNMD, 10 (October 1883), 610–612; H. B. Donkin, "Hysteria," in Daniel Hack Tuke, *Dictionary of Psychological Medicine* (London: J. and A. Churchill, 1892), I, 619; George M. Beard, "Neurasthenia, or Nervous Exhaustion," Bost J, 80 (April 29, 1869), 218.

7. Editors Table, "The Progress of Mental Science," *Popular Science Monthly, 25* (June 1884), 267.

8. James Jackson Putnam, "Analysis of a Case of Circumscribed Analgesia of the Skin After Typhoid Fever," American Neurological Association, *Proceedings, I* (1875), 37–40, 41, 47–50.

9. William James, "The Sense of Dizziness in Deaf-Mutes," *American Journal of Otology, 4* (October 1882), 252–253; James Jackson Putnam to Frances Morse, July 29, 1878, PP.

10. G. Stanley Hall to Charles Eliot Norton, June 8, 1879, The Norton Papers, Houghton Library, Harvard University.

11. John Hughlings Jackson, "The Evolution and Dissolution of the Nervous System," in James Taylor, ed., *Selected Writings of John Hughlings Jackson* (London: Hodder and Stoughton, 1932), I, 72.

12. William James, "Are We Automata," *Mind, 4* (January 1879), 1, 3.

13. Morton Prince, "Hughlings Jackson on the Connection Between the Mind and the Brain," *Brain, 14,* Parts II, III (1891), 253, 250–269.

14. John Charles Bucknill and Daniel Hack Tuke, *A Manual of Psychological Medicine,* 2nd ed., rev. and enl. (London: John Churchill, 1862), pp. 72–89; Daniel Hack Tuke, ed., *A Dictionary of Psychological Medicine* (London: J. and A. Churchill, 1892), pp. 372–382; James Mark Baldwin, ed., *A Dictionary of Philosophy and Psychology,* new ed., corrected (New York: The Macmillan Co., 1911), I, 550.

15. Moses Allen Starr, "Cortical Lesions of the Brain, a Collection and Analysis of the American Cases of Localized Cerebral Disease," *American Journal of the Medical Sciences, 87* (April 1884), 379–380, *88* (July 1884), 139; Lewellys F. Barker, "The Sense Areas and Association Centres in the Brain as Described by Flechsig," JNMD, 24 (June 1897), 355.

16. John Hughlings Jackson, "The Evolution and Dissolution of the Nervous System," *Popular Science Monthly, 25* (June 1884), 171–180; Sir William Broadbent, "Hughlings Jackson as Pioneer in Nervous Physiology," *Brain, 26* (Autumn 1903), 305–366.

17. J. S. Jewell, "The Varieties and Causes of Neurasthenia," JNMD, 7 (January 1880), 11; Charles L. Dana, "Clinical Lecture on Certain Sexual Neuroses," *Medical and Surgical Reporter, 65* (August 15, 1891), 241–245; D. H. Tuke, ed., *Dictionary of Psychological Medicine,* pp. 535, 784–786; Silas Weir Mitchell, *Doctor and Patient* (Philadelphia: J. B. Lippincott Co., 1889), p. 148; George M. Beard, *American Nervousness* (New York: G. P. Putnam's, 1881), pp. 9–10; Frederick Peterson, "A Nerve Specialist to His Patients," *Colliers, 42* (January 9, 1909), 11; Francis Xavier Dercum, *A Clinical Manual of Mental Diseases* (Philadelphia: W. B. Saunders Co., 1913), pp. 370–373; Moses Allen Starr, *Nervous Diseases, Organic and Functional,* 3rd ed. (New York: Lea Bros., 1907), pp. 765–767.

18. William Osler, *Aequanimitas* (Philadelphia: P. Blakiston's Son and Co., 1905), pp. 374–382.

19. Anonymous review of second edition of Thomas S. Kirkbride, *On the Construction, Organization and General Arrangements of Hospitals for the Insane with Some Remarks on Insanity and Its Treatment,* JNMD, 8 (April 1881), 336–347.

20. Morton Prince, "How a Lesion of the Brain Results in That Disturbance of Consciousness Known as Sensory Aphasia," JNMD, *12* (July 1885), 267; S. Weir Mitchell, *Lectures on Diseases of the Nervous System, Especially in Women,* 2nd ed., rev. and enl. (Philadelphia: Lea Bros., 1885), p. 14; Beard, *A Practical Treatise on Nervous Exhaustion,* pp. 28–30.

21. Mitchell, *Lectures on Diseases of the Nervous System,* pp. 81–84; Cesare Lombroso, "Paradoxical Anarchist," *Popular Science Monthly, 56* (January 1900), 312–315; Freud, "Five Lectures on Psychoanalysis," SE, XI, 10–12.

22. G. A. Devos, "Transcultural Diagnosis of Mental Health by Means of Psychological Tests," in A. V. S. de Reuch and Ruth Porter, eds., *Transcultural Psychiatry: Ciba Foundation Symposium* (Boston: Little, Brown and Co., 1965), pp. 328–356; Alexander Leighton, *An Introduction to Social Psychiatry* (Springfield, Ill.: Charles C. Thomas, 1965), pp. 4–18.

23. Mitchell, *Lectures on Diseases of the Nervous System,* p. 86.

24. *Ibid.,* p. 13.

25. *Ibid.,* p. 267.

26. *Ibid.,* pp. 56–57.

27. S. Weir Mitchell, "The Evolution of the Rest Treatment," JNMD, *31* (June 1904), 368–373; Mitchell, *Fat and Blood, and How to Make Them,* 2nd ed., rev. (Philadelphia: J. B. Lippincott, 1878), pp. 23–24.

28. Mitchell, *Doctor and Patient,* 3rd ed. (Philadelphia: J. B. Lippincott, 1889), p. 138.

29. Robert T. Edes, "The New England Invalid," Bost J, *133* (July 18,

1895), 56; George M. Beard, *The New Cyclopedia of Family Medicine,* rev. ed. (New York: E. B. Treat, 1896), pp. 703–705.

30. Leon Edel, ed., *The Diary of Alice James* (New York: Dodd, Mead and Co., 1964), pp. 151, 149, 66, 223.

31. Owen Wister, "S. Weir Mitchell, Man of Letters," in *S. Weir Mitchell, M.D., LL.D., F.R.S., 1829–1914. Memorial Addresses and Resolutions* (Philadelphia, 1914), p. 155.

32. Ernest Earnest, *S. Weir Mitchell, Novelist and Physician* (Philadelphia: University of Pennsylvania Press, 1950), p. 180.

33. Mitchell, *Doctor and Patient,* p. 10.

34. Quoted in Earnest, *S. Weir Mitchell,* p. 236; David M. Rein, *S. Weir Mitchell as Psychiatric Novelist* (New York: International Universities Press, 1952), p. 192.

35. S. Weir Mitchell, *Wear and Tear Or Hints for the Overworked,* 5th ed., thoroughly revised (Philadelphia: J. B. Lippincott Co., 1891), p. 74.

36. *Ibid.,* pp. 63–65.

37. George M. Beard, *Sexual Neurasthenia,* 5th ed. (New York: E. B. Treat, 1902), p. 294.

38. C. B. to James Jackson Putnam, November 9, 1874, PP; E. C. Seguin, Opinion in the Case of A. L. H., to James Jackson Putnam, undated, PP.

39. James Hendrie Lloyd, "Hysteria: A Study in Psychology," JNMD, *10* (October 1883), 604–606.

40. George M. Beard, "A New Theory of Trance and Its Bearings on Human Testimony," JNMD, *4* (January 1877), 1–47; Beard, *The Study of Trance, Muscle-Reading and Allied Nervous Phenomena in Europe and America with a Letter on the Moral Character of Trance Subjects, and a Defence of Dr. Charcot* (New York, 1882), pp. 12–13; Beard, *A Practical Treatise on Nervous Exhaustion (Neurasthenia) Its Symptoms, Nature, Sequences, Treatment* (New York: William Wood & Co., 1880), pp. 26–41.

41. Compare George M. Beard, "The Influence of the Mind in the Causation and Cure of Disease—The Potency of Definite Expectation," JNMD, *3* (July 1876), 429–436, and Beard, "Mental Therapeutics," JNMD, *4* (July 1877), 581–582, with Hack Tuke, *Illustrations of the Influence of the Mind upon the Body in Health and Disease Designed to Elucidate the Action of the Imagination* (Philadelphia: Henry C. Lea, 1873), pp. 21, 113–116.

42. Quoted in Ilza Veith, *Hysteria: The History of a Disease* (Chicago: University of Chicago Press, 1965), p. 214.

43. G. Stanley Hall, "Reaction Time and Attention in the Hypnotic State," *Mind, 8* (April 1883), 177–179, and note 9.

44. Allan McLane Hamilton, *Nervous Diseases: Their Description and Treatment* (Philadelphia: Henry C. Lea, 1878), p. 383; Allan McLane Hamilton, "The Pathogeny of Mental Disease," *The Medical Record* (New York), *81* (March 23, 1912), 551–561; Edward Cowles, "Neurasthenia and Its Mental Symptoms," Massachusetts Medical Society, *Medical Communications, 15* (1891), 287–288; Charles L. Dana, "Dr.

George M. Beard," *Archives of Neurology and Psychiatry, 10* (October 1923), 427–435; Dana, "Debate on the Therapeutic Value of Hypnotism," JNMD, *19* (February 1892), 161–162.

Chapter IV. The Crisis of the Somatic Style

1. Adolf Meyer, "The Dynamic Interpretation of Dementia Praecox," *Clark Lectures and Addresses,* pp. 148–149, 154–155.

2. Meyer, "A Review of Recent Problems of Psychiatry" (1904), CPM, II, 385; Boris Sidis, "The Nature and Principles of Psychology," AJI, *56* (July 1899), 52; August Hoch, "The Psychogenic Factors in the Development of Psychoses," *Psychological Bulletin, 4* (June 15, 1907), 161.

3. This discussion is freely based on Thomas S. Kuhn, *The Structure of Scientific Revolutions,* Phoenix ed. (Chicago: University of Chicago Press, 1964).

4. Bernard Sachs, "Advances in Neurology and Their Relation to Psychiatry," American Medico-Psychological Association, *Proceedings, 4* (1897), 149. Compare the confidence of A. M. Shew, "Progress in the Treatment of the Insane," AJI, *42* (April 1886), 429–451, and Eugene C. Riggs, "An Outline of the Progress in the Care and Handling of the Insane in the Last Twenty Years," JNMD, *20* (September 1893), 620–628, with Charles G. Hill, "How Can We Best Advance the Study of Psychiatry," AJI, *64* (July 1907), 1–8, and Peter M. Wise, Presidential Address, AJI, *58* (July 1901), 82–95.

5. S. Weir Mitchell, "Address Before the Fiftieth Annual Meeting of the American Medico-Psychological Association," JNMD, *21* (July 1894), 422–423.

6. Ira Van Gieson, "First Annual Report of the Pathological Institute of the State Hospitals," *New York State Commission in Lunacy, Eighth Annual Report* (1895–1896), pp. 979–983.

7. Compare William Hammond, *A Treatise on Diseases of the Nervous System* (New York: D. Appleton and Co., 1890), pp. 760–761, L. W. Baker, "Epilepsy," JNMD, *12* (January 1885), 27–34, and Archibald Church and Frederick Peterson, *Nervous and Mental Diseases,* 1st ed. (Philadelphia: W. B. Saunders, 1899), with Church and Peterson, 1911 ed., p. 628, and Moses Allen Starr, "Is Epilepsy a Functional Disease," JNMD, *31* (March 1904), 145–156, 259–263.

8. A. I. Noble, "The Curability of Insanity," AJI, *69:2* (April 1913), 722; Hack Tuke, *The Insane in the United States and Canada* (London: H. K. Lewis, 1885), pp. 100–188; Pliny Earle, "The Curability of Insanity: A Statistical Study," AJI, *42* (September 1885), 179–209; Charles W. Pilgrim, "The Care and Treatment of the Insane in the State of New York," American Medico-Psychological Association, *Proceedings, 18* (1911), 88–89; Charles P. Bancroft, "Is There an Increase Among the Dementing Psychoses?" AJI, *71:1* (July 1914), 59–73.

9. B.-A. Morel, *Traité des Dégénerescences de l'Espèce Humaine* (Paris: J. B. Baillière, 1857), pp. 343–346, pp. vi–ix, 352–354; see W. Grie-

singer, *Mental Pathology and Therapeutics* (London: The New Sydenham Society, 1867), pp. 150–156; R. von Krafft-Ebing, *Textbook of Insanity,* trans. Charles Gilbert Chaddock (Philadelphia: F. A. Davis Company, 1905), pp. 157–164, 359; John Charles Bucknill and Daniel Hack Tuke, *A Manual of Psychological Medicine* (Philadelphia: Lindsay and Blakiston, 1862), p. 268; A. Halipré, "Bénédict-Augustin Morel, 1808–1873," *Revue Médicale de Normandie* (Rouen, 1900), pp. 58–61.

10. Charles Rosenberg, *The Trial of the Assassin Guiteau* (Chicago: University of Chicago Press, 1968), pp. 62–70, 100–102, 144–150, 160–166.

11. William Alanson White, "The Physical Basis of Insanity and the Insane Diathesis," AJI, *50* (April 1894), 536.

12. Lloyd H. Rogler and August B. Hollingshead, *Trapped: Families and Schizophrenia* (New York: John Wiley and Sons, Inc., 1965), pp. 117–214; J. Delay P. Deniker and A. Green, "Le Milieu Familial des Schizophrènes," *L'Encéphale, 46:3* (1957), pp. 189–232; Morel, *Traité des Maladies Mentales* (Paris: Librairie Victor Masson, 1860), pp. 234–239, 527. The whole problem of the role of genetic factors in mental illness is still open.

13. Morel, *Traité des Dégénerescences,* pp. 4–5, 8–9; review of Morel, "Traité des Maladies Mentales," AJI, *17* (October 1860), 199.

14. Cesare Lombroso, *Crime, Its Causes and Remedies* (Boston: Little, Brown and Co., 1911), p. xvii; Moritz Benedikt, *Anatomical Studies Upon Brains of Criminals,* trans. G. P. Fowler (New York: William Wood, 1881), pp. vii–viii.

15. Krafft-Ebing, *Textbook of Insanity,* pp. 163–165, 279–284, 361–363.

16. Valentin Magnan, *Recherches sur les Centres Nerveux, Acoolisme, Folie des Héréditaires Dégénérés, Paralysie Générale, Médecine Légale* (Paris, G. Masson, 1893), pp. iii–v, 108–112, 118–228, 244–260.

17. Theodore H. Kellogg, *Textbook on Mental Diseases* (New York: William Wood, 1897), pp. 197–198.

18. Charles L. Dana, "On the New Use of Some Older Sciences, Being a Discourse on Degeneracy and Its Stigmata," *Medical Record* (New York), *46* (December 15, 1894), 737–741.

19. A. J. Rosanoff and Florence Orr, "A Study of Heredity in Insanity in the Light of Mendelian Theory," AJI, *68* (October 1911), 221–261; Abraham Myerson, "Psychiatric Family Studies," AJI, *73* (January 1917), 355–486, esp. pp. 356–361, and Abraham Myerson, *The Inheritance of Mental Diseases* (Baltimore: Williams and Wilkins Co., 1925), pp. 275–279.

20. W. Duncan McKim, *Heredity and Human Progress* (New York: G. P. Putnam's, 1901), p. 193; William Goodell, "Clinical Notes on the Extirpation of the Ovaries for Insanity," AJI, *38* (January 1882), 294–302.

21. Hubert Work, "Presidential Address. The Sociologic Aspect of Insanity and Allied Defects," American Medico-Psychological Association, *Proceedings, 19* (1912), 134–141; John Joseph Kindred, "Eugenics: Its

Relation to Mental Diseases," American Medico-Psychological Association, *Proceedings, 24* (1917), 441–442. Carlos F. MacDonald "Presidential Address," AJI, *71:*1 (July 1914), 9–10.

22. Richard S. Dewey, "Present and Prospective Management of the Insane," JNMD, *5* (January 1878), 82.

23. James V. May, "Immigration as a Problem in the State Care of the Insane," American Medico-Psychological Association, *Proceedings, 19* (1912), 186–196; Thomas W. Salmon, "The Relation of Immigration to the Prevalence of Insanity," AJI, *64* (July 1907), 53–71, esp. pp. 61, 66; Frank G. Hyde, "Notes on the Hebrew Insane," American Medico-Psychological Association, *Proceedings, 8* (1901), 133–135; Smith Ely Jelliffe, "Dispensary Work in Nervous and Mental Diseases," JNMD, *33* (April 1906), 237.

24. Smith Ely Jelliffe, "Glimpses of a Freudian Odyssey," Q, *2* (1933), 321–322; Jelliffe, review of William Hirsch, *Genius and Degeneration*, JNMD, *24* (June 1897), 382.

25. Henry Maudsley, "Insanity in Relation to Criminal Responsibility," *Alienist and Neurologist, 17* (April 1896), 175.

26. Kellogg, *Textbook on Mental Diseases*, p. 86.

27. J. B., review of Cesare Lombroso, *Genius and Insanity*, AJI, *48* (April 1892), 529.

28. Putnam to Dana, January 22, 1904, PP.

29. Remarks of S. Weir Mitchell on J. J. Putnam, "A Study in Heredity," JNMD, *34* (December 1907), pp. 771–772.

30. Meyer, "A Review of the Signs of Degeneration," CPM, II, 257; E. C. Spitzka, review of Max Nordau, *Degeneration*, AJI, *52* (July 1895), 106–117; William James, review of Cesare Lombroso, *Entartung und Genie, Neue Studien, Psychological Review, 2* (May 1895), 288–289; Franz Boas, review of C. Lombroso and W. Ferrero, *The Female Offender, Psychological Review, 4* (March 1897), 212–213, and "Remarks on the Theory of Anthropometry," *Papers on Anthropometry* (Boston: American Statistical Association, 1894).

31. G. Hughes, "The Stigmata of Degeneration, a Cursory Editorial Critique," *Alienist and Neurologist, 18* (January 1897), 65.

32. See note 19.

33. Myerson, *The Inheritance of Mental Diseases*, p. 275.

34. *Ibid.,* p. 279.

35. George W. Stocking, Jr., "Lamarckianism in American Social Science: 1890–1915," *Journal of the History of Ideas, 23* (April–June 1962), 239–256; Myerson, "Psychiatric Family Studies," AJI, *74* (April 1918), 498; review of J. Arthur Thompson, *Heredity, Nation, 87* (November 19, 1908), 499–500.

36. *State of New York, Commission on Lunacy, 19th Annual Report, October 1, 1906–September 30, 1907*, p. 91.

37. Thomas Clouston, *Clinical Lectures on Mental Diseases* (Philadelphia: Henry C. Leas's Son & Co., 1904), p. 43. Adolf Meyer, "Psychosis" (1902), CPM, II, 288, "A Few Trends in Modern Psychiatry," *ibid.,* pp.

291, 387–388; "Review of Recent Problems of Psychiatry," *ibid.*, p. 383.

38. Emil Kraepelin, *Lectures on Clinical Psychiatry* (New York: William Wood & Co., 1906), pp. 3, 21–29.

39. Clarence B. Farrar, "Dementia Praecox in France with Some References to the Frequency of This Diagnosis in America," AJI, *62* (October 1905), 266–267.

40. *Ibid.*, n. d.; G. Alder Blumer, "The History and Use of the Term Dementia," American Medico-Psychological Association, *Proceedings, 13* (1906), 213–223; for a historical review, see D'Orsay Hecht, "A Study of Dementia Praecox," JNMD, *32* (November 1905), 689–712, (December 1905), 762–790.

41. G. H. Hill, "Dementia Praecox," American Medico-Psychological Association, *Proceedings, 7* (1900), 286.

42. Comments of Edward C. Runge, *ibid.*, p. 288.

43. See note 40.

44. Edward Runge, "Psychic Treatment," American Medico-Psychological Association, *Proceedings, 8* (1901), p. 166.

45. *Ibid.*, p. 187.

46. J. T. W. Rowe, "Is Dementia Praecox the 'New Peril' in Psychiatry," AJI, *63* (January 1907), 385–393; Bernard Sachs, "Dementia Praecox," The New York Neurological Society, JNMD, *32* (January 1905), 38.

47. A. A. Brill, *Freud's Contribution to Psychiatry* (New York: W. W. Norton Co., 1944), pp. 17–18; William Alanson White, *The Autobiography of a Purpose* (New York: Doubleday, Doran and Co., 1938), pp. 52–79; A. A. Brill, *Lectures on Psychoanalytic Psychiatry* (New York: Alfred A. Knopf, 1949), pp. 1–8; Adolf Meyer, "Thirty-five Years of Psychiatry" (1928), AMP, II, 11.

48. William Alanson White, "Types in Mental Disease," JNMD, *33* (April 1906), 256.

49. Moses Allen Starr, "Is Epilepsy a Functional Disease," JNMD, *31* (March 1904), 145–156, 104–112.

50. Henry Stedman, "The Public Obligations of the Neurologist," JNMD, *33* (August 1906), 494–495; Morton Prince, "American Neurology of the Past—Neurology of the Future," JNMD, *42* (June 1915), 452.

51. John Eric Erichsen, *On Concussion of the Spine, Nervous Shock and Other Obscure Injuries of the Nervous System* (London: Longmans, Green and Co., 1875); G. L. Walton, "Contributions to the Study of the Traumatic Neuro-Psychoses," JNMD, *17* (July 1890), 432–449.

52. J. J. Putnam, "Typical Hysterical Symptoms in Men Due to Injury, and Their Medico-Legal Significance," JNMD, *11* (July 1884), 497–499; Putnam, "On the Etiology and Pathogenesis of the Post-Traumatic Psychoses and Neuroses," JNMD, *25* (November 1898), 786–799.

53. See note 28, and compare Dana, "The Partial Passing of Neurasthenia," JNMD, *31* (February 1904), 191–193, Bost J, *150* (March 24, 1904), 339–344, with Beard, *A Practical Treatise on Nervous Exhaustion (Neurasthenia) Its Symptoms, Nature, Sequences, Treatment* (New York: E. B. Treat, 1880), p. 52.

54. See note 49.

55. J. J. Putnam and George A. Waterman, "Certain Aspects of the Differential Diagnosis Between Epilepsy and Hysteria," Bost J, *152* (May 4, 1905), 509–516.

56. Joseph Babinski, *Ma Conception de l'Hysterie et de l'Hypnotisme (Pithiatisme)* (Chartres: Durand, Imprimerie, 1906); Charles K. Mills, "Hysteria, What It Is and What It Is Not," AJI, *66* (October 1909), 231–251.

57. William James, *Principles of Psychology* (Dover Publications, 1950), I, 29–30; Adolf Meyer, "Aphasia" (1905), CPM, I, 348.

58. Lewellys F. Barker, "The Sense Areas and Association Centres in the Brain as Described by Flechsig," JNMD, *24* (June 1897), 355.

59. Sigmund Freud, *On Aphasia, a Critical Study* (1891) trans. E. Stengel (New York: International Universities Press, 1953), pp. 55–73; Henri Bergson, *Matter and Memory,* 5th ed. (New York: Macmillan, 1912), p. 157; W. M. L. "A Theory of Subcortical Aphasia," JNMD, *19* (December 1892), 927; see also B. Onuf, "A Study in Aphasia," JNMD, *24* (January 1897), 86–87.

60. Henry Head, *Aphasia and Kindred Disorders of Speech* (New York: Macmillan Co., 1926), I, 133–140, 77–78.

61. Morton Prince, "Cerebral Localization from the Point of View of Function and Symptoms," JNMD, *37* (June 1910), 343.

62. Adolf Meyer, "Segmental-Suprasegmental Concept" (1898), CPM, I, 126–127.

63. J. J. Putnam, "The Value of the Physiological Principle in the Study of Neurology," Bost J, *151* (December 15, 1904), 641–647.

64. S. Weir Mitchell, "Address to the American Neurological Association," JNMD, *36* (July 1909), 385–386. See note 50 above.

65. JNMD, *35* (December 1908), 781–785; 389–396, 401–415.

66. Charles L. Dana, "The Future of Neurology," JNMD, *40* (December 1913), 754–755. See note 44, Chapter III.

67. Edward Cowles, "Progress in the Care and Treatment of the Insane During the Half-Century," AJI, *51* (July 1894), 17.

68. Ralph Lyman Parsons, "Psychotherapy," American Medico-Psychological Association, *Proceedings, 10* (1903), 374–376, 380.

69. E. C. "Obituary, John S. Butler, M.D.," AJI, *47* (July 1890), 95; Charles W. Page, "John S. Butler: The Man and His Hospital Methods," AJI, *57* (January 1901), 477–499.

70. James Jackson Putnam, "The Value of the Physiological Principle in the Study of Neurology," Bost J, *151* (December 15, 1904), 641–647.

71. James Hendrie Lloyd, "The Metaphysical Conception of Insanity," JNMD, *31* (June 1904), 374–385.

72. J. J. Putnam to Elmer Southard, November 16, 1906, PP.

73. Morton Prince, "American Neurology of the Past—Neurology of the Future," JNMD, *42* (June 1915), 453.

Chapter V. American Psychologists and Abnormal Psychology

1. Remarks of Lewellys F. Barker on James Jackson Putnam, "Personal Experience with Freud's Psychoanalytic Method," JNMD, 37 (October 1910), 632.

2. William James, The Principles of Psychology (New York: Dover Publications, 1950), I, vii; James to Meyer, April 29, 1901, February 2, 1899, AMP.

3. Hall, Adolescence (New York: D. Appleton and Co., 1904), I, ix.

4. Hall, Life and Confessions of a Psychologist (New York: D. Appleton and Co., 1923), pp. 377, 86.

5. Hall, Morale (New York: D. Appleton and Co., 1920), p. 4; Hall, Life and Confessions, pp. 219–223.

6. Hall to Norton, quoted in Dorothy Ross, "G. Stanley Hall, 1844–1895, Aspects of Science and Culture in the Nineteenth Century," Doctoral Dissertation, Columbia Department of History, Columbia University, 1965, p. 145.

7. Dorothy Ross, "G. Stanley Hall, the New Psychology and Psychoanalysis," Paper delivered before the Conference on the History of Medicine, New Haven, May 10, 1967, p. 4.

8. Lewis Terman, "Trails to Psychology," Carl Murchison, ed., A History of Psychology in Autobiography (New York: Russell and Russell, 1961), II, 312–317.

9. Hall, Life and Confessions, pp. 131–134.

10. Ibid., p. 407; Hall, Adolescence, II, 141.

11. Ibid., pp. 66–68.

12. Sara E. Wiltse, "A Preliminary Sketch of the History of Child Study in America," Pedagogical Seminary, 3 (October 1895), 190.

13. James Mark Baldwin, Mental Development in the Child and the Race, 2nd ed. (New York: Macmillan Co., 1903), pp. 4, 5.

14. Freud to Fliess, November 5, 1897 in Marie Bonaparte et al., eds., The Origin of Psychoanalysis: Letters, Drafts and Notes to Wilhelm Fliess 1887–1902 (New York: Doubleday Anchor Books, 1957), p. 231.

15. Stephen Kern, "Freud and the Emergence of Child Psychology: 1880–1910," Doctoral Dissertation, Department of History, Columbia University, 1970, pp. 45, 75, 85, 90–91, 98.

16. Jules Dallemagne, Dégénérés et Déséquilibrés (Bruxelles: H. Lamartin, 1894), pp. 525–526, cited in Kern, "Freud and the Emergence of Child Psychology," p. 45.

17. Earl Barnes, "The Development of Feelings and Ideas of Sex in Children," Pedagogical Seminary, 2 (1892), 199–203; Putnam to Freud, November 17, 1909, JPP, p. 86.

18. Sanford Bell, "A Preliminary Study of the Emotion of Love Between the Sexes," AJP, 13 (July 1902), 328.

19. *Ibid.*, p. 327.

20. Freud, "The Origin and Development of Psychoanalysis," *Clark Lectures and Addresses*, p. 28.

21. Kern, *Freud and the Emergence of Child Psychology*, pp. 44–51.

22. Colin Scott, "Sex and Art," AJP, 7 (January 1896), 198, 226.

23. James Leuba, "A Study in the Psychology of Religious Phenomena," AJP, 7 (April 1896), 309–385; "National Destruction and Construction in France as Seen in Modern Literature and in the Neo-Christian Movement," AJP, 5 (July 1893), 496–539.

24. W. I. Thomas, "The Sexual Element in Sensibility," *Psychological Review*, 11 (January 1904), 61–67.

25. George E. Dawson, "Psychic Rudiments and Morality," AJP, *11* (January 1900), 203.

26. Lewis Terman, "A Study in Precocity and Prematuration," AJP, *16* (April 1905), 145–183.

27. Hall, "A Study of Anger," AJP, *10* (July 1899), 524.

28. Edward L. Thorndike, review of Hall, *Adolescence, Educational Review*, 28 (October 1904), 217–227.

29. James, *The Varieties of Religious Experience* (New York: Longmans, Green and Co., 1902), p. 270.

30. Quoted in Ralph Barton Perry, *The Thought and Character of William James* (Boston: Atlantic Monthly Press, Little Brown and Co., 1935), II, 317.

31. *Ibid.*, p. 168.

32. James, *Principles of Psychology*, I, 22–23.

33. *Ibid.*, II, 548, 573, 579; and "What the Will Effects," *Scribner's Magazine*, 3 (January–June 1888), 240–250.

34. Perry, *The Thought and Character of William James*, II, 675; and Gay Wilson Allen, *William James* (New York: Viking Press, 1967), pp. 162–167, 212–214.

35. Quoted in Perry, *The Thought and Character of William James*, I, 786.

36. Quoted in Grace Foster, "The Psychotherapy of William James," R, *32* (July 1945), 307.

37. James, *Principles of Psychology*, II, 552ff (footnote).

38. Quoted in Hutchins Hapgood, *A Victorian in the Modern World* (New York: Harcourt Brace and Co., 1939), p. 77; *Principles of Psychology*, II, 43.

39. *Principles of Psychology*, II, 437.

40. "Words of Professor Royce at the Walton Hotel at Philadelphia, December 29, 1915," in "Papers in Honor of Josiah Royce on his Sixtieth Birthday," *Philosophical Review*, 25 (May 1916), 508–509.

41. Royce, "John Bunyan" and "Some Observations on the Anomalies of Self-Consciousness," in *Studies in Good and Evil: A Series of Essays upon Problems of Philosophy and Life* (New York: D. Appleton and Co., 1915, *c.* 1898), p. 190.

42. Robert S. Woodworth, "Josiah Royce: November 20, 1855–September 14, 1916," National Academy of Science, *Biographical Memoirs, 33* (New York: Columbia University Press, 1959), 387.

43. Hocking, "The Holt-Freudian Ethics and the Ethics of Royce," *Philosophical Review, 25* (May 1916), 479–506.

44. Julius Nelson, "A Study of Dreams," AJP, *1* (May 1888), 374, 380–381; George Trumbull Ladd, "Contribution to the Psychology of Visual Dreams," *Mind* (n. s.), *1* (April 1892), 299–304; Mary Calkins, "Statistics of Dreams," AJP, *5* (April 1893), 311–343; Freud, "Interpretation of Dreams," SE, IV, 32–33.

45. Joseph Jastrow, *The Subconscious* (Boston: Houghton Mifflin Co., 1906), pp. 179–182, 214–220.

46. James Rowland Angell, *Psychology* (New York: Henry Holt, 1904), p. 297.

47. E. L. Thorndike, "What Instruction in Educational Psychology Should be Given in a Professional Course for Teachers," *Teachers College Record, 6* (January 1905), 20–21.

48. E. L. Thorndike, *Educational Psychology* (New York: Lemcke and Buechner, 1903), p. 164.

49. Prof. James McKeen Cattell, "The Progress of Psychology," *Popular Science Monthly, 43* (October 1893), 785.

50. Henry Adams, *The Education of Henry Adams* (Boston: Houghton Mifflin Co., 1961), p. 434.

Chapter VI. American Psychotherapy

1. Morton Prince, *The Dissociation of a Personality* (New York: Longmans, Green and Co., 1905), p. 2.

2. *Ibid.*, pp. 519–520.

3. *Ibid.*, p. 406.

4. *Ibid.*, pp. 11–13.

5. *Ibid.*, pp. 214–215, 388–389.

6. *Ibid.*, p. 226.

7. *Ibid.*, p. 18.

8. Putnam, review of *The Dissociation of a Personality*, J Ab P, *1* (October 1906), 238.

9. Prince, "Thought Transference," Bost J, *116* (February 3, 1887), 107–112.

10. Putnam, "Remarks on the Psychical Treatment of Neurasthenia," Bost J, *132* (May 28, 1895), 505.

11. Clarence B. Farrar, "Psychotherapy and the Church," JNMD, *36* (January 1909), 11–24.

12. "Debate on The Therapeutic Value of Hypnotism," JNMD, *19* (February 1892), 161–162; Charles L. Dana, "The Nature of So-Called Double-Consciousness," *Science, 7* (April 2, 1886), 311–312 and "The Study

of a Case of 'Amnesia' or Double Consciousness," *Psychological Review, 1* (November 1894), 570–580.

13. W. P. Wilkin, "The Practical Use of Hypnotism in Public Clinics," *The Post Graduate,* New York, *11* (1896), 278–286; Moses Allen Starr, *Nervous Diseases, Organic and Functional* 2nd ed. (New York: Lea Bros., 1907), pp. 790–793; "Hypnosis," *Current Literature, 29* (October 1900), 389–390.

14. Lightner Witmer, "Mental Healing and the Emmanuel Movement," *Psychological Clinic, 2* (December 15, 1908), 220–221; John Kearsley Mitchell, *Five Essays,* ed. S. Weir Mitchell (Philadelphia: J. B. Lippincott and Co., 1859).

15. T. B. Keyes, "Hypnotic Anaesthesia," *American Journal of Obstetrics, 34* (September 1896), 369–371; W. X. Sudduth, "Suggestion as an Idio-Dynamic Force," JAMA (January 18, 1896), 106–111; William Lee Howard, "The Practical Use of Hypnotic Suggestion," *New York Medical Journal, 77* (April 18, 1903), 673–676.

16. Morton Prince, *The Nature of Mind and Human Automatism* (Philadelphia: J. B. Lippincott, 1885), pp. 58, 66.

17. *Ibid.,* pp. 156–158.

18. Morton Prince, "Hughlings Jackson on the Connection Between the Mind and the Brain," *Brain, 14* (1891), 250–260.

19. Freud, "Charcot" (1893), CPF, I, 22.

20. Ola Andersson, *Studies in the Prehistory of Psychoanalysis: The Etiology of Psychoneuroses and Some Related Themes in Sigmund Freud's Scientific Writings and Letters 1886–1896* (Sweden: Norstedt and Söner, Svenska Bokförlaget, 1962).

21. Quoted in Pierre Janet, *Psychological Healing* (New York: The Macmillan Co., 1925), I, 173.

22. Alfred Binet, *On Double Consciousness* (Chicago: Open Court Publishing Co., 1896), p. 6; Theodule Ribot, *The Diseases of Memory, an Essay in the Postive Psychology,* trans. William H. Smith (New York: D. Appleton & Co., 1887).

23. Janet, "Les Idées Inconscients et le Dédoublement de la Personalité," *Revue Philosophique, 22* (1886), 577–592; F. W. H. Myers, *Automatic Writing,* II, The Society for Psychical Research, *Proceedings,* III (London, 1885), 1–63, esp. p. 27.

24. James, *Principles of Psychology,* I, 394; Morton Prince, "The Unconscious," J Ab P, *3* (October–November 1908), 287–288; Janet, *The Mental State of Hystericals* (New York: G. P. Putnam's Sons, 1901), p. 67.

25. Janet, *L'Automatisme Psychologique* (Paris: Felix Alcan, 1889), p. 3.

26. Janet, *Psychological Healing,* I, 642, 657–658.

27. Breuer and Freud, "Preliminary Communication," SE, II, 12.

28. Quoted in Ralph Barton Perry, *The Thought and Character of William James* (Boston: Atlantic Monthly Press, Little Brown and Co., 1935), II, 123.

29. James, *The Varieties of Religious Experience* (New York: Longmans, Green and Co., 1902), pp. 234–235.

30. Morton Prince, "Some of the Revelations of Hypnotism, Post-Hypnotic Suggestion, Automatic Writing and Double Personality," Bost J, *122* (May 8, 1890), 463–467, (May 15), 475–476, (May 22), 493–495; Prince, "Remarks on Hypnotism as a Therapeutic Agent," *ibid.* (May 8, 1890), 448.

31. *Ibid.* (May 15, 1890), pp. 475–476.

32. John Holland Mackenzie, "The Production of the So-Called 'Rose Cold' by Means of an Artificial Rose," *American Journal of the Medical Sciences, 91,* n.s. (January 1886), 57.

33. Prince, "Association Neuroses: A Study of the Pathology of Hysterical Joint Affections, Neurasthenia and Allied Forms of Neuro-Mimesis," JNMD, *18* (May, 1891), 257–282.

34. Prince, "The Psychological Principles and Field of Psychotherapy," in Prince et al., *Psychotherapeutics* (Boston: Richard G. Badger, 1909), pp. 14–16; Prince, "Habit Neuroses as True Functional Diseases," Bost J *139* (December 15, 1898), 589–592; Prince, "The Educational Treatment of the Psychoneuroses," *ibid.* (October 6, 1898), 332–337.

35. Prince, "Sexual Perversion or Vice: A Pathological and Therapeutic Inquiry," JNMD, *25* (April 1898), 240.

36. *Ibid.*, p. 248.

37. *Ibid.*, p. 251.

38. *Ibid.*

39. Prince, "Fear Neurosis," Bost J, *139* (December 22, 1898), 616.

40. Prince, "The Educational Treatment of Neurasthenia and Certain Hysterical States," *ibid.* (October 6, 1898), 335.

41. Putnam, [The Establishment of Truth, undated, untitled] Harvard College essays, "St. Francis," June 15 [No.] 20; James to Putnam, January 17, 1879 PP; James, "Are We Automata," *Mind, 4* (January 1879), 20–21; Putnam to Charles Moorfield Storey, April 20 [1864] and September 7, 1867, in the possession of Charles Moorfield Storey; Putnam, "Neuralgia," in William Pepper, ed., *A System of Practical Medicine by American Authors* (Philadelphia: Lea Bros., 1886), V, 1221.

42. Putnam, "The Bearings of Philosophy on Psychiatry," *British Medical Journal* (October 20, 1906), 1023.

43. Putnam, "Remarks on the Psychical Treatment of Neurasthenia," Bost J, *132* (May 28, 1895), 506.

44. Putnam, "A Consideration of Mental Therapeutics as Employed by Special Students of the Subject," Bost J, *151* (August 18, 1904), 181.

45. Putnam, "The Bearing of Philosophy on Psychiatry," *British Medical Journal* (October 20, 1906), 1022.

46. Putnam, "The Treatment of Psychasthenia from the Standpoint of the Social Consciousness," *American Journal of the Medical Sciences, 135* (January 1908), 78.

47. Putnam, "Not the Disease Only but Also the Man," The Shattuck Lec-

ture, 1889, Massachusetts Medical Society, *Medical Communications, 18* (1899–1901), 58–59, 72.

48. Putnam, "The Treatment of Psychasthenia," *American Journal of the Medical Sciences, 135* (January 1908), 83, 79.

49. Putnam, "Neurasthenia," in Alfred Labbeus Loomis and William Gilman Thompson, eds., *A System of Practical Medicine by American Authors* (New York: Lea Bros., 1898), IV, 583, 572–573.

50. Putnam, "The Psychology of Health, IV," *Psychotherapy: A Course of Reading in Sound Psychology, Sound Medicine and Sound Religion* (New York: Centre Publishing Co., 1909), II, No. 1, p. 44.

51. Putnam, *American Journal of the Medical Sciences, 135* (January 1908), 87–88.

52. Putnam, "The Philosophy of Psychotherapy, II," *Psychotherapy: A Course of Reading,* III, No. 4 (1909), 28–38.

53. Putnam, "Not the Disease Only but Also the Man," p. 50; Putnam, "Remarks on the Psychical Treatment of Neurasthenia," Bost J, *132* (May 23, 1895), 510.

54. See note 51 above; Putnam, "The Treatment of Psychasthenia," *American Journal of the Medical Sciences, 135* (January 1908), 79.

55. See note 49.

56. *Ibid.*

57. James to Putnam, August 19, 1908, JPP, p. 74, Remarks of James to the Boston Society for Medical Improvement, Bost J, *132* (May 3, 1895), 516–517.

58. Prince to Hugo Münsterberg, October 17, 1905 and February 21, 1905, Hugo Münsterberg papers, Boston Public Library; Prince, "The Desirability of Instruction in Psychopathology in Our Medical Schools and Its Introduction at Tufts," Bost J, *159* (October 15, 1908), 498.

59. Compare the psychotherapists James Jackson Putnam (b. 1846), Morton Prince (b. 1854), Boris Sidis (b. 1868), Lewellys F. Barker (b. 1867), William Alanson White (b. 1870), Edward Wyllys Taylor (b. 1866) with the somaticists Bernard Sachs (b. 1858), Francis X. Dercum (b. 1856), Charles K. Mills (b. 1845), E. C. Spitzka (b. 1852), James Hendrie Lloyd (b. 1853), and S. Weir Mitchell (b. 1829).

60. Lewellys F. Barker, "The Psychic Treatment of the Functional Neuroses," *International Clinics,* 17th series, I (1907), 15–16.

61. Tom A. Williams, "Psychoprophylaxis in Childhood," in Morton Prince *et al., Psychotherapeutics* (Boston: Richard G. Badger, 1909), pp. 148, 164–165; Boris Sidis, *Genius and Philistine* [Address at Commencement of the Harvard Summer School in 1909] (Boston: Richard G. Badger, 1917), pp. 29, 15–16, 26.

62. Second Annual Report of Social Work at the Massachusetts General Hospital, October 1, 1906 to October 1, 1907 (Boston: The Fort Hill Press, undated), pp. 22–28.

63. Richard Cabot, "The Analysis and Modification of Environment," *Psychotherapy: A Course of Reading,* III, No. 3 (1909), 7; Cabot, "The American Type of Psychotherapy," *ibid.,* I, No. 1 (1908), 7–8.

64. Sidney I. Schwab, "The Use of Social Intercourse as a Therapeutic Agent in the Psychoneuroses, a Contribution to the Art of Psychotherapy," JNMD, *24* (August 1907), 497–503.

65. Putnam, "Personal Experience with Freud's Psycho-Analytic Method" (1910), Ad Psa, p. 41.

66. Sidney I. Schwab, "Neurasthenia Among Garment Workers," *Bulletin of the American Economic Association,* 4th series, No. 2 (April 1911), 265–270.

67. Richard Cabot, "The Use and Abuse of Rest," *Psychotherapy: A Course of Reading,* II, No. 2 (1909), 31–32.

68. Robert S. Woodworth, "Psychiatry and Experimental Psychology," American Medico-Psychological Association, *Proceedings, 13* (1906), 128–129.

69. Sidis, "The Psychotherapeutic Value of the Hypnoidal State," in *Psychotherapeutics,* pp. 124–130, 111, 113–114; William James, "The Energies of Men," *Philosophical Review, 16* (January 1907), 1–20; Sidis, *Philistine and Genius* (New York: Moffat, Yard and Co., 1911), p. 105.

70. Morton Prince, "The Psychological Principles and Field of Psychotherapy," *Psychotherapeutics,* p. 31.

71. Janet, *Psychological Healing,* I, 127, 118–119; Smith Ely Jelliffe, "Glimpses of a Freudian Odyssey," Q, *2* (1933), 323; Frank Hallock, "The Educational or Plain Talk Method in Psychotherapy," *Psychotherapy: A Course of Reading,* II, No. 4 (1909), 5–12; Paul Dubois, *The Psychic Treatment of Nervous Disorders,* trans. and ed. Smith Ely Jelliffe and William A. White (New York: Funk and Wagnalls, 1909), pp. 25–27, 242–243, 342.

72. Review of Thomas Clouston, *The Neuroses of Development, Brain, 18,* Part IV (1895), 600.

73. Janet, *The Major Symptoms of Hysteria* (New York: The Macmillan Co., 1907), pp. 12, 321–324.

74. Janet, *Les Obsessions et la Psychasthenie* (Paris: Felix Alcan, 1903), I, 621–622, 700–715, 495–497.

75. *Ibid.,* pp. 627–628.

76. *Ibid.,* pp. 606–613, 632–633; Janet, *The Mental State of Hystericals,* 526–527; *Les Névroses* (Paris: Ernest Flammarion, 1919), pp. 392–393.

77. Leonhard Schwartz, *Les Névroses et la Psychologie Dynamique de Pierre Janet* (Presses Universitaires de France, 1955).

78. Hugo Münsterberg, *Psychotherapy* (New York: Moffat, Yard and Co., 1909), p. x.

79. Hugo Münsterberg *et al., Subconscious Phenomena* (Boston: Richard G. Badger, 1910), pp. 9–15 and in J Ab P, *2* (April–May 1907), 25–44, (June–July 1907), 58–80; Prince, "Some of the Present Problems of Abnormal Psychology," *Psychological Review, 12* (March–May 1905), 118–143.

80. Prince, "The Psychological Principles and Field of Psychotherapy," *Psychotherapeutics,* pp. 13–18.

81. Janet, *Psychological Healing,* II, 733.

82. Morton Prince and Isador Coriat, "Cases Illustrating the Educational Treatment of the Psycho-Neuroses," J Ab P, *2* (October–November 1907), 166.

83. Edward Wyllys Taylor, "The Attitude of the Medical Profession Toward the Psychotherapeutic Movement," JNMD, *35* (June 1908), 402.

84. *Ibid.,* p. 414.

85. Sidis, *The Psychology of Suggestion* (New York: D. Appleton and Co., 1898), p. 15.

86. Schwab, "The Use of Social Intercourse as a Therapeutic Agent in the Psychoneuroses," JNMD, *34* (1907), 501.

87. Bronislaw Onuf (Onufrowicz), "Psychotherapy," JAMA, *50* (June 6, 1908), II, 1893.

88. Prince *et al., Psychotherapeutics,* p. 33; Joseph Collins, "Some Fundamental Principles in the Treatment of Functional Nervous Diseases, with Especial Reference to Psychotherapy," *American Journal of the Medical Sciences, 135,* n.s. (February 1908), 176.

89. Putnam, "The Work of Sigmund Freud" (1917), Ad Psa, p. 358.

90. Janet, *The Major Symptoms of Hysteria,* p. 6.

Chapter VII. Functional Psychiatry

1. William Alanson White, *The Autobiography of a Purpose* (New York: Doubleday Doran, 1938), pp. 57, 76; *Outlines of Psychiatry* (New York: Journal of Nervous and Mental Disease Publishing Co., 1907), pp. 16–17.

2. Emanuel Régis and A. Hesnard, "La Doctrine de Freud et Son École," *L'Encéphale, 8* (June 10, 1913), 542; Adolf Meyer, "Remarks on Habit Disorganization . . ." (1905), CPM, II, 421–431.

3. Edward Cowles, "Insistent and Fixed Ideas," AJP, *1* (February 1888), 224–225.

4. *Ibid.,* p. 237.

5. *Ibid.,* pp. 258, 256.

6. Cowles, "The Mechanism of Insanity," I, AJI, *46* (April 1890), 457–485, II, AJI, *47* (April 1891), 471–495; III, AJI, *48* (October 1891), 209–252.

7. Cowles, "Neurasthenia and Its Mental Symptoms," Massachusetts Medical Society, *Medical Communications, 15* (1891), 285–388; Meyer, "Remarks on Habit Disorganization," (1905), CPM, II, 421; Adolf Meyer, "Thirty-five Years of Psychiatry in the United States and Our Present Outlook," (July 1928), CPM, II, 6.

8. A. B. Richardson, "Habit in Insanity," AJI, *43* (April 1887), 422, 426.

9. Charles W. Page, "The Adverse Consequences of Repression," AJI, *49* (January 1893), 373, 374; Charles P. Bancroft, "Subconscious Homicide and Suicide: Their Physiological Psychology," AJI, *55* (October 1898), 263–273.

10. Boris Sidis, *Psychopathological Researches* (New York: G. E. Stechert, 1902), pp. 16–17.

11. Sidis and White, "Mental Dissociation in Functional Psychosis," *ibid.,* pp. 33–102.

12. Adolf Meyer, review of Sidis, *Studies in Psychopathology, Journal of Philosophy, Psychology and Scientific Methods,* 4 (November 7, 1907), 638.

13. *Psychopathological Researches,* p. 153.

14. Boris Sidis and Simon P. Goodhart, *Multiple Personality: An Experimental Investigation into the Nature of Human Individuality* (New York: G. Appleton and Co., 1905), pp. 83–226; Marcus Wyler, review of above, *Journal für Psychologie und Neurologie,* 7 (1906), 197–200.

15. Eunice E. Winters, "Adolf Meyer's Two and a Half Years at Kankakee, May 1, 1893–November 1, 1895," *Bulletin of the History of Medicine,* 40 (September–October 1966), 455.

16. Meyer, "Schedule for the Study of Mental Abnormalities in Children" (1895), CPM, IV, 339.

17. William James to Adolf Meyer, April 29 [?], AMP.

18. Adolf Meyer, CPM, II, 273.

19. Adolf Meyer, review of A. T. Schofield, "The Mental Factor in Medicine" (1902), CPM, II, 632–633.

20. "A Few Trends in Modern Psychiatry" (1904), CPM, II, 393; "A Few Remarks Concerning the Organization of Work in Large Hospitals" (1902), *ibid.,* p. 89.

21. Meyer, "A Few Trends in Modern Psychiatry," CPM, II, 403.

22. Meyer, "An Attempt at Analysis of the Neurotic Constitution," CPM, II, 321–330.

23. Meyer, "The Dynamic Interpretation of Dementia Praecox, CPM, II, 446; "Remarks on Habit Disorganization," *ibid.,* pp. 421–431.

24. Meyer, "Fundamental Conceptions of Dementia Praecox," CPM, II, 432–437.

25. Meyer, "Fundamental Conceptions of Dementia Praecox," JNMD, *34* (May 1907), 335.

26. G. Alder Blumer, "The Coming of Psychasthenia," JNMD *33* (May 1906), 338.

27. Isador Coriat to Adolf Meyer, May 10, 1905, AMP.

28. Henri Ellenberger, *The Discovery of the Unconscious* (New York: Basic Books, 1970); Stephen Kern, "Freud and the Emergence of Child Psychology: 1880–1910," Doctoral Dissertation, Department of History, Columbia University, 1970.

29. Jules Dallemagne, *Dégénérés et Déséquilibrés* (Bruxelles: H. Lamertin, 1894), p. 525, cited in Kern, "Freud and the Emergence of Child Psychology: 1880–1910, p. 45.

30. Freud, "Five Lectures on Psychoanalysis" (1909), SE, XI, 42; G. Stanley Hall, *Adolescence* (New York: D. Appleton, 1904), II, 95.

31. H. Oppenheim, *Letters on Psychotherapeutics* (New York: G. E. Stechert, 1907), *Diseases of the Nervous System, a Textbook for Students and*

Practitioners of Medicine, trans. Edward G. Mayer, 2nd American ed. (Philadelphia: J. B. Lippincott, 1904), pp. 704–707, 721, 734, 738–741; T. Ziehen, "Hysteria," in Archibald Church, ed., *Diseases of the Nervous System,* an authorized translation from "Die Deutsche Klinik" (New York: D. Appleton and Co., 1908), pp. 1091–1094, 1077.

32. Gilbert Ballet, *Traité de Pathologie Mentale* (Paris: Octave Doin, 1903), pp. 818–820; Jules Babinski, *Ma Conception de l'Hysterie et de l'Hypnotisme (Pithiatisme)* (Chartres: Imprimerie Durant, 1906), p. 12; Régis, *Précis de Psychiatrie,* 4th ed. (Paris: Octave Doin et Fils, 1909), pp. 892, 105.

33. Archibald Church and Frederick Peterson, *Nervous and Mental Diseases,* 5th ed. (Philadelphia: W. B. Saunders, 1905), pp. 578, 605; L. Harrison Mettler, *A Treatise on Diseases of the Nervous System* (Chicago: Cleveland Press, 1905), pp. 102–104; Moses Allen Starr, *Nervous Diseases, Organic and Functional* (New York: Lea and Febiger, 1909), pp. 824–826, 813–816, 841–847.

34. August Forel, "Fondation de la Société Internationale de Psychologie Médical et de Psychothérapie," *Informateur des Alienistes et des Neurologistes* (Supplement mensuel de *l'Encéphale,* Paris), 1910, V (February 25, 1910), 42–45; "La Psychologie et la Psychothérapie à l'Université, *Journal für Psychologie und Neurologie, 17,* Ergänzungsheft, 307, 314–315.

35. Review of Paul Sollier, "L'Hysterie et son Traitement," *Revue Neurologique, 9* (September 30, 1910), 914; Francis X. Dercum, *Rest, Mental Therapeutics, Suggestion* (Philadelphia: P. Blakiston's Son & Co., 1903), p. 107.

36. Stewart Paton, *Psychiatry: A Text-Book for Students and Physicians* (Philadelphia: J. B. Lippincott, 1905), pp. 3, 66–126, 146–166.

37. Remarks of Dr. Deady on Charles L. Dana, "The Limitation of Hysteria," American Neurological Association, *Proceedings, 32* (1906), 76.

Chapter VIII. Acquaintance and Conversion

1. Putnam to Freud, April 14, 1910, JPP, p. 98.

2. Freud, "Some Points for a Comparative Study of Organic and Hysterical Motor Paralyses" (1893), SE, I, 157–172.

3. Freud, "Heredity and the Aetiology of the Neuroses" (1896), SE, III, 141–185 and "My Views on the Part Played by Sexuality in the Aetiology of the Neuroses" (1906), SE, VII, 275–276.

4. Peter Amacher, *Freud's Neurological Education and Its Influence on Psychoanalytic Theory* (New York: International Universities Press, 1965), *passim.*

5. J. Mitchell Clarke, review of Breuer and Freud, *Studien über Hysterie, Brain, 19,* Parts II, III (1896), 401–414.

6. Meyer, "The Treatment of Paranoic and Paranoid States" (1913), CPM, II, 521.

7. Review of "The Structure of Living Nerves and Nerve Cells," JNMD, *9* (October 1882), 784–785.

8. Moses Allen Starr, "The Anatomy of the Nervous System," JNMD, *12* (July 1885), 386, and *13* (April-May 1886), 246; Bernard Sachs, *A Treatise on the Nervous Diseases of Children for Physicians and Students* (New York: William Wood and Co., 1895), pp. 536–539.

9. Freud, "A New Histological Method for the Study of Nerve Tracts in the Brain and Spinal Cord," *Brain, 7* (April 1884), 84–88; Dr. Henry S. Upson, "On Gold as a Staining Agent for Nerve Tissues," JNMD, *15* (November 1888), 685.

10. See note 5.

11. Dr. Beevor on S. Weir Mitchell, "Rest Treatment in Relation to Psychotherapy," American Neurological Association, *Proceedings, 34* (1908), p. 218.

12. James Jackson Putnam, "Personal Impressions of Sigmund Freud and His Work" (1909), Ad Psa, p. 3.

13. Erwin W. Runkle, "Psychological Literature, Studien über Hysterie von Dr. Jos. Breuer und Dr. Sigm. Freud," AJP, *10* (1898–1899), 592–593.

14. William James, review of Breuer and Freud, "Ueber den Psychischen Mechanismus Hysterischer Phänomene," *Psychological Review, 1* (March 1894), 199.

15. James Jackson Putnam, "Remarks on Psychical Treatment of Neurasthenia," Bost J, *132* (May 23, 1895), 510.

16. Robert T. Edes, "The New England Invalid," Bost J, *133* (July 18, 1895), 56.

17. August Forel, "Hypnotism and Cerebral Activity," *Clark University 1889-1899 Decennial Celebration* (Worcester, Mass., 1899), pp. 412–413.

18. Lewellys F. Barker, "On the Psychic Treatment of Some of the Functional Neuroses," *International Clinics,* 17th series, *1* (1907), 10–11; Isador Coriat, *Abnormal Psychology* (New York: Moffat, Yard & Co., 1910), p. 155.

19. Putnam, "Recent Experiences in the Study and Treatment of Hysteria at the Massachusetts General Hospital; with Remarks on Freud's Method of Treatment by 'Psycho-Analysis,' " J Ab P, *1* (April 1906), 26–41.

20. *Ibid.,* p. 27; Boris Sidis to William James, October 9, 1905, WJP.

21. Putnam, J Ab P, *1* (April, 1906), 30.

22. *Ibid.,* pp. 37–38.

23. *Ibid.,* p. 37.

24. *Ibid.,* pp. 36, 41.

25. Robert T. Edes, "The Present Relations of Psychotherapy," JAMA, *52* (January 9, 1909), 96.

26. Prince, "The Psychological Principles and Field of Psychotherapy," in Prince et al., *Psychotherapeutics* (Boston: Richard G. Badger, 1909), pp. 38–40.

27. *Ibid.,* p. 37.

28. Harry Linenthal and Edward Wyllys Taylor, "The Analytic Method in Psychotherapeutics. Illustrative Cases," Bost J, *155* (November 8, 1906), 542.

29. Bronislaw Onuf, "Psychotherapy," JAMA, *50* (June 6, 1908), 1892–1897; Francis Xavier Dercum, "An Analysis of Psychotherapeutic Methods," *Therapeutic Gazette* (Detroit), *32* (May 15, 1908), 305–316.

30. Leo B. Allen, "Psychotherapy," *University of Pennsylvania Medical Bulletin, 21* (May 1908), 76–80; L. Pierce Clark, "Freud's Method of Psychotherapy," JNMD, *35* (June 1908), 391–392; C. L. Allen, "Some Consideration with Regard to Present Popular Interest in Psycho and Religio-therapy; What Lessons Should It Convey to Physicians," *Southern California Practitioner, 24* (July 1909), 343–346; W. Jarvis Barlow, "Psychotherapy," *California State Journal of Medicine, 7* (August 1909), 274–279; C. Lull, "Remarks on Psychotherapy," *Alabama Medical Journal, 20* (1908), 395–400.

31. See note 19.

32. Sidney I. Schwab, "An Estimate of Freud's Theory of the Neuroses and Its Value to the Neurologist," *Interstate Medical Journal, 18* (September 1911), 938–948.

33. "Sigmund Freud's Foolish Conclusion," editorial, *Alienist and Neurologist, 20* (January 1899), 113–114; cf. also editorial, *Alienist and Neurologist, 17* (October 1896), 519–520.

34. Bronislaw Onuf, review of Freud, "The Warding-Off Neuro-Psychosen (Die Abwehr Neuro-Psychosen)," JNMD, n.s., *20* (February 1895), 129.

35. Hugh T. Patrick, "The 'Anxiety Neurosis'—Freud," *ibid.,* pp. 196–198.

36. "Etiology of Hysteria," JAMA, *27* (August 15, 1896), 393.

37. Havelock Ellis, "Hysteria in Relation to the Sexual Emotions," *Alienist and Neurologist, 19* (October 1898), 609.

38. Putnam, "Recent Experiences," J Ab P, *1* (April 1906), 28.

39. Adolf Meyer, "Interpretation of Obsessions," (1906), CPM, II, 633–635.

40. See note 11; "Discussion of S. Weir Mitchell, 'Rest Treatment in Relation to Psychotherapy,'" American Neurological Association, *Proceedings, 34* (1908), 215.

41. A. A. Brill, review of Freud, "Hysterical Fancies and Their Relations to Bisexuality," JNMD, *36* (May 1909), 311–313; Adolf Meyer, "A Discussion of Some Fundamental Issues in Freud's Psychoanalysis" (1909–1910), CPM, II, 604–617.

42. See note 20 and Sidis, review of "The Psychopathology of Everyday Life," J Ab P, *1* (June 1906), 103; Frederick Peterson, "Some New Fields and Methods in Psychology," *New York Medical Journal, 90* (November 13, 1909), 947.

43. Meyer, abstract and review of Jung, "Diagnostische Assoziationstudien," *Psychological Bulletin, 2* (June 15, 1905), 242–250; "Application of Association Studies," *ibid., 3* (August 15, 1906), 275–280.

44. Meyer, review of Sidis, *Journal of Philosophy, Psychology and Scientific Methods, 4* (November 1907), 638, "Nineteenth Annual Report of the State Commission in Lunacy, September 30, 1907, CPM, II, 151.

45. Jung, "On Psychophysical Relations of the Association Experiment," J Ab P, *1* (February 1907), 250.

46. William Alanson White, "The Theory of the Complex," *Interstate Medical Journal, 16* (1909), 256; Robert M. Yerkes and Charles S. Berry, "The Association Reaction Method of Mental Diagnosis," AJP, *20* (January 1909), 22–37; Hugo Münsterberg, *On the Witness Stand* (New York: Doubleday Page, 1909), pp. 78–84, 88–89.

47. Jung, *The Psychology of Dementia Praecox* (New York: Journal of Nervous and Mental Disease Publishing Co., 1909), pp. 35–38, 47.

48. Prince, "The Psychological Principles and Field of Psychotherapy," *Psychotherapeutics,* pp. 13, 17; *The Unconscious* (New York: Macmillan, 1914), pp. 266–267.

49. See notes 19 and 28; Freud, *The Interpretation of Dreams,* SE, IV, 19–21, 221; Mary Whiton Calkins, in Carl Murchison, ed., *A History of Psychology in Autobiography* (New York: Russell and Russell, 1961), I, 31–32, footnote 5.

50. Jung, *Memories, Dreams, Reflections* (New York: Vintage Books, 1961), pp. 146–147.

51. Frederick Peterson and A. A. Brill, preface to Jung, *The Psychology of Dementia Praecox* (New York: Journal of Nervous and Mental Disease Publishing Co., 1909), p. vi.

52. August Hoch, "The Psychogenic Factors in the Development of Psychoses," *Psychological Bulletin, 4* (June 15, 1907), 161–169.

53. August Hoch, "Constitutional Factors in the Dementia Praecox Group," American Medico-Psychological Association, *Proceedings, 17* (1910), 227–238; "The Psychogenic Factors in the Development of Psychoses," *Psychological Bulletin, 4* (June 15, 1907), 166.

54. Jung, *The Psychology of Dementia Praecox,* p. xx.

55. "The Dynamic Interpretation of Dementia Praecox" (1909), CPM, II, 455–458; Meyer, "The Growth of Scientific Understanding of Mentality and Its Relationship to Social Work" (1923), CPM, IV, 244; C. P. Oberndorf, *A History of Psychoanalysis in America* (New York: Grune and Stratton, 1953), pp. 81–86.

56. A. A. Brill, *Lectures on Psychoanalytic Psychiatry* (New York: Alfred A. Knopf, 1947), pp. 1–2.

57. Putnam, "A Consideration of Mental Therapeutics as Employed by Special Students of the Subject," Bost J, *151* (August 18, 1904), 179–183.

58. Boris Sidis, "Studies in Psychopathology," Bost J, *156* (March 14, 1907), 321–326; (March 28), 394–398; (April 4), 432–434; (April 11), 472–478.

59. "Discussion of S. Weir Mitchell, 'Rest Treatment in Relation to Psychotherapy,'" American Neurological Association, *Proceedings, 34* (1908), 216.

60. Schwab, "An Estimate of Freud's Theory of the Neuroses and Its Value to the Neurologist," *Interstate Medical Journal, 18* (September 1911), 938.

61. Bernard Hart, "The Psychology of Freud and His School," *Journal of Mental Science, 56* (July 1910), 451–452.

62. Trigant Burrow to Anastasia Burrow, October 19, 1909, *A Search for Man's Sanity, the Selected Letters of Trigant Burrow* (New York: Oxford University Press, 1958), pp. 24–25.

63. *Lectures on Psychoanalytic Psychiatry,* p. 10; Brill, "Psychological Factors in Dementia Praecox, an Analysis," J Ab P, *3* (October–November 1908), 223.

64. Ernest Jones, *Free Associations* (New York: Basic Books, 1959), pp. 187–193.

65. *Ibid.,* pp. 153, 138–146.

66. *Ibid.,* p. 161.

67. Ernest Jones, "Mechanism of a Severe Briquet Attack as Contrasted with That of Psychasthenic Fits," J Ab P, *2* (December 1907–January 1908), 218–227.

68. Cyril Greenland, "Ernest Jones in Toronto, 1908–1913, a Fragment of Biography," *Canadian Psychiatric Association Journal, 6* (June 1961), 132–139 and *11* (December 1966), 512–519.

69. Jones, *Free Associations,* pp. 189–190.

70. Putnam, "On Freud's Psycho-analytic Method and Its Evolution" (1912), Ad Psa, p. 121.

71. Hall to Freud, December 15, 1908, CUP.

72. Hall to Freud, April 13, 1909, CUP.

73. Hall to Freud, August 9, 1909, CUP.

74. Hall, *Adolescence* (New York: D. Appleton and Co., 1904), I, 177, 268–278, 285; II, 121–122; "The Needs and Methods of Educating Young People in the Hygiene of Sex," *Pedagogical Seminary, 15* (March 1908), 83.

75. Putnam, "The Relation of Character Formation to Psychotherapy," in Morton Prince *et al., Psychotherapeutics,* pp. 166–186.

76. Putnam, "Personal Impressions" (1909), Ad Psa, p. 30; Sandor Ferenczi, Introduction to Freud, *The Problem of Lay Analysis* (New York: Brentano's, 1927), pp. 16–18.

77. Putnam to Freud, November 17, 1909, JPP, pp. 86, 88.

78. Freud to Putnam, December 5, 1909, *ibid.,* pp. 89–90.

79. See Putnam, "Personal Impressions" (1909), Ad Psa, pp. 18, 3, 17.

80. Freud to Putnam, January 28, 1910, JPP, p. 92.

81. Putnam to James, June 25, 1910, *ibid.,* p. 79.

82. *Ibid.;* James to Putnam, June 4, 1910, *ibid.,* p. 77; Discussion of Putnam, "Personal Experience with Freud's Psychoanalytic Method," JNMD, *37* (November 1910), 657–674.

83. Putnam, "William James," *Atlantic Monthly, 106* (December 10, 1910), 837.

84. See note 82.

85. Schwab, "Some New Freudian Literature," *Interstate Medical Journal, 17* (September 1910), 697–700.

86. Putnam, "Personal Impressions" (1909), Ad Psa, p. 30; "Personal Experience," *ibid.,* pp. 31–53.

87. Putnam to editor, Bost J, *161* (July 1, 1909), 37.

88. Putnam, "On the Etiology and Treatment of the Psychoneuroses" (1910), Ad Psa, pp. 69–70.

89. *Ibid.,* pp. 64–65.

90. Freud to Putnam, September 29, 1910, JPP, p. 107.

91. Quoted in Ives Hendrick, ed., *The Birth of an Institute: Twenty-fifth Anniversary, The Boston Psychoanalytic Institute, November 30, 1958* (Freeport, Maine: Bond Wheelwright Co., 1961), p. 12.

92. William Alanson White, "The Theory, Methods, and Psychotherapeutic Value of Psycho-Analysis," *Interstate Medical Journal, 17* (September 1910), 644–645.

93. Ernest Jones, "The Action of Suggestion in Psychotherapy" (1910), *Papers on Psychoanalysis* (London: Bailliére, Tindall and Cox, 1913), pp. 241–282.

94. Putnam, "On the Etiology and Treatment" (1910), Ad Psa, p. 66.

95. Putnam, "Personal Experience" (1910), Ad Psa, p. 47.

96. Bernard Hart, "The Conception of the Subconscious," in Hugo Münsterberg *et al., Subconscious Phenomena* (Boston: Richard G. Badger, 1910), pp. 131, 112.

97. *Ibid.*

98. Jones, *The Life and Work of Sigmund Freud* (New York: Basic Books, 1955), II, 64.

99. Jones, "Psychoanalysis in Psychotherapy," in Prince *et al., Psychotherapeutics,* p. 103; Putnam, "Presidential Address Before the American Psychopathological Association," J Ab P, *8* (August-September 1913), 173–174.

100. Schwab, "An Estimate of Freud's Theory of the Neuroses and Its Value to the Neurologist," *Interstate Medical Journal, 18* (September 1911), 944.

101. William Alanson White, "The Theory of the Complex," *Interstate Medical Journal, 16* (April 1909), 254–256.

102. Remarks of Smith Ely Jelliffe on a paper by August Hoch, "Psychogenetic Factors in Some Paranoic Conditions, with Suggestions for Prophylaxis and Therapy," JNMD, *34* (October 1907), 673.

103. Putnam, "Personal Impressions" (1909), Ad Psa, p. 16; "On the Etiology' and Treatment" (1910), *ibid.,* p. 64.

104. Putnam, J Ab P, *8* (August-September 1913), 173, 179.

105. A. A. Brill, *Freud's Contribution to Psychiatry* (New York: W. W. Norton, 1944), p. 33.

106. Putnam, "Personal Experience" (1910), Ad Psa, p. 32.

107. Ernest Jones, remarks on Putnam, "Personal Experience," JNMD, *37* (October 1910), 636.

108. Jones, *Free Associations,* p. 189; Remarks of Dr. Edward B. Angell on Putnam, "Personal Experience," JNMD, *37* (October 1910), 631.

109. Putnam, "On the Etiology and Treatment" (1910), Ad Psa, pp. 57—58.

110. Putnam, Presidential Address, J Ab P, *8* (August-September 1913), 174.

111. Jones, "Psychoanalysis in Psychotherapy," in Prince *et al., Psychotherapeutics,* p. 104.

112. Burrow to Anastasia Burrow, February 23, 1910, *A Search for Man's Sanity,* p. 31.

113. Freud, "Sexuality in the Aetiology of the Neuroses" (1898), SE, III, 275.

114. Putnam, "On the Etiology and Treatment" (1910), Ad Psa, p. 55; Remarks of Charles L. Dana and Lewellys Barker on Putnam, "Personal Experience," JNMD, *37* (October 1910), 631—633.

115. *Ibid.,* p. 638.

116. Putnam, "On the Etiology and Treatment" (1910), Ad Psa, pp. 62—63.

117. Prince, "The Mechanism and Interpretation of Dreams. A Reply to Dr. Jones," J Ab P, *5* (February-March 1910), 349—353.

118. Remarks of Ernest Jones on Putnam, "Personal Experience," JNMD, *37* (October 1910), 636.

119. Jones, *Freud,* II, 75.

Chapter IX. Mind Cures and the Mystical Wave: Popular Preparation for Psychoanalysis

1. *Boston Evening Transcript,* Saturday, September 11, 1909, Part Three, p. 3.

2. The Rev. Elwood Worcester, The Rev. Samuel McComb, and Isador Coriat, *Religion and Medicine: The Moral Control of Nervous Disorders* (New York: Moffat, Yard, 1908), pp. 145—150.

3. Worcester, *Life's Adventure: The Story of a Varied Career* (New York: Charles Scribner's Sons, 1932), pp. 285—288.

4. James to Theodore Flournoy, September 28, 1909, in Henry James, ed., *The Letters of William James* (Boston: Atlantic Monthly Press, 1920), II, 327—328; James to Mary Calkins, September 19, 1909, in Ralph Barton Perry, *The Thought and Character of William James* (Boston: Atlantic Monthly Press, Little Brown and Co., 1935), II, 123.

5. William James, *The Varieties of Religious Experience* (New York: Longmans Green and Co., 1902), p. 96; Richard C. Cabot, "The American Type of Psychotherapy," *Psychotherapy: A Course of Reading, 1,* no. 1 (1908), 6.

6. Allan McLane Hamilton, "Mental Medicine, the Treatment of Disease by Suggestion," *Century, 46,* n.s. (24 July 1893), 430—435; William James, "The Hidden Self," *Scribner's Magazine, 7* (March 1890), 361—373; James Mark Baldwin, "Among the Psychologists of Paris," *Nation, 55* (July 28, 1892), 68; W. R. Newbold, "Suggestion in Therapeutics," *Popular Science Monthly, 49* (July 1896), 342—353.

7. John G. Green, "The Emmanuel Movement, 1906–1929," *New England Quarterly,* 7 (September 1934), 507–508; Morton Prince, "An Objection to Newspaper Headlines," Bost J, *122* (June 26, 1890), 649.

8. Frederick Peterson, "The New Divination of Dreams," *Harper's Monthly Magazine, 115* (August 1907), 448–452; Peterson, "The Nerve Specialist to His Patients," *Colliers, 42* (January 1909), 11.

9. The *Reader's Guide* listed 40 articles on hypnotism from 1890 to 1899, and 71 from 1900 to 1909; it listed 8 articles on diseases of the nervous system (which included treatment) from 1890 to 1900, while there were 48 between 1900 and 1909. Similarly, the number of articles on insanity increased from 10 between 1890 to 1900 to 85 from 1900 to 1909.

10. Worcester, *Life's Adventure,* pp. 294–296; *The Reader's Guide* listed 15 articles on nervous diseases (including treatment) in women's magazines from 1905 to 1909, compared with 14 in all other periodicals.

11. Hereward Carrington, "To Become Beautiful by Thought," *Good Housekeeping, 49* (August 1907), 221; Clarence B. Farrar, "Psychotherapy and the Church," JNMD, *36* (January 1909), 10–18; William James, *Talks to Teachers on Psychology; and to Students on Some of Life's Ideals,* introduction by John Dewey (New York: Henry Holt, 1910), pp. 200, 217, 223; Annie Payson Call, "How Women Can Keep from Being Nervous," *Ladies' Home Journal, 25* (March 1908), 8.

12. H. A. Bruce, "Insanity and the Nation," *North American Review, 187* (January 1908), 70–79; "Increasing Insanity," *Current Literature, 36* (May 1904), 547–548.

13. Worcester, *Religion and Medicine,* p. 171; Charles W. Pilgrim, "Insanity and Suicide," AJI, *63* (January 1907), 349–360; Bruce, "The New Mind Cure Based on Science," *American Magazine, 70* (October 1910), 773–778.

14. Clarence B. Farrar, "Psychotherapy and the Church," JNMD, *36* (January 1909), 12.

15. "A Disease of Civilization," *Nation, 81* (September 7, 1905), 196; Joseph Collins, "The General Practitioner and the Functional Nervous Diseases," JAMA, *52* (January 1909), 92.

16. Quoted in Clifford Beers, *A Mind That Found Itself,* 25th ed. (Garden City, New York: Doubleday, Doran and Co., 1944), p. xiv; Perry, *The Thought and Character of William James,* II, 318–319; Ernest Jones, review of *A Mind That Found Itself,* J Ab P, *5* (1910–1911), 41; reviews of Beers, *A Mind That Found Itself, Dial, 44* (May 1, 1908), 278; *Literary Digest, 36* (April 4, 1908), 489; *Independent, 65* (September 17, 1908), 663; *Outlook, 88* (March 21, 1908), 654; *Review of Reviews, 37* (March 1908), 383; *New York Times* (June 13, 1908), 335.

17. Review, *A Mind That Found Itself, Nation, 86* (March 19, 1908), 265–266.

18. Franklin C. Sanborn, "Progress in the Treatment of Insanity over the Past Half-Century," National Conference of Charities and Corrections, *Proceedings,* 1909, pp. 66–67, esp. p. 71.

19. F. B. Wines, Conference Sermon, "The Healing Touch," National Conference of Charities and Corrections, *Proceedings,* 1900, p. 18; J. M. Buck-

ley, "How to Safeguard One's Sanity," *Century, 60,* n.s. *38* (July 1900), 375–378.

20. Worcester, *Religion and Medicine,* pp. 125–132; Paul Dubois, "The Method of Persuasion," *Psychotherapy: A Course of Reading, 3,* No. 2 (1909), 35–38.

21. "A Disease of Civilization," *Nation, 81* (September 7, 1905), 196; H. Addington Bruce, "The New Mind Cure Based on Science," *American Magazine, 70* (October 1910), 775; "The Age of Nerves," *Living Age, 267* (November 19, 1910), 505–507; H. Addington Bruce, "Masters of the Mind," *American Magazine, 71* (November 1910), 74–77.

22. "The Autobiography of a Neurasthenic," *American Magazine, 71* (December 1910), 223–231, esp. pp. 231, 228; James Jackson Putnam, "The Nervous Breakdown," *Good Housekeeping, 99* (November 1909), 597–598.

23. Robert Herrick, *The Master of the Inn* (New York: Charles Scribner's Sons, 1909), pp. 6, 9, 10–11, 17–18, 23–24, 29.

24. Quoted in Frank Dekker Watson, *The Charity Organization Movement, a Study in American Philanthropy* (New York: Macmillan, 1922), p. 33.

25. Worcester, *Religion and Medicine,* p. 139.

26. Review, Gelett Burgess, *The White Cat, Bookman, 25* (August 1907), 568; Gelett Burgess, *The White Cat* (New York: Grossett and Dunlap, 1907), pp. 326–327, 356–385, 141–150, 172, 226.

27. Quoted in David M. Rein, *S. Weir Mitchell as a Psychiatric Novelist* (New York: International Universities Press, 1952), pp. 85–87.

28. Agnes Repplier, "The Nervous Strain," *Atlantic Monthly, 106* (August 1910), 198–201.

29. Worcester, *Religion and Medicine,* p. 124.

30. Ray Stannard Baker, "The Spiritual Unrest, II, The New Mission of the Doctor," *American Magazine, 67* (January 1909), 232; Edward Wakefield, "Nervousness, the National Disease of America," *McClure's, 2* (February 1894), 302.

31. Richard Cabot, "The American Type of Psychotherapy," *Psychotherapy: A Course of Reading, 1,* No. 1 (1908), 7, 11, 13; Richard Cabot, "The Analysis and Modification of Environment," *Psychotherapy: A Course of Reading, 3,* No. 3 (1909), 7; Ray Stannard Baker, *New Ideals in Healing* (New York: Frederick Stokes, 1909), p. 84.

32. Worcester, *Religion and Medicine,* pp. 158–159, 4; Worcester, *Body, Mind and Spirit* (Boston: Marshall Jones Co., 1931), pp. 77–78, 195–196.

33. Worcester, *Life's Adventure,* pp. 136, 92–96.

34. Worcester, *Body, Mind and Spirit,* pp. 7–11, 39–43, 47–54, 130–134, 150–152, 193–195.

35. Worcester, *Religion and Medicine,* p. 42; Worcester and Samuel McComb, *The Christian Religion as a Healing Power: A Defense and Exposition of the Emmanuel Movement* (New York: Moffat, Yard, 1909), pp. 114–118.

36. Irving Babbitt, "Bergson and Rousseau," *Nation, 95* (November 14,

1912), 452–455; "Bergson's Reception in America," *Current Opinion, 54* (March 1913), 226; William James, "The Philosophy of Bergson," *Hibbert Journal, 7* (April 1909), 562–577.

37. Walter Lippmann, "The Most Dangerous Man in the World," *Everybody's, 27* (July 1912), 100–101.

38. Henri Bergson, "The Birth of a Dream," *Independent, 76* (October 30, 1913), 200, 203; Alvan F. Sanborn, "Bergson: Creator of a New Philosophy," *Outlook, 103* (February 15, 1913), 357; Bergson, "Such Stuff as Dreams are Made Of," *Independent, 76* (October 23, 1913), 163.

39. Bergson, "The Birth of a Dream," *Independent, 76* (October 30, 1913), 203.

40. H. Addington Bruce, "The Soul's Winning Fight with Science," *American Magazine, 77* (March 1914), 21–26.

41. William James, "The Powers of Men," *American Magazine, 65* (November 1907), 56–65; William James, *The Energies of Men* (New York: Henry Holt, 1911), p. 4.

42. George Santayana, "William James," *Character and Opinion in the United States* (London: Constable and Co., 1920), p. 82.

43. "Hypnosis," *Current Literature, 29* (October 1900), 390; John Duncan Quackenbos, *Hypnotic Therapeutics in Theory and Practice* (New York: Harper and Brothers, 1908), pp. 10–11, 69–70, 147–149, 181–182; "Hypnotism in Practice," *Current Literature, 27* (January 1900), 25–26; "Reciprocal Influence in Hypnotism," *Harper's Monthly Magazine, 103* (June 1901), 110–112; for a hostile review see *Popular Science Monthly, 57* (December 1900), 214–216; see also A. MacDonald, "Alcoholic Hypnotism," *American Journal of Sociology, 5* (November 1899), 383, and "Hypnotism for the Gangster," *Literary Digest, 44* (May 11, 1912), 991, for glowing accounts of hypnotic therapy for alcoholics and delinquents.

44. A. A. Brill, *Freud's Contribution to Psychiatry* (New York: W. W. Norton, 1944), pp. 22–23.

45. Horatio W. Dresser, *A History of the New Thought Movement* (London: George C. Harrap, 1919), pp. 161, 165, 306–307.

46. Max Eastman, *The Enjoyment of Living* (New York: Harper and Brothers, 1948), pp. 241–244, 260–263, 273.

47. H. Addington Bruce, "Insanity and the Nation," *North American Review, 187* (January 1908), 77; H. Addington Bruce, "The Origin and Development of Mental Healing," *Outlook, 92* (August 28, 1909), 1039–1047.

48. "How One Girl Lived Four Lives," as told by John Corbin, *Ladies' Home Journal, 25* (November 1908), 11.

49. "The Greatest Modern Discovery," *Current Literature, 45* (September 1908), 304–307; Frederick Peterson, "The New Divination of Dreams," *Harper's Monthly Magazine, 115* (August 1907), 451–452.

50. Ray Stannard Baker, "The Spiritual Unrest, I, Healing the Sick in Churches," *American Magazine, 67* (December 1908), 199.

51. James Jackson Putnam, "The Service to Nervous Invalids of the Physician and the Minister," *Harvard Theological Review, 2* (April 1909),

238–239, 249; "The New Crusade in Behalf of 'Religious Therapeutics,' " *Current Literature, 44* (March 1908), 289–292; Putnam to Rev. Elwood Worcester, September 12, 1908, PP.

Chapter X. The Repeal of Reticence

1. Granville Stanley Hall, "The Needs and Methods of Educating Young People in the Hygiene of Sex," *Pedagogical Seminary, 15* (March 1908), 91.

2. Earl Barnes, "Books and Pamphlets Intended to Give Sex Information," *Studies in Education* (Stanford: Stanford University Press, 1897), I, 301–308.

3. A. C. McClanahan, "A Plea for Physiology of the Sexual System," *New York Medical Journal, 64* (July 4, 1896), 15; Sir James Paget, *Selected Essays and Addresses* (New York: Longmans, Green and Co., 1902), p. 34.

4. Mrs. Kate Gannett Wells, "Why More Girls Do Not Marry," *North American Review, 152* (February 1891), 175–181.

5. A. D. Rockwell, "Sexual Erethism, Its Neurotic Origin and Treatment," *New York Medical Journal, 58* (August 19, 1893), 201–202; William Hammond, quoted in Joseph W. Howe, *Excessive Venery, Masturbation and Continence* (New York: E. B. Treat, 1883), pp. 86–87.

6. *The National Purity Congress: Its Papers, Addresses and Portraits* (New York: American Purity Alliance, 1896), p. 158.

7. *Ibid.*, p. 318.

8. *Ibid.*, pp. 124, 250–268.

9. Jane Addams, *A New Conscience and an Ancient Evil* (New York: The Macmillan Co., 1912), pp. 185–186, 196.

10. David Jay Pivar, "The New Abolitionism, the Quest for Social Purity, 1872–1900," Doctoral Dissertation, University of Pennsylvania, 1965, pp. 299–303.

11. Prince Morrow, "The Relation of Social Diseases to the Family," *American Journal of Sociology, 14* (1908–1909), 634.

12. "The Havoc of Prudery," *Current Literature, 50* (February 1911), 174–175.

13. William A. Neilson, ed., *Charles W. Eliot: The Man and His Beliefs* (New York & London: Harper and Bros., 1926), II, 652–663.

14. Havelock Ellis, *The Task of Social Hygiene* (Boston and New York: Houghton Mifflin Co., 1912), p. 301.

15. Jane Addams, *A New Conscience*, pp. 79–80.

16. Quoted in William Trufant Foster, ed., *The Social Emergency: Studies in Sex Hygiene and Morals* (Boston: Houghton Mifflin, 1914), pp. 133–134.

17. *The National Purity Congress, Papers*, p. 401; Report of the Vice Commission of Minneapolis to His Honor, James C. Haynes, Mayor, Minneapolis, 1911, p. 104.

18. George J. Kneeland, *Commercialized Prostitution in New York City,* Introduction by John D. Rockefeller, Jr. (New York: Century Co., 1913), p. 79.

19. Jane Addams, *A New Conscience,* p. 172.

20. *Independent,* 74 (April 3, 1913), 752; Edward L. Bernays, *Biography of an Idea; Memoirs of Public Relations Counsel Edward L. Bernays* (New York: Simon and Schuster, 1965), pp. 53–62.

21. Eugène Brieux, *Damaged Goods,* trans. J. Pollock, in *Three Plays by Brieux* (New York: Brentano's, 1911), p. 241.

22. G. Stanley Hall, "The Pedagogy of Sex," *Educational Problems* (New York & London: D. Appleton & Co., 1911), I, 478; James Foster Scott, *The Sexual Instinct* (New York: E. B. Treat, 1899), pp. 82–99; Ira S. Wile, *Sex Education* (New York: Duffield & Co., 1912), p. 57; W. J. Hardman, "The Duty of the Medical Profession to the Public in the Matter of Venereal Diseases and How to Discharge It," JAMA, 47 (1906), 1246–1248.

23. H. D. Sedgwick, "A Gap in Education," *Atlantic Monthly, 87* (January 1901), 70.

24. Winfield S. Scott, *From Youth into Manhood,* 13th ed. (New York: Association Press, 1919), pp. 49–54, 68–69.

25. Ralph Reed, "Sexual Education of the Child," *The Medical Record* (New York), *81* (April 6, 1912), 657–662. Psychoanalytic books did not appear in book lists in the American Social Hygiene Review nor was psychoanalysis mentioned except in passing in social hygiene literature.

26. Joseph Ishill, ed., *Havelock Ellis, In Appreciation* (Berkeley Heights, New Jersey: The Oriole Press, privately printed, 1929).

27. Arthur Calder-Marshall, *The Sage of Sex: A Life of Havelock Ellis* (New York: G. P. Putnam's Sons, 1959), pp. 166, 172; Havelock Ellis, *My Life* (Boston: Houghton Mifflin Co., 1939), pp. 370–371.

28. *The Little Review 2* (March 1915),15.

29. *Nation, 71* (August 23, 1900), 157; *Psychological Review, 6* (1899), 134–145; *New York Medical Journal, 73* (March 9, 1901), 435–437.

30. Hall, *Adolescence* (New York: D. Appleton and Co., 1904), II, p. 141.

31. JNMD, *32* (May 1905), 351.

32. Freud to Havelock Ellis, September 12, 1926, *The Letters of Sigmund Freud* (New York, Toronto, London: McGraw-Hill, 1960), pp. 370–371; Freud to Wilhelm Fliess, January 3, 1899, *The Origins of Psychoanalysis* (Garden City, New York: Doubleday Anchor, 1957), pp. 274–275.

33. Havelock Ellis, *The Philosophy of Conflict and Other Essays in War-Time* (Boston and New York: Houghton Mifflin Co., 1919), p. 200.

34. Ellis, *Studies in the Psychology of Sex* (Philadelphia: F. A. Davis Co.), VI, *Sex in Relation to Society* (1911), pp. 190, 199–200, 524.

35. Smith Ely Jelliffe, review of Ellis's *Studies in the Psychology of Sex,* VI, *Sex in Relation to Society,* JNMD, *39* (May 1912), 355–356.

36. Quoted in Ellis, *Studies,* VI, *Sex in Relation to Society,* note 3, p. 370, and p. 119; pp. 429–435.

37. Ellis, *Studies,* III, *Analysis of the Sexual Impulse; Love and Pain; The Sexual Impulse in Women* (Philadelphia: F. A. Davis Co., 1905), pp. 44, 202–214.

38. *Ibid.,* pp. 15–17.

39. Ellis, *Studies,* II, *Sexual Inversion* (1901), pp. 193–202.

40. Review of Ellis, *Studies in the Psychology of Sex, New York Medical Journal,* 78 (November 7, 1903), 920.

41. Calder-Marshall, *The Sage of Sex,* p. 154.

42. Ellis, *Studies,* VI, *Sex in Relation to Society,* p. 512; *The Task of Social Hygiene,* p. 276.

43. Ellis, *Studies,* VI, *Sex in Relation to Society,* p. 116.

44. Ellis, "Morality as an Art," *Atlantic Monthly, 114* (November 1914), 700–707; *Studies,* VI, *Sex in Relation to Society,* pp. 177; 216, 213, 514.

45. *Ibid.,* p. 428.

46. Ellis, "Psychoanalysis in Relation to Sex," in *The Philosophy of Conflict and Other Essays in War-Time,* pp. 215–216.

47. *Ibid.,* p. 210.

48. Wren Jones Grinstead, "Reading for Teachers of Sex Hygiene," *School Review, 22* (April 1914), 249–253; Foster, *The Social Emergency,* p. 203. Maurice A. Bigelow, *Sex Education* (New York: Macmillan, 1916), p. 240.

49. "Is the Granting of Freer Divorce an Evil?" *American Journal of Sociology, 14* (1908–1909), 791, 787.

50. Ellen Key, *Love and Ethics* (New York: B. W. Huebsch, 1911), p. 69; and *Love and Marriage,* trans. Arthur G. Chater (New York & London: G. P. Putnam's Sons, 1911), pp. 14, 32–36; William E. Carson, *The Marriage Revolt: A Study of Marriage and Divorce* (New York: Hearst's International Library, 1915), pp. 387–389.

51. *Boston Evening Transcript,* September 10, 1909, p. 4; Jones, *Freud,* II, p. 57.

52. Emma Goldman, *Anarchism and Other Essays* (New York: Mother Earth Publishing Association, 1910), p. 192; *Mother Earth, 6* (February 1912), 98, 354–355; 8 (April 1913), 63; 4 (January 1910), 344–351, 369–370.

53. Quoted in Hutchins Hapgood, *A Victorian in the Modern World* (New York: Harcourt, Brace & Co., 1939), p. 279; Joseph Ishill, ed., *A New Concept of Liberty from an Evolutionary Psychologist: Theodore Schroeder; Selections from His Writings* (Berkeley Heights, New Jersey: The Oriole Press, privately printed, 1940), pp. xxiv–xxviii.

54. Theodore Schroeder, "The Evolution of Marriage Ideals," *Arena, 34* (December 1905), 578–589; James S. Van Teslaar, "Is Sex the Basis of Religion?" *New York Times,* April 4, 1915, Section V, p. 2; R, *1* (1913–1914), 148.

55. *American Journal of Urology,* 11 (October 1915), 391–405; Editorial, "Freud and the Continence Advocates," *Medical Critic and Guide,* 15 (March 1912), 98.

56. *Nation,* 98 (January 22, 1914), p. 81; Bigelow, *Sex Education,* pp. 51–52, 240.

57. Joyce Kilmer, "The Happy Eugenists," *New York Times,* November 17, 1912, p. 672; Schroeder, "The Impurity of Divorce Suppression," *Arena,* 33 (February 1905), 142–146.

58. Agnes Repplier, "The Repeal of Reticence," *Atlantic Monthly,* 113 (March 1914), 297–304.

59. Irving Babbitt, "Bergson and Rousseau," *Nation,* 95 (November 14, 1912), 452–455; Paul Elmer More, "A Naughty Decade," *Nation* (May 21, 1914), 600.

60. "Morbid Pedagogism," *Nation,* 93 (July 27, 1911), 80–81; R. M. Barrington, "G. Stanley Hall's *Educational Problems,*" *Bookman,* 34 (September, 1911), 88–90.

61. *Nation,* 96 (May 15, 1913), 503–505.

62. "Brieux' *Damaged Goods,*" *Outlook, 104* (May 31, 1913), 226.

Chapter XI. Opposition and Debate: Science and Sexuality

1. William McDonald, "Progress of Psychiatry in 1910, America," *Journal of Mental Science,* 57 (July 1911), 517–518.

2. Putnam, "Personal Experience with Freud's Psychoanalytic Method," JNMD, 37 (October 1910), 637.

3. Jones, *The Life and Work of Sigmund Freud* (New York: Basic Books, 1955), II, 115.

4. Compare Francis X. Dercum, American Neurological Association, *Proceedings* (1908), 215, and W. Spielmeyer, review of Freud, "Brüchstuck einer Hysterie-Analyse," *Centralblatt für Nervenheilkunde und Psychiatrie,* 29 (April 15, 1906), 322–324; Bleuler, letter to editor on Spielmeyer, *ibid.* (June 1, 1906), 460–461; and Spielmeyer, reply to Bleuler, *ibid;* with Gustav Aschaffenburg, "Die Beziehungen des sexuellen Lebens zur Entstehung von Nerven und Geisteskrankheiten," *Münchener Medizinische Wochenschrift, 53;* 2, No. 37 (September 11, 1906), 1793–1798.

5. Robert S. Woodworth, review of "Nervous and Mental Disease Monographs," *Science* n.s., 38 (December 26, 1913), 929; Bernard Hart, "The Psychology of Freud and His School," *Journal of Mental Science,* 56 (July 1910), 451–452.

6. Pierre Janet, "Psychoanalysis," J Ab P, 9 (1914–1915), 1–35, 153–187; Charles W. Burr, "A Criticism of Psychoanalysis," AJI, 71 (October 1914), 234; Bernard Hart, *The Psychology of Insanity* (Cambridge: at the University Press, 1912), pp. 166–167.

7. Francis X. Dercum, "The Role of Dreams in Etiology," JAMA, 56 (May 13, 1911), 1373–1377.

8. See, for example, Dercum's anachronistic discussion of Freud's view of sexuality. *Ibid.*

9. A. Friedländer, "Hysteria and Modern Psychoanalysis," J Ab P, 5 (February-March 1911), 313.

10. See note 7 and Dercum, "An Evaluation of Psychogenic Factors in the Etiology of Mental Disease, Including a Review of Psychoanalysis," JAMA, 62 (March 7, 1914), 754; Charles W. Burr, "The Prevention of Insanity and Degeneracy," AJI, 74 (January 1918), 409–424, and "A Criticism of Psychoanalysis," AJI, 71 (October 1914), 241–242.

11. *Ibid.*

12. Dercum, JAMA, 62 (March 7, 1914), 752.

13. Smith Ely Jelliffe, "Predementia Praecox: The Hereditary and Constitutional Features of the Dementia Praecox Makeup," in Adolf Meyer *et al., Dementia Praecox, a Monograph* (Boston: Richard G. Badger, 1911), p. 23; E. Stanley Abbott, "Meyer's Theory of the Psychogenic Origin of Dementia Praecox. A Criticism," AJI, 68 (July 1911), 15–22.

14. Remarks, "Symposium on the Pathogenesis of Morbid Anxiety," J Ab P, 6 (June–July 1911), 172; Ernest Jones, "Remarks on Dr. Morton Prince's Article: The Mechanism and Interpretation of Dreams," J Ab P, 5 (February–March 1911), 331 and "Psycho-Analysis in Psychotherapy," Morton Prince *et al., Psychotherapeutics* (Boston: Richard G. Badger, 1909), p. 103.

15. Morton Prince, "The Mechanism and Interpretation of Dreams," J Ab P, 5 (October–November 1910), 158–159.

16. Ernest Jones, "The Action of Suggestion in Psychotherapy," J Ab P, 5 (December 1910–January 1911), 217–254.

17. Jones to Putnam, October 13 [1910], Prince to Putnam, October 21 and November 22, 1910, PP.

18. Putnam to Jones, January 5, 1911, JPP, pp. 249–250.

19. Jones, J Ab P, 5 (February–March 1911), 334.

20. Prince, *ibid.,* pp. 349–350, 338, 336.

21. *Ibid.,* pp. 347, 343–344.

22. *Ibid.,* pp. 353, 336.

23. Jones to Putnam, February 19, and February 27, 1911, JPP, pp. 257, 260.

24. Prince to Putnam, November 22–26, 1910, March 3, 1911, JPP, pp. 325, 333, 323–324.

25. Jones, J Ab P, 7 (April–May 1912), 85–86.

26. Robert S. Woodworth, "Some Criticisms of the Freudian Psychology," J Ab P, 12 (August 1917), 174–175; Bernard Hart, "The Psychology of Freud and His School," *Journal of Mental Science,* 56 (July 1910), 451.

27. Adolf Meyer, remarks, "Symposium on Pathogenesis of Morbid Anxiety," J Ab P, 6 (June–July, 1911), 169; Frederick Lyman Wells, "A Critique of Impure Reason," J Ab P, 7 (June–July 1912), 91–92; Woodworth, J Ab P, 12 (August 1917), 183.

28. *Ibid.,* p. 178.

29. Samuel A. Tannenbaum, "Some Current Misconceptions of Psychoanaly-

sis," J Ab P, *12* (February 1918), 396–397; Freud, "Five Lectures on Psychoanalysis," SE, XI, 29–32, 38.

30. Schwab, "Symposium on Freud's Theory of the Neuroses and Allied Subjects," JNMD, *38* (August 1911), 496; Prince, J Ab P, *5* (February–March 1911), 347; Dercum, "The Role of Dreams in Etiology," JAMA, *57, Part 2* (May 13, 1911), 1376.

31. Woodworth, J Ab P, *12* (August 1917), 179; discussion of Prince, "The Psychopathology of a Case of Phobia, a Clinical Study," J Ab P, *8* (December 1913–January 1914), 331; Putnam to Freud, January 5, 1913, JPP, pp. 153–154.

32. Freud, "Inhibitions, Symptoms and Anxiety" (1914), SE, XX, 77–174; Prince, "A Clinical Study of a Case of Phobia," J Ab P, *7* (October–November 1912), 276; Jones, *Freud,* III, 355.

33. Robert Sessions Woodworth, *Dynamic Psychology* (New York: Columbia University Press, 1918), pp. 168–176; J Ab P, *12* (August 1917), 189–190.

34. John Broadus Watson, "Behavior and the Concept of Mental Disease," *Journal of Philosophy, Psychology and Scientific Methods, 13* (October 26, 1916), 589–597; William Alanson White, "The Behavioristic Attitude," *Mental Hygiene, 5* (January 1921), 1–18; Knight Dunlap, *A System of Psychology* (New York: Charles Scribner's Sons, 1912), pp. 327–330, 361, and "The Pragmatic Advantages of Freudo-Analysis," R, *1* (February 1914), 149–152.

35. Woodworth, J Ab P, *12* (August 1917), 181–182, 176; Alfred Reginald Allen, "Presidential Address to the American Psychopathological Association," J Ab P, *9* (1914–1915), 283.

36. Brill, Comments on Prince, "A Psychopathological Study of a Case of Phobia," J Ab P, *8* (December 1913–January 1914), 342.

37. Schwab, "An Estimate of Freud's Theory of the Neuroses, and Its Value to the Neurologist," *Interstate Medical Journal, 18* (September 1911), 948.

38. Edward Wyllys Taylor, "Suggestions Concerning a Modified Psychoanalysis," J Ab P, *12* (December 1917), 371–374; Dercum, JAMA, *62* (March 7, 1914), 755.

39. Prince, J Ab P, *5* (February–March 1911), 351–352; Dercum, JAMA, *62* (March 7, 1914), 754; Woodworth, *Science,* n.s., *38* (December 26, 1913), 930; G. Stanley Hall, "Anger as a Primary Emotion and the Application of Freudian Mechanisms to Its Phenomena," J Ab P, *10* (1915), 82.

40. E. Pumpian-Mindlin, *Psychoanalysis as Science* (Palo Alto: Stanford University Press, 1952); Sidney Hook, ed., *Psychoanalysis: Scientific Method, and Philosophy* (New York: Grove Press, 1959).

41. Freud, " 'Wild' Psycho-Analysis" (1910), SE, XI, 223.

42. "Three Essays on Sexuality" (1905), SE, VII, 176; "Sexuality in the Aetiology of the Neuroses" (1898), CPF, I, 244.

43. Jones, *Freud,* II, 63; I, 317.

44. Beatrice Hinkle, "Jung's Libido Theory and the Bergsonian Philosophy," *New York Medical Journal, 99* (May 30, 1914), 1080–1086; Samuel

A. Tannenbaum, "Sexual Abstinence and Nervousness," *American Journal of Urology, 9* (June 1913), 290–322.

45. Putnam, *Human Motives* (Boston: Little, Brown and Co., 1915), pp. 85–94; A. A. Brill, *Psychoanalysis: Its Theories and Practical Application* (Philadelphia: W. B. Saunders, 1913), pp. 20, 90–93, 268.

46. Frederick Lyman Wells, "On Formulation in Psychoanalysis," J Ab P, *8* (October–November 1913), 223.

47. Charles W. Burr, "A Critique of Psychoanalysis," American Medico-Psychological Association, *Proceedings, 21* (1914), 305–306.

48. Pierre Janet, J Ab P, *9* (1914–1915), 160, 179–180.

49. *Ibid.,* pp. 170–172, 166.

50. Charles W. Burr, AJI, *71* (October 1914), 154–155, 163, 244; Alfred R. Allen, "Presidential Address to the American Psychopathological Association," J Ab P, *9* (1914–1915), 284.

51. Albert Carrier, "What Shall We Teach the Public Regarding Venereal Diseases," JAMA, *67* (October 20, 1906), 1250–1251; Francis X. Dercum, ed., *A Textbook on Nervous Diseases by American Authors* (Philadelphia: Lea Bros. & Co., 1895), p. 55.

52. Havelock Ellis, *Studies in the Psychology of Sex, II, Sexual Inversion* (Philadelphia: F. A. Davis Co., 1901), 188–190; see also review of Krafft-Ebing, *Psychoathia Sexualis,* AJI, *50* (July 1893), 93–95; William Lee Howard, "Sexual Perversion," *Alienist and Neurologist, 17* (January 1896), 6; Irving G. Rosse, "Sexual Hypochondriasis and Perversion of the Genesic Instinct," JNMD, *19* (November 1892), 795–811.

53. G. Stanley Hall, J Ab P, *10* (1915–1916), 82.

54. Boris Sidis, "Fundamental States in Psychoneuroses," J Ab P, *5* (February–March 1911), 322–323.

55. Frederick Peterson, "Some New Fields and Methods in Psychology," *New York Medical Journal, 90* (November 13, 1909), 947; Janet, J Ab P, *9* (1914–1915), 164.

56. Woodworth, J Ab P, *12* (August 1917), 175.

57. Hall, J Ab P, *10* (1915–1916), 82.

58. Southard, J Ab P, *10* (1915–1916), 276.

59. Meyer, "A Discussion of Some Fundamental Issues in Freud's Psychoanalysis" (1909), CPM, II, 605; Dercum, JAMA, *62* (March 7, 1914), 755; Lewellys F. Barker, "The Psychic Treatment of the Functional Neuroses," *International Clinics,* 17th series (1907), I, pp. 10–11.

60. Jones, *Free Associations,* p. 192.

61. Janet, J Ab P, *9* (1914–1915), 186–187.

62. Dercum, JAMA, *62* (March 7, 1914), 755.

63. James Hendrie Lloyd, "The So-Called Oedipus Complex in Hamlet," JAMA, *56* (May 13, 1911), 1377; A. Friedländer, J Ab P, *5* (February–March 1911), 299, 302.

64. Charles L. Dana, JNMD, *37* (September 1910), 631–632; Burr, JNMD, *38* (1911), 494–495.

65. Boris Sidis, *Symptomatology, Psychogenesis and Diagnosis of Psycho-*

pathic Diseases (Boston: Richard G. Badger, 1914), pp. vi–vii; Frederick W. Peterson, "Credulity and Cures," JAMA, *73* (December 6, 1919), 1740.

66. Ernest Jones to Alfred Reginald Allen, September 5, 1911, *American Journal of Psychiatry, 123:2* (August, 1966), 233.

67. See notes 59, 30.

68. Putnam, "Personal Impressions" (1909), Ad Psa, pp. 17, 22.

69. Burr, American Medico-Psychological Association, *Proceedings, 21* (1914), 317; White, "Psychoanalytic Parallels," R, *2* (April 1915), 177.

70. *New York Times,* April 5, 1912, p. 8; Putnam, "Comments on Sex Issues from the Freudian Standpoint" (1912), Ad Psa, pp. 136–137.

71. Putnam to Freud, June 4, 1912, JPP, p. 140.

72. Freud to Putnam, June 25, 1912, *ibid.,* p. 143.

73. Schwab, JNMD, *38* (August 1911), 495–496.

74. Jones, J Ab P, 6 (June–July 1911), 172, 126–134.

75. Prince and Putnam, "A Clinical Study of a Case of Phobia: A Symposium," J Ab P, 7 (October–November 1912), 303, 298.

76. Prince, "The Psychopathology of a Case of Phobia, a Clinical Study," J Ab P, *8* (December 1913–January 1914), 334–335.

77. Prince, J Ab P, *5* (February–March 1911), 342–343.

78. Janet, *Psychological Healing,* I, 184.

79. Prince to Putnam, October 21, 1910, PP.

80. Prince, J Ab P, 7 (October–November 1912), 301.

81. C. P. Oberndorf, "A Case of Hallucinosis induced by Repression," J Ab P, 6 (February–March 1912), 441.

82. Prince, *The Creed of Deutschtum and other Essays* (Boston: Richard G. Badger, 1918), pp. 252–260, 307–308.

83. Dercum, JAMA, *62* (March 7, 1914), 756.

84. Frederick P. Gay, *Elmer Ernest Southard, the Open Mind* (Normandie House, 1938), pp. 248–262; for Putnam's favorable reactions to what must have been the Armory Show on tour in Boston, see Putnam to Susan Blow, March 5, 1913, PP.

85. Dercum, JAMA, *62* (March 7, 1914), 754; Woodworth, J Ab P, *12* (August 1917), 175.

86. J. Victor Haberman, "A Criticism of Psychoanalysis," J Ab P, 9 (1914–1915), 269.

87. William S. Walsh, *Psychotherapy* (New York: D. Appleton & Sons, 1912), pp. 480–481, 595–596; Adolf Meyer, "Conditions for a Home of Psychology in the Medical Curriculum," J Ab P, 7 (December 1912–January 1913), 324.

88. Samuel A. Tannenbaum, "Freud's Apprehension Neurosis," *American Medicine, 17* (December 1911), 642–645; Brill, "Freud's Compulsion Neurosis," *ibid.,* p. 647.

89. Brill, Preface to Freud, *Selected Papers on Hysteria,* 2nd edition (New York: Journal of Nervous and Mental Disease Publishing Co., 1912), p. ix.

90. Editorial, "Freudism on Dangerous Ground," *Medical Review of Reviews, 18* (March 1912), 150.

91. Tannenbaum, "Sexual Abstinence and Nervousness," *American Journal of Urology, 9* (June 1913), 321; editorial, "The Awakening of a New Conscience," *Medical Review of Reviews, 19* (May 1913), 289–291.

92. Editorial, "The Psychoanalyst and the Puritan Strain," *The Medical Record* (New York), *89* (March 4, 1916), 425–426.

93. Janet, J Ab P, 9 (1914–1915), 181.

94. Burr, American Medico-Psychological Association, *Proceedings, 21* (1914), 304–315.

95. Dercum, JAMA, *62* (March 7, 1914), 756.

Chapter XII. *The American Psychoanalytic Organizations*

1. Putnam to Jones, January 5, 1911, JPP, p. 249.

2. Franz Alexander, Sheldon T. Selesnick, "Freud-Bleuler Correspondence," *Archives of General Psychiatry, 12* (January 1965), 1–9.

3. Freud, " 'Wild' Psychoanalysis" (1910), SE, XI, 226–227; "On the History of the Psycho-analytic Movement" (1914), SE, XIV, 43–44.

4. Ernest Federn and Herman Nunberg, eds., *Minutes of the Vienna Psycho-analytic Society,* II, 1908–1910 (New York: International Universities Press, 1967), 463–471.

5. Freud to Putnam, June 16, 1910, JPP, pp. 100–101.

6. Jones to Putnam, July 12, 1910, *ibid.,* p. 222.

7. Putnam to Freud [late July, 1910], *ibid.,* p. 103.

8. Jones to Putnam, August 14, 1910, *ibid.,* p. 225.

9. Jones to Putnam, September 9, 1910, *ibid.,* p. 227.

10. Prince to Putnam, November 22, 1910, *ibid.,* p. 327.

11. Jones to Putnam, December 11, 1910, *ibid.,* pp. 246–247.

12. Among these were Brill; George H. Kirby, thirty-six, the clinical director and the son of a physician from North Carolina; Maurice Karpas, thirty-two, who had emigrated with his family at the age of thirteen from St. Petersburg and whom Brill had known in night school; C. P. Oberndorf, age twenty-nine, the son of a merchant from Selma, Alabama, who became the first historian of the American movement; L. Bish; Frederic J. Farnell, twenty-six, from Providence, Rhode Island; Ernest M. Poate, twenty-seven, son of a minister from Rushford, New York; J. Rosenbloom; William C. Garvin, thirty-eight, state hospital psychiatrist, and Charles Ricksher, thirty-two, who had graduated from the Johns Hopkins Medical School in 1905 and had studied in Zurich with Jung in 1907. Others included E. W. Scripture, forty-six, a psychologist and physician, former director of the Yale Psychological Laboratory and then an associate in psychiatry at Columbia University, and Samuel Tannenbaum, thirty-seven, an immigrant from Hungary who had graduated from Columbia University Medical School and who practiced psychotherapy and general medicine. New York Psychoanalytic Society, Minutes (Courtesy of

Edmund Brill) February 12, 1911; "The New York Psycho-analytic Society," J Ab P, 6 (April–May 1911), 80; "The American Psycho-Analytic Association," *ibid.* (October–November 1911), 328; C, *2* (1912), 233, 236, 241–242. Information about members comes from a variety of sources, obituaries, biographical material, medical school records. L. Bish probably was the popular medical writer, Louis E. Bisch.

13.   "The American Psycho-Analytic Association," J Ab P, 6 (October–November 1911), 328; Jones, *The Life and Work of Sigmund Freud* (New York: Basic Books, 1955), II, 87–88.

14.   New York Psychoanalytic Society, Minutes, April 25, 1911, June 27, 1911; C, *2* (1911), 236; see also Bertram Lewin and Helen Ross, *Psychoanalytic Education in the United States* (New York: W. W. Norton, 1960), pp. 8–9. For the Boston Society see J Ab P, 9 (April–May 1914), 71; Z, *2* (1914), 404.

15.   "Discussion" of G. Stanley Hall, "The Application of Freudian Mechanisms to Other Emotions," J Ab P, *10* (1915), 275.

16.   Ralph Reed, "From Mesmer to Freud, a Review of Psychotherapy," *Lancet-Clinic* (Cincinnati), *103:14* (April 2, 1910), 354–367, and Jones to Putnam, January 23, 1911, JPP, p. 254.

17.   J. McKeen Cattell, "Families of American Men of Science," *Popular Science Monthly,* 86 (May 1915), 504–515; "The American Society of Naturalists, Homo Scientificus Americanus," *Science,* n.s., 17 (April 10, 1903), 561–570; "Families of American Men of Science," *Scientific Monthly,* 4 (March 1917), 248–262; *ibid.,* 5 (March 1917), 368–377.

18.   Freud, "The Future Prospects of Psychoanalytic Therapy" (1910), SE, XI, 144–145.

19.   C. P. Oberndorf, *A History of Psychoanalysis in America* (New York: Grune and Stratton, 1953), pp. 117–118; "New York Psychoanalytic Society," *International Journal of Psychoanalysis, 2* (1921), 151.

20.   Michael Balint, "On the Psychoanalytic Training System," J, *39,* Part 3, (1948), 167–173, "Analytic Training and Training Analysis," *ibid., 35,* Part 2 (1954), 157–162; Rudolph Ekstein, "A Historical Survey on the Teaching of Psychoanalytic Technique," *Journal of the American Psychoanalytic Association,* 8 (1960), 500--516; J, 4 (October 1923), 520.

21.   H. W. Frink, "Report of the Psychotherapeutic Clinic of the Cornell Dispensary," *New York Medical Journal,* 94 (September 30, 1911), 671–672.

22.   Adolf Stern, "Neurotic Manifestations in Children," *The Medical Record* (New York), *89* (February 26, 1916), 361–363.

23.   New York Psychoanalytic Society, Minutes, May 23, 1911, December 23, 1919.

24.   Z, *2* (1914), 413; Z, *6* (1920), 187–188.

25.   Z, *2* (1914), 409; Z, *5* (1924), 115–116; C, *3* (1913), 102–103; Z, *4* (1916–1917), 124.

26.   Percy Hickling, "Psychoanalysis in Its Relation to Psychiatry," *Washington Medical Annals, 15* (1916), 342.

27. Cyril Greenland, "Ernest Jones in Toronto: 1908–1913," *Canadian Psychiatric Association Journal,* 6 (June 1961), 132–134; Part II, *ibid., 11* (December, 1966), 512–519.

28. Jones to Putnam, April 9, 1910, JPP, p. 217.

29. Charles W. Burr, "A Criticism of Psychoanalysis," American Medico-Psychological Association, *Proceedings, 21* (1914), 304.

30. Freud, "The Question of Lay Analysis" (1926), SE, XX, 250; "Introduction to Oskar Pfister, *The Psychoanalytic Method"* (1913), SE, XII, 330–331.

31. "Discussion of Lay Analysis," J, 8 (April 1927), 221; James Oppenheim, *American Types, a Preface to Analytic Psychology* (New York: Alfred A. Knopf, 1931), p. 123; interview, Moritz Jagendorf, New York, May 25, 1960.

32. White, *Mechanisms of Character Formation: An Introduction to Psychoanalysis* (New York: Macmillan Co., 1916), p. 257.

33. Max Eastman, "Exploring the Soul and Healing the Body," *Everybody's Magazine, 32* (June 1915), 743; Sandor Lorand, "A. A. Brill," J, 29, Part I, (1948), 2–3.

34. Oberndorf, *A History of Psychoanalysis in America,* pp. 134–135.

35. Smith Ely Jelliffe, "Glimpses of a Freudian Odyssey," 2 (April 1933), 325–327; New York Psychoanalytic Society, Minutes, November 25, 1913; William Alanson White to G. Stanley Hall, February 11, 1913, CUP.

36. Oberndorf, *A History of Psychoanalysis in America,* p. 110.

37. C, 2, (1912), 675–677.

38. C. P. Oberndorf, *A History of Psychoanalysis in America,* p. 136; J, 1 (1920), 211–214.

39. Freud, "On the History of the Psychoanalytic Movement," (1914), SE, XIV, 32; "Introductory Lectures on Psychoanalysis, (1916–1917), SE, XVI, 423.

40. Jones, *Freud,* II, 152–167.

Chapter XIII. The American Interpretation of Psychoanalysis

1. A few saw the pessimistic elements in Freud during the Great War, but most American analysts ignored them. See Joseph Jastrow, "The Psycho-Analyzed Self," *Dial, 62* (May 3, 1917), 395–398, and the *Chicago Daily News,* May 10, 1918, p. 6.

2. William Alanson White, R, 5 (October 1918), 448; "Definition by Tendency," *ibid., 15* (October 1928), 373–383.

3. White, "Extending the Field of Conscious Control," *ibid.,* 7 (April 1920), 148–162; Freud, "Instincts and Their Vicissitudes" (1915), SE, XIV, 134–139; Louville Eugene Emerson, abstract of Freud, "Impulses and Their Mutations," R, 6 (July 1919), 345–346.

4. Z, 2 (1914), 481.

5. Jones, *The Life and Work of Sigmund Freud* (New York: Basic Books, 1957), III, 29; Edward Bernays, *Biography of an Idea: Memoirs of Public Relations Counsel, Edward L. Bernays* (New York: Simon and Schuster, 1965), pp. 261–267; Bruce Barton, "You Can't Fool Your Other Self," *American Magazine, 92* (September 11, 1921), 11–13, 68–72.

6. Putnam, "Symposium Before the New England Hospital Medical Society," October 19, 1911, PP.

7. Garry R. Austin, "An Analysis of Certain Characteristics of Recent Widely Distributed Psychology Books for the Lay Reader," Doctoral Dissertation, Northwestern University, School of Education, 1951, p. 50; Putnam, introduction to Horace W. Frink, *Morbid Fears and Compulsions* (New York: Moffat, Yard and Co., 1918), pp. vii–viii; Jones, *Free Associations* (New York: Basic Books, 1959), p. 231.

8. Putnam to Ernest Jones, July 21, 1915, JPP, pp. 290–291. see Brill, *Fundamental Conceptions of Psychoanalysis* (New York: Harcourt, Brace & World, 1921); Isador Coriat, *What Is Psychoanalysis?* (New York: Moffat, Yard and Co., 1917).

9. Freud to Putnam, December 5, 1909, JPP, pp. 90–91.

10. Freud, SE, XVI, 434.

11. *American Journal of Urology, 11* (October 1915), 391–405.

12. Freud, "Civilized Sexual Morality and Modern Nervousness," CPF, II, 76–99, 92.

13. Jelliffe, review of Carpenter, *Love's Coming of Age,* JNMD, 44 (September 1916), 288.

14. John B. Watson and K. S. Lashley, "A Consensus of Medical Opinion upon Questions Relating to Sex Education and Venereal Disease Campaigns," *Mental Hygiene, 4* (October 1920), 769–847.

15. Coriat, *What Is Psychoanalysis?,* pp. 72–74; White, *Mechanisms of Character Formation: An Introduction to Psychoanalysis* (New York: Macmillan, 1916), p. 99; Frink, *Morbid Fears and Compulsions,* pp. 146–147.

16. Freud to Putnam, May 14, 1911, JPP, pp. 121–122.

17. Jones, *Freud,* II, 103.

18. Monroe A. Meyer, review of Ernest Jones, *Papers on Psychoanalysis,* 3rd ed., *Mental Hygiene, 8* (January 1924), 266–267; Ernest Jones, "The Value of Sublimating Processes for Education and Re-Education" (1911), *Papers on Psychoanalysis* (London: Baillière, Tindall and Cox, 1913), pp. 417–419, 424–428.

19. Oskar Pfister, *The Psychoanalytic Method* (New York: Moffat, Yard and Co., 1917), pp. 315–316.

20. Freud, "Turnings in the Ways of Psycho-Analytic Therapy" (1937), CPF, II, 395.

21. Freud, "Civilization and Its Discontents" (1930), SE, XXI, 83–84; "Formulations on the Two Principles of Mental Functioning" (1911), SE, XII, pp. 222–223; *Introductory Lectures on Psycho-Analysis* [1916–1917], SE, XVI, 356–357.

22. Compare SE, XII, 222–223, with Frink, *Morbid Fears,* pp. 93–94, 550.

23. Brill, *Fundamental Conceptions of Psychoanalysis,* pp. 29–30, 187–188; Putnam, "Personal Impressions of Sigmund Freud and His Work" (1909), Ad Psa, pp. 24, 9; Smith Ely Jelliffe, *The Technique of Psychoanalysis* (New York: Journal of Nervous and Mental Disease Publishing Co., 1918), p. 52; James J. Walsh, *Psychotherapy* (New York: D. Appleton, 1912), p. 481.

24. Frink, *Morbid Fears,* pp. 4–5.

25. White, *The Mental Hygiene of Childhood* (Boston: Little Brown, & Co., 1919), pp. 27–32, 119–121, 186; Putnam, *Human Motives* (Boston: Little, Brown, & Co., 1915), pp. 85–101.

26. Freud, *The Interpretation of Dreams* (1900), SE, IV, 257.

27. Otto Rank, *The Myth of the Birth of the Hero: A Psychological Interpretation of Mythology,* trans. F. Robbins and Smith Ely Jelliffe (New York: Nervous and Mental Disease Publishing Co., 1914), pp. 63–68, 74–76.

28. Smith Ely Jelliffe, *The Technique of Psychoanalysis* (New York: Nervous and Mental Disease Publishing Co., 1918), pp. 49–53, 55–61; White, *Mechanisms of Character Formation,* pp. 155, 162.

29. Brill, *Psychoanalysis: Its Theories and Practical Application* (Philadelphia: W. B. Saunders, 1913), pp. 282–287.

30. William Alanson White, "Underlying Concepts of Mental Hygiene," *Mental Hygiene, 1* (1917), 12–13; Samuel Haber, *Efficiency and Uplift: Scientific Management in the Progressive Era* (Chicago: University of Chicago Press, 1964), Chapter V.

31. Mabel Dodge Luhan, *Intimate Memories,* III, *Movers and Shakers* (New York: Harcourt Brace, 1936), p. 506.

32. See note 3.

33. Brill, *Fundamental Conceptions of Psychoanalysis,* p. 329; Frink, *Morbid Fears,* p. 93.

34. Trigant Burrow, "Character and the Neuroses," R, *1* (February 1914), 123; "Psychoanalysis and Society," J Ab P, 7 (December 1912–January 1913), 340–346, 350–351; "The Psychoanalyst and the Community, JAMA, 62 (June 13, 1914), 1876–1878.

35. White, R, 5 (October 1918), 445–446.

36. Frink, *Morbid Fears,* p. 518; Jones, "Psychoanalysis in Psychotherapy," in Prince *et al., Psychotherapeutics* (Boston: Richard G. Badger, 1909), pp. 96–97; White, *Outlines of Psychiatry,* 5th ed. (New York: Nervous and Mental Disease Publishing Co., 1915), pp. 39–40.

37. Bruce Barton, "You Can't Fool Your Other Self," *American Magazine, 92* (September 11, 1921), 70, 72; Jelliffe, *The Technique of Psychoanalysis,* pp. 66, 14; Putnam to Frances Morse, September 1, 1911, PP.

38. Freud, "Civilization and Its Discontents," SE, XXI, 143, 113.

39. Freud, *Introductory Lectures,* SE, XVI, 431–432, 438–439, 447; Jung, *The Psychology of Dementia Praecox* (1907) (New York: Journal of Nervous and Mental Disease Publishing Co., 1909), pp. 34–35, 73.

40. A. A. Brill, "Studies in Paraphrenia," *New York Medical Journal, 110* (November 15, 1919), 792–798.

41. Discussion of A. A. Brill, "Schizophrenia and Psychotherapy," *American Journal of Psychiatry, 86* (November 1929), 538.

42. Coriat, "Some Statistical Results of the Psychoanalytic Treatment of the Psychoneuroses," R, *4* (April 1917), 209–216.

43. Brill, *Fundamental Conceptions of Psychoanalysis,* p. 21; *Psychoanalysis: Its Theories and Practical Application,* pp. 279–281.

44. Pfister, *The Psychoanalytic Method,* authorized translation by Charles Rockwell Payne (New York: Moffat, Yard and Co., 1917), pp. 535, 560–561; Fritz Wittels, "Brill, the Pioneer," R, *35* (October 1948), 397–398.

45. Freud to Oskar Pfister, July 12, 1909, in Heinrich Meng and Ernst L. Freud, eds., *Psychoanalysis and Faith: The Correspondence of Sigmund Freud and Oskar Pfister* (New York: Basic Books, Inc., 1963), p. 27; Freud to Lou Andreas-Salomé, July 30, 1915, in Ernst L. Freud, ed., *The Letters of Sigmund Freud* (New York: McGraw-Hill Book Co., 1964), p. 311; Pfister, *The Psychoanalytic Method,* pp. 560–561.

46. William Alanson White, "Psychoanalytic Tendencies," AJI, *73* (April 1917), 605; Freud, "The Ways of Psycho-Analytic Therapy" (1919), CPF, II, 400–402.

47. Freud, preface to August Aichorn, *Wayward Youth* (1925), CPF, V, 99; introduction to Oskar Pfister, *The Psychoanalytic Method,* pp. vi, vii.

48. Oberndorf, *A History of Psychoanalysis in America,* p. 123; "New York Psychoanalytic Society," Meeting, January 25, 1921, J, *2* (March 1921), 151.

49. Freud, "On Beginning the Treatment" (1913), SE, XII, 124.

50. *Boston Evening Transcript,* September 10, 1909, p. 4; Jelliffe, *The Technique of Psychoanalysis,* 2nd ed. (1920), pp. 138, 67–68; Brill, *Psychoanalysis: Its Theories and Practical Application,* pp. 138–187, and "A Few Remarks on the Technique of Psychoanalysis," *Medical Review of Reviews, 18* (April 1912), 252.

51. Freud, "Studies on Hysteria," SE, II, 302–305.

52. Freud, "On the Dynamics of Transference" (1912), SE, XII, 108; "Observations on Transference Love" (1915), SE, XII, 170.

53. Freud, *Introductory Lectures,* SE, XVI, 444–445.

54. Brill, "A Few Remarks on the Technique of Psychoanalysis," *Medical Review of Reviews, 18* (April 1912), 253; William Alanson White, "The Mechanism of Transference," R, *4* (October 1917), 373–381; A. Stern, "On the Nature of Transference in Psychoanalysis," *New York Medical Journal, 107* (1918), 398–402; C. P. Oberndorf, "Resistance and Transference in Psychoanalysis," *The Medical Record* (New York), *114* (1918), 542–546.

55. White, "The Mechanism of Transference," R, *4* (October 1917), 373–381.

56. Jelliffe, *The Technique of Psychoanalysis,* pp. 70, 80–81.

57. John B. Watson, "Behavior and the Concept of Mental Disease," *Journal of Philosophy, Psychology and Scientific Methods, 13* (October 26, 1916), 590.

58. *Ibid.*, p. 596.

59. John B. Watson, "The Psychology of Wish-Fulfilment," *Scientific Monthly,* 3 (November 1916), 487.

60. Frink, *Morbid Fears,* pp. 196–198.

61. Alfred Adler, *The Neurotic Constitution,* trans. Bernard Glueck (New York: Moffat, Yard and Co., 1916), pp. 1–6, xiv, 29–30, 43–67.

62. Quoted in White, *Mechanisms of Character Formtion,* pp. 249, 258–259, 42–43; Jelliffe, *The Technique of Psychoanalysis,* pp. 78–81.

63. André Tridon, *Psychoanalysis and Love* (Garden City, New York: Garden City Publishing Co., 1922), title page.

64. White, *Mechanisms of Character Formation,* p. 256.

65. Jones, Freud, II, 91, 144–145, 150.

66. Jung, *The Theory of Psychoanalysis* (New York: Journal of Nervous and Mental Disease Publishing Co., 1915), p. 2, passim.

67. White, *Mechanisms of Character Formation,* pp. 219, 232–238.

68. White, *Mechanisms of Character Formation,* p. 275.

69. *Ibid.*, p. 225.

70. Frink, *Morbid Fears,* pp. 552–553.

71. *Ibid.*, pp. 55–61, 511–512; see note 14.

72. Putnam, "Elements of Strength and Elements of Weakness in Psycho-Analytic Doctrines" (1918), Ad Psa, p. 453.

73. Discussion of Trigant Burrow, "The Meaning of Psychoanalysis: An Apologia," J Ab P, *11* (1916–1917), 412–413.

74. White, "Psychoanalytic Tendencies," AJI, *73* (April 1917), 603–604.

75. Brill, *Fundamental Conceptions of Psychoanalysis,* pp. 315–327; Coriat, "A Note on the Anal Character Traits of the Capitalistic Instinct," R, *11* (October 1925), 435–437.

76. John T. MacCurdy, "Ethical Aspects of Psychoanalysis," *Johns Hopkins Hospital Bulletin,* 26 (May 1915), 169–173; Frink, *Morbid Fears,* pp. 42–43, 62–66, 71.

77. White, *Mechanisms of Character Formation,* pp. 58–60, 310–313; Putnam, "The Interpretation of Certain Symbolisms" (1918), Ad Psa, pp. 413, 438–446.

78. Coriat, *The Meaning of Dreams* (Boston: Little, Brown, and Co., 1915), p. 65.

79. Coriat, *What Is Psychoanalysis?,* pp. 79–80.

80. Putnam, *Human Motives* (Boston: Little, Brown and Co., 1915), pp. 106–107; White, *The Mental Hygiene of Childhood,* pp. 187–188.

81. Jones, "The Therapeutic Action of Psychoanalysis" (1911), *Papers on Psychoanalysis,* p. 307.

82. White, *Principles of Mental Hygiene* (New York: The Macmillan Co., 1917), pp. 170–185; "The Social Significance of Mental Disease," *Archives of Neurology and Psychiatry,* 22 (November 1929), 873–900; A. A. Brill, "Alcohol and the Individual," *New York Medical Journal, 109* (May 31, 1919), 928–930.

83. John T. MacCurdy, "Ethical Aspects of Psychoanalysis," *Johns Hopkins Hospital Bulletin, 26* (May 1915), 173.

84. Frink, *Morbid Fears,* pp. 135–136.

85. Brill, "What's Wrong with Cities?," *Saturday Evening Post, 204* (October 17, 1931), p. 44.

86. Freud, "Three Essays on Sexuality" (1905), SE, VII, 222–224; White, *The Mental Hygiene of Childhood,* pp. 133–151; Brill, *Psychoanalysis: Its Theories and Practical Application,* p. 280.

87. Putnam, *Human Motives,* pp. 96–97; Brill, *Psychoanalysis: Its Theories and Practical Application,* p. 38.

88. White, *The Mental Hygiene of Childhood,* pp. 94–111, 116–118, 133–141; Brill, *Fundamental Conceptions of Psychoanalysis,* pp. 280–282; Coriat, *What Is Psychoanalysis?,* pp. 84–86.

89. Coriat, *What Is Psychoanalysis?,* p. 77; Jelliffe, *The Technique of Psychoanalysis,* pp. 97–117.

90. *Ibid.,* pp. 158–160.

Chapter XIV. American Psychoanalysts

1. Freud, "On the History of the Psychoanalytic Movement" (1914), SE, XIV, 31.

2. Jones, obituary, James Jackson Putnam, Ad Psa, p. 457.

3. Edward W. Taylor, "James Jackson Putnam: His Contributions to Neurology," *Archives of Neurology and Psychiatry, 3* (March 1920), 313.

4. Putnam, "The Philosophy of Psychotherapy," *Psychotherapy: A Course of Reading, 3,* No. 3 (1909), 15–16.

5. Putnam, Presidential Address to the American Psychopathological Association, J Ab P, *8* (August–September 1913), 174, 179; Putnam to Jones, September 14, 1910, JPP, p. 229.

6. Putnam, *Human Motives* (Boston: Little, Brown, 1915), pp. 65–66, 150, 168.

7. Putnam, "On the Etiology and Treatment of the Psychoneuroses" (1910), Ad Psa, pp. 70–72, 77–78.

8. Putnam, *Human Motives,* p. 150; Putnam, "On Freud's Psycho-Analytic Method and Its Evolution" (1911), Ad Psa, p. 111.

9. Putnam, *Human Motives,* pp. 169–170, 110–111.

10. Putnam, review of William Healy, *The Individual Delinquent,* R, *2* (October 1915), 469–472.

11. White, review of *Human Motives,* R, *3* (January 1916), 116; Richard C. Cabot, review of *Human Motives, Survey, 36* (June 1916), 292; *Dial, 59* (August 15, 1915), 114; *Nation, 101* (September 16, 1915), 362.

12. Freud to Putnam, July 8, 1915, in Jones, *The Life and Work of Sigmund Freud* (New York: Basic Books, 1953–1957), II, 416, 180; III, 389–390.

13. Freud, "An Autobiographical Study" (1925), SE, XX, 51.

14. Oskar Pfister, *Some Applications of Psychoanalysis* (New York: Dodd, Mead, 1923), pp. 178–179; Jones, *Freud,* II, 86; Freud, "Lines of Ad-

vance in Psycho-Analytic Therapy" (1919), SE, XVII, 165; Ferenczi, "Philosophie und Psychoanalyse" (Bermerkungen zu einen Aufsatze des H. Prof. Dr. James J. Putnam von der Harvard Universität U.S.A.), *Imago, 1* (1912), 519–526; Antwort auf die Erwiderung des H. Dr. Ferenczi, *ibid.,* 527–530.

15. Putnam, "The Work of Sigmund Freud," Ad Psa, p. 360; Putnam, "The Work of Alfred Adler, Considered with Especial Reference to That of Freud" (1915), *ibid.,* pp. 314, 333; Putnam, "The Work of Sigmund Freud" (1917), *ibid.,* pp. 349, 353, 360.

16. *Ibid.,* pp. 362–363.

17. *Ibid.,* pp. 355–359.

18. Putnam, "Elements of Strength and Elements of Weakness in Psychoanalytic Doctrines" (1919), Ad Psa, pp. 451–452.

19. Putnam, "On Freud's Psycho-Analytic Method and Its Evolution" (1911), *ibid.,* p. 122.

20. Putnam, "The Work of Sigmund Freud" (1917), *ibid.,* p. 348.

21. *Dictionary of American Biography, 8,* 283; *Boston Evening Transcript,* November 7, 1918, p. 12.

22. Coriat, "Some Personal Reminiscences of Psychoanalysis in Boston," R, *32,* (January 1945), 6; George B. Wilbur, obituary, Isador Coriat, R, *30* (October 1943), 479–483.

23. Coriat, *Repressed Emotions* (New York: Brentano's, 1920), pp. 43–46.

24. Coriat, "The Future of Psychoanalysis," R, *4* (October 1917), 383–384.

25. Coriat, *What Is Psychoanalysis?* (New York: Moffat, Yard and Co., 1917), pp. 86, 76; Coriat, "Anal-Erotic Character Traits in Shylock," J, *2* (September–December 1921), 354–360; Coriat, "A Note on the Anal Character Traits of the Capitalistic Instinct," R, *11* (October 1924), 435–437.

26. Coriat, *The Hysteria of Lady Macbeth,* 2nd ed. (Boston: Four Seas Co., 1920), originally published 1912; Coriat, "A Note on the Sexual Symbolism of the Cretan Snake Goddess," R, *4* (July 1917), 367–368.

27. Ray Lyman Wilbur, introduction to William Alanson White, *William Alanson White: The Autobiography of a Purpose* (New York: Doubleday, Doran, 1938), pp. xvii, 107, 176–177.

28. Elizabeth Shepley Sergeant, "A Specialist in Human Beings," *Harper's Magazine, 154* (March 1927), 480; Jelliffe, obituary, William Alanson White, JNMD, *85* (May 1937), 629–634.

29. White, *Forty Years of Psychiatry* (New York and Washington: Nervous and Mental Disease Publishing Co., 1933), pp. 149–151; White, "The Theory, Methods and Psychotherapeutic Value of Psychoanalysis," *Interstate Medical Journal, 17* (1910), 654–655.

30. White, "Psychoanalytic Tendencies," AJI, *73* (April 1917), 603–605.

31. White, *Autobiography,* pp. 112–113.

32. Remarks of Elmer Ernest Southard on White, "Psychoanalytic Tendencies," AJI, *73* (April 1917), 606.

33. White, *Autobiography,* p. 192; White, *Crimes and Criminals* (New York: Farrar and Rinehart, 1933), pp. 134, 178–180, 232–245.

34. White, *Mechanisms of Character Formation* (New York: Macmillan, 1920), preface.

35. White, *Forty Years of Psychiatry,* pp. 57–61.

36. White, "Definition by Tendency," R, *15* (October 1928), 381; White, review of James Hay, Jr., *Mrs. Marden's Ordeal,* R, *5* (October 1918), 447–448.

37. Seminar, Karl Menninger and the Topeka Psychoanalytic Institute, Topeka, Kansas, May 15, 1960; interview, Bernard Glueck, June 10, 1960, Chapel Hill, North Carolina; James Hay, Jr. and Florenz K. Buschman, "Dr. William Alanson White," *The World's Work, 59* (May 1930), 44–46. Hay was the author of America's first psychoanalytic novel, *Mrs. Marden's Ordeal,* published by Little, Brown and Co., Boston, in 1918.

38. White, *Forty Years of Psychiatry,* pp. 101–102; review of White, *The Mental Hygiene of Childhood, Dial, 67* (September 20, 1919), 264; John T. MacCurdy, review of White, *Mental Hygiene of Childhood, Mental Hygiene, 3* (October 1919), 711–712.

39. Max Eastman, *The Enjoyment of Living* (New York: Harper, 1948), p. 491.

40. Jelliffe, "Glimpses of a Freudian Odyssey," Q, *2,* (1933), 319, 322; Karl Menninger, "Smith Ely Jelliffe," in Menninger, *A Psychiatrist's World: The Selected Papers of Karl Menninger* (New York: Viking Press, 1959), pp. 826–828; C. P. Oberndorf, obituary, Smith Ely Jelliffe, J, *26* (1945), 186–189; A. A. Brill, obituary, Smith Ely Jelliffe, JNMD, *106* (1947), 221.

41. Jelliffe, "Sigmund Freud and Psychiatry: A Partial Reappraisal," *American Journal of Sociology, 45* (November 1939), 340.

42. Jelliffe, "Glimpses of a Freudian Odyssey," Q, *2* (1933), 325.

43. Eastman, *The Enjoyment of Living,* p. 492.

44. Jelliffe, discussion, J Ab P, *8* (December 1913–January 1914), 350.

45. Jelliffe, "Women and the Old Immorality," *Forum, 77* (February 1927), 189, 197–198.

46. Jelliffe and Louise Brink, "Compulsion and Freedom, the Fantasy of the Willow Tree," R, *5* (July 1918), 256–257; Jelliffe and Brink, *Psychoanalysis and the Drama* (New York and Washington: JNMD Publishing Co., 1922), pp. 4–8, 59–60, 112–113.

47. Quoted in Karl Menninger and George Devereux, "Smith Ely Jelliffe, Father of Psychosomatic Medicine," R, *35* (October 1948), 351; Karl Menninger, "Smith Ely Jelliffe and Peter Bassoe, 1866–1945 and 1874–1945;" Menninger, *A Psychiatrist's World,* pp. 826–828; Jelliffe, "Sigmund Freud and Psychiatry: A Partial Reappraisal," *American Journal of Sociology, 45* (November 1939), 326–340.

48. C. P. Oberndorf, M. A. Meyer, A. Kardiner, obituary, Horace W. Frink, Q, *5* (1936), 601–603; obituary, Horace W. Frink, *New York Times,* April 19, 1936, Section 2, p. 11; A. A. Brill, obituary, Horace W. Frink, JNMD, *84* (August 1936), 239–240.

49. Frink, *Morbid Fears and Compulsions,* pp. 50–55.

50. *Ibid.*, pp. 226, 502–530.

51. *Ibid.*, p. 549.

52. *Boston Evening Transcript,* May 11, 1918, Part 3, p. 6; James Harvey Robinson, review of *Morbid Fears and Compulsions, New Republic, 17* (November 30, 1918), 140.

53. Max Eastman, "A Significant Memory of Freud," *New Republic, 104* (May 19, 1941), 694.

54. A. A. Brill to Smith Ely Jelliffe, December 4, 1940, courtesy of Edmund Brill.

55. *Ibid.*

56. A. A. Brill, "The Adjustment of the Jew to the American Environment," *Mental Hygiene, 2* (January 1918), 219–231.

57. Paula S. Fass, "A. A. Brill—Pioneer and Prophet," Master's Thesis, Political Science, Columbia University, 1969, pp. 29–33.

58. Brill, "The Adjustment of the Jew to the American Environment," p. 224.

59. A. A. Brill, *Psychoanalysis: Its Theories and Practical Application,* dedication.

60. A. A. Brill, "Reminiscences of Freud," Q, 9 (1940), 177–178.

61. Theodore Dreiser, "The Mercy of God," *American Mercury, 2* (August 1924), 459.

62. A. A. Brill, "The Adjustment of the Jew to the American Environment," 222.

63. "Dance Man Epidemic," *New York Times,* April 6, 1914, Section IV, p. 8.

64. A. A. Brill, "The Psychology of the Jew," *American Hebrew and Jewish Tribune,* January 13, 1933, p. 170; letter to author from Edmund Brill, November 29, 1964.

65. "The Psychopathology of Reformers," *New York Times,* July 31, 1921, Section III, p. 7; Brill, "Alcohol and the Individual," *New York Medical Journal, 109* (May 31, 1919), 928–930; A. A. Brill as told to Selma Robinson, "What's Wrong with Cities," *Saturday Evening Post, 204* (October 17, 1931), 44.

66. A. A. Brill, "The Adjustment of the Jew to the American Environment," pp. 229–230.

67. A. A. Brill, preface to Freud, *Three Contributions to the Theory of Sex,* 4th ed. (New York and Washington: JNMD Publishing Co., 1930), p. xiii; Brill, "What's Wrong with Cities," p. 44.

68. A. A. Brill, "Diagnostic Errors in Neurasthenia," *Medical Review of Reviews, 36* (March 1930), 122.

69. Brill, introduction to *The Basic Writings of Sigmund Freud* (New York: Modern Library, 1938), p. 3.

70. Jones, *Freud,* III, 37–38; Jones, *Freud,* II, 175–176.

71. Quoted in "Damage and Defense," *Time, 27* (May 25, 1936), 31; Frederick Peterson, "Credulity and Cures," JAMA, 73 (December 6, 1919), 1740.

72. Brill, "Prof. Freud and Psychiatry," R, *18* (July 1931), 242.

73. Frederick Peterson to A. A. Brill, June 18, 1934, A. A. Brill to Frederick Peterson, June 28, 1934, in the possession of Edmund Brill.

74. Brill, "Masturbation, Its Causes and Sequellae," *American Journal of Urology and Sexology, 12* (May 1916), 214–222.

75. Quoted in Karl Menninger, "Contributions of A. A. Brill to Psychiatry," (1948), in Menninger, *A Psychiatrist's World,* pp. 832–833; see also Brill, "Reflections on Euthanasia," JNMD, *84* (July 1936), 1–12; review of Freud, *The Discomforts of Civilization,* JNMD, *72* (August 1930), 113–124.

Chapter XV. Popular Psychoanalysis

1. These figures are based on corrections of the *Readers' Guide to Periodical Literature* and on *Recent Social Trends in the United States: Report of the President's Research Committee on Recent Social Trends* (New York: McGraw, Hill Book Co., 1933), I, 395, 414.

2. Chicago *Daily News,* May 10, 1918, p. 6; Edwin Tenney Brewster, "Dreams and Forgetting," *McClure's Magazine, 39* (October 1912), 714–719.

3. See the *Index Medicus.*

4. Max Eastman, *The Enjoyment of Living* (New York: Harper and Bros., 1948), p. 491.

5. Quoted in J. Victor Haberman, "Psychic Therapy, Clinical Psychology and the Layman Invasion," *The Medical Record* (New York), *87* (April 24, 1915), 680.

6. Max Eastman, "Exploring the Soul and Healing the Body," *Everybody's Magazine, 32* (June 1915), 741.

7. Floyd Dell, "Speaking of Psychoanalysis, the New Boon for Dinner Table Conversationalists," *Vanity Fair, 5* (December 1915), 53.

8. Lucian Cary, "Escaping Your Past," *Technical World, 23* (August 1915), 733.

9. *Ibid.,* p. 813.

10. "The Establishment of a Psychoanalytic Priest in a Confessional of Science," *Current Opinion, 62* (April 1917), 260.

11. Max Eastman, "Exploring the Soul," *Everybody's Magazine, 32* (June 1915), 750; Freud, "The Future Prospects of Psychoanalytic Therapy" (1910), SE, XI, 148.

12. Pearce Bailey, "The Wishful Self," *Scribner's Magazine, 58* (July 1915), 115–121.

13. Peter Clark Macfarlane, "Diagnosis by Dreams," *Good Housekeeping, 60* (February 1915), 127.

14. *New York Times,* March 2, 1913, Section 5, p. 10; W. M. A. Maloney, "Modern Means of Investigating Mental Processes," *Scientific American Supplement, 74* (November 9, 1912), 290.

15. Max Eastman, "Exploring the Soul," *Everybody's Magazine, 32* (June 1915), 746.

16. Rev. Samuel McComb, "The New Interpretation of Dreams," *Century Magazine, 84* (September 1912), 665.

17. Floyd Dell, *Vanity Fair, 5* (December 1915), 53.

18. Max Eastman, *Everybody's Magazine, 32* (June 1915), 742.

19. Peter Clark Macfarlane, *Good Housekeeping, 60* (February 1915), 125, 132.

20. *Ibid.* (March 1915), pp. 280–283.

21. Havelock Ellis, "The Psychoanalysts," *Bookman, 46* (September 1917), 54–55.

22. Peter Clark Macfarlane, "Diagnosis by Dreams," *Good Housekeeping, 60* (March 1915), 286.

23. "The Modern Medicine Man," *Unpopular Review, 10* (July–September 1918), 134.

24. Peter Clark Macfarlane, "Diagnosis by Dreams," *Good Housekeeping, 60* (March 1915), 283.

25. *Ibid.,* pp. 283, 286.

26. *Ibid.,* p. 283.

27. See note 17.

28. Edwin Tenney Brewster, "Dreams and Forgetting: New Discoveries in Dream Psychology," *McClure's Magazine, 39* (October 1912), 715.

29. "Freud's Discovery of the Lowest Chamber of the Soul," *Current Literature, 50* (May 1911), 513.

30. "The Modern Medicine Man," *Unpopular Review, 10* (July–September 1918), 132.

31. Edward Sapir, "Psychoanalysis as a Pathfinder," review of Oskar Pfister, *The Psychoanalytic Method, Dial, 63* (September 27, 1917), 267–269.

32. Rev. Samuel McComb, "The New Interpretation of Dreams," *Century Magazine, 84* (September 1912), 664.

33. Arthur B. Reeve, "The Dream Doctor," *Cosmopolitan, 55* (August 1913), 334; "The Psychic Scar," *Cosmopolitan, 64* (April 1918), 130; "The Soul-Analysis," *Cosmopolitan, 60* (1916), 875.

34. Charles F. Oursler, "Behind the Madman's Dreams," *Technical World, 21* (April 1914), 207.

35. Alfred Booth Kuttner, "What Causes Slips of the Tongue," *New York Times,* October 18, 1914, Section V, p. 10; "Psychoanalysis: Getting at the Facts of Mental Life," *Scientific American, Supplement, 71* (April 22, 1911), 256.

36. See note 28.

37. Stephen S. Colvin, "Real Mind Reading," *Independent, 71* (December 7, 1911), 1258.

38. Hugo Münsterberg, "The Third Degree," *McClure's Magazine, 29* (October 1907), 614–622.

39. Max Eastman, "Mr. -er-er- Oh! What's His Name?," *Everybody's Magazine, 33* (July 1915), p. 100.

40. Peter Clark Macfarlane, "Diagnosis by Dreams," *Good Housekeeping, 60* (February 1915), 126; J. P. Toohey, "How We All Reveal Our Soul Secrets," *Ladies' Home Journal, 34* (November 1917), 97.

41. Florence Kiper Frank, "The Psychoanalyst," R, 4 (October 1917), 459.
42. Jeanne d'Orge, "The Interpreter (Sixteen Years)," *The Little Review, 3* (March 1916), 16, quoted in Florence Kiper Frank, "Psycho-Analysis, Some Random Thoughts Thereon," *The Little Review, 3* (June–July 1916), 15.
43. H. Addington Bruce, "Selfishness and Your Nerves," *Good Housekeeping, 63* (October 1916), 39.
44. Max Eastman, "Exploring the Soul and Healing the Body," *Everybody's Magazine, 32* (June 1915), 743.
45. "The Modern Medicine Man," *Unpopular Review, 10* (July–September 1918), 139.
46. James Hay, Jr., *Mrs. Marden's Ordeal* (Boston: Little, Brown and Co., 1918), pp. 306, 126, 268–270.
47. Morton Prince, "Roosevelt as Analyzed by the New Psychology," *New York Times,* March 24, 1912, Magazine Section, Part 6, pp. 1–2.
48. "A Scientific Vivisection of Mr. Roosevelt," *Current Literature, 52* (May 1912), 518–522.
49. Ernest Jones, "Psycho-Analyse Roosevelts," C, 2 (1912), 674–677.
50. Carl Jung, "America Facing Its Most Tragic Moment," *New York Times,* September 29, 1912, Section 5, p. 2; *New York Times,* February 12, p. 6, February 14, p. 12, February 17, p. 10, 1916.
51. "Dance Man Psychic Epidemic," *New York Times,* April 26, 1914, Section 4, p. 8.
52. Robert S. Woodworth, "Followers of Freud and Jung," *Nation, 103* (October 26, 1916), 396.
53. Wilfrid Lay, " 'John Barleycorn' Under Psychoanalysis," *Bookman, 45* (March 1917), 47–54.
54. Quoted in W. David Sievers, *Freud on Broadway: A History of Psycho-analysis and the American Drama* (New York: Hermitage House, 1955), pp. 50–51.
55. *Nation, 96* (May 15, 1913), 503–505.
56. Warner Fite, "Psycho-Analysis and Sex-Psychology," *Nation, 103* (August 10, 1916), 127–129.
57. Samuel E. Eliot, Jr., *Nation, 103* (September 7, 1916), "Psycho-Analysis," 219.
58. Samuel A. Tannenbaum, *ibid., 103* (September 7, 1916), "Psycho-Analysis Debated," pp. 218–220; "Professor Fite's Reply," *ibid.,* p. 219.
59. Christine Ladd Franklin, "Freudian Doctrines," *Nation, 103* (October 19, 1916), 373–374.
60. Horace M. Kallen, "The Mystery of Dreams," *Dial, 55* (August 1, 1913), 78–80.
61. Ernest R. Groves, *Moral Sanitation* (New York: International Committee of the YMCA, 1916), pp. 17–20, 39; interview, Mrs. Ernest Rutherford Groves, Chapel Hill, North Carolina, June 11, 1960.
62. Ira S. Wile, review of Groves, *Moral Sanitation, Survey, 37* (December 23, 1916), 342; William Alanson White, R, 4 (January 1917), 124.

63. Groves, "Freud and Sociology," R, *3* (July 1916), 241–253.

64. Groves, "Sociology and Psychoanalytic Psychology: An Interpretation of Freudian Hypothesis," *American Journal of Sociology, 23* (July 1917), 116; "The Development of Social Psychiatry," R, *22* (January 1935), 1–9.

65. Wilfrid Lay, *Man's Unconscious Conflict: A Popular Exposition of Psychoanalysis* (New York: Dodd, Mead and Co., 1917), pp. 20, 220, 324, 77.

66. *Ibid.,* pp. 11, 130, 231.

67. *Boston Evening Transcript,* March 28, 1917, Part 2, p. 8; White, review of Lay, *Man's Unconscious Conflict,* R, *4* (October 1917), 471.

68. Interview, B. W. Huebsch, New York, May 25, 1960.

69. Holt, *The Freudian Wish and Its Place in Ethics* (New York: B. W. Huebsch, 1915), pp. 3–4, vi–vii.

70. *Ibid.,* pp. 130–131.

71. *Ibid.,* p. 124.

72. *Ibid.,* p. 145.

73. *Ibid.,* p. 39.

74. John B. Watson, "Does Holt Follow Freud?," *Journal of Philosophy, Psychology and Scientific Methods, 14* (February 15, 1917), 85, 86; James Jackson Putnam, *Harvard Theological Review, 10* (January 1917), 94; *Nation, 102* (January 20, 1916), 82.

75. Alfred Kuttner, "Freud's Contribution to Ethics," *New Republic, 5* (November 27, 1915), 102, 103.

76. Joseph Jastrow, "The Psychoanalyzed Self," *Dial, 62* (May 3, 1917), 397; Elmer Ernest Southard, "Sigmund Freud, Pessimist," J Ab P, *14* (August 1919), 197–216.

77. Letters from George Allen and Unwin, Ltd., December 30, 1960; Dodd, Mead and Co., December 10, 1959; Random House, Inc., December 11, 1959; Liveright Publishing Corp., March 14, 1960; The Macmillan Co., March 10, 1960; W. B. Saunders Co., March 2, 1960.

78. Havelock Ellis, "The Psychoanalysts," *Dial, 46* (September 7, 1916), 54.

79. Max Eastman, "Exploring the Soul," *Everybody's Magazine, 32* (June 1915), 749.

80. Putnam, introduction to *Morbid Fears and Compulsions* (New York: Moffatt, Yard and Co., 1918), p. 1; William Alanson White, R, *5* (October 1918), 448.

81. Edward Sapir, "Psychoanalysis as a Pathfinder," *Dial, 63* (September 27, 1917), 267.

82. Jones, *The Life and Work of Sigmund Freud* (New York: Basic Books, 1955), II, 60.

Chapter XVI. A New Model in Neurology and Psychiatry

1. Morton Prince, *Clinical and Experimental Studies in Personality,* revised and enlarged by A. A. Roback (Cambridge, Mass.: Sci-Art Publishers, 1929, 1939), pp. 10–11.

2. American Medico-Psychological Association, *Proceedings, 31* (1914), 318–322.

3. *Ibid.,* pp. 323–324.

4. Freud to Putnam, December 5, 1909, JPP, p. 90.

5. William Alanson White, *Outlines of Psychiatry* (New York: Journal of Nervous and Mental Disease Publishing Co., 1907), pp. 205–209, 216–219.

6. White, *Outlines of Psychiatry,* 3rd ed. (New York: Journal of Nervous and Mental Disease Publishing Co., 1911), pp. 247–257.

7. White and Jelliffe, eds., *The Modern Treatment of Nervous and Mental Diseases by American and British Authors* (Philadelphia: Lea and Febiger, 1913), I, 333–342, esp. p. 339.

8. White, *Outlines of Psychiatry* (1907), pp. 9–13, 22.

9. White, *Outlines of Psychiatry* (1911), pp. 20–22, 35–41, 124–125, 185.

10. White and Jelliffe, *The Modern Treatment of Nervous and Mental Diseases,* I, p. 48.

11. *Ibid.,* pp. 592–596, 602–606.

12. *Ibid.,* pp. 614–661.

13. *Ibid.,* pp. 814, 815.

14. *Ibid.,* p. 334.

15. *Ibid.,* p. xiii.

16. *Ibid.,* p. 335. See, for example, *The Medical Record* (New York), *84* (September 20, 1913), 541; William G. Spiller, JNMD, *40* (July, 1913), 483–485; *Chicago Medical Recorder, 35* (May, 1913), 305–306.

17. Smith Ely Jelliffe, and William A. White, *Diseases of the Nervous System: A Text-Book of Neurology and Psychiatry* (Philadelphia and New York: Lea and Febiger, 1915), p. iii.

18. *Ibid.,* pp. 702, 709.

19. *Ibid.,* p. 701.

20. Bost J, *179* (August 15, 1918), 242; Meyer Solomon, *Chicago Medical Recorder, 37* (November, 1915), 683; Elmer E. Southard, "General Psychopathology," *Psychological Bulletin, 13* (June 15, 1916), 242–245.

21. John C. Burnham, *Psychoanalysis and American Medicine* (New York: International Universities Press, 1967), pp. 150–152; *Announcements of the Medical School of Harvard University,* 1912–1913, p. 50; *ibid.,* 1914–1915, p. 64; *Johns Hopkins University Medical School,* Summer 1914–1915, p. 107; *Bulletin of the University of Nebraska, Annual Catalogue of the College of Medicine,* 1917–1918; *University of California, Medical Department, San Francisco, Announcement of Courses,* 1917–1918, p. 87.

22. Adolf Meyer, "Objective Psychology or Psychobiology, with Subordination of the Medically Useless Contrast of Mental and Physical" (1915), CPM, III, 42–43; Edwin G. Boring, "The Course and Character of Learning in Dementia Praecox," *St. Elizabeth's Hospital Bulletin,* No. 5 (1913), pp. 51–79.

23. Frink, remarks to the New York Neurological Society on Brill, "Facts and Fancies in Psychoanalytic Treatment," JNMD, *50* (August 1919), 234; L. E. Emerson, "Psychoanalysis and Hospitals," R, *1* (July 1914), 285; Julius Grinker, "Progress in Neurology and Psychiatry During the Last 25 Years," *International Clinics, 4,* 25th series, 1915, 194–195; Freud, *Introductory Lectures on Psychoanalysis* (1916–1917), SE, XVI, 423.

24. Edward R. Fisher, "Psychotherapy: Its Uses and Abuses," *New York Medical Journal, 92* (December 24, 1910), 1266–1268; William G. Somerville, "The Psychology of Hysteria," AJI, *73* (April 1917), 640.

25. Jelliffe, JNMD, *43* (February 1916), 200.

26. Frederick Tilney, "Address of the Incoming President to the New York Neurological Society," January 2, 1917, JNMD, *45* (April 1917), 367.

27. Francis X. Dercum, "An Analysis of Psychotherapeutic Methods," *Therapeutic Gazette* (Detroit), *32* (May 15, 1908), 305–316; "An Evaluation of the Psychogenic Factors in the Etiology of Mental Disease," JAMA, *62* (March 7, 1914), 752; *Rest, Suggestion and Other Therapeutic Measures in Nervous and Mental Diseases* (Philadelphia: P. Blakiston's Sons and Co., 1903), pp. 249–318; *ibid.,* 2nd ed., 1917, pp. 90, 264, 334–354; Dercum, *A Clinical Manual of Mental Diseases* (Philadelphia: W. B. Saunders, Co., 1913), pp. 351, 356–357, 366–367; *ibid.,* 1917 ed., pp. 197–199, 382–411; American Medico-Psychological Association, *Proceedings, 31* (1914), 319.

28. AJI, *74* (January 1917), 496.

29. Charles L. Dana, *Textbook of Nervous Diseases* (New York: William Wood, 1915), pp. 489–490, 513.

30. Dana, "Psychiatry and Psychology," *The Medical Record* (New York), *91* (February 17, 1917), 267.

31. Compare Dana, *Textbook of Nervous Diseases,* 1920 ed., pp. 503–504, 71–73, 512, with 1908 ed., pp. 87–90.

32. Moses Allen Starr, *Nervous Diseases Organic and Functional* (New York and Philadelphia: Lea and Febiger, 1913), pp. 863–877, 887–891; 1909 ed., pp. 840–841.

33. Archibald Church and Frederick Peterson, *Nervous and Mental Disease* (Philadelphia: W. B. Saunders, 1911), pp. 716, 763–764; *ibid.,* 1908 ed., pp. 578–582, 612.

34. Massachusetts General Hospital, Ward G, Nerve, *Records, 4* (September 21, 1911), p. 130.

35. Mary O'Malley, "Psychoses in the Colored Race, a Study in Comparative Psychiatry," AJI, *71*:1 (October 1914), 309–337.

36. Helen C. Kuhlmann, "The Father Complex," AJI, *70* (April 1914), 905–939.

37. Levin, "Is Psychoanalysis of Any Value in Understanding and Treating Our Hospital Patients?," *State Hospital Quarterly, 3* (November 1917), 11, 20–21.

38. *Ibid.,* pp. 50, 54.

39. *Ibid.,* pp. 57–58; Wagner; "Presidential Address, Recent Trends in Psychiatry," AJI, 74 (July 1917), 11–14.

40. *State Hospital Quarterly, 3* (November 1917), 59–60.

41. Schwab, "The Newer Concepts of the Neuroses: An Estimate of Their Clinical Value," *American Journal of the Medical Sciences, 154,* n. s. (September 1917), 338–351.

42. *Ibid.,* pp. 347, 341.

43. Taylor, "Suggestions Regarding a Modified Psychoanalysis," J Ab P, *12* (December 1917), 372.

44. *Archives of Neurology and Psychiatry, 6* (December 1921), 595–633, esp. pp. 620–621, 629–630.

45. Unpublished lecture, 1924, AMP.

46. Isador Coriat, "Some Statistical Results of the Psycho-Analytic Treatment of the Psycho-Neuroses," R, 4 (April 1917), 209–216.

47. David H. Keller, "A Psycho-analytic Cure of Hysteria," *Institutional Quarterly* (Springfield, Ill.), 8 (March 31, 1917), 78–82; Massachusetts General Hospital, Ward G, *Records, Nerve, 4* (January 21, 1912).

48. Garfield Tourney, "A History of Therapeutic Fashions in Psychiatry, 1800–1966," *American Journal of Psychiatry, 124* (December 1967), 784–796.

49. Henry S. Munro, *Handbook of Suggestive Therapeutics, Applied Hypnotism, Psychic Science,* 3rd ed. (St. Louis: C. V. Mosby Co., 1912), pp. 292–308., 257–270.

50. Carl Renz, "Clinical Use of Psychotherapy Illustrated by Cases from Private Practice," *California State Journal of Medicine, 9* (November 1911), 478; Cecil E. Reynolds, "Mental Conflicts and Their Physical Homologues," *Southern California Practitioner, 30* (April 1915), 115.

Chapter XVII. Conclusion: "Civilized" Morality and the Classical Neuroses

1. Ernest Jones, *The Life and Work of Sigmund Freud* (New York: Basic Books, 1955), II, 57; Freud, "On the History of the Psycho-Analytic Movement" (1914), SE, XIV, 31.

2. G. Stanley Hall, *Educational Problems,* I (New York: D. Appleton & Co., 1911), 477.

3. Hall, Preface to American edition, Freud, *A General Introduction to Psychoanalysis* (Garden City, New York: Garden City Publishing Co., 1943), p. 5.

4. Putnam, "The Work of Sigmund Freud" (1917), Ad Psa, p. 360.

5. Freud, "Fragment of an Analysis of a Case of Hysteria" (1905), SE, VII, 3–122.

6. Horace W. Frink, *Morbid Fears and Compulsions: Their Psychology and Psychoanalytic Treatment* (New York: Moffat, Yard and Co., 1918), pp. 444–495.

7. Mrs. Kate Gannett Wells, "Why More Girls Do Not Marry," *North American Review, 152* (February 1891), 177–178.

8. Horace W. Frink, *Morbid Fears and Compulsions: Their Psychology and Psychoanalytic Treatment*, pp. 522–553.

9. Edward J. Kempf, "The Psychoanalytic Treatment of Dementia Praecox, Report of a Case," R, 6 (January 1919), 20–54, 58.

10. Dudley Ward Fay, "The Case of Jack," R, 7 (October 1920), 333–351.

11. A. A. Brill, *Psychoanalysis: Its Theories and Practical Application* (Philadelphia: W. B. Sanders, 1913), pp. 136–145.

12. Putnam, "Sketch for a Study of New England Character" (1917) Ad Psa, pp. 373, 388.

13. Francis X. Dercum was a "cultured gentleman." DAB, *21* (1944), 240–241.

14. Smith Ely Jelliffe, "Modern Art and Mass Psychotherapy," Bost J, *179* (November 14, 1918), 609–613.

15. Charles W. Burr, "The Prevention of Insanity and Degeneracy," AJI, *74* (January 1918), 409, 414, 422–423.

16. Elsie Clews Parsons, "Sex Morality and the Taboo of Direct Reference," *Independent, 61* (August 16, 1906), 391–392.

17. Alfred Kinsey *et al., Sexual Behaviour in the Human Female* (Philadelphia: W. B. Saunders Co., 1953), pp. 303, 357–358, 387–400, 333–339, 686.

18. Havelock Ellis, "Sexual Problems, Their Nervous and Mental Relations," in William A. White and Smith Ely Jelliffe, eds., *The Modern Treatment of Nervous and Mental Diseases by American and British Authors* (Philadelphia: Lea and Febiger, 1913), I, 100–143, and Jones, "The Treatment of the Neuroses, Including the Psychoneuroses," *ibid.,* p. 411.

19. President's Research Committee on Social Trends, *Recent Social Trends in the United States,* (New York: McGraw-Hill Book Co., 1933), I, 415–423; Freda Kirchway, ed., *Our Changing Morality* (New York: Albert and Charles Boni, 1924), pp. vi–viii, 41–90; Phyllis Blanchard, *The Adolescent Girl: A Study from the Psychoanalytic Viewpoint* (New York: Moffat, Yard and Co., 1920), p. 224.

20. *Recent Social Trends,* I, 408–414.

21. The present writer has surveyed major pulp magazines of the 1920's and found no trace of psychoanalysis. See also Geoffrey H. Steere, "Freudianism and Child-Rearing in the Twenties," *American Quarterly, 20* (Winter 1968), 759–767.

22. Simon Patten, *The New Basis of Civilization* (New York: The Macmillan Co., 1912), pp. 11–17, 25–26, 147–158, 164.

23. Putnam to Freud, April 14, 1913, JPP, p. 160.

24. Mabel Dodge Luhan, *Intimate Memories,* III, *Movers and Shakers* (New York: Harcourt, Brace and Co., 1936), pp. 253, 393.

25. Harry Woodburn Chase, "Psychoanalysis and the Unconscious," *Pedagogical Seminary, 17* (September 1910), 290–291; Philippe Ariès, *Centuries of Childhood* (New York: Alfred A. Knopf, 1962), pp. 100–103.

26. Sidney I. Schwab, "The Newer Concepts of the Neuroses: An Estimate of Their Clinical Value," *American Journal of the Medical Sciences, 154* (September 1917), 346.

27. Edward Wyllys Taylor, "James Jackson Putnam, His Contributions to American Neurology," *Archives of Neurology and Psychiatry, 3* (March 1920), 314.

28. Freud to Lou Andreas-Salomé, July 30, 1915, in Ernst L. Freud, ed., *The Letters of Sigmund Freud* (New York: Basic Books, Inc., 1960), p. 311.

Bibliographical Essay

Manuscripts

The papers of James Jackson Putnam in the Francis A. Countway Library of Medicine, Boston, are the most valuable manuscript collection available. It includes not only material for the history of American psychoanalysis but the whole development of neurology and psychotherapy in the United States, as well as sidelights on the social and intellectual life of Boston-Cambridge. See also the valuable notebooks and a few letters of Isador Coriat and the correspondence of Louville Eugene Emerson, especially with Putnam, also in the Countway. The papers of Morton Prince in The Countway and the Massachusetts Historical Society, Boston, are scanty, although containing significant correspondence with patients. The Freud Archives as well as the Brill Archives, both in the Library of Congress, are closed. The papers of G. Stanley Hall at Clark University concern chiefly Hall's early life, although some of his later correspondence has survived. Some of the papers of William Alanson White and Smith Ely Jelliffe probably were burned in a fire at Jelliffe's house on Lake George. Jelliffe's scrapbooks for the period to about 1909 are in the possession of his daughter, Mrs. Carel Goldschmidt, Mt. Kisco, N.Y. Major insights are provided by the papers of Adolf Meyer at the William Henry Welch Library of Medicine, Baltimore, and those of Lydiard Horton, Columbia University Library. Occasional sidelights occur in the papers of Edward Bradford Titchener at the John M. Olin Library, Cornell University, and the papers of Hugo Münsterberg in the Boston Public Library.

Published Sources

The published correspondence of Freud with Karl Abraham, Oskar Pfister, and others and Ernst Federn, ed., *Minutes of the Vienna Psychoanalytic Society,* 2 vols. (New York: International Universities Press, 1962, 1967), are indispensable. Jung's absorbing *Memories, Dreams, Reflections* includes letters written during the trip to America in 1909. Freud's *Autobiographical Study,* SE, XX, and his *History of the Psychoanalytic Movement,* SE, XIV, more formal and polemical, are also invaluable.

Major Secondary Works

The Life and Work of Sigmund Freud, 3 vols. (New York: Basic Books, 1953–1957), by Ernest Jones, remains the most important study, although its biases require correction. Revision already is underway and the relation of Freud to the whole context of European psychiatry and culture has been superbly reinterpreted by Henri Ellenberger in *The Discovery of the Unconscious* (New York: Basic Books, 1970). However, Ellenberger underplays the opposition that occurred on sexual grounds. The most brilliant interpretation of Freud from the perspective of his views after 1914 is Philip Rieff, *Freud: The Mind of the Moralist* (New York: Viking Press, 1959). For a view of Freud filtered through Anglo-American ego psychology, see Paul Roazen, *Freud: Political and Social Thought* (New York: Alfred A. Knopf, 1968).

No comprehensive chronological study of the development of Freud's thought yet exists, but see the beginnings in Walter A. Stewart, *Psychoanalysis: The First Ten Years* (New York: The Macmillan Co., 1967) and Bartlett Stoodley, *The Concepts of Sigmund Freud* (Glencoe, Ill.: The Free Press, 1959). The best introduction to Freud's intellectual development is the editorial matter of James Strachey in *The Complete Psychological Writings of Sigmund Freud,* 24 vols. (London: The Hogarth Press and the London Institute of Psychoanalysis, 1953–1966). A painstaking study of Freud's first psychological departures is Ola Andersson, *Studies in the Prehistory of Psychoanalysis* (Sweden: Norstedt and Söner, 1962). Freud's base in neurological theory is described in Peter Amacher, *Freud's Neurological Education and Its Influence on Psychoanalytic Theory* (New York: International Universities Press, 1965). The most sophisticated history of psychoanalytic thought after Ellenberger is Dieter Wyss, *Depth Psychology: A Critical History,* trans. Gerald Onn (New York: W. W. Norton, 1966), although without Ellenberger's attempts to place theory in a social matrix.

Freud and America

Most recent historians have begun with C. P. Oberndorf, *A History of Psychoanalysis in America* (New York: Grune and Stratton, 1953), an essential, reminiscent account by an early participant. The most comprehensive study of Freud's early influence in America, which avoids deliberately any use of Freud's writings, is John C. Burnham, "Psychoanalysis in American Civilization Before 1918," Doctoral Dissertation, Stanford University, 1958. For the impact on medicine, see Burnham's excellent monograph and bibliography, *Psychoanalysis in American Medicine: 1894–1918: Medicine, Science, and Culture* (New York: International Universities Press, 1967), which remedies the major defect of the dissertation. See also Burnham, "Psychology, Psychoanalysis and the Progressive Movement," *American Quarterly,* 12 (1960), 457–465. For the best treatments of the response of intellectuals, see Fred Matthews, "Freud Comes to America: The Impact of Freudian Ideas on American Thought, 1909–1917," Master's Thesis, University of California, Berkeley, 1957; and "The Americanization of Sigmund Freud," *Journal of American Studies, 1* (April 1967), 39 62.

Freud's influence on American sociology, psychiatry, anthropology, etc. are discussed with invaluable reminiscences by Brill, Jelliffe, Kroeber, and others in the *American Journal of Sociology, 45* (November 1939). See also David Rapaport and David Shakow, *Freud's Influence on American Psychology* (New York: International Universities Press, 1964); Ernst Kris *et al.,* "Freud's Theory of the Dream in American Textbooks," J Ab P, 38 (1943), 319–334. For a compilation of excerpts from the *American Psychoanalytic Review,* see Murray H. Sherman, *Psychoanalysis in America: Historical Perspective* (Springfield, Ill.: C. C. Thomas, 1966); see also Franz Alexander, ed., *Psychoanalytic Pioneers* (New York: Basic Books, 1966), which includes European and American psychoanalysts. Frederick J. Hoffman, *Freudianism and the Literary Mind* (Baton Rouge: Louisiana University Press, 1957) is still the best study of Freud's impact on writers, chiefly in the early period.

Chapter I. The Clark University Conference

The Clark Conference is best studied through the sources cited in the notes, especially the scrapbooks at Clark University and the *Boston Evening Trancript.* See also Virginia Payne, "Psychology and Psychiatry at the Clark Conference of 1909," Dissertation for the degree, Doctor of Medicine, University of Wisconsin, 1950, which includes an unpublished letter of Carl Jung. For a summary which exaggerates the impact of the conference, see William A. Koelsch, "Freud Discovers America," *Virginia Quarterly Review,* 46 (Winter 1970), 115–132. The most perceptive account is in Dorothy Ross, *G. Stanley Hall: The Psychologist as Prophet* (Chicago: University of Chicago Press, 1971).

Chapter II. American "Civilized" Morality

Only recently have scholars seriously begun to study the history of American sexual attitudes and mores; the bulk of scholarly attention hitherto has focused on British Victorianism. By far the most significant essay is Peter J. Cominos,

"Late Victorian Sexual Respectability and the Social System," *International Review of Social History,* 8 (1963), 18–48, 216–250, which draws an original analogy between economic and sexual man. Stephen Marcus has contributed important insights, especially concerning male ambivalence and class relations in his study of pornography, *The Other Victorians* (New York: Basic Books, 1966).

For American sexual morality, see the pioneering remarks in Henry F. May, *The End of American Innocence* (New York: Alfred A. Knopf, 1959). The literature on the American woman is burgeoning. Three important books that deal primarily with the moral revolution of the early twentieth century also contain insightful discussions of the nineteenth-century background: David M. Kennedy, *Birth Control in America: The Career of Margaret Sanger* (New Haven: Yale University Press, 1970), with a useful bibliographical essay; William L. O'Neill, *Divorce in the Progressive Era* (New Haven: Yale University Press, 1967); and O'Neill, *Everyone Was Brave* (Chicago: Quadrangle, 1969). For nineteenth-century sexual hygiene, see Stephen Nissenbaum, "Careful Love: Sylvester Graham and the Emergence of Victorian Sexual Theory in America, 1830–1840," Doctoral Dissertation, University of Wisconsin, 1968. Leslie Fiedler, *Love and Death in the American Novel* (New York: Criterion Books, 1960) contains rambling, but useful discussions of incestuous and homosexual themes and the absence of adult sexual love in nineteenth-century American literature. A breezy, unsystematic introduction to the American woman is Andrew Sinclair, *The Better Half* (New York: Harper and Row, 1967). No adequate history of the American family exists, and the most ambitious, Arthur W. Calhoun, *A Social History of the American Family from Colonial Times to the Present* (Cleveland: Arthur M. Clark Co., 1917–1919), is impressionistic and out of date.

Chapter III. The Somatic Style

The history of American psychiatry and especially neurology in the later nineteenth century remains to be written. The most comprehensive study is Barbara Sicherman, "The Quest for Mental Health in America, 1880–1917," Doctoral Dissertation, Columbia University, 1968. For an illuminating and pioneering account of psychiatric theory and institutional practice, see Gerald Grob, *The State and the Mentally Ill* (Chapel Hill, University of North Carolina Press, 1965). For the emergence of somatic determinism in psychiatric theory, see the important study by Charles Rosenberg, *The Trial of the Assassin Guiteau* (Chicago: University of Chicago Press, 1968). Albert Deutsch, *The Mentally Ill in America* (New York: Doubleday, Doran and Co., Inc., 1937) remains a stimulating account. See also Ruth B. Caplan, *Psychiatry and the Community in Nineteenth Century America* (New York: Basic Books, 1970). Uneven but also useful is J. K. Hall, ed., *One Hundred Years of American Psychiatry* (New York: Columbia University Press, 1944).

A hopeful summary from the early nineteenth century to the present is Garfield Tourney, "History of Biological Psychiatry in America," *American Journal of Psychiatry,* 1969, *126:*1, 29–42. A lucid secondary account is Charles Rosnberg's "The Place of George M. Beard in Nineteenth Century Psychiatry," *Bulletin of the History of Medicine, 36* (May–June 1962), 245–259.

An indispensable guide to the medical literature up to about 1915 is the *Index Catalogue of the Library of the Surgeon General's Office, United States Army.* For a useful summary of a major issue, see E. H. Hare, "Masturbatory Insanity: The History of an Idea," *Journal of Mental Science, 108* (January 1962), 2–25. Norman Himes, *Medical History of Contraception* (Baltimore: Williams and Wilkins, 1936), includes material on the attitude of American physicians in the nineteenth century.

Neurological discoveries are described by Fielding H. Garrison, "History of Neurology," in Charles L. Dana, *Text-book of Nervous Diseases for the Use of Students and Practitioners of Medicine,* 10th ed. (New York: William Wood and Co., 1925), xv–lvi. See also Walter Riese, "History and Principles of Classification of Nervous Diseases," *Bulletin of the History of Medicine, 18* (December 1945), 465–512. For American neurology, see Smith Ely Jelliffe and Frederick Tilney, eds., *Semi-Centennial Volume of the American Neurological Association* (American Neurological Association, 1924); William J. Morton, "Neurological Specialism," JNMD, *10* (October 1883), 618–629; Charles K. Mills, "Neurology in Philadelphia, 1874–1904," JNMD, *31* (June 1904), 353–367; Frederick G. King, "An Historical Sketch of Neurology," *New York Medical and Physical Journal, 3* (April, May, June 1824), 141–162; Charles L. Dana, "Early Neurology in the United States," JAMA, *90:2* (May 5, 1928), 1421–1424; Anon., "Neurology and Psychiatry in America," JAMA, *32* (June 3, 1899), 1236–1237; F. W. Langdon, "Neurologic Progress and Prospects," JAMA, *41* (July 18, 1903), 145–161; Roland P. Mackay, "The History of Neurology in Chicago," *Illinois Medical Journal, 125* (January–June 1964) 51–58; 142–146; 256–259; 341–344; 539–544; 636–642; *126* (July 1964), 60–63.

The most illuminating contemporary discussion of localization is Frederick Albert Lange, *The History of Materialism,* 2nd ed., 1873, authorized translation by Ernest Chester Thomas (New York: Harcourt Brace & Co., 1925), Second Book, pp. 111–161, 162–201. A definitive contemporary account is Jules Soury, *Le Système Nerveux Central, Structure et Fonctions, Histoire Critique des Théories et des Doctrines* (Paris: George Carre, et C. Naud, 1899); see also Walter Riese and Ebbe C. Hoff, "A History of the Doctrine of Cerebral Localization," *Journal of the History of Medicine, 5* (1950), 50–71; *ibid., 6* (1951), 439–470. David Krech provides an excellent survey, "Cortical Localization of Function," in Leo Postman, ed., *Psychology in the Making* (New York: Alfred A. Knopf, 1962), pp. 31–72. For an astute account of the "physiologizing" of cerebral function, see Robert M. Young, Cambridge University, Whipple Science Museum, "The Functions of the Brain: Gall to Ferrier, 1808–1886," *Isis, 59* (Fall 1968), 521–568. Edwin G. Boring's classic account is Chapter XII, *A History of Experimental Psychology* (New York: Appleton-Century, Crofts, 1950). For Pliny Earle and curability, see Franklin B. Sanborn, *Memoirs of Pliny Earle, M.D.* (Boston: Damrell and Upham, 1898); Isaac Ray, "Recoveries from Mental Disease," *Alienist and Neurologist, 1* (April 1880), 131–142. George Mora has provided an important guide in "The History of Psychiatry: A Cultural and Bibliographical Essay," R, *52* (1965), 154–183. The best account of theories of hysteria to the early Freud with perceptive sections on Weir Mitchell and Charcot is Ilza Veith, *Hysteria: The History of a Disease* (Chicago: University of Chicago Press, 1965).

Chapter IV. The Crisis of the Somatic Style

There is no comprehensive treatment of the transition from a somatic to a psychological style in neurology and psychiatry. Several histories, such as Gregory Zilboorg and George W. Henry, *History of Medical Psychology* (New York: W. W. Norton, 1941), and Franz Alexander and Sheldon T. Selesnick, *The History of Psychiatry: An Evaluation of Psychiatric Thought and Practice from Prehistoric Times to the Present* (New York: Harper and Row, 1966), embalm a belligerent Freudianism. Contemporary texts and articles contain the most useful accounts. The best brief discussion of degeneracy theory is in Erwin Ackerknecht, *A Short History of Psychiatry* (London: Hafner Publishing Co., 1959).

See also Arthur Major Fink, *Causes of Crime: Biological Theories in the United States* (Philadelphia: University of Pennsylvania Press, 1938); and George Mora, "One Hundred Years from Lombroso's First Essay, 'Genius and Insanity,'" *American Journal of Psychiatry, 121* (December 1964), 562–571. Mark Haller, *Eugenics* (New Brunswick: Rutgers University Press, 1963), is best for the period after 1910. A definitive history by a True Believer is Georges Genil-Perrin, *Histoire des Origines et de l'Évolution de l'Idée de Dégénérescence en Médicine Mentale* (Paris: Alfred Leclerc, 1913). See also Frederick Peterson, "Some of the Principles of Craniometry," *The Medical Record* (New York), *33* (June 23, 1888), 681–689. For immigration and insanity, see George H. Kirby, *A Study in Race Psychopathology* (New York: Journal of Nervous and Mental Disease Publishing Co., 1912). The standard self-image of the psychiatrists is contributed by Henry M. Hurd, *The Institutional Care of the Insane in the United States and Canada,* 4 vols. (Baltimore: The Johns Hopkins University Press, 1916–1917). A useful discussion of the vagaries of psychiatric classification is in the Appendix to Karl Menninger, *The Vital Balance* (New York: The Viking Press, 1963). Hans Kind has contributed a helpful survey on "The Psychogenesis of Schizophrenia: A Review of the Literature," *International Journal of Psychiatry, 3* (May 1967), 383–417.

Chapter V. American Psychologists and Abnormal Psychology

Literature on the history of American psychology is rapidly growing; among the most useful accounts are Gardner Murphy, *An Historical Introduction to Modern Psychology,* rev. ed. (New York: Harcourt, Brace & Co., 1949); Robert I. Watson, *The Great Psychologists from Aristotle to Freud* (Philadelphia: Lippincott, 1968); Calvin Hall, *Psychology, an Introductory Textbook* (Cleveland: H. Allen, 1960). A lively summary remains Edna Heidbreder, *Seven Psychologies* (New York: The Century Co., 1933). For G. Stanley Hall and the origins of American psychology in addition to the cited work of Dorothy Ross, see Frank McAdams Albrecht, Jr., "The New Psychology in America, 1880–1895," Doctoral Dissertation, The Johns Hopkins University, 1961, especially useful for the relation between psychology and psychical research. See Alan Gauld, *The Founders of Psychical Research* (New York: Schocken Books, 1968), for Richard Hodgson and the American branch. For a theory of role hybridization in the genesis of scientific psychology, see Joseph Ben-David and Randall Collins, "Social Factors in the Origins of a New Science," *American Sociological Review, 31* (August 1966), 451–465. On child psychology, in addition to Stephen Kern's cited work, see Bernard Wishy, *The Child and the Republic* (Philadelphia: University of Pennsylvania Press, 1967); James Dale Hendricks, "The Child-Study Movement in American Education, 1880–1910: A Quest for Educational Reform Through a Systematic Study of the Child," Doctoral Dissertation, Indiana University, 1968. For a survey of the early sexologists, see Annmarie Wettley, *Von der "Psychopathia Sexualis"* (Stuttgart: Enke, 1959). The most relevant Royce essays are "Psychology and the Consulting Psychologist," National Education Association, *Proceedings* (1898), 554–570, and "The Recent Psychotherapeutic Movement in America," *Psychotherapy: A Course of Reading,* I, No. 1 (1909), pp. 17–36, (New York: Centre Publishing Co., 1909); *The Philosophy of Loyalty* (New York: Macmillan, 1908); *The Problem of Christianity* (New York: The Macmillan Co., 1913). See also James Harry Cotton, *Royce on the Human Self* (Cambridge, Mass.: Harvard University Press, 1954). For a comprehensive study of Thorndike, see Geraldine Joncich, *The Sane Positivist: A Biography of Edward L. Thorndike* (Middletown, Conn.: Wesleyan University Press, 1968). A useful survey of clinical psychology from 1890 to 1959 and its relation to psychotherapy is John M. Reisman, *The Development of Clinical Psychology* (New York: Appleton-Century, Crofts, 1966).

Chapter VI. American Psychotherapy

No comprehensive history of the origins of American psychotherapy exists but see the relevant sections in Walter Bromberg, *Man Above Humanity: A History of Psychotherapy* (Philadelphia: J. B. Lippincott, Co., 1954). The best accounts of Morton Prince are Merrill Moore, "Morton Prince, M. D., 1854–1929, a Biographic Sketch and Bibliography" (Boston, Massachusetts, 1938); William Sentman Taylor, *Morton Prince and Abnormal Psychology* (New York: D. Appleton, 1928); Henry Murray, *The Harvard Psychological Clinic, 1927–1967* (Cambridge, Mass., 1957); and "Morton Prince, Sketch of His Life and Work," J Ab P, 52 (May 1956), 291–295. See also A. A. Roback, *A History of Psychology and Psychiatry* (New York: Philosophy Library, 1961); and "Morton Prince, 1845–1929," *American Journal of Orthopsychiatry, 10:1* (January 1940), 177–184; Roback *et al., Problems in Personality: Studies Presented to Dr.*

Morton Prince (New York: Harcourt, Brace, 1925); and *Freudiana, Including Unpublished Letters from Freud, Havelock Ellis, Bernard Shaw, Romain Rolland et alie,* presented by A. A. Roback (Cambridge, Mass.: Sci-Art Publishers, 1957). For the reception of *The Dissociation of a Personality,* see Otto Marx, "Morton Prince and the Dissociation of a Personality," *Journal of the History of the Behavioral Sciences,* 6 (April 1970), 120–130. For the vexed issue of multiple personality, see Corbett H. Thigpen and Henry Cleckley, "A Case of Multiple Personality," J Ab P, 49 (January 1954), 135–151; the historical summary in W. S. Taylor and M. F. Martin, "Multiple Personality," J Ab P, 39 (1944), 281–300; C. P. Oberndorf, "Co-Conscious Mentation," Q, 10 (1941); and the skeptical view of P. L. Harriman, "A New Approach to Multiple Personality," *American Journal of Orthopsychiatry,* 13 (1943), 638–644. Pierre Janet, *Psychological Healing,* includes American developments beginning with the early mind cures. A thorough study of Boris Sidis is badly needed, but see Herbert S. Lengfeld, "Boris Sidis," *Dictionary of American Biography,* XVII; and H. Addington Bruce, "Boris Sidis, An Appreciation," J Ab P, 18 (April 1923–March 1924), 274–280; the note on Boris Sidis, Jr. in Norbert Wiener, *Ex-Prodigy* (New York: Simon and Schuster, 1953), pp. 131–132; and the correspondence on Sidis's medical degree in the Putnam papers. For the Boston ambiance, see Martin Green, *The Problem of Boston: Some Readings in Cultural History* (New York: W. W. Norton, 1966); and for Charles W. Eliot's campaign against the new class distinctions, see Barbara Solomon, *Ancestors and Immigrants* (Cambridge, Mass.: Harvard University Press, 1956); and Dibgy Baltzell, *Philadelphia Gentlemen: The Emergence of a National Upper Class* (Glencoe, Ill.: Free Press, 1958) and *The Protestant Establishment* (New York: Random House, 1964). An illuminating discussion is in Ellenberger, *The Discovery of the Unconscious,* pp. 126–147. For Putnam, see the notes in Chapter XIV. For the beginnings of psychiatric social work, see Roy Lubove's excellent *The Professional Altruist: The Emergence of Social Work as a Career, 1880–1930* (Cambridge, Mass.: Harvard University Press, 1955).

Chapter VII. Functional Psychiatry

The most comprehensive treatments are in Sicherman and Grob, already cited. Stephen Kern treats European and American developments in child psychology as part of a single international movement, but a comparative study that includes neurology and psychiatry would be illuminating. The tentative comparison offered here is based on Ellenberger, Kern, and major textbooks.

Chapter VIII. Acquaintance and Conversion

See Burnham, *Psychoanalysis and American Medicine* and "Psychoanalysis in American Civilization before 1918," for the importance of personal contact in understanding Freud's theories. For Jones's later assessment of his former colleagues, see "Reminiscent Notes on the Early History of Psychoanalysis in English-speaking Countries," *International Journal of Psychoanalysis, 26* (1945), 8–16; and for his contemporary view of the British and American literature, see "Bericht über die neuere englische und amerikanische Literatur zur klinsichen Psychologie und Psychopatholgie," *Jahrbuch für Psychoanalyse,* II (1910), 316–346; For word association tests, see Hugo Münsterberg, *On the Witness Stand* (New York: Doubleday Page, 1909), pp. 78–84, 88–90, and Robert M. Yerkes and Charles S. Berry, "The Association Reaction Method of Mental Diagnosis," *American Journal of Psychology, 20* (1909), 22–37.

Chapter IX. *Mind Cures and the Mystical Wave: Popular Preparation for Psychoanalysis*

Contemporary periodicals remain the best source. A stimulating account of the New Thought is Donald B. Meyer, *The Positive Thinkers* (New York: Doubleday, 1965).

Chapter X. *The Repeal of Reticence*

The best treatments are May, Kennedy, and O'Neill cited earlier. See also Peter Fryer, *The Birth Controllers* (New York: Stein and Day, 1966). Emma Goldman has been thoroughly studied in Richard Drinnon, *Rebel in Paradise: A Biography of Emma Goldman* (Chicago: University of Chicago Press, 1961). Christopher Lasch has provided provocative but uneven sketches of Jane Addams and some important mandarins in *The New Radicalism in America: The Intellectual as a Social Type* (New York: Alfred A. Knopf, 1966). James R. McGovern argues from magazines and novels, but without using Kinsey's evidence, that the American woman was moving toward new roles and sexual norms by 1912 in "The American Woman's Pre World War I Freedom in Manners and Morals," *Journal of American History, 55* (September 1968), 315–333. For prostitution, see Egel Feldman, "Prostitution, the Alien Woman and the Progressive Imagination," *American Quarterly, 19* (Summer 1967), 192–206; Roy Lubove, "The Progressive and the Prostitute, *The Historian, 24* (May 1962), 308–330; and Robert E. Riegel, "Changing American Attitudes Toward Prostitution, 1800–1920," *Journal of the History of Ideas, 24* (July-September 1968), 437–452.

Chapter XI. *Opposition and Debate: Science and Sexuality*

Ernest Jones's view of the early opposition to psychoanalysis has been sharply challenged by Henri Ellenberger in *The Discovery of the Unconscious,* pp. 454–456, 796–822. For an incomplete survey of the early reaction to some of Freud's work that suggests far more partial acceptance than Jones would admit, see Ilse Bry and Alfred H. Rifkin, "Freud and the History of Ideas: Primary Sources, 1886–1910," *Science and Psychoanalysis,* 5 (1962), 6–36. A thorough study of Freud's reception in Europe and a social history of the European movement is badly needed, but see Michel David, *La Psicoanalisi nella Cultura Italiana* (Torino: Boringhiesi, 1966) and *Letteratura e Psicoanalisi,* (Milano: Mursi, 1967). See also the forthcoming study by Hannah Decker, "The Reception of Psychoanalysis in Germany, 1893–1907," Doctoral Dissertation, Columbia University, 1971. On the scientific issue, see also Michael Scriven and Herbert Feigel, eds., *The Foundations of Science and the Concepts of Psychology and Psychoanalysis* (Minneapolis: University of Minnesota Press, 1956). Richard La Pierre, who in my opinion fails to grasp Freud's meaning, blames him for destroying the Protestant ethic in *The Freudian Ethic* (New York: Duell, Sloan and Pearce, 1959).

Chapter XII. The American Psychoanalytic Organizations

For a chatty account of the schisms, with a useful essay on Havelock Ellis, see Vincent Brome, *Freud and His Early Circle* (London: Heinemann, 1967). Oskar Pfister, *Zum Kampf um die Psychoanalyse* (Leipzig, Vienna, and Zurich: International Psychoanalytic Press, 1920), partly translated as *Some Applications of Psychoanalysis* (New York: Dodd, Mead, 1923); and Ludwig Binswanger, *Sigmund Freud: Reminiscences of a Friendship* (New York: Grune and Stratton, 1957) are important biographical records. M. M. Wangh, ed., *Fruition of an Idea: Fifty Years of Psychoanalysis in New York* (New York: International Universities Press, 1962), is a skimpy, anniversary sketch of the New York Institute. Ives Hendricks, *The Birth of an Institute* (Freeporte, Maine: The Wheelwright Co.), is an ambitious, uneven account of psychoanalysis in Boston, with anecdotal material. Bertram D. Lewin and Helen Ross, *Psychoanalytic Education in the United States* (New York: W. W. Norton, Co., 1962), an otherwise scholarly study, gives no direct information about the social origins of psychoanalysts. The Putnam papers and the psychoanalytic journals are the best sources. One of the most suggestive interpretations remains Eric Fromm, *Sigmund Freud's Mission* (New York: Harper, 1959), which argues that Freud was driven by inner needs and ambitions to found a quasi-religious world movement. A revealing retrospective is C. P. Oberndorf, "History of the Psychoanalytic Movement in America," R, *14* (July 1927), 281–297.

Chapter XIII. The American Interpretation of Psychoanalysis

For a jaundiced view of American popularizers and eclectics, see Horace W. Frink, *Sammelreferat über die amerikanische psychoanalytische Literatur in den Jahren 1920–1922,* Z, *10* (1924), 57–86, 180–198. The best accounts of the American interpretation are those of Burnham and Matthews cited earlier, although in my opinion psychoanalysis was far more ambiguous and explosive than Burnham suggests. For an amusing interview with Freud, see George Sylvester Viereck, "Cheerful Humility Marks Freud at Seventy," Baltimore *Sun*, August 28, 1927, Part II, Section 2, p. 8. A popular version of Jung's psychology by a journalist and lay analyst is James Oppenheim, *American Types: A Preface to Analytic Psychology* (New York: Alfred A. Knopf, 1923). For a serious interpretation, see Beatrice Hinkle, *The Recreation of the Individual* (New York: Harcourt, Brace, 1923). The best presentation of Adler is in Heinz L. Ansbacher and Rowena Ansbacher, *The Individual Psychology of Alfred Adler* (New York: Basic Books, 1956); see also Heinz L. Ansbacher, "The Significance of the Socio-Economic Status of the Patients of Freud and Adler," *American Journal of Psychotherapy, 13* (April 1969), 376–382; see also the slight biography by Phyllis Bottome, *Alfred Adler: Apostle of Freedom* (London: Faber and Faber, 1939); for Adler's philosophy, see his *What Life Should Mean To You* (Boston: Little Brown & Co., 1922); Jung's work to 1920 gives the best account of his growing differences with Freud, especially *The Theory of Psychoanalysis* NMDMS No. 19 (New York: Journal of Nervous and Mental Disease Publishing Co., 1915), and *Two Essays on Analytical Psychology* (New York: Dodd, Mead and Co., 1928). Princeton University Press plans to publish the Freud–Jung correspondence, hitherto unavailable. The best interpretation of Jung and Janet is in Ellenberger, *The Discovery of the Unconscious*. For Jung's later view of his differences with Freud and Adler, see *Modern Man in Search of a Soul* (New York: Harcourt, Brace, 1933). For a stimulating interpretation of behaviorism, see Lucille Birnbaum, "Behaviorism: John Broadus Watson in American Social Thought, 1913–1933," Doctoral Dissertation, University of California, Berkeley, 1964. See also John C. Burnham, "On the Origins of Behaviorism," *Journal of the History of the Behavioral Sciences, 4* (April 1968), 143–151; David Bakan, "Behaviorism and American Urbanization," *ibid, 2* (January 1966), 5–28; and W. Harrell and R. Harrison, "The Rise and Fall of Behaviorism, *Journal of General Psychology, 18* (1938), 367–421.

Chapter XIV. American Psychoanalysts

Published obituaries and contemporary journals are the most important sources apart from the Putnam papers. For a fascinating study of A. A. Brill, see Paula Fass, "A. A. Brill—Pioneer and Prophet," Master's Thesis, Columbia University, 1969; and for Putnam, Russell Vasile, "James Jackson Putnam—From Neurology to Psychoanalysis: A Study of the Reception and Promulgation of Freudian Psychoanalytic Theory in America, 1895–1918," Senior Thesis, Princeton University, 1970. For an insightful link between the self-improving strain of American Protestantism and psychoanalysis, see Howard M. Feinstein, "The Prepared Heart: Puritan Theology and Psychoanalysis," *American Quarterly, 22* (Summer 1970), 166–176.

Chapter XV. Popular Psychoanalysis

There is almost nothing on this vital subject apart from Matthews and Burnham. David Sievers' *Freud on Broadway* is an indiscriminate compendium that classifies almost every play dealing with the family or sexuality as "Freudian." The best study of semi-popular literature is Hoffmann, *Freudianism and the Literary Mind,* already cited. Mark Sullivan, *Our Times* (New York: Charles Scribner's Sons, 1932), IV, 166–176, is an early and exaggerated assessment of the Freudian impact. For an impressionistic account that is a testimonial to Freud as a liberator, see Sidney H. Ditzion, *Marriage, Morals and Sex in America: A History of Ideas* (New York: Bookman Associates, 1953). Richard Weiss, *The Myth of Success* (New York: Basic Books, 1969), sees the early interest in psychoanalysis as a reflection of the concern with mind cures.

Chapter XVI. A New Model in Neurology and Psychiatry

For changes in psychiatry, see Sicherman and Burnham, and for a sketch of the earlier pessimism, see Burnham, "Psychoanalysis in American Civilization Before 1918."

Chapter XVII. Conclusion: "Civilized" Morality and the Classical Neuroses

The essays of Talcott Parsons, especially those on the superego, have been suggestive. See *Family, Socialization and Interaction Process* (Glencoe, Ill.: Free Press, 1960), and *Social Structure and Personality* (New York: Free Press, 1965). Alan Wheelis, *The Quest for Identity* (New York: W. W. Norton, 1958), and Kenneth Keniston, *The Uncommitted* (New York: Harcourt, Brace & World, 1965) are pioneering accounts of the internalization of social change. See also the suggestions in Ellenberger and Veith. The most comprehensive attempt to integrate psychoanalytic insights into a theory of historical change is Gerald Platt and Fred Weinstein, *The Wish to be Free* (Berkeley: University of California Press, 1969). The Mabel Dodge Luhan papers in the Beinecke Library, Yale University, contain a wealth of information regarding her relationships to Jelliffe and Brill.

Index